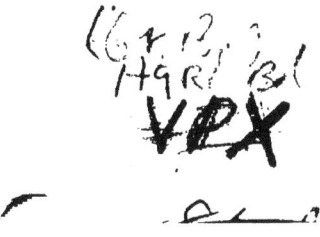

SOLD BY MOZLEY, AND DRURY, DERBY; BRADLEY, CHESTER-
FIELD; AND TODD, SHEFFIELD.
Printed by B. M'Millan, Bow-Street, Covent-Garden.

1817.
[*Price Eighteen Shillings.*]

Plate III. to face page 293 Vol.

GENERAL VIEW

OF THE

AGRICULTURE

OF

DERBYSHIRE;

WITH

OBSERVATIONS ON THE MEANS OF ITS IMPROVEMENT.

DRAWN UP FOR THE CONSIDERATION OF

THE BOARD OF AGRICULTURE,

AND INTERNAL IMPROVEMENT.

VOL. III.

CONTAINING A FULL ACCOUNT OF

The various Breeds of Live Stock, their food, management, uses, and comparative advantages; their Houses, Stalls, &c.; with accounts of the preparation of Cheese, Butter, Bacon, &c.

Rural details regarding modes and prices of Labour, Cottages, prices of Provisions, Fuel, &c.

Politico-Economical details, regarding Game, Roads, Rail-ways, Canals, Fairs, Markets, Weights and Measures, various Manufactures, Commerce; Parish Maintenance of the Poor and their own Benefit Societies, &c.; and regarding the increase of the Population, their healthiness, modes of living, &c.

The Obstacles to Improvements, and facilities for their adoption; with a concluding brief recapitulation of the various hints and suggestions, of Measures calculated for Improvement, scattered through these Volumes.

ILLUSTRATED BY A MAP OF ROADS, CANALS, &c.
AND TWO PLATES.

BY JOHN FAREY, SEN.

MINERAL SURVEYOR,

OF HOWLAND-STREET, FITZROY-SQUARE, LONDON.

LONDON:

PRINTED FOR SHERWOOD, NEELY AND JONES,
PATERNOSTER-ROW.

SOLD BY MOZLEY, AND DRURY, DERBY; BRADLEY, CHESTERFIELD; AND TODD, SHEFFIELD.

Printed by B. M'Millan, Bow-Street, Covent-Garden.

1817.

[*Price Eighteen Shillings.*]

ADVERTISEMENT.

The circulation of the original Quarto County Reports, with wide Margins, having occasioned a very general desire, that the several Counties might be re-surveyed, and more copious Reports printed in Octavo, the same has accordingly been done.

It is proper at the same time to add, that the Board does not consider itself responsible for every statement contained in the Reports thus re-printed, and that it will thankfully acknowledge any additional information which may still be communicated.

PREFACE.

THE first Chapter of this Report, Volume I. treating of the Surface, Divisions, Climate, and Soils, of the County of Derby, and entering very fully into the description of its *Strata* and its *Mineral Concerns*, and those of its Environs, has now been five years before the Public, and has, I understand, obtained a pretty extensive circulation amongst Persons interested in the important objects which it embraces.

The Second Volume, Chapters Two to Thirteen, treating of Estates, Buildings, Occupancy, Implements, Inclosing, Arable Cultivation and Crops, Grass Lands, Gardens, Woods, Plantations, and Waste Lands; and of Improvements by Draining, Liming, Manuring, Irrigating, Embanking, &c. within the County, has now been more than three years on sale.

On presenting at so late a period, the Third and concluding Volume of this Report, an apology seems necessary from me, for the great and unprecedented delay which has occurred in completing it, and which I hope and trust, that my many kind Friends and the Public, will accept, on learning more particularly, the circumstances which have occurred to prevent it.

In the Summer of 1807, forming an insufficient estimate, of the force of *prejudice* in favour of preconceived *Theories* concerning the structure of the Earth and its Mineral contents, by one class of Persons, the Writers and Readers on these subjects; and of the habitual *indifference* by another class, the Owners and Operative Men concerned with Mineral Property, to every thing coming from those who have taken pen in hand, or attempt to generalize, on whatever grounds;—through this mistake, I was too sanguinely led to think, that the delay of engraving and publishing Mr. Smith's general Map of the Strata of England, &c. (I. 116), and the want of publicity to any Specimens of more *minute Mineral Maps* of Districts, made on the principles which he had discovered and taught, were the chief, if not *the only impediments* which required removing, for seeing myself, and many others treading in the same steps, very fully and *profitably* employed, in practically exploring, and in *arranging and publishing* the details, of the vast Mineral Treasures which these Islands possess, in their Coals, Iron, Limestones, Sandstones and useful Clays, in particular, and of such details regarding the curious structures and states of these Islands, as might throw new light on Geology as a Science.

And in consequence, with a zeal and industry which has not often been surpassed, as I am persuaded, I laboured incessantly, to the near completion of my large and minute Mineral Map of all Derbyshire, and of almost as many acres of its Environs (I. v. &c.), and to the publication of Vol. 1. of this Report.

But alas! I soon then found, after having devoted a length of *time*, and given thereby a neglect and check to my increasing professional Business, which my circumstances and Family had not warranted; that, on the contrary, the prejudice and indifference before alluded to, were so general, so well understood, and so fully appreciated, by all those concerned in, or able with effect, *to bring out and sell* Maps and Works, of the kind I had contemplated, that those in particular, which went sufficiently into minute and long details, *to render them of any material use*, could not be published at all in this Country!

In this dilemma, I was rather cruelly led and persuaded, into a further sacrifice of my *time*, to no inconsiderable amount, in preparing, partly from the materials of a minute Mineral Survey I had made for Sir Joseph Banks, Bart. (with his consent, II. ix), and partly from the materials of my General Survey of the County, to prepare two Maps, a Section of more than Seven Miles in length of Surface, and a Memoir of great length, fully explaining these, and the principles and modes on which they had been made; with the hopes, that *their publication* might draw the more general attention and literary patronage to the subject, which seemed wanting, for permitting the accomplishment of the design I had formed in 1807, as above-mentioned;—but here again I was entirely disappointed, as is stated in the " Philosophical Magazine," Vol. 42, p. 55, Note, by the very improper and interested views of *a Party*, alike hostile to the cause of real Geological Science, as to those Persons engaged in Practical and un-theoretical investigations of the facts of the British Stratification*, *by refusing to publish*, and yet claiming the right to detain my Manuscripts!: but which last, they were

---

* I am sorry to add, that the very meritorious exertions of my friend *Mr. Elias Hall* (II. x, and Phil. Mag. Vol. 34, p. 340) have met just a similar fate, through the same *Party*;— no sooner was one of his *Models* shewn, before several of its Leaders, than without waiting to examine a single particular of its laborious details, advantage was taken, of an injudicious use of rather too glaring colours, in Mr. Hall's first attempt, to raise *a laugh*, by the far-fetched and contemptible joke, that "a tray of Guts and Garbage in a Fishmonger's or Poulterer's Shop," rather than any thing else, was called to the mind! by viewing this elaborate Model of the Stratified Hills and Dales, of a tract of Country!—Such was the conduct of the Heads of a Geognostic Society, and such the reward of meritorious labours, out of its pale; and my Friend remains, and is likely to do, a considerable loser by what he has done.

soon induced to abandon, and the Papers remain yet useless on my hands.

My friend *Mr. Smith*, has seen equal cause to complain of the same malignant influence, as is stated in p. 337, Vol 45 of Phil. Mag.; the principal part of his hard-earned materials, having been surreptitiously obtained, and the progress of a rival Map of our Strata, unblushingly announced, by the same Party, *while his was engraving and colouring!*—and since its completion, an extensive sale of it has been prevented, such as was necessary *to repay the outlay* of Engraving and Colouring, independent of any profit to Mr. Cary, or remuneration to Mr. Smith, by the sinister arts employed to under-value the same, and to dissuade, even the few who are disposed *to purchase a Map of the Strata*, to wait for the improved Map of a *ci-devant* Member of Parliament! who, it may be proper to observe, entered on this pursuit years after Mr. Smith had, not only commenced, but brought his Map almost to its present state of approach to perfection, had freely shewn it to hundreds, and was soliciting Subscribers' *names,* for its publication.

Such very discouraging results from my Derbyshire labours, as have been mentioned above, and others, were thought by most of my Friends and Acquaintance, to justify, and require on my part, after the publication of Volume I. the utmost compression or abridgment of the materials remaining on my hands for this Report, which could accord with the literal terms, of my original engagement with the Board, and which would have produced only one small Volume, in place of the two which I have since produced.

But considering, that provided the Board would allow me, while other County Reports were yet unfinished, and the much-wanted *General Report* on the Kingdom, was not ordered or provided for by Parliament, to lay by, and resume this Report, at such times as my professional Business, prevented or permitted my application to it; in such case, I say, I thought it more just, to the many Gentlemen and Practical Men of the County, who had furnished me with information, to the Country, to the Board, and to myself, to do as I have since been enabled to do, by the kind indulgence of the Board.

I had also another motive, for wishing the course I have been permitted to pursue for lengthening this Report, in order that the results, of no ordinary industry and pains on my part, during the ten years in which I conducted the Bedfordshire affairs of its late and greatly revered Duke, and since, to investigate and come to conclusions of my own, in various departments of Rural Management and Political

Economy, that such might not be lost to the Public, as they had, and were likely, in every pecuniary sense to prove to me and my Family.

These, I humbly presume to hope, will be accepted as sufficient apologies for my apparent tardiness in producing these Volumes; the very multifarious subjects of which, and the importance of great numbers of them, seemed to require the present extension; at the same time, that the candour and utmost allowance of my Readers are, I am sensible, necessary, for the imperfect manner in which many of them have been handled; I have, however, endeavoured to do the best my materials and information would allow; and in sincerity and disinterestedness to offer, various hints and suggestions for future Improvements, (most of which are briefly recapitulated in the concluding Chapter) for which I solicit the like candid reception.

Amongst such very numerous and different kinds of details as these Volumes furnish, the utmost care has been insufficient, to guard against various mistakes and errors, which the communications of my Friends and others, and my continued attention and research in the affairs of Derbyshire and its Environs, have brought to light, since the printing*; the most material of which will be noticed in the *Errata*,

---

* See Phil. Mag. Vol. 42, p. 113, and Vol. 45, p. 161.

which follows the Contents of this Volume; for others, which my Readers may yet discover, I must request their candid allowance, and shall at all times be extremely obliged, by any corrections or additions which are given or sent to me, and which I will carefully make and insert in my own copy, in order, that if perchance a second Impression should be called for (which with the first Volume I have been flattered by thinking is not very improbable) I, or my Sons after me, may give every possible degree of accuracy and perfection, to the very ample details regarding one of the finest of the English Counties, which my work furnishes.

The "Plan" prescribed by the Board, has as closely as possible been followed in my arrangement throughout; and my *Indexes* and References will, I may hope, by their convenience and utility, correspond with the labour they have occasioned.

JOHN FAREY, Sen,

June 5, 1816.

CONTENTS OF VOLUME III.

CONTAINING

CHAPTERS XIV. TO XVIII. AND CONCLUSION, AND APPENDIX.

CHAP. XIV. LIVE STOCK, pages 1 to 183.

SECT. 1. CATTLE, pages 1 to 87.
   1. Breeds, 1 to 20.
      1. Old Long-horned, 2.
      2. Short-horned, 3.
      3. New Long-horned, 4.
        Lists of Bull-letters, of this Breed, 6, 7.
        Account of Mr. Mundy's Sale, 11.
      4. Devon, 14.
      5. Hereford, 15.
      6. Scotch, 16.
      7. French, or Alderney, 16.
      8. Welsh, 16.
      9. White Breed, 16.
      10. Long-horn and Short-horn Cross, 17.
      11. Old Long-horn and New Long-horn Cross, 18.
      12. Long-horn and Devon Cross, 18.
      13. Short-horned and Devon Cross, 19.
      14. ———————— French Cross, 19.
      15. ———————— White Cross, 20.
      16. ———————— Lincoln Cross, 20.
      17. Devon and French Cross, 20.
      18. Scotch and White Cross, 20.
   2. Food for Neat Cattle, 21.
      1. In Winter,
        Roots, 21.
        Oil-cake, 22.
        Green Crops, 22.
      2. In Summer,
        Natural and sown Grasses, 22.
        Soiling, 22.

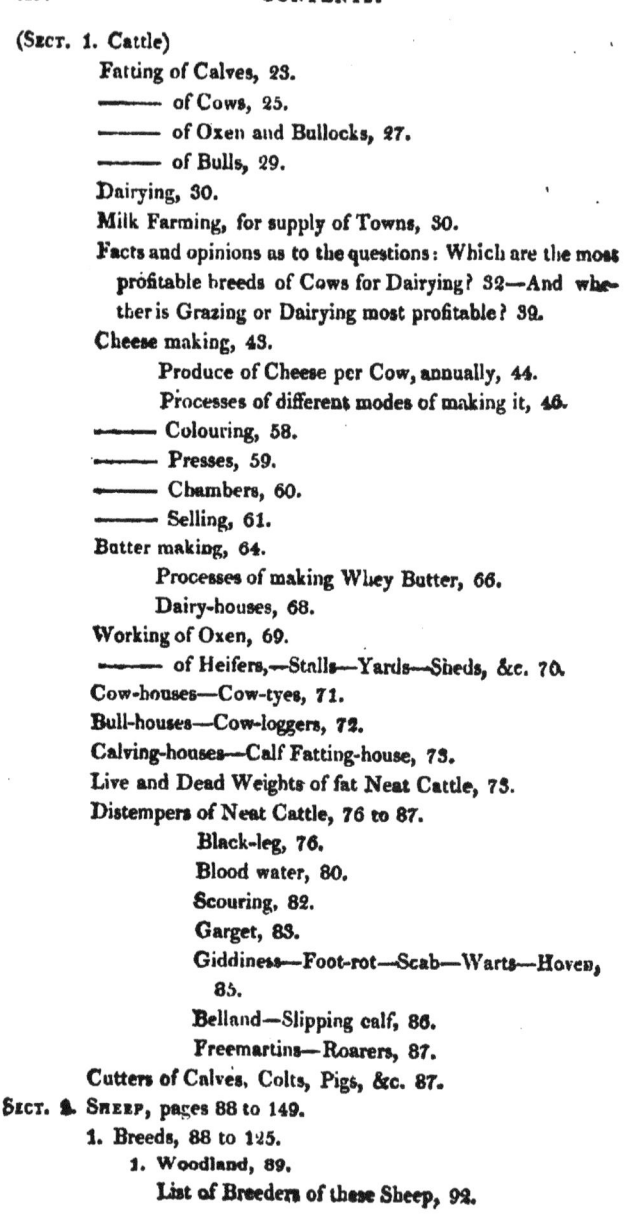

CONTENTS.

(SECT. 1. Cattle)
    Fatting of Calves, 23.
    ——— of Cows, 25.
    ——— of Oxen and Bullocks, 27.
    ——— of Bulls, 29.
    Dairying, 30.
    Milk Farming, for supply of Towns, 30.
    Facts and opinions as to the questions: Which are the most profitable breeds of Cows for Dairying? 32—And whether is Grazing or Dairying most profitable? 39.
    Cheese making, 43.
        Produce of Cheese per Cow, annually, 44.
        Processes of different modes of making it, 46.
    ——— Colouring, 58.
    ——— Presses, 59.
    ——— Chambers, 60.
    ——— Selling, 61.
    Butter making, 64.
        Processes of making Whey Butter, 66.
        Dairy-houses, 68.
    Working of Oxen, 69.
    ——— of Heifers,—Stalls—Yards—Sheds, &c. 70.
    Cow-houses—Cow-tyes, 71.
    Bull-houses—Cow-loggers, 72.
    Calving-houses—Calf Fatting-house, 73.
    Live and Dead Weights of fat Neat Cattle, 75.
    Distempers of Neat Cattle, 76 to 87.
        Black-leg, 76.
        Blood water, 80.
        Scouring, 82.
        Garget, 83.
        Giddiness—Foot-rot—Scab—Warts—Hoven, 85.
        Belland—Slipping calf, 86.
        Freemartins—Roarers, 87.
    Cutters of Calves, Colts, Pigs, &c. 87.
SECT. 2. SHEEP, pages 88 to 149.
    1. Breeds, 88 to 125.
        1. Woodland, 89.
            List of Breeders of these Sheep, 92.

CONTENTS.

(SECT. 2. Sheep. Breeds)
  2. Old Limestone, 96.
  3. Old Leicester, 96.
  4. Forest, 96.
  5. New Leicester, 97.
   Lists of Tup-letters, of this Breed, 98, 100.
  6. South-Down, 110.
  7. Ryeland, 112.
  8. Portland, 112.
  9. Merino, 113.
   A Table of Calculations on *Crossing* Sheep, 120.
  10. Spanish, 123.
  11. Gritstone, 123.
  12. Old Limestone and New Leicester Cross, 123.
  13. New Leicester and Northumberland Cross, 124.
  14. New Leicester and South Down Cross, 125.
  15. Merino and Woodland Cross, 125.
  16. Merino and South Down Cross, 125.
  17. Merino and Ryeland Cross, 125.
 Food for Sheep, 126.
 Sheep Walks, 126.
 Cotting of Sheep, 127, with a *Plate* facing p. 128.
 Fatting of Sheep, 128.
 Mutton Hams—Fat Lamb, 129.
 Live and Dead Weights of Fat Sheep, 129.
 Washing of Sheep, 134, with a *Plate* facing p. 144.
 Shearing of Sheep—Shearing Lambs—Wool-Chambers, 136.
 Sale of Wool—Wool-Fairs, 137.
 Distempers of Sheep, 140 to 149.
  Rot, 140.
  Red water—Resp—Brown Ger—Scouring, 145.
  Turn, 146.
  Rickets—Foot-rot, 147.
  Scab—Maggots, 148.
  Ticks—Bud, 149.
 Goats, 149.
SECT. 3. HORSES, pages 150 to 161.
  Breeds,
   List of Stallion-letters, 151.

(Sect. 3. Horses)
  Training of Nags, 155.
  Racing, or Blood Horses, 155.
  Work, 156.
  Food—Oats, 157.
   Soiling, 158.
   Littering, 159.
  Distempers of Horses, 159.
Sect. 4. Asses, 161.
Sect. 5. Mules, 163.
Sect. 6. Hogs, pages 164 to 176.
 1. Breed, 164.
  Roasting Pigs, 168.
  Food, 168.
  Sties—Troughs, 171.
  Live and Dead Weights of Fat Hogs, 172.
   Bacon, 173.
   Hams, 175.
  Distempers of Hogs, 176.
Sect. 7. Rabbits, 176.
  Warrens, 177.
Sect. 8. Poultry, 177.
  Turkies, 178.
  Geese—Fowls, 179.
  Ducks, 180.
Sect. 9. Pigeons, 181.
  Dove-Cotes, 181.
Sect. 10. Bees, 183.

CHAP. XV. RURAL ECONOMY, pages 184 to 197.

Sect. 1. Labour, pages 184 to 194.
  Hired Servants—Labourers, 184.
  Piece Work, or Job Work, 189.
  Cutting of Hedges—Dibbling—Draining—Hoeing—
   Mowing—Planting—Reaping, 189.
  Spreading of Dung—Stone digging—Taking up Potatoes
   —Thatching—Thrashing—Trenching—Walling, 190.
  Letting of Work, 191.

(Sect. 1.)
      Cottages attached to Farms, 194.
      Cottagers keeping Cows, and renting Land, 194.
Sect. 2. Prices of Provisions, 195.
Sect. 3. Fuel, 196.
      Coal—Wood—Peat, 196.
      Candles—Gas-Lights, 197.

CHAP. XVI. POLITICAL ECONOMY, pages 198 to 632.

    Game, pages 198 to 206.
      Black Game—Dotterels—Wild Ducks—Wild Geese—Grouse—Partridges—Pheasants—Pigeons, 199.
      Wood-Pigeons—Reeves and Ruffs—Snipes—Woodcocks, 200.
      Deer, 200.
      Foxes—Hares, 200.
      Packs of Hounds kept, 200.
      Rabbits, 202.
      Fish, 202 to 206.
        Castrating of Carp, 204.
        Angling permitted, paying for the Fish, 205
        Salmon, 205.
        Trout, 206.
    On *Improved Communication* by Roads, Rail-ways, and Canals; with a *Map* facing p. 193.
Sect. 1. Roads, pages 206 to 281.
      Turnpike Roads, 208, 228; and *Map* facing p. 193.
      Mail Coach Routes, 210, 228.
      Canals and Navigations, 212; *Map*, facing p. 193.
      Roman Roads and Stations, 216.
      Principles of setting out Roads formerly, were defective, 219.
      New lines of, wanted, 225.
      Turnpike Roads were commenced on wrong principles, 228, 232.
      Improvements of them commenced, 230.
      Further ones recommended, 235.
      Post Towns, a List of, 249.

CONTENTS.

(SECT. 1. Roads)
    Materials for making them, 250.
        Gravel, 250.
        Limestone, 258.
        Rider-stones—Chert, 254.
        Crowstone—Cank—Gritstone, 255.
        Burnt-stone, 256.
        Bricks or Burnt Clay—Iron Slag or Cinders, 257.
        Lead-Slag—Gypsum—Coals—Wood, 258.
    Expence of Roads, 259.
        Mode of letting repairs of Roads, 260.
        List of Road-makers, 261.
        Statute duty, 262.
        Parish or Township Roads, 263.
        Bye-sets—Ploddings—Scotching-stones, 264.
        Farm Ways, and private Roads, 271.
        Foot-paths, 272.
        Pack-Horses—Pack-Asses, 274.
    Waggons and Carts—Prices or Rates of Carriage, 274.
    Concave Roads, 275.
    Convex Roads, 276.
    Flat Roads, 277.
    Waving or undulating Roads, 279.
    Application of Water, 280.
    Fords—Ferrys, 282.

SECT. 2. Iron RAIL-WAYS; and *Map* facing p. 193.
    The Author's Article, and intended Work on these, and Canals, &c. 283, 456.
    Public Rail-way, instead of private ones, wanted in Tyne and Wear Dales, 285.
    Rail-way branches to Public Canals, 287.
    Wooden Rail-ways, where used, 288.
    Cost of laying Rail-ways, 289.

SECT. 3. CANALS, pages 290 to 456; and *Map* facing p. 193.
    Their advantages to the Public, exceed those to their Proprietors, 291.
    Adelphi Canal, described, 294.
    Ankerbold and Lings Rail-way, 295.
    Ashby-de-la-Zouch Canal, 297.
        Estimate for the Ticknal Rail-way, 301.

CONTENTS.

(SECT. 3. Canals)
 Ashover and Chesterfield, proposed Rail-way, 304.
 Barnsley Canal, 306.
 Baslow and Brimington, proposed, 309.
 Baslow and Chesterfield, proposed, 311.
 Belper, proposed, 312.
 Belper and Morley-Park Rail-way, 313.
 Birmingham and Fazeley Canal, 313.
 Breedon, proposed Rail-way, 316.
 Bridgewater's Canal, 316.
 Chesterfield, 317.
  Unfair restrictions imposed by Mill-owners, 322.
 Chesterfield and Swarkeston, proposed Canal, 328.
 Commercial, proposed, 329.
 Congleton Rail-way, 331.
 Coventry Canal, 331.
 Cromford, 336.
  Section of Strata in Butterley Tunnel, 343.
 Cromford and Bakewell, proposed Canal, 352.
 Dearne and Dove Canal, 353.
 Derby, 355.
 Derwent (Derby) late Navigation, 360.
 Dilhorn, proposed Canal, 360.
 Don Navigation, 361.
 Erewash Canal, 363.
 Grantham, 366.
 Greasley's, 368.
 High Peak Junction, proposed, 369.
 Huddersfield Canal, 371.
  The long Tunnel, described, 372.
 Idle Navigation, 375.
 Leicester Navigation, 376.
 Leicester and Melton Mowbray Navigation, 381.
 Leicestershire and Northamptonshire Union Canal, 382.
 Loughborough Navigation, 384.
 Macclesfield, proposed Canal, 386.
 Manchester, Ashton and Oldham, 389.
 Mersey and Irwell Navigation, 391.
 Newcastle Underline, Canal, 391.
 ———————————— Junction, 392.

CONTENTS.

(SECT. 3. Canals)
  North-eastern, proposed Canal, 393.
  Nottingham Canal, 397.
  Nutbrook, 400.
  Peak Forest, 402.
   The large Inclined Plane, described, 404.
  Rochdale Canal, 410.
   Unfair advantages claimed by Mill-owners, 412.
  Stafford Rail-way, 415.
  Staffordshire and Worcestershire Canal, 415.
  Swadlingcote and Newhall, proposed Rail-way, 418.
  Ternbridge and Winsford, proposed Canal, 419.
  Tinsley and Grindleford-Bridge, proposed, 420.
  Trent, lower, Navigation, 421.
  Trent, upper, late Navigation, 428.
  Trent and Mersey Canal, 430.
   The large Inclined Planes, described, 436.
   The large Reservoir, described, 441.
  Wilden and King's-Bromley, proposed Canal, 449.
  Woodeaves Canal, 451.
  Wyrley and Essington, 451.
SECT. 4. Fairs, alphabetical List of them, 457.
SECT. 5. Markets, alphabetical List of them, 459.
  Markets for Corn, 459.
  Auction and Ticket Sales, 460.
SECT. 6. WEIGHTS AND MEASURES, pages 460 to 475.
  Land Measure, 461.
   The Author's proposals, in 1795, for New Decimal Weights, Measures and Money; raised from the *Acre*, the *Pound* avoirdupoise, and the *Pound* sterling, 461.
   Building—Roads—Fencing—Draining—Trenching—Corn Measures, 469.
   Reaping—Flour—Hay—Potatoes—Butter—Cheese—Liquids, 471.
   Wood—Bark—Charcoal—Coals, 472.
   Lime, 473.
   Gypsum—Stone—Ironstone—Lead-Ore, 474.
   Lead—Wool—Cotton, 475.
SECT. 7. Price of Products, compared with Expences, 475.
SECT. 8. Manufactures, pages 476 to 509.

(SECT. 8. Manufactures)
- 1. Species of Manufactures, 478 to 496.
    - 1. Trades, &c. depending on Animal Products of the County, 479.
    - 2. ——————————— Animal Substances Imported, 481.
    - 3. ——————————— Vegetable Productions of the County, 482.
    - 4. ——————————— Vegetable Substances Imported, 483.
    - 5. ——————————— Mineral Products of the County, 488.
    - 6. ——————————— principally on Mineral Substances Imported, 493.

    Chief articles of Export from the County, 497.
    Great manufacturing progress in Glossop Parish, 498.
- 2. Earnings of Manufacturing Labourers, 499.
- 3. Rise of Rents occasioned by Manufacturers, 500.
    Injurious effects of Child labour and Apprenticeship, 501.
    Manufacturers grown Rich, and Land-owners Poor, 505.
- 4. Poor Rates, increased by Manufactures, 500, 502, 506.
    Great abuses of Apprentice and Poor Laws, 508, (and 502.)

SECT. 9. Commerce, 509.
- The Circulating Medium is Paper-money, 510.
- Frequent occurrence of forged Bank-notes, 511.
- List of Bankers in the County, 512.
- Silver Tokens—Soho Halfpence, &c. 512.
- Effects of Commerce on Agriculture, 512.
- Commercial Men, often are good Farmers, 513.

SECT. 10. The Poor, pages 514 to 664.
- The constitution of Society, as to *Rich* and *Poor* People, 514.
- The *wants* of each, mutually guarantee their *rights*, and point out their several Duties, 515, 521.
- *Charity* in the Poor, towards each other, is most important, 518.

(SECT. 10. The Poor)

  Church-dependance, fatally paved the way for Parish-dependance, by the Poor, 518.
  Origin of the Apprentice-Law Monopolies, 520 Note.
  ———— Poor-Laws system, 521.
  ———— Work-House system, 522.
  ———— Houses of Industry instead of Work-houses, 523 Note.
  Parish-dependance, more mischevious than Church-dependance had been, 524.
  The *Rights* of the Poor, quite mistaken by many Persons, 525.
  The Agricultural and Rural Labourers, are insufficiently paid, 526.
  Necessary preliminary steps (in increasing *their* pay, &c.) to abolishing the Poor-laws, 527 Note.
  Journeymen of established Trades, are sufficiently paid, or more, 528.

1. State of the Poor, 528.
  Going the Rounds, or House-row, 529.
  State of Education—Schools, 529.
  Hints for rendering them more useful, 530, 569 Note.
2. Annual Receipt and Expenditure on behalf of the Poor, 531; with a folding Table facing, p. 529.
  What are the reasons of such very different *Pound rates* to the Poor, in different places? 532.
  Nominal Rents, are very generally used *in Rating* for the Poor, 534, 539, 544.
  Table of proportions of Agriculturists, Traders, &c. Paupers, and Rentals, 528.
  Suggestions for improved Poor Returns, 540.
3. Sums raised by Rates, 543.
4. Work-houses, 544 to 564.
  A Table of Parish Work-houses, of Paupers, and their cost each, 545.
  1. Management of Parish Work-houses, 547.
  Expences per head, of Paupers in Parish Work-houses, 547.
5. Houses of Industry, 548.

CONTENTS. xxiii

(SECT. 10. The Poor)
  Subscription Poor-houses, pages 549 to 564.
  A Table of Subscription Poor-houses, Paupers, and cost of each, 549.
  History of the Ashover Poor-house, 550.
  A List of its Subscribing Parishes and Places, 551.
  Particulars of the other Subscription Poor-houses in the County, 551.
 1. Management of Subscription Poor-houses, 552.
  Rules in Ashover, for the Overseers of subscribing Parishes, &c. 552.
  ——— for Paupers, 554.
  ——— for the Master and Mistress, 555.
  Bill of Fare, for Ashover Poor-house, 557.
  Some accounts of Ilkeston and Rosliston Houses, 558.
 2. Expences per Head, in Subscription Poor-houses, 558.
  Abstract of 4 years very minute accounts, at Ashover Poor-house, 560, 561.
6. Box Clubs, or *Friendly Societies*, pages 564 to 578.
 1. Men's Friendly Societies, 564.
  Table of Societies, No. of Members, and their proportions to the Population, 565.
 2. Women's Friendly Societies, 566.
  Tables of them, as above, 566.
  Rules of a Women's Society in Glossop, 567.
  Great good results from Friendly Societies, 569.
  Respect paid to them by the Inhabitants, 570.
  Publican's Clubs, engender *Combinations* amongst Workmen, &c. 573.
  Means suggested, of encouraging Friendly Societies, 574.
  Saving Banks, are very useful 571 Note, 575.
  Flour Clubs, 578.
  Alms Houses, 578.
  Soup Shops—Cheap Bread, 579.
SECT. 11. Population, pages 579 to 621.
  Mr. Pilkington's Enumeration of Houses in 1788, p. 579.

(SECT. 11. Population)

    Parliamentary Enumerations, &c. in 1801 and 1811, p. 580; with a folding Table of Abstracts, &c. facing p. 577.

    Ecclesiastical and Civil divisions of the County compared, 581 Note.

    Deductions from the Population Table, 584.

    Proportions of Agricultural and Manufacturing People, 587, (535).

    Returns of Family Occupations, are very erroneous, 588.

    Directions suggested, for better determining the Employs of the People, 590.

    Area of the County and Kingdom, compared with their fertility and population, 592 Note.

    Towns, their correct Population is important to be known, 593.

    An Alphabetical List of 700 British Towns, and their People, 594.

    Its errors, and their sources, 604.

1. Tables of Births, Burials, and Marriages, 605.

    A summary of Parish Registers, in Derbyshire, 607.

    Abstracts of Parish Registers of the Hundreds, &c. 608.

    Deductions from these, and modes of calculating the Population, 610.

2. Has the progress of Population depended solely on Food, &c.? 612.

    A List of Places, increased in their Population, 613.

    Causes of these increases, 615.

    A List of Places, decreased in their Population, 616.

    Causes of this, 617.

    Has the facility of erecting Cottages, increased the People? 617.

3. Is the County over or under Peopled? 618.

    Bastardy, its mischievous effects on Society, 619.

    New Laws on the subject, should be enacted, 620.

4. Healthiness of the County, 621.

    Ague—Hooping-cough, 621.

    Stone—Insanity—Thick Throat, 622.

    Sore Eyes—Belland, 623.

    Asthmas and Consumptions, 624.

(Sect. 11. Population)
- 5. Food of the People, and mode of living, 624.
  - Soured Oat Bread, 624.
  - Fat Mutton—Milk, 625.
- 6. Customs, Opinions, Amusements, &c. of the People, 625.
  - Rush-bearing, 625.
  - Decorated Wells—Hare Supper—Superstitions, 626.
  - A proper mode of discouraging astrological imposture, 627.
  - Silly Stories and notions entertained, 627.
  - Wakes—Bull-baiting—Badger-baiting, 628.
  - Bear-baiting—Cock-fighting—Squirrel-hunt—Bonfires, 629.
  - Foot-ball—Single-stick—Gig-Fairs—Mountebanks, 630.
  - Gipsies—Fortune-tellers—The Stocks—Pounds, 631.
  - Village Prisons—Useless Gallows in Ilkeston, 632.

CHAP. XVII. OBSTACLES TO IMPROVEMENTS, pages 633 to 649.

Sect. 1. Obstacles relative to Capital, 633.
- Small Farmers are mostly deficient in Capital, 633.
- And cannot pay such Rents per Acre, as larger Occupiers, 634.

Sect. 2. Obstacles relative to Prices, 635.
- Importations of Corn have depressed prices, 636.
- Small Farmers have mostly found prices too low, 635.
- Others, less so, 636.

Sect. 3. Obstacles relative to Expences, 636.
- Rents—Labour—Roads, 636.

Sect. 4. Obstacles relative to want of power to Inclose, 637.
- Mineral Rights, and claims on the Lands, 637.
- Obstacles relative to want of proper Leases, 637.
  - Are produced by our fluctuating currency, 638.
  - Restrictive Covenants, 638.

Sect. 5. Obstacles relative to Tithes, 638.
- Lay Persons, and non-resident Clergymen, exact most Tithes, 639.

(SECT. 5. Obstacles relative to Tithes)
>Curates are often most inadequately paid, 639.
>Exonerated Lands, occasion Tithe rates of Counties, to vary greatly, 646.

SECT. 6. Obstacles relative to Poor's Rates, 640.
>Means, adopted for lessening the Poor's Rates in Derby, 641.

SECT. 7. Obstacles relative to a want of disseminated knowledge, 641.
>1. Agricultural Libraries, 641.
>Reading Societies, 642.
>2. Cheap Publications, 642.
>Agricultural Books—Farmer's and Agricultural Magazines, Farmer's Journal, 642.

SECT. 8. Enemies to Improvement, 643.
>1. Red or Wire Worm, 643.
>2. Slugs—Cockchafers, 643.
>Grubs—Worms—Flies—Caterpillars, 644.
>Insects—Ants—Crickets—Wasps, 645.
>3. Rats—Mice—Squirrels—Moles, 646.
>4. Sparrows—Bird-clacks or Windmills, 646.
>Tom-tits—Wood-Pigeons—Rooks—Swallows, 647.

CHAP. XVIII. MISCELLANEOUS ARTICLES.

SECT. 1. Agricultural Societies, 649.
>1. Those established in the County, 649.
>Derby—Repton—Hayfield, 649.
>Saltersbrook, 650.
>General List of South-British Agricultural Societies, 651.
>Earl Chesterfield's Premiums to Tenants, 654.
>Mr. Thomas's Sheep-shearing, 655.
>Philosophical Societies at Derby and Chesterfield, 655.
>Coal-Masters', and Iron-Masters' Societies, 655.
>2. Where Agricultural Societies are wanting, 656.
>Chapel-en-le-Frith (or Castleton), Chesterfield, Measham, 656.

Sect. 2. Provincial Terms, 656.
>An extensive Glossary has been compiled by the Author, for separate publication, 657.
>Local Terms are sufficiently definite, when understood, 658.
>Mineralogical, or Geological and Mineral Surveying *Terms*, should be distinct, 659 Note.

CONCLUSION.

Means of Improvement, and the Measures calculated for that purpose, 661.
Recapitulation of various hints and suggestions offered for effecting Improvement, by the Author, or others, thro' these Volumes, 661.

APPENDIX.

A Meterological Table, from a Journal kept at Derby by the late Mr. Thomas Swanwick, during 11 years, 685.
Deductions from the same, 686.
Form for the Agreements, by which *Cottages* are annexed to Collieries, or other Works, 688.
Form of ditto, by which *Work-people* are bound to Cotton-Mills, or other Works, 689.

DIRECTIONS TO THE BINDER,
FOR PLACING THE PLATES, &c.

Plate I. A Lambing Fold .... to face page 128
II. A Sheep-wash 144
III. A Map of Improved Communication 193
Folding Table of Poor-Rates Returns 529
Ditto of Population Returns 577

AGRICULTURAL SURVEY

OF

DERBYSHIRE.

CHAP. XIV.

LIVE STOCK.

SECT. I.—CATTLE.

NEAT Cattle, Beasts, or Cow Stock, for the purposes of the Dairy chiefly, form a principal feature in the economy of the Derbyshire Farms: without the County possessing any original or distinct *Breed* of these Animals, notwithstanding that some few Breeders of the County call their stock by the name of the *New Derbyshire* Long-Horn Breed, or New Long-Horn, as will be further mentioned herein.

I noticed in the County, besides a great variety of mixed or uncertainly crossed Animals, nine different *Breeds* of Cow Stock, and nine Crosses of these, and others, viz.

1. Old Long-Horn (Old L H).
2. Short-Horn (S H).
3. New Long-Horn (New L H).
4. Devon.
5. Here-

5. Hereford.
6. Scotch.
7. French.
8. Welsh.
9. White.
10. Long and Short-Horn.
11. Old and New Long-Horn.
12. Long-Horn and Devon.
13. Short-Horn and Devon.
14. Short-Horn and French.
15. Short-Horn and White.
16. Short-Horn and Lincoln.
17. Devon and French.
18. Scotch and White.

And I have arranged the Notes, made during my Survey, in this order, viz.

1. *Old Long-Horn*, Lancashire or Westmoreland; a useful sort of Cows for the Dairy, with large Bags, bred in places to the north-west of this County, as well as within it, prevailed almost generally in this County formerly, and did so to a very great degree, until about the beginning of the present Century. The owners of herds of this breed are,

Mr. William Garman of Persal Pits, in Croxall.
Mr. William Gould of Hanson-grange, in Thorpe.
Mr. Charles Greaves of Rowlee, makes Cheese or Butter, only on a small scale, rears all his Calves, the Cow Calves generally kept until they have had two Calves, and some longer; the Oxen are sold off at three to four year old, in September or October, and fetch 8*l.* to 12*l.* each, about 9*l.* on the average. In 1808 Mr. G. adopted a Cross of these Cows with a Devon Bull, that was out of a Holderness Cow.
Mr. Timothy Greenwood of New Haven, keeps 10 Cows of this breed, for making both Cream for his Inn, and Whey Butter, and Cheese.

Mr.

Mr. Joshua Lingard of Blackwell, keeps 10 Cows, and rears Calves; makes a little Butter and Cheese for sale, beyond the family consumption.

Mr. John Pearsall of Foremarke, keeps 30 Cows of this breed, for Cheese-making.

Mr. John Redsham of Dalbury.

Mr. George Toplis of Brassington.

The late Mr. John Wall of Weston Underwood,
&c. &c.

2. *Short-Horn*, Holderness, Yorkshire or Durham Cows, are brought to the Fairs at Derby and other places in this County, from places N E of it*, and many are bred in it: they are a most useful kind of Dairy Cows: kept as follows, viz.

Mr. John Blackwall of Blackwall, tried these Cows in his Dairy, but finding them too large for his cold hilly pastures, he was introducing a Devon cross.

Francis Bradshaw, Esq. of Barton Blount.

Earl Chesterfield of Bradby Park: see an account of the Milk, Cream, Butter and Cheese from Cows of this breed, farther on.

Sir Henry Crewe, Bart. of Calke, has an excellent herd of these Cows, originally purchased in Durham.

Edward Coke, Esq. of Longford, lately, but has now preferred the Devons and Herefords.

Mr. William Etches of Sturston.

Mr. William Garman of Persal Pits, in Croxall.

---

* Short-Horn Bulls, bred from the stock of C. and of Robert Collins, are kept to let, or use for hire, by Mr. Parker of Langton, near New Malton, Yorkshire.—*S.*

Mr. William Gould of Hanson-grange, part of his Dairy Cows.

Thomas Hassall, Esq. of Hartshorn, keeps 25 Cows of this breed, originally from Durham: uses his own Bull.

Abraham Hoskins, Esq. of Newton Solney.

Mr. Robert Lea of Burrow Fields, in Walton, finds these Cows not bear wintering on Straw till near the time of their calving, so well as New Long-Horns.

Earl Moira of Donnington Park, on the confines of Leicestershire: see Mr. Pitt's Report, p. 30 and 234.

Mr. Robert Needham of Perry-Foot, has a Dairy of these Cows: and annually grazes several.

Samuel Oldknow, Esq. of Mellor, keeps eight Cows of this breed, for milking.

Sacheverel Chandos Pole, Esq. (the late) of Radburne, kept these Cows, but found them too delicate for his wet clayey Lands.

Mr. James Robinson of Pyegrove, in Glossop, keeps these Cows, for milking.

Mr. John Smith of Repton, buys in a few Cows of this breed, at the Derby Spring Markets.

Mr. William Smith of Swarkestone Lows, keeps 53 of this breed, for Cheese-making, purchased originally in Durham.

Mr. George Toplis of Brassington.

The late Mr. John Wall of Weston Underwood.

Sir Robert Wilmot, Bart. of Chaddesden.

&c. &c.

3. *New Long-Horn*, New Derbyshire, Bakewell's, Dishley, or Rollright breed of Cows. Mr. William Pitt,

Pitt, in his Leicestershire Report, pages 216 to 223, has given several particulars of the early History of this improved Breed, originally from Craven, in Yorkshire, as is supposed, and first selected from thence, and improved, by the late Mr. Webster of Canley, the late Mr. Phillips of Garrendon, and Mr. Fenwick of Westmoreland, and whence the late Mr. Robert Bakewell of Dishley, the late Mr. Fowler of Rollright, and Thomas Prinsep, Esq. of Croxall, derived their chief aids, in raising the very celebrated herds of Cattle, which these three last Gentlemen possessed, and which I need not therefore repeat. Mr. William Cox of Culland informed me, that about the year 1760, the late Thomas Coke, Esq. German Pole, Esq. and Littleton Points Meynel, Esq. were instrumental in introducing improved Long-Horn'd Cattle into this County, from different persons' herds, and distributing them among their Tenants. What foundation there may be for the report, which I heard, when at Honey Hill in Chilcote, that Mr. Prinsep's stock had in part originated there, I am unable to say, from the reserve of that Gentleman, on almost all points of my enquiries, when I called on him at Croxall: and which, to the credit of the County I may state, is quite a solitary instance, among the many scores of persons of all ranks, to whom I have had occasion to apply for information: fortunately, Mr. P. is situated so very near to the edge of the County, as to have been claimed by Mr. Pitt as a *Staffordshire* Man, and in whose Report, p. 132, an account and plate of Mr. P.'s Stock will be found, and in Mr. Joseph Plymley's Shropshire Report, p. 242, the dimensions of one of Mr. P.'s Bulls is given, as Staffordshire bred. Some other particulars of Mr. P.'s Stock have also appeared in Mr. Pitt's Leicestershire Report, p. 290, in Mr. Parkinson's Rutland Report, p. 122, and the Agricultural Magazine, IV. p. 334, &c., which makes it the less to be regretted, that I am enabled to say but little about them.

The following is a List of *Bull-letters*, or Professional Breeders of Neat Cattle, of the improved or *New Long-Horn* kind, not now or formerly residing in Derbyshire, viz.

Astley, Richard, of Oddstone Hall, near Market Bosworth N, Leicestershire.

Bakewell, Robert (the late), of Dishley Grange, near Loughborough N N W, Leicestershire.

Clarke, Thomas (the late), of Lockington, near Loughborough N W, Leicestershire.

Dyott, Richard, of Freeford, near Lichfield S E, Staffordshire.

Fenwick, — (the late), of   in Westmoreland.

Fowler, — (the late), of Rollright, near Chipping Norton N, Oxfordshire.

Grundy, —, of   near Atherstone.

Hacket, —, of Nailston, near Ashby-de-la-Zouch S E, Leicestershire.

Harris, William, of Overseal, near Ashby-de-la-Zouch W, Leicestershire.

Honeyborn, Robert, of Dishley Grange (successor to Mr. Bakewell), near Loughborough N W, Leicestershire.

Kerby, Thomas, of Ibstock, near Ashby-de-la-Zouch S E, Leicestershire.

Knowles, Samuel, of Nailston, near Ashby-de-la-Zouch S E, Leicestershire, had a Sale 16th of January 1805.

Meek,

Meek, Richard, of Dunstall, near Burton W, Staffordshire.—Sale in October, 1811.
Paget, Thomas (late of) Ibstock near Ashby-de-la-Zouch S S E, Leicestershire: his stock was sold off on the 14th of November, 1793: see Mr. Pitt's Leicester Report, p. 222.
Parkinson, William, (the late) of Quarndon near Loughborough S E, Leicestershire.
Phillips, — (the late) of Garrendon near Loughborough, W. Leicestershire.
Phillips, T. M. of Garrendon near Loughborough W, Leicestershire.
Vernam, — of
Webster, — (the late) of Canley near Coventry S W, Warwickshire.
Wright, — of   in Leicestershire.

A List of Derbyshire *Bull-letters*, or Breeders of Improved or New Long-Horn Cows.

Cox, William, of Culland, in Brailsford (until 1809).
Harrison, Richard, of Ash, in Sutton on the Hill.
Harvey, Thomas, of Hoon Hay, in Marston on Dove.
Jowett, Thomas and Robert, of Draycot in Sawley.
Moore, George, of Appleby.
Mundy, Francis N. C. of Markeaton in Mackworth, (until 1808).
Prinsep, Thomas, of Croxall (until 1811).

I shall now proceed to my Notes on the various Herds or Dairies of *New Long-Horn* Cows in the County, viz.

Mr. John Bancroft of Synfin, keeps 25 Cows; hires Mr. R. Harrison's Bulls.

Mr. John Blackwall of Blackwall, has part of his Dairy Cows of this breed.

The late Mr. Francis Bruckfield of Alton in Worksworth, kept 15 or 16 Cows, derived from the Stock of F. N. C. Mundy, Esq. and of Sir Robert Wilmot of Chaddesden.

Earl Chesterfield of Bradby Park: see an account of the Milk, Cream, Butter, and Cheese from Cows of this breed, further on.

Mr. George Clay of Arleston, keeps 20 Cows, originally from the Stock of Mr. Robert Bakewell of Dishley, but since somewhat crossed by Short-horns: occasionally buys in Cows, breeds and uses his own Bulls; his present one, is a Son of a Bull that belonged to the late Mr. Thomas West of Twyford.

Mr. Samuel Cocker of Ilkeston, keeps 16 to 20 Cows; has bought his Bulls of Messrs. Thomas and Robert Jowett, but has intention of using a larger Bull, from Mr. Grundy of Leicestershire.

Mr. William Cocks of Sandiacre.

Edward Coke, Esq. of Longford, formerly, but Devons and Herefords are now kept.

Mr. Edward S. Cox, of Brailsford, keeps 40 or 50 Cows, derived from his Father's Stock, Mr. William Cox,

Mr. William Cox of Culland, kept 30 very fine longhorn Cows, pickt out of large numbers, and prior to November 1809, he sold and let Bulls of his breed, principally to distant Breeders: the Farmers round, as Mr. C. told me, bought any sort of Cows, which they thought would give Milk, and parted again with such, as on trial did not answer in their Dairies. I was in hopes of being able to have given here, the particulars, with the Buyers and Prices at

Mr.

Mr. Cox's Sale in 1809, but have not been favoured with a copy as requested.

Mr. William Eaton of Sutton on the Hill, keeps 35 Cows.

Mr. Henry Fletcher of Killis Farm, keeps and much approves this breed of Cows.

Mr. Robert C. Greaves of Ingleby.

Mr. Richard Harrison of Ash, keeps 50 Cows in Milk, his Father Mr. John H. derived his Stock originally from Dishley and Rollright; he has hired Bulls of Mr. Samuel Knowles of Nailston: he breeds all that he uses; usually raises 15 or 16 Cows and 3 or 4 Bulls annually.

Mr. Thomas Harvey of Hoon Hay, keeps 10 Cows of this breed for Butter-making, for sale in the neighbourhood.

Mr. John Holland of Barton Fields, kept this breed of Cows, but was changing them in 1809, for others more disposed to Milk.

Eusebius Horton, Esq. of Catton.

Abraham Hoskins, Esq. of Newton Solney, kept these Cows, but now crosses with the Short-horn.

Mr. Thomas Jowett, Sen. of Draycot, keeps 25 to 30 Cows, originally from the Stock of the late Mr. William Parkinson of Quarndon: raises most of his Calves.

Messrs. Thomas and Robert Jowett of Draycot, keep Cows of this breed, and 4 or 6 Bulls to let, at 10 to 15 Guineas, from May-day to 1st of September. This breed, was originally derived by their late Father Mr. Robert J. from the Stock of the late Mr. Thomas Clarke of Lockington, Mr. Thomas Kerby of Ibstock, and F. N. C. Mundy, Esq. of Markeaton.

Mr.

Mr. Robert Lea of Burrow Fields, keeps 20 Cows or more, of this breed: rears 14 or 15 Calves annually.

Mr. James Matthews of Loscoe Farm in Repton, keeps 12 or 14 Cows, bred from the Stock of Mr. Richard Meek of Dunstall, and Thomas Prinsep, Esq. of Croxall.

George Moore, Esq. of Appleby, keeps Cows and Bulls of this breed to let.

Mr. Thomas Moore of Lullington.

Mr. Benjamin Mousley of Honey Hill in Chilcote, keeps 25 Cows of this breed; he has hired Bulls of Mr. Richard Harrison of Ash, Mr. Thomas Harvey of Hoon Hay, Mr. Hasket of Nailston, and one from George Moore, Esq. of Appleby, that came from Mr. Prinsep's.

Francis Noel Clarke Mundy, Esq. of Markeaton Park, kept 25 Cows of this breed, which were very greatly admired, and of which some account has already appeared in the " Agricultural Magazine," Vol. V. p. 334, and Vol. VII, p. 332, and in Mr. Parkinson's Rutland Report, p. 122: he also kept and let a good many Bulls; but finding, as he expressed himself to me, the rage decline for his particular breed, and himself growing older, on the 21st of April 1808, he sold off the principal part of them, by auction: unfortunately, a deep Snow fell in the preceding night, and it snow'd and rain'd most of the day of sale, owing to which, probably, it was but very poorly attended. The particulars of the Sale are as follows, viz.

CATTLE—MR. MUNDY'S SALE.

Lot	Description.	Age.	Purchasers.	Residence.	Price. £ s. d.
	Dairy Cows.				
1	Rosemary, by Shakespear (a son of Mr. Paget's Shakespear): bulled by Lot 40	7	Bought in		21 0 0
	Bull Calf, off ditto, six weeks old, by Lot 40		Mr. Storer	Weston, Derbyshire	4 4 0
2	Primrose, by Shakespear: bulled by Lot 40	6	Mr. Jervis	Pinner House, Warwickshire	27 6 0
3	Broken-horns, by ditto: ditto	7	Bought in		25 4 0
	Cow Calf, off Lot 3, by Lot 40		Lord Huntingfield	Suffolk	6 6 0
4	Garland, by Mr. Prinsep's Shakespear	4	Wm. Cox, Esq.	Culland, Derbyshire	31 10 0
	Bull Calf, two days old, off Lot 4, by Lot 38	5	Mr. Storer	Weston, ditto	3 3 0
5	Buttercup, by Mr. Prinsep's Shakespear	5	Mr. Thos. Harvey	Hoon Hay, ditto	31 10 0
6	Rose, by Shakespear: bulled by Lot 40	10	Mr. Thos. Shorthose	Eggington, ditto	21 0 0
7	Birds-eye, by ditto: in-calf to Lot 40	8	Rich. Astley, Esq.	Oddstone, Leicestershire	26 5 0
*8	Marygold, by ditto: bulled by Lot 40	7	Rich. Leaper, Esq.	Derby	36 15 0
	Bull Calf, six weeks old, off Lot 8, by Mr. Harrison's Bull, bred by Mr. Knowles		Mr. Hickingbottom	Aston, Derbyshire	12 12 0
9	Posy, by Shakespear	10	Mr. Wild	West Leake, Notts	27 6 0
10	Tease, by Pine-apple (sold to Lord Talbot): bulled by Lot 40	4	Bought in		40 19 0
11	Tansey, by Marrowfat, very ill	4	Not bid for		
12	Pink, off Lot 9, by the late Mr. Healey's Bull: bulled by Lot 40	4	Sir R. Wilmot, Bart.	Chaddesden, Derbyshire	43 1 0
13	Turtle, by Marrowfat, off a daughter of Lot 17	5	Lord Huntingfield	Suffolk	27 6 0

* Lot 8 won the Héfer Prize at Derby.

Totley,

CATTLE—MR. MUNDY'S SALE.

Lot	Description.	Age.	Purchasers.	Residence.	Price.
					£ s. d.
14	Totley, by a son of Shakespear (off Mouse, the dam of Marrowfat and May-fly); bulled by Lot 39	6	Lord Huntingfield	Suffolk	36 15 0
	Cow Calf, one month old, off Lot 14, by Lot 40		Bought in		4 14 6
*15	Minikin, by Mr. Pole's Bull (bought of Mr. Walton), off Lot 15; bulled by Lot 39	4	Mr. William West	Twyford, Derbyshire	27 6 0
	Bull Calf, off Lot 15, by Lot 40	¼	B. Heathcote, Esq.	Littleover, ditto	8 8 0
16	Thistle, by Shakespear, off Old Truelove (sold into Ireland for 40 guineas): this Cow produced 17lbs. of Butter in one week!	10	Mr. Wild	West Leake, Notts	36 15 0
17	Truelove, by Shakespear, off Old Truelove: in-calf to Lot 40	9	Bought in		37 16 0
*18	May-fly (the dam of Lots 38, 39, and 40), by a Bull sold into Ireland, and repurchased 5 years afterwards: bulled by Lot 40	12	Rich. Astley, Esq.	Oddstone, Leicestershire	26 5 0
	Bull Calf, off Lot 18, by Mr. Harrison's Bull, bred by Mr. Knowles, four months old		Lord Huntingfield	Suffolk	17 17 0

Heifers, Stirks, and Yearlings.

Lot	Description.	Age.	Purchasers.	Residence.	Price.
19	Heifer, off Lot 2, by Marrowfat	3	John Toplis, Esq.	Wirksworth, Derbyshire	17 17 0
20	Butterfly, off Lot 7, by Lot 38: in-calf to Lot 40	3	Mr. Jervis	Pinner House, Warwickshire	17 17 0
21	Heifer, off Lot 17: barren	3	Mr. Robert Briggs	Thurlston, Derbyshire	17 17 0
22	Rosetta, off Lot 6, by Lot 39: in calf to Lot 40	3	Mr. Stych	Barton-under-Needwood, Staffordshire	31 10 0

* Lots 15 and 18, won the Heifer Prizes at Derby.

CATTLE—MR. MUNDY'S SALE.

Lot	Description	Buyer	Location	£	s	d
*23	Pheasant's-eye, off Lot 9, by Lot 38; bulled by Lot 40	3 Rich. Astley, Esq.	Oddstone	30	9	0
24	Hemlock, by Lot 39; in-calf to Lot 40	3 Mr. Lydale	Near Coventry	33	12	0
25	Heifer Stirk, by Marrowfat	2 Mr. Jervis	Pinner House	12	12	0
26	Ditto, off Lot 16, by Lot 38	2 Bought in		16	16	0
27	Ditto, by Marrowfat	2 Mr. Jervis	Pinner House	16	16	0
28	Ditto, off Lot 2, by Mr. Jowett's Bull	2 Rich. Astley, Esq.	Oddstone	26	5	0
29	Ditto, off Mab (daughter of Shakespear and Mouse), by Lot 39	2 R. Honeyborn, Esq.	Dishley, Leicestershire	24	0	0
30	Ditto, off Lot 5, by Lot 38	2 Wm. Cox, Esq.	Cullard	17	17	0
31	Ditto, off Lot 6, by Lot 39	2 Mr. Stych	Barton-under-Needwood	22	1	0
32	Heifer Calf, off Lot 10, by Lot 40	1 Bought in		10	10	0
33	Ditto, off Lot 9, by Prizer (which won at Derby)	1 Ditto		7	7	0
34	Ditto, off Lot 14, by Lot 40	1 Ditto		10	10	0
35	Ditto, off Lot 9, by Lot 40	1 Ditto		14	14	0
36	Ditto, off Lot 3, by Prizer	1 Ditto		14	14	0
37	Ditto, off Lot 18, by Prizer	1 R. Honeyborn, Esq.	Dishley	15	15	0

Bulls.

Lot	Description	Buyer	Location	£	s	d
38	Madcap, off Lot 18, by Shakespear	5 Bought in		30	9	0
39	Goldenrod, off Lot 18, by Ditto	6 Mr. Hinkley	Ash, Derbyshire	42	0	0
*40	Sweetwilliam, off Lot 18, by Marrowfat (a son of Shakespear and Mouse)	3 Bought in		84	0	0
				£1093	11	6

* Lot 23, won the Heifer Prize at Derby.—Lot 40, won the Bull Prize at Derby, on the day next after the Sale.

Mr.

Mr. Philip Oakden of Bentley Hall, kept 30 Cows of this breed, but now has them crossed with the short-horn.

Sacheverel D. Pole, Esq. (the late) of Radburne, kept these Cows, but found they did not stand well to their bulling; had in 1809 20 Cows of various breeds, selected for Milk.

Thomas Prinsep, Esq. of Croxall. Of the Sale of this Gentleman's famous Stock in 1811, I am unable to give the particulars, as I wished to do, because the same has not been sent to me, as requested by Letter.

Mr. Thomas Rowbottom of Ley-fields in Doveridge, keeps 30 Cows of this breed.

Mr. Samuel Rowland of Mickleover, keeps 40 Cows; hired Bulls of F. N. C. Mundy, Esq.

Mr. John Smith of Linton.

Mr. John Smith of Repton, keeps 24 Cows of this breed.

Mr. Robert Stone of Boylstone, keeps 36 Cows.

Mr. William West of Twyford.

Sir Robert Wilmot, Bart. of Chaddesden, keeps 8 Cows of this breed.

&c. &c.

4. *Devon.*—This useful breed of Cattle, are in the hands of a few persons in the County, viz.

Earl Chesterfield of Bradby Park, keeps Cows of this breed; has had Bulls from Mr. Emanuel Pester of Yeovil, Somersetshire; raises some Oxen for Work. Mr. Francis Blaikie, his Lordship's Bailiff, stated, that in March their Devon Bull thickens in the Neck, similar to what a Buck does in October.

See

See an account of the Milk, Cream, Butter, and Cheese from Cows of this breed, further on.

Edward Coke, Esq. of Longford, keeps 12 Cows of this breed, from Mr. Emanuel Pester's stock, I believe: these and Herefords are preferred by Mr. Coke to five other breeds, which have been tried at Longford; hay and straw only, given them in the winter: rears all the Bullocks he can.

Samuel Frith, Esq. of Bank Hall in Chapel-en-le-Frith.

Mr. Thomas Harvey of Hoon-Hay, keeps 6 Milking Cows of this breed; two calves are supplied with Milk by one of these Cows, till weaned at the end of August; he keeps them on straw when dry, till they spring for Calving.

Samuel Oldknow, Esq. of Mellor, keeps 8 Cows of this breed, for milking; has crossed these with the Short-Horn.

&c. &c.

5. *Hereford*, or Middle-Horn, as Mr. Blaikie proposes to call them, are bred and kept by

Edward Coke, Esq. of Longford, who keeps 12 Cows of this breed, originally purchased in Herefordshire, and has a Bull from the Duke of Bedford's at Woburn: these and Devons are preferred by Mr. Coke, to five other breeds, which have been tried at Longford: hay and straw only, given them in the winter: rears all the Bullocks he can.

Sir Henry Crewe, Bart. of Calke.

Sacheverel C. Pole, Esq. (the late) of Radburne.

Mr. William Smith of Swarkestone Lows.

&c. &c.

See the *Dead-weight*, &c. of an Ox of this breed, further on.

6. *Scotch*

6. *Scotch* or Highland.

Edward Coke, Esq. of Longford, formerly, but now Devons and Herefords are preferred there, to any others.

&c. &c.

7. *French* or Alderney.

Mr. John Bainbrigge of Hales Green in Shirley.

Mr. Joseph Butler of Killamarsh.

Earl Chesterfield of Bradby Park. See an account of Milk, Cream, Butter and Cheese from Cows of this breed, further on.

Sir Henry Crewe, Bart. of Calke, has kept 4 or 5 Cows of this breed, of a yellow dun colour, for 11 or 12 years past; their quantity of Milk small, but very rich.

Earl Moira of Donnington Park.

&c. &c.

8. *Welsh*.

Edward Coke, Esq. of Longford, formerly, but now uses only Devons and Herefords.

9. *White* Breed.

William Drury Lowe, Esq. of Locko Park, keeps 12 large white-pol'd Cows, and a Bull, from Lord Vernon's: the Cows give 16 quarts of Milk per day, in the height of summer.

Edward Miller Mundy, Jun. Esq. of Walton, has also of this breed.

Lord Vernon (the late) of Sudbury, for several years kept 14 to 20 of these white Cows, polled, with black ears and a tuft on their head, and bred 6 or 8 yearly: they are extremely gentle, are good Milkers, and have probably a mixture of the Holderness

in

in their breed. They are, however, tender animals, and don't stand well to the Bull, particularly of their own breed, and the young Cows often prove barren: on which accounts they are now crossed by short-horn Bulls, in his Lordship's Park.

Lord Waterpark of Doveridge, keeps 8 Cows, white spotted, rather smaller than Lord Vernon's breed, his Bull was from Clement Kinnersley, Esq. of Sutton in Scarsdale.

&c. &c.

10. *Long-horn and Short-horn Cross.*—Mr. Richard Parkinson, as he informs us in his Rutlandshire Report, p. 121, made an excursion from thence thro' Leicestershire and parts of this County, on purpose to examine the Neat Cattle, and compare and weigh the merits of the different Breeds, the result of all which is (page 123), an advice to the Farmers of Rutland to adopt a cross between a Long-horn Bull and a Short-horn Cow, as preferable to any other sort of stock. Mr. P. does not, however, mention having seen the Stock of the Cross he recommends, in the Dairies of any of the following Gentlemen, viz.

The late Mr. Fletcher Bullivant of Stanton Ward, for 7 years or more before his death, in 1812, kept 50 well selected milking Cows of this cross; bred from a Short-horn Bull, and the cross of his daughters, without finding any declining of Milk.

Abraham Hoskins, Esq. of Newton Solney, keeps 40 milking Cows of this cross. His New Long-horn Bulls, bought of Mr. Thomas Jowett of Draycot, of Mr. William Harris of Overseal, and Mr. Richard Meek of Dunstall, and he has proceeded to the

1st, 2d, and 3rd crosses, with Short-horn Cows and their progeny, but no further, selecting the best milkers to breed from: he raises 15 Cow Calves annually.

Mr. Thomas Moore of Lullington, has some of his large dairy of Cows thus crosst, the others are New Long-horn.

Mr. Philip Oakden of Bentley Hall, has part of his 30 Cows, crossed with a Short-horn Bull, formerly he kept only New Long-horn.

Mr. Samuel Rowland of Mickleover, keeps 40 Cows, prefers the first cross of a Long-horn Bull and Short-horn Cow, but to go much further would occasion the form to degenerate, he thinks, without an increase of Milk; they would be tenderer, eat more, and not go so well thro' the winter: he rears 14 Cow Calves annually.

&c. &c.

11. *Old Long-horn and New Long-horn Cross.*

Mr. John Webb of Barton Lodge, keeps 25 Cows, bred from old Long-horn Cows, by a Bull of Mr. William Cox's of Culland; he finds them hardy, have a great propensity to fatten, and give a fair quantity of Milk.

12. *Long-horn and Devon Cross.*

Earl Chesterfield of Bradby Park, has Cows of a Cross between new Long-horn Cows and Devon Bulls, which are found very hardy, are kind feeders, and the Meat of a good quality. See an account of Milk, Cream, Butter and Cheese from Cows of this Cross, further on.

Mr.

Mr. Charles Greaves of Rowlee, is crossing his Long-horn Cows with a Devon Bull, or rather Short-horn and Devon Bull, as it was bred out of a Holderness Cow; in 1809 he was rearing 10 Calves of this cross.

13. *Short-horn and Devon Cross.*

Mr. John Blackwall of Blackwall, procured in 1809 a two-year old Devon Bull, to cross his Short-horn Dairy Cows, in order to lessen their size, and increase the quantity of their Milk.

Earl Chesterfield of Bradby Park, occasionally buys into his Dairy (of 22 Cows) large new-milkt Short-horn Cows, and such as turn out the best milkers, are crossed by a small handsome Devon Bull; the progeny very closely resemble the Hereford or Middle-horn, are of a large size, more so than the Long-horn and Devon, mentioned above; are a hardy and valuable kind of Stock, kind Feeders, and the Meat of an excellent quality. See an account of Milk, Cream, Butter, and Cheese from Cows of this Cross, further on.

Mr. Charles Greaves of Rowlee, has a Bull of this Cross, as mentioned above.

Samuel Oldknow, Esq. of Mellor, is rearing this cross from his Devon Bull and Short-horn Cows, as better adapted than either of these breeds, for supplying Milk to the families of his numerous work-people, and to his Cotton-Mill Apprentices.

14. *Short-horn and French Cross.*

Mr. John Bainbrigge of Hales-green, has Cows of this Cross in his Dairy.

Earl Chesterfield of Bradby Park, has tried this Cross, with good effect, for Milking.

Sir Henry Crewe, Bart. of Calke, has tried and much approves this Cross, for feeding Bullocks in particular.

15. *Short-horn and White Cross.*

Lord Vernon (the late) of Sudbury, lately used this Cross with part of his White Cows, in order to remedy their want of prolific powers, as above-mentioned.

16. *Short-horn and Lincoln Cross.*

Mr. Isaac Bennet, Jun. of Over Haddon near Bakewell, has a dairy of Cows of this Cross, of a kind of dun colour; his Bulls, from Lincolnshire Bulls and Short-horn Cows, to cross with Lincoln Cows, bought in for the purpose, has used this Cross since 1798; rears 20 annually, and fattens all himself.

17. *Devon and French Cross.*

Earl Chesterfield of Bradby Park, keeps Cows of this Cross, between a Devon Bull and French or Alderney Cows, which prove very valuable Stock, of a moderate size; the Cows much improved in symmetry, hardy, with a great propensity to early fatness, even on indifferent food, and the Meat very rich: see an account of Milk, Cream, Butter, and Cheese from Cows of this cross, and the *dead-weight* of a fat Cow of the same, further on.

18. *Scotch and White Cross.*

Mr. Walter Plimley of Styd Hall, in Cubley, keeps Cows of a cross between the White Cattle of Lord Vernon's and the Scotch breed.

Food for Neat Cattle in Winter.—The chief dependance for the support of Cattle in the Winter Months is on *Hay*, in the Peak Hundreds in particular, and their Winter being rather lengthened, as observed Vol. I. p. 96, is therefore a considerable disadvantage. In the autumn of 1807, I witnessed the fall of Snow at Wensley, on the 1st of November, and on the 30th of April following, vast quantities of it were laying drifted in the hollow lanes and under the Walls at Matlock Bank, so as materially to impede my progress, but this was deemed such a winter for length, as few persons living had seen. Mr. John Dakin, of Croslow-bank, N of Alsop, foddered his Cows during 29 weeks this winter! The present spring (1812), is I fear too nearly a parallel to it in backwardness, judging from the state of vegetation near London, and South of it: and the Report from the N E extremity of Yorkshire, near the Coast, where a considerable fall of Snow happened on the 7th of May!

Roots.—Various instances of the consumption of Turnips by Neat Stock have been mentioned, Vol. II. p. 138, of Swede Turnips, Vol. II. p. 147[*] (see also the Farmer's Journal, Vol. VI. p. 54), and of Potatoes, Vol. II. p. 155; and violently as the latter have been declaimed against as Cattle food, by some persons, it cannot be doubted, but their more extended use would prove a great national advantage, by providing one of the very best resources for the occasional use of the Inhabitants, in scarcities of Corn like those of 1795, 1800, and the present time (May 1812).

---

[*] Mr. Richard Harrison of Ash, has found, that Swede Turnips given to Cows that suckle their Calves, cause the latter to Scour, which was omitted to be mentioned, Vol. II. p. 147.

Oil-cake, seems but little used in the district; I heard of its being given to Cattle, only by Mr. Isaac Bennet of Over Haddon, and Mr. Joseph Greaves. For Mr. Curwen's use of it, on his Schoon Farm in Cumberland, see the Note p. 23.

Green Crops.—Instances of the consumption of Cole or Rape by Neat Cattle, are mentioned, Vol. II. p. 140; Mr. George Clay of Synfin, gives this vegetable with Straw, to his milking Cows, which supports their Milk excellently, from October till Christmas: this use of Cabbages have been instanced, Vol. II. p. 113; and of Thousand-leaved Cabbage, Vol. II. p. 148. Mr. Robert Stone of Boylstone, in Spring, gives Winter Tares, and Oats, and Hay to his new Milk'd Cows, in Stalls.

The *Summer Food* of Neat Cattle is natural Grass, and sown or artificial Grasses: great advantage being experienced in some few Dairies, by the early or Spring use of Watered or Irrigated Grass, as observed in Section 14 of Chap. XII.; I hope that, at no very distant period, this most beneficial practice will be far more general.

Soiling or Summer stall-feeding of Cattle is practised occasionally, by a few, as mentioned in treating of the different cultivated Crops in Chapter VII.: Mr. John Blackwall of Blackwall, and Francis Bradshaw, Esq. of Barton Blount, during the hottest months, tye up their Grazing and Dairy Cattle in the day time, and feed them with mown Vetches or Spring Tares, turning them out in the evenings. Mr. Samuel Rowland of Mickleover, mentioned his conviction, of the great advantage of this practice, in economising food and increasing manure, as well as preventing *gadding,*
but

but finds labour too dear near to Derby, to admit of its adoption. I have been almost shocked to see the irritation, almost approaching to madness, among many excellent dairies of Cows, from the *Gad-fly** in hot weather, and the furious manner in which they run, with their large bags swinging, as if in danger of being torn from their bellies; surely, almost any cost should be incurred, to prevent this mischief. I did not hear of the *currying* of Milking Cows here, when stall-fed, as is practised about Leeds in Yorkshire, with great success, as I am told. Mr. John Aveson of Glossop, beds his stall Cows with Shudes or Oat Shillings†, which are too generally thrown away, elsewhere.

Fatting of Calves, Cows, Oxen, and Bulls, come, according to the Printed Plan, to be treated of here, and which I have complied with, for the sake of uniformity in the arrangement of the different County Reports, altho' *Dairying* and *Working* of Cattle, would more properly have preceded fatting of them.

My travelling Notes on the *Fatting of Calves*, are as follows, viz.

Earl Chesterfield of Bradby Park, suckles his Calves

---

* Or Breeze-fly (*Œstrus Bovis*) which perforates and inserts its eggs in the backs of the Cows, that afterwards produce the Warrel Worms or maggots, under the skin, which so dreadfully torment these animals, when neglected.

† That very spirited and able Improver, John Christian Curwen, Esq. of Workington Hall, Cumberland, stews, by means of steam, the Husks or Chaff of Wheat and other grain from the thrashing Fanners, and finds 28 lb. of the same, with 4 lb. of oil-cake, and 112 lb. of Turnips, at the cost of 9*d.* with one pennyworth of Wheat Straw, an ample daily allowance, to each of his Milking-Cows: see Transactions of the Society of Arts, Vol. XXX. p. 60: it would be worth while perhaps to try the effect of steamed Shudes of Oats, as food for Cattle.

till six or nine weeks old: Short-horns, at nine weeks, weigh about 36lb. per quarter.

Mr. Richard Harrison of Ash, suckles New Long-horns to four or six weeks old, which weigh 20 to 25lb. per quarter.

Abraham Hoskins, Esq. of Newton Solney, suckles his Bull Calves six weeks, when they average about 24lb. per quarter.

Mr. Robert Lea of Burrow Fields, suckles his Calves till three to six weeks old.

Mr. Joshua Lingard of Blackwell, feeds Calves with new Milk, for 14 or 21 days, but never lets them suckle, after the Cows have been in part milked, as some do, because it teaches the Cows to hold their Milk, and the Calves are soon taught to drink out of a Kit. After this, two quarts of Linseed boiled over a slow fire in an iron digester in two gallons of water, makes a kind of a jelly, which when cold is stirred into the Cheese Whey, on which the Calves are then fed, and are not liable to scour or run-out.

Mr. Richard Phillips of Somersall-Herbert, a small Farmer, feeds his Calves with Whey and some Milk, till four or five weeks old.

Mr. Samuel Rowland of Mickleover, suckles one-third of his drait Calves, in the Spring, to six weeks old, and sells them at 3*l.* on the average: the remainder are fatted on the top or cream of the scalded Cheese Whey, with an allowance of two quarts of new Milk each, daily, and sells, at seven to twelve weeks old, at four guineas on the average. This practice of giving the heat-raised *Whey Cream* to fatting Calves, is common about Weston Underwood, and other parts of the County.

Mr. Robert Stone of Boylstone, gives the top of his Whey to fatting Calves.

Fatting

Fatting of Cows.

Mr. Isaac Bennet of Over Haddon, feeds several three or four-year old Cows, annually, summered on Grass, and on hay and Oil-cake in the winter; sells at 35*l.* to 38*l.* a head.

Mr. Joseph Butler of Norbrigs, before turning out Cattle in the Spring to graze, gives each two doses of gentle purgatives, which he thinks saves two months time, in the fattening of a Cow or an Ox.

Earl Chesterfield of Bradby Park, fats a good many Cows: in 1805, a beautiful small French Cow, killed here, at the weight of only 25 stone, or 17¼ score lb., had within her 116lb. of loose fat: being barren, she fed in one Summer, on Grass, to such an abundant degree, as to be almost all fat, with little lean, and was scarcely eatable on that account.

In 1809, his Lordship fatted a most perfect Heifer, as to symmetry, of the Devon and French Cross, owing to her not proving a breeder, but a *Roarer* or Common-Buller, which is a kind of disease, accompanied by a falling down of the gristles of the rump, and the appearances of a springing Cow, and not unfrequently occurs to the master Cows of large herds, brought on, it is supposed, by frequently riding of the other Cows, that go to Bull, and from which they rarely recover, so as to breed, tho' for feeding they seem none the worse. See farther particulars of the feeding and *dead weights* of this Cow, further on.

Mr. George Clay of Arleston, fats his Cows at five or six years old, to 8 score or 8¼ score lbs. weight.

Mr. Edward S. Cox of Brailsford, used to graze great numbers of Scotch and other Heifers, but, in his small Enclosures, they proved so subject to the *Garget* in their Bags, that he now feeds only Oxen.

Sir

Sir Henry Crewe, Bart. of Calke, at seven years old, in general, fats his French Cows to six score per quarter: they prove disposed to fat, and the flesh kind.

Mr. Joseph Gould of Pilsbury, usually fattens 40 Cows in the year.

Mr. Richard Harrison of Ash, at different ages, according as they decline in Milk, drys his Cows in the middle of September, and fats them on the Aftermaths and Hay and Turnips; if kept till Christmas they usually weigh 8 or 9 score lbs. per quarter.

Abraham Hoskins, Esq. of Newton Solney, fats his Cows, at four to eight or nine years old, their weight $9\frac{1}{4}$ score on the average: he used to give them Hay and Brewer's *Grains* from Burton, but has substituted Swede Turnips for the Grains.

Mr. Thomas Jowett, Sen. of Draycot, fats his Cows, at three to eight years old, according as they are found to milk, gives them no Corn (but sometimes Potatoes), and brings them to 11 to 15 stone, or $7\frac{1}{4}$ to $10\frac{1}{4}$ score lbs. per quarter.

Mr. Robert Lea of Burrow Fields, fats his Cows, at four to ten years old, as they go barren, on Grass, to 7 to 11 score lbs. per quarter.

Mr. Thomas Lea of Stapenhill, buys and grazes a good many barren Cows, and Scotch Heifers.

Mr. Joshua Lingard of Blackwell, fats a few Cows, and some Oxen, of his own rearing, and sells them off before Christmas at the latest.

William Drury Lowe, Esq. of Locko Park, fats his White Cows, at six or seven years old, to 15 stone or $10\frac{1}{4}$ score lbs. per quarter.

Mr. James Matthews of Loscoe Farm in Repton, fats his Cows, at four to six years old, to 7 to 9 score lbs. per quarter.

Mr.

Mr. John Pearsall of Foremarke, fats his Cows, at different ages, according to their value as Milkers, to 8 to 10 score lbs. per quarter.

John Radford, Esq. on his Farms in Smalley and Great Hucklow, fattens a considerable number of Spayey Heifers, which he has worked, and Cows.

Mr. Samuel Rowland of Mickleover, annually fattens about 10 Cows, they are bulled early in the preceding year, and dried at the end of August; keeps them from the Bull, and feeds them in the winter with Turnips and Potatoes, clean washed and given raw in the field, when the weather is open, and at other times in open sheds: five or six .of the most backward, are finished with a month's Grass next Spring.

Mr. William Smith of Foremarke Park, in 1808 fatted a Freemartin or barren Heifer, to $10\frac{1}{4}$ score per quarter.

Lord Vernon (the late) of Sudbury, annually feeds some of his White Cows, which fat kindly, and come to 8 or 10 score lbs. per quarter.

Mr. John Webb of Barton Lodge, usually grass feeds 30 or 40 Cows.

Fatting of Oxen or Bullocks.

Mr. Isaac Bennet of Over Haddon, fattens the Shorthorn and Lincoln cross Bullocks of his own breeding; summered on Grass and fed on Hay and Oilcake in the winter. They sell for near 40*l.*

Messrs. Francis Bradshaw and Son of Newton-grange, grass feed several Oxen.

Earl Chesterfield of Bradby Park, usually buys some Hereford Oxen at Leicester Fair, on the 12th of May, and after working them two years, fats them on Grass,

Grass, Hay and Turnips: the dead-weight of an Ox of this breed killed here, will be found further on. His Lordship also annually fats Devon Oxen, bred and worked by himself.

Edward Coke, Esq. of Longford, rears as many Hereford and Devon Oxen as his stock of 24 Cows will admit, and fats most of them.

Mr. Edward S. Cox of Brailsford, was fatting some New Long-horn Oxen, and six large Hereford Oxen, in 1809, two of which last, were three-year olds, bred by Edw. Coke, Esq. of Longford: he had also a great many Galloway Scotch Oxen fatting, which were very promising.

Sir Henry Crewe, Bart. of Calke, works 12 Hereford Oxen and fats them, and breeds and fats Short-horn and French Cross Bullocks, of 9 or 10 score lbs. per quarter, which are disposed to fatten, and have good flesh. Sir Henry also fats some Scotch Oxen, annually.

The Duke of Devonshire at Chatsworth, usually fats some Scotch Oxen.

Thomas Hassall, Esq. of Hartshorn, works and fats Hereford Oxen.

Mr. John Holland of Barton Fields, usually fattens some Scotch Oxen.

Abraham Hoskins, Esq. of Newton Solney, used to fat some Oxen annually on Brewer's *Grains* from Burton, and Hay, but has substituted Swede Turnips for the Grains.

Mr. Thomas Jowett, Sen. of Draycot, rears and fats New Long-horn Oxen, at four years old, to 56 stone or $39\frac{1}{4}$ score lbs.

William Drury Lowe, Esq. of Locko Park, usually fats

fats 30 or 40 Hereford and Shropshire Oxen (having worked a few of them), on Grass, and on Hay, Turnips and Cabbages, in the Stall.

Earl Moira of Donnington Park, on the borders of Leicestershire, fats Short-horn and Scotch Oxen, as mentioned by Mr. Wm. Pitt in his Survey of that County, p. 234.

Mr. William Pickering of Mackworth, feeds a few Oxen, in part on Potatoes.

John Radford, Esq. on his Farms at Smalley and Great Hucklow, feeds about 100 Oxen and Cows and Spayed Heifers, annually, the Oxen chiefly Herefords.

Mr. William Smith of Swarkestone Lowes and Foremarke Park, works and fattens several Hereford Oxen, and feeds also some Scotch Oxen.

Lord Vernon (the late) of Sudbury, in September or October, buys in 40 or 50 Scotch Oxen, keeps them on the Aftermaths and on poor Hay thro' the Winter, and summers them in the low Meadows by the Dove: about two-thirds of them are killed into the House, and the others sold fat, in Summer and Autumn, and usually pay 6*l.* or 7*l.* a head for their keep. His Lordship also breeds and fattens some large White Oxen annually.

Sir Robert Wilmot of Chaddesden, fattens several Scotch Oxen.

Fatting of Bulls.—It is an ancient custom with the Vernon Family, at Sudbury, to buy and keep two Bulls in the Park, to be killed and given to the Poor of the place, on Old Christmas-day.

Dairy-

Dairying.

It has already been observed, that Cheese-making forms a principal feature in the Farming business of this County: before however I proceed to this subject, I will mention a few particulars relating to the supply of the Inhabitants with

Milk.—Around Derby, Chesterfield and others of the larger Towns, there are numbers of Cows kept, and their Milk sent twice a day, in small conical tubs or barrels, slung on the sides of Asses or Poneys, to supply the regular Milk-sellers or Hawkers, and the Inhabitants. Messrs. Strutts of Belper, in order to ensure a constant supply of Milk to the Inhabitants, and make it the interest of the Cow-keepers, to keep up their stock of Milking Cows through the Winter, engage for a sufficiency of Milk, at $1\frac{1}{2}d.$, $2d.$, $2\frac{1}{2}d.$ and even $3d.$ per quart, during different periods of the year, according to the expense and difficulties of procuring the article, and a person serves it out to their numerous Work-people in the Cotton Works, and keeps accounts until the end of the week, when they pay for it out of their wages. This regular supply of Milk, is found of the utmost benefit to the Poor of Belper, and it were well that the system of Milk Farming were more universally spread, in populous districts: indeed in every district the Farmers might supply the local Poor [*].

Samuel

---

[*] This was an object which the late excellent Duke of Bedford had much at heart to accomplish, on his Bedfordshire Estates; and but a very short time before his death, he directed a circular Letter of Queries on the subject, to be sent to all his considerable Dairy Tenants, intending from the results of their answers, to take some effectual measures for

organising

Samuel Oldknow, Esq. of Mellor, milks 14 Cows for supplying his Cotton-Mill Work-people and Tenants with Milk, as mentioned above.

In Glossop, where the hands employed in between 50 and 60 Cotton-mills, and in many other Manufactories, are very numerous, a large portion of the Grass Land is appropriated to raising Milk and some Butter, and but very little Cheese is made: Mr. James Robinson of Pyegrove, keeps Short-horn Cows for this purpose, and sells new Milk at 2*d.* to 3*d.* per quart, and skimmed Milk at half these prices: Butter at 1½*d.* per pound (16 ounces), standing price the year round. His Cows average 7 lb. of Butter per Week for 20 Summer Weeks, and about one half of this in Winter: at the Winter prices, not more than half the Summer quantity of Milk is sold.

Much of the Grass Land in Beighton, Eckington, and Norton, is assigned to the supply of Sheffield with Milk; the Farmers send their Milk, night and morning,

---

organising a system, for supplying *all the Poor constantly with Milk*, at the lowest prices at which it could be afforded. A similar and regular retail supply of Coals and Wood for *Fuel*, at their lowest prices, to all the Poor of his district, was equally an object with this truly great Man to accomplish, preparatory to effective steps, not less essential to their true interests and happiness, for suppressing totally the practice, unfortunately become almost general, of pilfering Wood from his Park, Plantations and Woods (see Vol. II. 307, Note), as well as from the Hedges, Rails, &c. or wherever else it could be secretly laid hold of: notwithstanding that near 2000 individuals were twice or thrice in the year supplied with Underwood, Faggots, Billets, Roots, &c. from His Grace's Woods, Plantations, &c. at very moderate prices, and *credit* for the whole given, until the conclusion of the next Harvest, annually; a system attended with immense trouble to his agents, and expense, independent of a considerable portion of the amount being lost, on the deaths of the parties, &c. that were thus trusted, among the very lowest classes.

in barrels slung on a Horse or Ass, to Agents, who for 1*d*. in the Shilling, or 4*s*. per week, for disposing of seven Gallons daily, employ Hawkers or Milk-carriers to sell it, and collect the Money; the usual sale price of new Milk being 10*d*. per gallon, and of skimmed Milk at 6*d*. per gallon; at home they sell skimmed Milk at 4*d*. per gallon. Mr. John Milnes of Ashover informed me, in 1812, that the selling of new Milk even at 7*d*. per gallon, was there a profitable concern*.

Among the Dairy Farmers of the County and others, two questions seem to be frequently agitated, viz. *which is the most profitable breed of Cows* for this purpose? and *whether is dairying or grazing most profitable?* Without offering, as others have done, any decided opinions of my own on either of these questions, I shall record all the principal facts, and opinions of practical Men, which were stated to me, during my Survey, whence others may draw their own conclusions.

With respect to the first question, as to *the Cows most profitable in the Dairy*, Mr. William Cox of

---

* In a late Communication to the Society of Arts, by John Christian Curwen, Esq. of Workington Hall in Cumberland, see Vol. XXX. p. 65, of their Transactions, he states, that by the use of steam-cooked *warm* food, two-thirds of the number of Cows now kept to supply our Towns with *Milk*, would give a larger quantity and better quality of Milk, than at present: and the Cows at the same time, by the means that he uses on his Schoon Farm, would be maintained in a state of flesh nearly fit for the Butcher. In another place Mr. C. has stated, that his Cows give 3000 to 4000 Wine quarts of Milk per annum, the average of his Dairy during four years, being 3739 quarts, sold at 2*d.* per quart, which, with the calf, produced 39*l*. 3*s*. 2*d*. each Cow per annum. Their cost of food being 10*d*. per day, or 15*l*. 4*s*. 2*d*. per annum, which with 3*l*. for interest of Money, risk, and insurance, &c. left 12*l*. of annual profit and upwards, per Cow, exclusive of her Calf.

Culland

Culland, stated, that since 1795 he had paid particular attention to Milking, as well as Fattening properties in the New Long-horn Cattle that he had selected and bred from, and had in 1809 30 Milking Cows, out of which he would pick 20, to match against any dairy of 20 Cows in the County, for quantity and quality of Milk, or he would match the whole 30 against any man's Cows of his own breeding, in the County. One of his Cows, of which he shewed me a Painting, formerly gave 12 quarts per meal, or 24 per day, and a sister of her's, nearly as much, but she did not hold out so well, yet he sold her to Mr. Richard Meek for 60 guineas.

Mr. Richard Harrison of Ash, stated, that he usually milks 50 New Long-horn Cows in the Season, and begins Cheese-making about the middle of April, and in the height of the season makes 8 Cheeses per day, of 18lb. to 20lb. each, usually at New Michaelmas 5 Cheeses, and generally leaves off Cheese-making about the end of November.

Thomas Prinsep, Esq. of Croxall, merely said to me, that " he would shew a Dairy of Cheese with any Man."

F. N. C. Mundy, Esq. of Markeaton, seemed to admit, that Cows of his New Long-horn breed seldom were great Milkers, but at the same time shewed a strong disposition to fatten; and mentioned, that some years ago he had two Sister Cows, Nos. 16 and 17 in his subsequent Sale, p. 12, only a year different in age, which were kept exactly alike, and calved nearly about the same time; that during the productive season of their milking, Thistle was constantly low of flesh, and Truelove as invariably fit for the Butcher: and that the Milk of these Cows being kept and skimmed and churned separate, Thistle was found to produce 17lb.

17 lb. of Butter per week, and Truelove only 5 lb. per week!

Mr. Thomas Jowett, Sen. of Draycot, said, that Flesh is had at the expense of Milk, in a great measure.

Mr. Robert Charles Greaves of Ingleby, said, that new Long-horns were less in fashion than they had been, for that when the grass comes to maturity, it disposes them so to feed, that by August or soon after, they become dry, or nearly so: but that fleshy Cows have some advantage over poor milking Cows, at the commencement of Winter. That he was in favour of *small Cows*, as consuming less in Winter, when unproductive: and, continuing four or five years in the Dairy, their disadvantage compared with larger Cows, come only into consideration at the end of that time.

Mr. James Matthews of Loscoe Farm, said, that his new Long-horn Cows gave but little Milk, comparatively.

Sacheverel S. Pole, Esq. (the late) of Radburne, found them not give a sufficiency of Milk.

Mr. George Toplis of Brassington, keeps Cows of various sorts, and whether long or short-horn'd, prefers fleshy ones, and is positive, that they milk better, than rough flesh't poor Cows; he sells his Cows in the Spring, sometimes at 25 to 28 guineas a head.

Mr. William Gould of Hanson-grange, says, that old Long-horn or Short-horn Cows, which are most disposed to feed, are not worse Milkers on that account, tho' he believes it to be otherwise with the new Long-horn.

Mr. Robert Lea of Burrow Fields, finds the Short-horn Cows to give more Milk than Long-horn, but of inferior quality.

Samuel Oldknow, Esq. of Mellor, finds Short-horn Cows

Cows give rather more milk than Devons; and a cross between these breeds, to answer best with him.

Mr. Thomas Harvey of Hoon-hay, finds the Devons good Milkers, and inclined to fatten.

Edward Coke, Esq. of Longford, prefers Devons and Herefords, to five other sorts of Cows that had been tried there; they give 8 quarts of Milk at a Meal, and make an hundred weight more Cheese each in the Season, than the usual average of the neighbourhood.

Sacheverel C. Pole, Esq. (the late) of Radburne, don't find the Herefords there, to milk well.

The Earl of Chesterfield, in the months of May and June 1807 and 1808, caused the following very interesting comparative experiments to be made on his Farm in Bradby Park, with seven of the different Breeds and Crosses of Cows which have been mentioned above; when alike kept, on Red Marl and Gravelly Lands, of middling quality, well watered. His Lordship first ascertained the quantity and produce of three milkings of one Cow*, of each of the seven Breeds, as follows, viz.

---

* A large Dairy of *Short-born* Cows, on the Schoon Farm of John Christian Curwen, Esq. of Workington-Hall in Cumberland, gave on the average of three years, 3945 wine quarts of Milk per Cow, annually: his previous Dairy of *Long-born* Cows had in one year given 3123 quarts of Milk each Cow, which are nearly as 29 to 23; and hence it seems probable, that Mr. Curwen's Long-horns were better milkers, compared with the Short-horns, than was the case with these two breeds on Bradby Farm.

No. page 1.	Breeds and Crosses.	Cows' Names.	Milk of three Meals.		Cream.		Butter.		Milk.		Pressed Cheese Curd.	
			qts.	pts.	qts.	pts.	lb.	oz.	qts.	pts.	lb.	oz.
2	Short-horn	Poll	29	0	2	0½	2	6½	29	0	8	5
3	New Long-horn ...	Lark	19	0½	2	0	1	10	19	0½	7	3¼
4	Devon	Marquis	16	1	1	1	1	12	16	1	5	9¼
7	French	Lily	19	0½	1	1	1	9	19	0½	8	8½
12	New Long-horn and Devon Cross	Beauty	28	0	2	0	1	13	28	0	9	0
13	Short-horn and Devon Cross	Young Poll	25	0	2	0½	2	0	25	0	8	3¼
17	Devon and French C.	Tidy	12	0	1	0½	1	5	12	0	5	0

Each of the three Crossed Cows, Nos. 12, 13, and 17, were by a Devon Bull. Column four shews the aggregate quantity of *Milk* at three milkings, from each of the Cows; and the two following, the measure of *Cream* and averdupoise weight of *Butter*, therefrom, churned at once. Column seven, shews a similar quantity of Milk from three other Milkings of the same Cows; and the last column, the weight of pressed *Cheese-curd* procured thereform, by the usual cheese-making processes of the District.

His Lordship further caused one quart of Milk to be taken, at the same milking, from five different Cows of each of the seven Breeds above-mentioned, and the produce of these, when mixed, in Butter and Cheese-curd, to be ascertained, viz.

Number.	Breeds and Crosses.	Butter from 5 quarts of Milk.	Proportions for 3 Milkings of 1 Cow. Milk.		Butter.		Pressed Cheese Curd from 5 Quarts of Milk.		Proportions for 3 Milkings of 1 Cow. Milk.		Curd.	
		oz.	qts.	pts.	lb.	oz.	lb.	oz.	qts.	pts.	lb.	oz.
2	Short-horn	7	29	0	2	8½	2	4	29	0	13	0¾
3	New Long-horn	6¼	19	0½	1	10	2	6	19	0½	9	2¼
4	Devon	8¼	16	1	1	11¾	2	9½	16	1	8	9
7	French	9¾	19	0½	2	4¾	2	4	19	0½	8	10½
12	New Long-horn and Devon Cross	8	28	0	2	12¾	2	9½	28	0	14	8½
13	Short-horn and Devon Cross	8¼	25	0	2	11¼	2	10	25	0	13	2
17	Devon and French Cross	9	12	0	1	5½	2	4	12	0	5	6½

The third column, shews the Butter, and the sixth column the pressed Cheese-curd, severally obtained from five quarts of the Milk of each Breed: in columns four and seven I have set down the aggregates of three milkings of a Cow of each breed from the former Table; and in columns of five and eight, I have calculated the weights of Butter and of Cheese-curd, which such three milkings ought to give, according to the produce of these in columns three and six, from five quarts: in order, that by comparing column six in the first, and column five in the second Table, it may be seen, how far the quantities of Butter from equal quantities of Milk from one Cow and from five Cows of each Breed, are uniform and consistent; and, by comparing the last columns in each of the above Tables, it may be seen, whether equal quantities of Cheese-curd were had from equal quantities of Milk of one Cow, and of five Cows. These comparisons exhibit differences so great, both in produce of Butter and of Cheese, between the one Cow in Table 1, and the five Cows in Table 2, in several instances, that I am am at a loss to account for them. Whence can it arise, that the five Cows' Milk yielded both Butter and Curd, at rates near one-fourth larger than* the single Cows? These differences seem to indicate, I think, that several such series of experiments as these are wanting, to come at any accurate and general comparison, of the values of the different Breeds of Cows and Crosses, to the Dairy Farmer; ascertaining the quantity of food of similar kinds consumed by each Cow during the experiments; their live weights frequently taken, might also prove useful, in calculating their comparative merits, as Dairy Cows.

His Lordship caused some Notes to be made, of the *food* consumed by each of the seven different kinds of Cows experimented on, as above, sufficient for placing them in rotation according to the quantity of food they eat, as in the following Table, viz.

Order as to Food consumed.	Nos. p. 1,	Breeds and Crosses.	Calculated Weekly Produce per Cow.			
			Milk.	Cream.	Butter.	Pressed Curd.
			qts. pts.	qts. pts.	lb. oz.	lb. oz.
1st	2	Short-horn	135 6½	10 1	11 8½	
2d	13	Short-horn & Devon Cross	116 1½	10 1		
3d	3	New Long-horn	90 0	9 0¼	7 9¼	
4th	12	New Long-horn and Devon Cross	130 1½	9 0¼		
5th	3	Devon	77 0	7 0	8 1	
6th	17	Devon and French Cross	56 0	5 1½	6 4¼	24 4¾
7th	7	French	90 0	7 0		40 1¾

In the four last columns of this Table, I have calculated the quantities for 14 milkings, or one Week, from

---

* The totals of Butter being 15lb. 0½ oz. and 12½lb. 8 oz.; and of Curd 72lb. 7½ oz. and 51 lb. 13¾ oz.

the several produces of three meals, in the preceding Tables. The reason of the several blanks left in the two last columns, are, the disagreement between the produces for one and for five Cows, in these instances: the weekly quantities of Butter and Curd here given, are calculated from the mean, of the nearly consistent results, in Table 1 and 2. It must be observed with regard to all these Tables, that separate quantities of Milk were used to obtain the Cheese-curd, from those which furnished the Cream and Butter: it being *Cream Butter* and not *Whey Butter*, which is here spoken of. The average weekly produces per Cow, from all the above experiments, are 100 quarts of *Milk*, yielding $8\frac{1}{4}$ quarts of *Cream*, and near 10lb. of *Butter*, or yielding $41\frac{1}{4}$ lbs. of pressed *Cheese-curd*, besides Whey, capable of yielding some Butter.

With respect to the 2nd Question (p. 32), *whether Grazing or Dairying is most profitable?* the late Mr. Francis Bruckfield, who was Secretary to the Derby Agricultural Society, used to mention, that Francis Bradshaw, Esq. of Barton Blount, had made a calculation of the expense of keeping a Dairy Cow, with the interest of Money for the original purchase, and the risk of death, and found the amount so very nearly what he made of her Cheese, that Calves and the feeding of Pigs were all the remuneration he received for his attention and time. I was unfortunate in not meeting with Mr. Bradshaw at home, to make more particular enquiry on this head; but happening to mention this disappointment to Mr. George Nuttall, Land Surveyor, late of Matlock, he furnished me with the following calculations made in 1809, in support, as he says, of Mr. Bradshaw's opinion as above, viz.

"*Keep*

"*Keep of a Cow for a Year, by a Person not occupying Land, at Matlock.*

	£	s.	d.
Summer Ley, from 20th of May to 11th October, 22 weeks,	3	3	0
Edish, from 11th of October to 22d of November, 6 weeks, at 4s.	1	4	0
In the Cow-house to the 20th of May following, 24 weeks, at 2 cwt. of Hay per week, 48 cwt. at 4l. per ton,	9	12	0
	£13	19	0

					£	s.	d.
Attendance, at 2d. per day,	£3	0	10				
Rent of Cow-house,	0	10	6				
Annual decrease of Capital, supposing the Cow to have cost 15l., to continue 7 years, with interest at 3l. per ann.	3	0	0				
					6	11	4
					£20	10	4

1st, To sell *Milk*.

	£	s.	d.
22 Weeks' Milk in the Pasture, at 14 quarts per day, or 2s. 4d. per day, or 16s. 4d. per week,	17	19	4
6 Weeks' ditto at Edish, at 10 quarts per day, or 1s. 8d. per day, or 11s. 8d. per week,	3	10	0
10 Weeks' ditto in the Cow-house, at 3 quarts per day, or 6d. per day, or 3s. 6d. per week,	1	15	0
Carry forward,	£23	4	4
14 Weeks			

	£	s.	d.
Brought forward,	23	4	4
14 Weeks Dry, or the Milk given to fatten the Calf,	0	0	0
Price of the Calf at a Month old,	2	10	0
Dung made during the Winter, 2 tons at 7s.	0	14	0
	£26	8	4

2d, To make *Cheese*.

	£	s.	d.
25 Weeks making Cheese, 3cwt. at 65s.	9	15	0
3 Weeks' Milk, at the end of the Edish, at 10s.	1	10	0
10 Weeks in the Cow-house, as above,	1	15	0
14 Weeks dry, as above,	0	0	0
Calf, ditto,	2	10	0
Dung, ditto,	0	14	0
Whey Butter, at 3lb. per Week, at 1s. per pound, for 25 weeks,	3	15	0
10 Quarts of Whey per day, at $0\frac{1}{4}d.$ per Gallon for 25 weeks,	0	18	3
	£20	17	3

If the Cow-keeper occupied Land, the expenses would be,

	£	s.	d.
3 Acres of Pasture Land, at 63s.	£9	9	0
Poors' Rates and other Taxes,	1	5	0
Hay-getting, &c. 20s. per acre,	1	10	0
Fencing and incidental expenses	0	6	0
Attendance, Cow-house and Interest, as above,	6	11	4
	£19	1	4"

Conversing on this subject with Mr. Francis Bradshaw of Newton-grange, he said, that the land there (Shale Limestone, Vol. I. p. 303) was good enough to fatten Oxen, which he found rather more profitable than Cheese-making, but that the reverse is the case on cold Pastures, as observed Vol. II. p. 190.

Mr. George Toplis of Brassington, stated in 1808, that Cheese-making there (with Cows of various breeds), paid better than Grazing, or Sheep-keeping.

Mr. William Smith of Swarkestone Lows, stated, that the strong Red Marl Lands, as about Ash, Brailsford, Etwall, &c. are better adapted to Dairying than to Grazing; feeding Cows, being there very apt to disorders in their bags: the Garget has prevented Mr. Edward S. Cox from fatting Cows on his Farm in Brailsford, as observed p. 25.

I was informed, that Mr. William Greatorex of Foulbrook Farm, near Derby, several years ago, had an Old Long-horn Cow, which at 12 years old he offered for sale at 6 Guineas, but being bid no more than 5 Guineas, he resolved to keep her, and that some time after she had been upon good Land, she gave 16 quarts of Milk per day, and which produced when separately churned, 16 lbs. of Butter (of 17 oz.) per week: and this probably is the Cow to which Mr. Thomas Brown alludes, in the original quarto Report, page 22.

Mr. William Cox of Culland, informed me, that Mr. Philip Burton of Churchfield Farm in Brailsford, makes 20*l*. per head of the Cows that he keeps, viz. by Calves, Pigs, Butter and Cheese, the latter being at 70*s*. per cwt.

Eighteen Cows belonging to Sir John Borlase Warren, Bart., in the absence of the family from Stapleford,

ford, Notts. some years ago, thro' the management of the Dairy-maid, produced, as I was informed, in Butter, Cheese, Calves and Pigs, 18*l.* each, on the average, within the year.

In conformity with the practice of this County, of making Cheese first and Butter afterwards, I shall here treat of these products of the Dairy, in this order; which practice of making Cheese from the new Milk, and Butter afterwards from the Whey, is either entirely unknown, or very little practised, I believe, in the greater part of England, tho' here so well established and approved.

On Cheese-making.

Respecting *the kind of Land* most proper for Cheese-making, it has often been said to me, that rather poor Land makes the fattest Cheese, as observed Vol. II. page 191, tho' less of it, than on richer soils. That old Sward makes more and better Cheese than new Lands, as observed in the same place. That Sheep kept along with Cows, lessen the produce of the Dairy, by picking most of the best Grasses, as observed Vol. II. p. 190; and that Dairy Cows kept upon Seeds or artificial Grasses, are seldom productive. Most of the above positions are however strongly controverted by others, the last of them in particular, by Mr. Samuel Cocker of Ilkeston Hall, who changes his mowing and dairying Lands alternately, instead of always feeding the same lands, according to the almost general custom; and says, that new Seeds or Lands under artificial Grasses, make very good Cheese with him, tho' requiring a little extraordinary care, as such Cheese is apt to swell at two to four weeks old, and the sides of

them

them to grow convex; but which subside again, without injury to the quality or appearance of the Cheese.

With respect to *the kinds of Cows* best adapted for Cheese-making, a great many facts and opinions have already been stated at p. 33: and some further information will be gained, from the following tabular account of the various *quantities of Cheese*, reported to me at different Dairies which I visited, as being their average annual sale from each Cow, independent of family consumption, in most instances: the breeds of Cows kept, being annexed, in the abbreviations used in page 1, viz.

5 cwt. per Cow (of 120 lb.).—Edward Coke, Esq. of Longford, Herefords.
 Mr. William Cox of Culland, New L. H. (sometimes 3 cwt.).
 Mr. William Smith of Swarkestone Lows, S. H. (net).
 Mr. John Webb of Barton Lodge, Old and New L. H. cross (sometimes 4 cwt.).

4½ cwt.—Mr. Philip Burton of Brailsford.
 Mr. Robert Stone of Boylestone, Old L. H.
 This is said by some to be the average produce of the Old L. H.

4 cwt.—Mr. William Eaton of Sutton on the Hill, New L. H.
 Mr. Richard Harrison of Ash, New L. H.
 Thomas Hassall, Esq. of Hartshorn, S. H.
 Mr. John Holland of Barton-field, New L. H.
 Abraham Hoskins, Esq. of Newton Solney, L. and S. H. (net).
 This is said to be the general Sale, around Bradburne and Longford.

3¼ cwt.

3¼ cwt.—Mr. John Smith of Repton, New L. H.
  Mr. John Wall (the late) of Weston Underwood, S. H. and Old L. H.
3½ cwt.—Mr. John Bainbrigge of Hales Green, French, &c.
  Mr. John Bancroft of Synfin, New L. H.
  Mr. George Clay of Arleston, New L. H.
  Mr. Philip Oakden of Bentley Hall, L. and S. H. cross.
  Mr. Samuel Rowland of Mickleover, New L. H.
  Mr. John Smith of Linton, New L. H.
3¼ cwt.—Mr. Richard Phillips of Somersall Herbert, various.
3 cwt.—Mr. John Blackwall of Blackwall, New L. H.
  Mr. Francis Bruckfield (the late) of Alton, New L. H.
  Mr. Thomas Jowett, Sen. of Draycot, New L. H.
  Messrs. Thomas and Robert Jowett of ditto, ditto.
  Mr. Robert Lea of Burrow Fields, New L. H. and S. H.
  Mr. Thomas Moore of Lullington, New L. H.
  Mr. Benjamin Mousley of Chilcote, New L. H.
  Mr. John Redsham of Dalbury, Old L. H.
  Mr. Francis Robinson of Melborne, various.
  Mr. Thomas Rowbottom of Ley Hill in Doveridge, New L. H.
  This is said by some, to be the average produce of the New L. H.
2¼ cwt.—Fletcher Bullivant, Esq. (the late) of Stanton Ward, L. and S. H. cross.
2 cwt.—Mr. Samuel Cocker of Ilkeston Hall, New L. H. (a large Family).
  Mr. James Matthews of Loscoe in Repton, New L. H. (ditto).

The

The *processes of Cheese-making* seem very differently conducted by different persons; Mr. Thomas Brown, in his original 4to. Report, describes three processes, but without mentioning any person's names where he saw them practised; these I shall extract, and then give my own Notes, of some of the processes that I saw or had described to me.

" *In the first Dairy*, the Mistress says, the colder the Milk is when put together for making Cheese, the better; that when she finds it sufficiently cold, she puts a sufficient quantity of Rennet to make it 'come' in an hour; it is then stirred or broke down with the hand very small, and left to settle about half an hour; then the Whey is got from the Curd as much as possible, and gathered into a firm state in the Cheese-pan; then a Vat is placed over the Pan, and the Curd broke *slightly* into it; it is then pressed by the hand in the Vat over the Pan, whilst any crushings will run from it (the more it is crushed the better); a small quantity of the Curd is then cut off round the edge of the Vat, and broke small in the middle of the Cheese; and after a little more pressing it is turned in the Vat, and the same method of cutting the edge off is again observed; afterwards, a clean dry Cloth is put under and over the Cheese in the Vat, and it is put in press for one hour; then it is again turned in the Vat, and pressed ten hours, when it is taken out, and salted on both sides (a Cheese of 12 lbs. will require a large handful of Salt on each side); it is then put in the Vat, wrapped in another clean dry Cloth, returned to the Press, and kept there two or three days, turning it every ten or twelve hours; the last time it is turned, it is put into a dry Vat without a Cloth, to take away any impressions.

" This

"This Dairy-woman's Cheese never heaves, but is in general dry, sound Cheese, and is in perfection at a year and a half or two years old.

"*The second* Dairy-woman pursues the same mode as the first, till the breaking the Curd in the Vat, which the first does *slightly*, but she breaks it very much (otherwise, she says, the Cheese is subject to be unsound); in crushing it over the Pan, they make use of a thick board of a half circular form, which covers half the Vat; they kneel upon the Board and press the other side by hand, frequently changing the Board from one side to the other; she presses it in this manner very much, but does not cut the edges off round the Vat; she has a Cloth under it in the Vat from the beginning, and turns it once in making. After being in press about an hour, she puts it into clean Water, rather warmer than new Milk, wrapped up in the Cloth it was made in; it remains in the Water about three hours (she thinks Water is better than Whey for scalding, it making the Cheese milder); it is then put in press again for an hour, and a dry clean Cloth is then put to it, and it is pressed for another hour; then salted and put in press again; she always presses them two days, and turns them every 12 hours; the last 12 hours they are pressed in a Vat without a Cloth, to take away any impressions; she always uses a Tin Girth, which she puts round the inside of the Vat, to prevent the Cheese from being pressed over the edge of the Vat; when the Cheese is in the chamber, it is turned every day for a fortnight or three weeks; after that time, once in two days; it is rubbed, whilst soft, twice a week with a Linen Cloth, and after that time, once a week or fortnight with a Hair Cloth, which keeps it very clean, and makes it look well.

"The

"*The third* Dairy-woman makes her Cheese exactly the same as the second, except scalding it, which she does in Whey, and much hotter than the second, viz. nearly boiling; she also colours it in the following manner, viz. by rubbing a piece of the best Spanish Arnotta upon a smooth stone, or bottom of an earthen pot, into a small quantity of Milk, till she has sufficient for the Arnotta to colour the whole Cheese; she then puts the Arnotta through a fine lawn Sieve into the Cheese-tub at the same time she puts the Rennet in. If in turning the Cheese, when in press, she finds it not firm, she rubs a little Salt upon it; the richer and better the Milk is, the more Salt the Cheese requires."

At the late Mr. John Wall's at Weston Underwood, I took Notes from Mrs. Wall's process of Cheese-making, as follows, viz. 21 Cows, partly of Short and partly of the Long-horn'd breeds, are milked, night and morning, and except about a gallon which is set bye each Meal for Cream and Butter for the family consumption (6 lbs. weekly), all the remainder of the new Milk is *Sied* or strained thro' a fine hair Sieve into the Cheese *Pan*, which is a large and stout bason-shaped Pan of Brass, kept very bright and clean, having a stout top edge, and a strong iron rim outside near its bottom, from whence three short iron legs proceed, for it to stand upon.

A piece of good *Earning-skin* (which will be further described below) about two inches square, having been soaked 12 hours in a tea-cup full of cold Whey, this solution, called *Rennet*, is poured into the Pan; and a dish-full of Milk being taken from the Pan, a *Colouring* Cake of prepared Arnotta (which will be mentioned below) is rubbed or grated into it, upon a piece
of

of hard Tile, until this dish of Milk, when well stirred, is judged to be high enough coloured, to give the Cheese the proper hue, in which the experience of the Dairy-maid must direct her, in suiting the fancy of her customers.

The Milk in the Pan is then stirred gently round, three or four times, by a wooden dish, and in very cold weather this is done near to the fire: the Pan is then covered over by a Cloth, to keep out dust and flies, and it stands for an hour and a half, or longer in cold weather, when, often, the Cheese is the better for this delay, tho' the trouble is much increased; in very hot weather, when the Milk is too warm at the time of putting in the Rennet, either cold water, or cold Milk of a previous Meal, is put into it; and in very cold weather, or when the milking is protracted by any cause, some Milk, heated, or warm water, is added in the Pan, to produce the proper warmth. Water thus added, is found to make the Whey poor, but not to affect the quality of the Cheese. The *Curd* being by this time formed, *come* or hard-come, as they express it, the contents of the Pan are stirred round with a dish, to break the fleaks of Curd in the Whey into small particles, when it is said to be *broke-down*, which operation should be moderately and gently performed, if rich and fat Cheese is expected.

The Pan is then left for half an hour, for the Curd to *settle* or subside in the Whey*, and all the clear or *green Whey* without Curd, is then laded off the top

---

* Those who don't regard the richness of their Cheese, commence the gathering and lading of the Whey, as soon as the Curd is broken down: and they also break it much smaller into the Vat, by which the cream or fat presses out into the Whey, and enriches it, at the expense of the Cheese.

with a dish; when the process of *gathering* commences, which is intended to expedite the settling, by very gently moving the spread hands downwards in the Whey, and slightly compressing all such parts of the sunk Curd, as feel less firm or more mixed with Whey than the other parts: the Dairy-maid also causes her fingers to meet when the hands are held edge-ways up, and sinking them at the further side of the Pan, very gently draws them towards her, in such a manner that the Curd is drawn together before them and consolidated, the Whey escaping over or thro' her fingers: and as often as any green Whey collects at top, the same is laded off, by gently sinking a brass dish into it, so that the Whey may flow over its edges, its bottom at the same time assisting in settling the Curd beneath; which operations are repeated, perhaps eight or ten times, as often as clear Whey rises, and at length the Curd lays in a lump seven inches thick, or more perhaps, on the middle of the bottom of the Pan, which is best made hollow or dishing, for collecting what is now called the *quick Curd*. This Curd is now cut with a knife into pieces or lumps about 7 inches square, and these are shoved away from the middle of the bottom of the Pan, so that a tin bason or dish can be turned upside down thereon, to prevent the lumps of Curd from slipping again into the Whey, as it collects under the bason, during the *draining* of the Curd.

The Cheese *Vat*, a shallow cylinder of turned wood, of 16 or 17 inches diameter inside, and of the shape of the intended Cheese, is then brought, and the lumps of Curd are gently lifted into it, upon the *Cheese-cloth*, previously spread over it, and when there, the pieces of Curd are carefully opened by the hand, into pieces larger than walnuts, using as little violence

as

as possible in so doing*. The Vat being thus heaped up with pieces of Curd, the Cheese-cloth, which is of the open canvas called strainer, and has one end left out when it is laid into the Vat, has then its end turned over the heap of Curd, and its corners are tucked in all round within the edge of the Vat, now standing on

---

* The importance of these precautions will appear, on considering, that in Milk, the *butyraceous* or fatty, and the *caseous* or curdy principles, exist separately, and can only be procured therefrom, separately, by the use of Rennet, and by the agitation of Churning; the mixture of these substances being only mechanical, in the former case; the many minute cells of the newly formed curd, being filled with mixtures of Cream and Whey: and hence it happens, that perfect and very repeated breakings of Cheese Curd, made from *new* Milk, and suddenly and violently pressing of it in the intervals, will produce Cheese, as horny and perfectly devoid of fat or richness, as any that is made in the Southern Butter Counties, from *skimmed* Milk, as often skimmed, as more Cream will rise; as is well known to some few Servants and others in this County, who are doomed to eat such-like lean and bad Cheese, from which all the Butter has been extracted, in the form of *crushings*, while the Cheese was making, even from new Milk.

This distinct nature and existence of the matter of Cheese and of Butter, in new or in partially skimmed Milk, will also explain the reason, why very good Butter can be made from Cheese Whey, and that where Butter and Cheese are both intended to be made from the same Milk, *either of these may be first extracted*, yet there can be little doubt of the advantages attending the making of Cheese first, it being impracticable by any known process, I believe, to retain or fix all the Butter in the Curd, which any given quantity of new Milk will produce, or near the whole of it, perhaps: and probably, where the Milk has been *once* skimmed, for producing Butter of a choice quality, a very skilful and dexterous Dairy-maid, would afterwards make Cheese from such skimmed Milk, quite as good in quality, and not less in quantity, than the average of that produced by the ordinary processes on *new* Milk, especially in the Counties and districts where Butter has been the chief object of study and attention, and Cheese is only made for the Servants' use, unless occasionally from new Milk, as an article of Luxury, and absurdly called Cream Cheese.

the *Cheese Ladder*, or frame laid across the top of the Pan, and the hands are used to compress it at first very evenly and gently. A flat round board is then laid on the top of the Curd*, and a moveable Screw Press (which will be mentioned, p. 59) is then brought down to act upon this board, at first very easily, and increasing by slow and regular degrees, during a quarter of an hour; the *Crushings* or *white Whey* which is pressed out, dripping over the edges of the Vat, into the Pan beneath.

At the end of this time, the Cloth is unfolded, and a knife is used to *pare* off such parts of the Curd as have protruded beyond the edges of the Vat, and this is done in a sloping direction all round, an inch and half at top, within the upright of the sides; these parings are then broken rather small, on to the top of the Cheese, the Cloth is folded over and tucked as before, and the Vat is returned under the Screw Press, and the progressive pressing of it is continued, for another quarter of an hour: at the end of which time the Cheese is emptied out of the Vat, by reverting it, over a table, and striking the Vat against a wall to start the Cheese, if it sticks therein; the Cloth is then unfolded from the Cheese, and the top and bottom cor-

---

* In many small Dairies, the Dairy-maid continues yet to use a half-round Board; kneeling upon it to compress the Curd, as Mr. Brown has mentioned, see p. 47; but, independently of this occupying her whole time, instead of now and then giving a turn to the screw handle, while going about her other work, the weight of the Dairy-maid brought at once on the half Board, causes too great a gushing or protruding motion in the fresh Curd, which her hands or fists are employed, in a constant kneading operation, to counteract, whereby the fat is too much discharged from the cells of the Curd, into the crushings or thrustings, and towards the latter part of the operation, her weight is often less, than might be beneficially applied, in fitting the Cheese for the large Press.

ners

ners of it are pared as before, which repeated paring, facilitates the escape of the Whey from the cavities in which it is apt to collect within the Cheese; the other end of the Cloth, to what was used before, is now laid across the Vat, and the Cheese is turned over into it, with what was before the upper side, now downwards; which last operations require to be dexterously and carefully performed, to avoid breaking or cracking the Cheese, which would greatly injure it.

The parings having been broken small on to the top of the Cheese, the Cloth is folded over it and tucked as before, and it is placed again under the Screw Press, to sustain a harder pressure than before, during half an hour: at which period the Cheese is again turned out of the Vat, and a dry Cloth being substituted for the wet one, it is replaced, and the corners neatly tucked; when it is removed to the *Cheese Press* (which will be further mentioned, p. 60), and remains there two hours, when it is taken from the Vat and Cloth, and *scalded*, by putting it loose into boiling hot water or Whey, and letting it rest therein until cold.

The Cheese is then returned to the Vat in a dry Cloth, and is again pressed for about two hours; a fresh and dry Cloth is then applied, and the pressing continued for two hours more, when it is ready for *salting* (no salt having yet been applied), which is done, by strewing about one-tenth of an inch thick of fine Salt on to the bottom and top and edges of the damp Cheese. It is then returned to the Vat, without a Cloth, and is pressed during three days, turning it every 12 hours, and applying a little fresh Salt each time. The Cheese is then placed for some time in strong brine, as much as it will swim in, and is then washed with a scrubbing-brush and hot water, to clear all the remaining Salt from the surface, when it is ready for *drying*, in an airy and shady *Cheese-chamber* (see p. 60), turning it twice a day for a fortnight, and then once a day thro' the remainder of the Summer: during which period, it is once a week rubbed well all over with a coarse Hair Cloth.

The usual weights of Mrs. Wall's Cheeses are 20 lbs. but towards the Autumn their thickness is reduced, to about 17 lbs. each, that they may dry the readier, and go off with the rest in September, when the Dealers lay in their Stocks. Three to four cwt. (of 120 lbs.) is her usual produce per Cow, between the turning out to Grass in April, and the failing of Grass in November.

When the quantity of Curd from a Meal of Milk, exceeds the quantity that a Vat will contain, the surplus Curd is preserved in cold water, until the next Meal: and if less is made, than will be properly pressed by the board, under the Screw Press, owing to its resting on the sides of the Vat, when the Cheese is first taken out to be pared, it is changed *to* a shallower Vat. Deep Vats are made to hold their proper quantity of Curd, by at first using a *garth* or hoop of tin or thin wood, that will just slide into the Vat, into which its edge is just entered, and which presses into the Vat along with the Curd, during the progressive action of the Screw Press.

The *Whey* laded from the Cheese Pan, is often reserved for the Pigs, as will be mentioned in Section VI. of this Chapter: tho' often, the *Cream* is first raised therefrom, either for making *Butter*, as will be mentioned further on, or for the fatting Calves, given with a portion of Whey or Milk, as has been stated in page 24.

Mr. John Blackwall of Blackwall, stated, that the
proper

proper heat of the Milk at the time of putting in the Rennet, is 84° of Fahrenheit: if it be much hotter, the Curd will be tough, and if much colder it will be tender, and mix with the Whey.

Mr. William Cox of Culland, stated, that the Milk should be warm from the Cows, when put together in the Cheese-pan, which, as well as all the other vessels and utensils, should be kept very clean and sweet. That the Rennet should be untainted, and used in proper quantities only, excess of it being injurious to the Cheese: that an hour or rather more after, the Curd should be broke down, in which, and in the subsequent operations, the same should be broken or crushed as little as possible. This Gentleman dairies on 100 acres of Sward, of a century old or more; his pastures are clean, and without hard-irons, thro' the occasional eating of Sheep in the spring, as observed, Vol. II. p. 195.

When at Mr. William Smith's at Swarkestone Lows, I made Notes of the processes in Mrs. Smith's well-managed Dairy, as follows, viz. In hot weather, the heat of the Milk is often too great for receiving the Rennet, until after standing an hour in a leaden vessel, or in one surrounded by water, to cool sufficiently. The weight of a Guinea of *Derby Cake*, for colouring each double Gloucester Cheese of 25 lb. weight, is dissolved over night in boiling water, and is strained, before putting into the Cheese-pan. After the coloured Milk and Rennet has stood covered up, 1½ or 2 hours, and a jelly-like curd is formed throughout it, a tin skimming-dish is used to cut or slash it, in every direction, which occasions the Whey to begin to separate and the Curd to sink. After half an hour allowed for it to settle, the clear Whey is laded off with a wooden Bowl, into wooden vessels, used for raising the Cream from it; the Curd is then gently broken up, into the remaining Whey, by the hands, and it then stands half an hour to settle and be laded again. The gathering now commences, and this forms the Curd into separate lumps, from which the Whey being laded, these are put into the Vats and slightly pressed, in the Screw-press, for 10 minutes. The contents of the Vat are then cut out into 4 or 5 pieces with a knife, which pieces are laid to drain, and are afterwards broken small into the Vat again: which last process, tho' it somewhat impares the fatness of the Cheese, is found essential (on the Swarkestone Red Marl Land, I. p. 148), for preventing the swelling and bursting of the Cheeses.

The Vat is now returned to the Screw-press, the edges are pared, if necessary, and it is then placed under the lightest of the four *Cheese-presses* which Mrs. S. uses in succession. In four or five hours afterwards it is turned, into a dry cloth in the Vat, and removed into the next heaviest Press, where it remains until next morning, when the Cheeses are salted. In this operation, about a pound of salt is rolled very fine, and rubbed on to each Cheese; next day they are rubbed with a little more salt, returned to the Vat, and removed to a heavier Press, and on the third day the salt is washed off (some Dairy-maids scald them): the Cheeses are now taken out of the Vats, and removed to a shelf in the Cheese-chamber, and are turned every day, and at 14 to 20 days old, they are removed to the floor of the Chamber, and are there turned 3 or 4 times per week. Mr. Smith used 58 Short-horn or Durham Cows, in 1808, and sold 11 tons of Cheese from them, besides supplying part of his family consumption.

Mr. Richard Harrison at Ash, used 50 new Derbyshire Long-horn Cows in 1809; began to make
Cheese

Cheese in the middle of April; in the height of the season made 8 Cheeses per day, of 18lb. or 20lb. each; at New Michaelmas 5 such Cheese; and he left off about the middle of November.

Mr. Joshua Lingard of Blackwell, keeps 10 old Long-horn Cows, for making Butter and Cheese, the Curd for which, he causes to be broke very perfectly into the Vat.

Mr. Joseph Gould of Pilsbury, has his Curd broke very small into the Vat.

Mr. William Redshaw of Longford; here, after the Curd in the Vat has been 15 or 20 minutes under the Screw-press, the same is turned out, and is entirely broken again into the Vat, and pressed by the hands of the Dairy-maid.

In my enquiries respecting Cheese-making in Derbyshire, I heard nothing of stabbing or skewering the Curd, soon after being put into the Vat, and afterwards, as is mentioned in Mr. Henry Holland's Cheshire Report, p. 279 and 280, respecting the practices of that County.

Shottle and Aldwark, two villages in this County, have long been famous for particularly rich and fine *Toasting Cheese;* made at the former place by Mr. William Statham, and others, and at the latter by Mr. Walter Buxton, and others.

Mr. Robert Stone of Boylstone, milks 36 Long-horn Cows, and makes *Stilton Cheese* of very good quality, during the Summer, and Butter in the Winter and Spring.

Rennet.—The Maw-skins, Earning-skins*, or sto-

---

* Called also in different places, the Bag, Keslop, Rendle, Yerning, &c.

machs of sucking Calves, whence the Steep or Rennet, for turning the Milk and producing Cheese-curds is prepared, are generally, in this County, cleaned, salted and dried by the Butchers, who sell them in this state, at Derby and the other Markets. Salted Mawskins will keep a long time, and are thought by some, to be better, for laying in brine 12 months or more, but are not so good, when kept a long time in a dried state: the manner of using them, has been mentioned, pages 48 and 55.

The late Mr. Hunter and Sir Everard Home, have ascribed the effect of Rennet, principally to the gastric juice which it contains; and the latter Gentleman communicated a Paper to the Royal Society on this subject, in January 1813.

Colouring.—Owing to the necessity of colouring so large a quantity of Milk, for each Cheese, the consumption and expense of Spanish Arnatto, which either is, or ought to be, the sole article employ'd, is exceedingly great: Mr. Fenna has calculated, that between 5 and 6000*l.* is thus expended annually in Cheshire, and from the double Gloucester Cheeses of this County, being much higher coloured than the Cheshire Cheeses, it seems not improbable, that in this County another sum nearly as large may be expended, on a matter of *mere fancy*, and that adds nothing to the value of the Cheese, in which part of it is infused, but more probably, detracts from its wholesomeness, in a degree, to say nothing of its effect, on the Butter which is made from this coloured Whey, and on the Animals, that are at length to take this colouring matter into their stomachs, along with the Whey. It were much to be wished, that this silly and useless

expen-

expenditure of a foreign article, were discontinued, and the colouring of Cheese, ranked only with the childish colourings of Sugar-plumbs and Sweetmeats.

Prepared *Arnatto*, in Balls, or Derby Cakes, as they are called, is sold at the Shops at 7s. or 8s. per pound: the manner of using it has been mentioned, pages 48 and 55. In many of the smaller Dairies, colouring of their Cheese has been discontinued, on account of the expense of the Arnatto. At Derby Fair in October 1811, coloured Cheeses were said to be less valued, as such, than they formerly had been.

Cheese-Presses.—Two kinds of these are now in use, one an iron Screw-press, for the first pressings of the Curd, and a larger kind, actuated by a heavy stone Weight, which is used in the subsequent operations. The former of these presses have only been in use a few years, tho' now become very common, and are kept for sale in the Ironmongers' Shops in Derby and other places. This press consists of a plate or templet, to be fixed to the ceiling of the Kitchen, just over the place found most convenient for setting the Cheese-pan, when in use: to this plate a long screw is attached by a joint, which admits of the screw and all its apparatus being turned up against the ceiling, and there fastened by a strap or hook, when not in use, and yet it can almost instantly be let down, and by turning the nut-handle, the end of the Screw can be brought to press on the board, laid on the Curd in the Vat (as mentioned, pages 52 and 56), with any required degree of force.

In fixing up these Screw-presses, care must however be taken, that the beams and joists to which they are attached, are strong enough, and the weights of Walls

on

on their ends, or of straps of iron purposely fixed from the walls below, are sufficient, to counteract the strain, and prevent the floor or walls and roof above, from being broken or lifted by the action of the screw. It will rarely happen, that the weight of a chimney can be made to counteract this upward thrust, with any safety, owing to the danger of breaking the joints of the brick or stone-work in it, and risking the firing of the house.

The larger Cheese-presses, are usually formed by a large hewn grit-stone (which are prepared for the purpose at many of the Quarries, mentioned Vol. I. p. 416), and has a screw and nut handle, for suspending it when not intended to act; which screws are commonly sold at the Ironmongers', and the country Carpenter and Mason, can together fit up this machine.

The bed beneath the press, is usually a large stone or block of wood, having a circular ring and cross within it, deeply engraven therein, for collecting and carrying off the white Whey or pressings, into a vessel set to catch them. The dimensions of the stone of Mr. Wall's Press, mentioned p. 53, was 24 × 22 × 19 inches. Mr. Smith, and many other large Dairy-men, have three or four presses, with stones of different dimensions and weights, for the progressive pressing of the Cheeses, as already mentioned, p. 56.

Cheese-Chambers.—In the new Farm premises of Abraham Hoskins, Esq. at Newton Solney, a very complete Cheese-Chamber has been constructed; the Floor is of Plaster (Vol. II. p. 16), covered by rolls of clean drawn Wheat Straw, neatly tied up in long cylinders of about an inch and a half in diameter, with old tarred rope strands, and laid close by the side of each other: these keep the Cheeses from actual contact

tact with the floor, and facilitate their gradual drying. Near to the floor, is a row of small holes thro' the outer Walls, for admitting fresh air freely.

In Earl Chesterfield's new Farm Premises at Bradby Park (see Plate I. Vol. II.), there is a very complete Cheese-chamber.

Mr. William Smith of Swarkestone Lows, has a spacious Chamber with a Plaster Floor, neatly covered with drawn Straw.

Mr. William Cox of Culland, has a large Chamber, in an out-house, with a Plaster Floor, on which the Cheeses lie, without any intermediate bedding: so has Mr. Robert Harrison of Ash, and this I believe to be rather a general practice.

It is very pleasing to see, towards the end of Summer, the neatness, order and regularity, with which the large stocks of Cheeses, in the above-mentioned Chambers, and others in the County, are kept, and rubbed, and turned at stated periods.

Mites in Cheese, are said to be destroyed or prevented, by water in which Elder Leaves have been steeped.

Selling of Cheese.—Since the making of the Trent and Mersey, the Derby and the Erewash and other *Canals*, the trade in this staple commodity of Derbyshire Farming has been much changed, and is now principally conducted as follows, viz. at several of the Wharfs on these Navigations, large Cheese-Warehouses have been built, and an experienced person appointed as the Clerk of each, whose business it is, to receive the Cheeses from the Farmers' teams, who deliver them at appointed times, rejecting and returning any which are cracked or damaged, or not sufficiently dried;

dried; and to stack up, and from time to time to turn and rub, and attend to the stock under his care, taking out all such as crack, or shew symptoms of decay, to be disposed of in the neighbourhood, for present consumption; and when the Factors or Dealers, on whose account the Cheeses are sent in, make sales or contracts for quantities of Cheese, it is the business of these Clerks to see that none but perfect Cheeses, and such as will bear the carriage, are weighed or sent off from the Warehouses: by which means the uniformity and credit of the commodity is upheld, and all parties seem benefited.

Several Cheese *Factors* reside in or near the County, and are considerable Dairy-men themselves, in some instances; some of these buy 2000 or more Tons of Cheese annually, principally on commission for London Dealers, or for those who have the Government Contracts, I believe. In the month of August the Factors usually travel round the County, and call on the small Dairy-men, to examine their Cheeses, made from four to two weeks before the time of this visit, and after feeling the latter-made Cheeses, and tapping several of the earlier ones, if the Dairy be new to them, they usually bargain for, and mark those Cheeses that they accept. The small Dairy-men, being obliged to sell for want of money, generally accept the prices thus offered by the Factors, and within two or three days after, they usually deliver it, at the Warehouses at Derby, Shardlow, Horninglow, &c.; where it is weighed, by the long-hundred of 120 lbs., and a check given for it by the Clerk: with these men, it is a ready-money trade, though some few give six or eight weeks credit. Between the larger and more opulent Dairy-men and the Factors, a strange practice seems to prevail, that of

selling

selling their Cheese at the period above mentioned, or soon after, but *without fixing any price!* Mr. Richard Harrison of Ash, informed me, that he thinks full half the Cheese of the County is delivered into the possession of the Factors, in September, without any price being fixed, until the time of payment, usually two or three months afterwards; the Factors in the mean time advancing money *on account*, in numerous instances. The prices are subsequently fixed, or pretended to be so, by *the prices* at Derby St. Luke's Fair, in the middle of October, or others at Burton, &c. before or after this period*.

Mr. William Smith of Swarkestone Lows, after justly reprobating this childish and absurd mode of dealing for Cheese, said, that he and many whom he knew, never would comply with it: that at the Derby Fair of 1808, the average price was about 70s. per cwt. but that on keeping his a week longer, he sold at 81s.; his Cheeses being remarkably neat and perfect. Mr. Hoskins in the same season sold his coloured Cheese at 82s., yet Mr. Richard Phillips of Somersall Herbert, and numerous other little Dairy-men, had sold to the Factors in September, at 60s. per cwt., as I was informed.

---

* When I was in the County of Durham in October last (1812), I was surprised to find, that this unusual and absurd mode of dealing for Cheese, had prevailed in that County and in the North Riding of Yorkshire adjoining; the delivery taking place in September, and the prices depending on the *subsequent* Cheese Markets at Yarm: but which practice, as leading to monopoly and frauds, the Farmers were then invited to resist and discontinue, by Resolutions printed and posted in Sunderland and other Market Towns.

On Butter-making.

With respect to the *kinds of Land* and *of Cows* most proper for Milking, I must refer to what is said respecting Cheese in pages 43 and 44, and on the general subject of Dairying to page 30, &c. Few if any Dairies of Cows in this County, are kept expressly for Butter-making, as in many other districts is common; but here, Cheese (or Milk in some few instances) is the principal, and Butter only a secondary consideration, and very little if any more of the latter article is made, than the consumption of the County requires, or that of its immediate neighbourhood.

The produce of *Cream* from different breeds of Cows at Bretby has been stated at page 36; my further Notes, as to the raising and management of Cream for Butter-making or for Table use, are as follows, viz.

Mr. Samuel Cocker of Ilkeston Hall, sets his Whey in shallow Leaden Vessels (such as are common in the South of England) for raising Cream, for Whey-Butter.

Mr. Joshua Lingard of Blackwell, sets his Milk in Earthen Pancheons or shallow brown glazed Pans (which is the most prevailing practice of the County, I. 450), and if his Cows are eating of Turnips at the time, he puts about a pint of boiling water to each gallon of Milk, at the time of setting it, which is found to aid the *throw* or separation of the Cream, and to prevent the taste of Turnips in the Butter. Mr. L.'s Milk, stands two to four Meals, and yet he skims it *only once*, at the end of that time; and to every four gallons of Cream, he adds half an ounce of Nitre, dissolved in water, when in the Cream Pot.

Mr.

Mr. William Smith of Swarkestone Lows, sets his Whey in shallow wooden Tubs or Kivers, for raising its Cream.

Bache Thornhill, Esq. of Stanton in the Peak, has stout grit-stone stands in his Dairy for placing the Milk-pancheons upon, in which the Milk stands three and sometimes four Meals, and is only skimmed at the end of that time, for Butter: the Cream is found to *cast* or gather better in earthen pans, than in any thing else, and the solidity and firmness of their supports, is here thought to be important.

The elegant yellow Milk Vessels used at Earl Chesterfield's, will be mentioned in speaking of his Dairy-house, further on, p. 69. In some southern parts of the County, shallow Cisterns of Swithland Slate, are in use, for setting of Milk and Whey, as mentioned Vol. I. pages 153 and 434.

Mrs. Wall of Weston Underwood, described the practice of several of her neighbours, in raising *Whey Cream*, or the fleetings or top of the Whey, to be, by boiling the Whey in a brass Pan, and when it begins to skim over, they pour in a quart or two of cold sour Whey or Butter-milk, and stir the whole gently with a tinned ladle; the Cream then immediately begins to rise, and is skimmed off: it is then stirred again, and skimmed, and so on, as long as any Cream rises. Into every two gallons of this Cream, when cold, two ounces of Saltpetre is usually put to dissolve, and be churned with it, and which they find a preventative against the taste of Turnips in Butter.

On the Shale in Ashover and some other places, where there are Sallow and Willow Trees in the hedges, the Farmers, before they turn out their Dairy Cows into the Edishes of such fields in the Autumn, strew some slacked and powdered lime, over the leaves of all such, that the Cows could reach, in order to prevent their eating them, to injure the taste of their Milk and Butter.

The Rosam or wild Garlick growing in Cow Pastures, has been mentioned Vol. II. p. 194, as giving a taste to Butter, which is not disrelished in some places, when not too prevalent: some Pastures on Grit-stone, about Dronfield, produce this weed.

Mr. Thomas Brown, in the original 4to. Report, describes the processes of three different Dairy-maids in making and managing *Whey Butter*, as follows, viz.

"*The first* gathers no Butter from the green Whey, but from the crushings which she sets up in Pans for 24 hours, then skims off the Cream or thick part, and immediately boils it; then before it is cold, she puts it into an earthen Pan, in which she collects a week's Cream for churning; she likes her Cream to be sour, and for that purpose saves a little of the last week's, which she puts in the bottom of the Pot, and adds the next week's Cream to it; she also puts her Cream into the Pot before it is cold, and in Winter sets it near the fire; in Summer she churns her Milk and Whey Butter together; she also makes her Milk Cream as sour as possible. It has been recommended to her by an experienced Mistress, to put a small quantity of Saltpetre dissolved in water into her Milk which she sets up for Cream, as a good thing for making the Milk throw up a greater quantity of Cream, making it churn easier, and giving the Butter a better colour; it was also told her, it was a good thing for preventing the Butter from having any disagreeable taste, such as arise from the Cows eating Turnips, Cabbage, &c. She has tried the experiment, by putting about the size of a hazel nut of

Saltpetre

Saltpetre into as much Cream as made 6lbs. of Butter, and approves of it very much; but did not put it into the Milk as advised, thinking it would spoil it for family uses.

"*The second* sets up all her Whey, as well green as white, for Cream, which she gathers every 24 hours, into large earthen Cream Pots; she boils hers only twice a week, and differs from the first by wishing to keep it as sweet as possible, and for that purpose changes it into sweet Pots twice a week; she gathers a small quantity of Milk Cream, which she churns together with the Whey Cream; her Butter is very good, and very little inferior to real Milk Butter.

"*The third* gathers all her Whey, as well green as white, into a large brass Kettle, over a fire, and as the top or thick part begins to rise, she takes it off with a brass skimmer full of holes; when her Whey is near boiling, she puts into it about a quart of cold spring water, and as much cold sour Butter-milk, which makes it throw up more top or thick curdy Cream; she sets it by till cold, then puts it into her Cream Pots for churning; her Cream Pots have all tap-holes at the bottom, by which she twice a day draws off the thin Whey that has settled there. She likes to keep her Cream sweet, and for that purpose sets it in a cold place, and changes it into clean sweet Cream Pots every two days. In Summer she churns her Milk Cream and Whey Cream together, once a week; her Butter is very good, and from her Cheese being coloured, it always looks well, and she sells it for the best price in the market, and mostly to the same persons who have bought it before. Another Mistress of a Dairy, equal for her good management and civility, informs me, that she sets both her green and white Whey by, for skimming, 24 hours;

24 hours; when skimmed, she puts about a gallon or six quarts of boiling clean water to a Pancheon (an earthen vessel that holds about six quarts) of Cream, and stirs them together well; the water will settle to the bottom, and when cold she takes the Cream from it, and puts it into Cream Pots for churning; she stirs it well in the Cream Pots once or twice a day, and changes it into clean ones twice a week, to keep the Cream as sweet as possible. This is a very much approved mode of making Whey Butter."

Mr. John Blackwall of Blackwall, stated, that the proper *heat* at which Cream should be put into the Churn is 60° of Fahrenheit.

Mr. William Green of Strind's Inn, Yorkshire, and others in the north-east part of this County adjoining, use finely powdered Salt instead of brine, for flavouring their Butter in a due degree, for immediate use.

Whey Butter, when well made and fresh, appeared to me in all respects as good, and not to be distinguished from Cream Butter; it has been said, that it will not keep good quite so long as Butter made from Cream, but in the large Towns in particular, it seems in equal request with any other.

The *Churns* used in this County have been described Vol. II. page 68, and several Makers of them mentioned, to which I wish now to add the name of Mr. William Candee of Brockhurst in Ashover, which is there omitted.

Dairy-houses —In the Plan and particulars of the Earl of Chesterfield's Farm Premises in Bradby Park, Vol. II. page 10, the situation and general arrangement of his Lordship's Dairy establishment will be seen. It remains here further to mention, that considerable
pains

pains have been taken in the construction of this Dairy, by ventilation in hot weather, and the use of well-disposed flues in Winter, to preserve as nearly as possible an even temperature in it, at all times, by which the sourness of Cream in Summer, and disagreeable bitter of that which has been frozen in Winter, are in a great measure prevented, and other advantages are gained. The Milk here is set to raise its Cream, in very handsome yellow dishes, with lips for pouring it out, which are made at the Pottery at Wooden Box, near Ashby Wolds, Vol. I. p. 449. Several other neat and well-contrived Dairies are to be found in this County.

Working of Oxen and Heifers.

Oxen or Bullocks are much less generally used for draught in the County now, than they were at a distant period, as observed Vol. II. p. 95; yet their use has probably increased somewhat, of late years: my Notes on the subject are as follows:

The Earl of Chesterfield of Bradby Park, works several Hereford and Devon Oxen, the Herefords are usually bought in at Leicester Fair on the 12th of May, worked two years, and then fatted on Grass, Hay and Turnips: the Devons are bred on the Farm, and worked two or three years before fatting as above. Sometimes four Oxen with a Horse before them, are used in ploughing: they do the carriage of Dung, Hay, ground-work, &c. Harrowing, &c. In Summer they are fed upon Grass, without Corn, and in Winter on Straw, the tops, outsides and bottoms of Hay-stacks, and the refuse Hay from the Stables, which has been browsed and pulled down among the litter by the Horses, of which they seem particularly fond

fond, and will toss the litter about with their horns, to pick out, even that which may have been wetted with the Horses' urine.

Sir Henry Crewe, Bart. of Calke, works 12 Hereford Oxen, in all the labour of his Farm and Park.

The Duke of Devonshire has usually eight Oxen in work on his Farm at Chatsworth, half of which are used on alternate days, and rested on the others, by which treatment they continue to improve in condition, until put up to fatten.

Thomas Hassall, Esq. of Hartshorn, works six Hereford Oxen, single in Collars, for three or four years, and then fats them.

William Drury Lowe, Esq. of Locko Park, has four Hereford and Shropshire Oxen in work, at which they are kept three or four years, before fatting.

Mr. William Smith of Foremarke Park, usually has some Oxen in work, and uses them with Horses, in his double-shared Plough.

Heifers.—The practice of Mr. Honeyborn, which Mr. Pitt commends in his Leicestershire Report, pages 232 and 245, is followed in this County by Mr. John Radford, on his Farm in Great Hucklow, where four teams of Spayed Heifers are kept in regular work, and he calculates on considerable savings by the practice, principally on account of the difference in the expense of keeping them and Horses.

Stalls, Yards, Sheds, &c.—In Chapter III. Sect. 2, on Farm Buildings, Vol. II. p. 9, the Farm Offices of Messrs. Francis Bradshaw, Esq. of Barton Blount, the Earl of Chesterfield of Bradby, Joseph Gould of Pilsbury, Timothy Greenwood of Newhaven, Abraham

ham Hoskins, Esq. of Newton Solney, William Drury Lowe, Esq. of Locko-Park, Robert Stone of Boylstone, and Sir Robert Wilmot, Bart. of Chaddesden, have been mentioned (and one of them delineated and described), as being worthy, as others also are, of notice, for their many improved conveniences for conducting the farming business; and within the County there are doubtless many others deserving of my commendation, either for their general design, or the novelty or utility of some of their particular parts, but of which I did not happen to obtain notes: what further occurs to me under these heads is as follows, viz.

Cow-houses.—These at Newton Solney, for tying up 34 Milking Cows and 12 fatting Beasts, and at Bradby for 14 Milking Cows, seem very complete: at the latter place they are white-washed annually. At Boylstone 36 Cows are tied up, with a board between every pair: the Cows are let out into the Yard in the middle of the day, and at different times to water, there being none in their stalls. Mr. Joseph Gould's Cow-stalls in his new Farm Premises at Pilsbury, are well contrived.

A sort of *Cow-tyes* formerly were general in the County, called a Sole and Bosquin, formed almost entire of wood, viz. a flat piece of wood which lay across the Cow's neck, into holes near the ends of which, the ends of a bent piece of tough ash were inserted, forming a bow or collar that embraced her neck; the bow had knobs at its ends to prevent its drawing out, and one of these had a hole and mortice connected, by means of which the sole was taken off, tho' the spring of the bow prevented this happening by accident: two small wooden hoops or bosquin-rings were connected together by an iron pin with two heads, so as to form a swivel, and one of these rings was put on to the bow of the sole, and the other on to a smooth upright piece of wood called a bosquin, which was set into the ground at bottom, near to the rack, and the other end was spiked to a beam or runner of the Cowhouse above: many of these ancient Cow-tyes are still in use in the Peak Hundreds, where others of the ancient wooden implements remain yet in use; but Chaintyes, with a ring and bosquin are now in the most general use. The fixed upright Cow-fatners, which are almost general in the southern counties, that Mr. Parkinson has drawn and described in his Rutland Report, p. 183, are little known in this County, I believe.

Here I may, perhaps with as much propriety, mention, as in any other place, that *Cow-loggers* are here very commonly used, for Cows that are turned out into lanes and commons, or on such as are apt to break their pastures; consisting for the most part of a long stout stick, slung before the Cow's neck by a chain or rope, round it, and balanced, so as to hang horizontal, to prevent their bursting through hedges. In Skegby I saw a Cow loggered by a large forked stick, which trailed between her legs; the fork embraced her neck, and was suspended therefrom by a chain. In the Cow-paddocks of Edward Coke, Esq. at Longford, the Cows all had neat brass knobs or *tips*, screwed on to their horns, and gave them a neat appearance.

Bull-houses.—In the Earl of Chesterfield's Farmery, there is a well-contrived House for two Bulls, (No. 12, p. 10, Vol. II.) which are kept tied up: the passage behind the stalls is of the same width as each stall, and a tall gate hung to the post of the partition of the stalls,

prevents

prevents the possibility of the two Bulls getting together, in case they get loose: at the same time, that either of these bulls can be shut close up, while the other is letting out or cleaning, with the greatest ease and safety. Mr. Robert Stone's, and most others of the newly-erected Premises, have proper Bull-houses in them.

Calving-house.—This very necessary appendage to the Farm Premises, has been provided with every convenience for two Cows at the same time, in the Earl of Chesterfield's Farmery; p. 10, Vol. II.

Calf Fatting-house.—This at Bradby, is furnished with a strong and pretty close latticed floor, of Oak bars, above the paved floor, for draining away all moisture, and supplying fresh air to the Calves. Mr. Stone and many others, have also been very attentive to cleanliness in the construction and use of this part of their Farm Premises.

Live and Dead Weights of Fat Neat Cattle.—Considering how much Weighing-machines are spread in this County, as observed, Vol. II. p. 65, I was somewhat surprised to meet with no registers kept, of the *live-weights* of Oxen or Cows, during the progress of their fattening, not even a recorded instance of such weight, immediately before slaughtering: a very desireable species of information, which is even rarely found in the numerous documents collected by the *Smithfield Club* in London (a Society suggested by a Farmer of this County, as observed, Vol. II. p. 362, Note) since 1798, and published for several years in Mr. Arthur Young's " Annals of Agriculture," and since in Mr. V. Griffiths's " Agricultural Magazine," annually,

annually, respecting the dead weights of the fat Oxen and Cows that are yearly exhibited in December, for its premiums*.

The following account of the feeding and produce of a Hereford Ox (called Merryman), is extracted from the Earl of Chesterfield's Farm Book at Bradby Hall, December the 18th, 1802, viz.

---

* Mr. Layton Cooke, in his very useful " *Tables* adapted to the use of Farmers and Graziers," (which no experimental Agriculturist should be without), has, partly from these documents, and partly from others collected by Lord Somerville at his Cattle Shows, and in His Majesty's Victualling-yard at Deptford, the yard of William Mellish, Esq. &c. found the component parts of a ripe or extra fatted *Ox*, properly fasted before killing, and weighed when cold, to be on the average as follows, viz.

Carcass or quarters, skirts and kidneys (meat .6 and bone .1) ...	.7000
Loose Fat	.0900
Hide and Horns	.0550
Head, Brains, and Tongue	.0230
Feet	.0140
Heart, Lights, Sweetbreads, and Bladder	.0084
Tripe (without Fat), Feck, Reed, Liver, Gall, and Melt	.0256
Entrails and contents	.0362
Blood	.0278
Loss, by evaporation	.0200
Live Weight	1.0000

The proportionate weight of the Carcass of an Ox moderately fat, he has found to be .65, or sixth-tenths and five-hundredths of the live-weight, and that of an Ox merely marketable for want of more perfect fattening, to be .6 or six-tenths of the live weight. From these data, very convenient and copious Tables are calculated by Mr. Cooke, for finding the weight of carcass and other valuable parts of an *Ox* or *Cow* by inspection, and their value, where the live weight has been ascertained. See the Farmer's Journal for May 3, 1813, p. 251.

The proportional weight of the Carcass of fatted *Calves* he has found to lie between .56 and .64 parts of the live-weight considered as 1.00, as in the case of Oxen above.

		£	s.	d.
To the prime cost of Merryman Ox,		18	0	0
Lattermath in 1801,		1	11	6
Summer's Grass, 1802,		3	3	0
House-feed 40 weeks, at 8s. per week,		16	0	0
Total Cost,		£38	14	6
By-Carcass, 1235 lb. (or 15 score 9 lb. per quarter) at 7d.		36	0	0
Hide 136 lb. at 6d.		3	8	0
Tallow 180 lb. (12 st. 12 lb.) at 6d.		4	10	0
Offal,		0	10	0
		44	8	0
Net profit to the Grazier of this Ox,		£5	13	6

Another extract from the same Book, relates to the beautiful Devon and French Heifer mentioned p. 25, as follows, viz. She was out of an Alderney Cow by a Devon Bull, calved in May 1805, was kept in store order until April 1808, when not proving a breeder, she was put to grazing. During the following winter, she was kept on Hay and Turnips, and in the Summer and Autumn of 1809, was fed on Clover, Hay, Turnips, Grains, and ground Buck-wheat. When slaughtered on the 18th of December, 1809, her dead-weight was as follows, viz.

Carcass, { Fore-quarter, 245 lb. / Ditto, 245 / Hind-quarter, 245 / Ditto, 237 } 972 lb. or 12 score, 3 lb. per quarter.

Tallow, cake, and rough fat, 150 lb. or 10 stone 10 lb. Hide, 69 lb.

The grain of the meat was beautiful, and a more perfect Carcass was perhaps never seen. The bone of the

the leg girted as follows, viz. below the Knee $3\frac{7}{4}$ inches, and below the Hock $3\frac{7}{4}$ inches.

Distempers.—The diseases of Neat Cattle seem now much better understood and attended to, than they were formerly, and the losses therefrom are proportionally diminished; principally owing, to the more liberal education, and to the more frequent employment of the professional Cow or *Cattle Doctors*, in attending Cattle under disease, or in administering preventive medicines: the names and address of these Gentlemen are as follows, viz.

Mr. Henry Bowyer (surgeon) of Brailsford, near Ashburne, S E.

Mr. George Brough of Upper Town in Ashover, near Chesterfield, S W.

Messrs. Thomas and George Draper (Vet. Surg.) of Derby; and of Castle Donnington, near Loughborough, Leicestershire, N W.

Mr. James Horsley of Melborne, near Derby, S.

Mr. William Robinson of King's Newton, near Derby, S.

&c. &c.

The most prevailing, as well as that disease which is attended with the greatest loss of Neat Cattle in this County, affects only young Beasts or rearing Calves, and has been mentioned and written for me, with all the following Names, viz.

Black-leg, Black-quarter, Foul, Hian, Hiand, Hion, Hyan, Hyon, Iron, Murrain, Quarter-evil, Spade, Speed, &c. I shall go thro' my travelling Notes in the order of the places, and mention such particulars

as

as to the symptoms and treatment of this terrible disorder, as were stated to me.

At Blackwell, Mr. Joshua Lingard stated, that the Black-leg sometimes happens to his rearing Calves, attacking those that are in the best condition; the causes of it are unknown, and it is without a remedy: it is not contagious.

Bretby: Mr. Francis Blaikie, the Bailiff on Bradby Farm, describes this disorder as a sudden and incurable disease, or mortification of some membrane under the skin of the leg, the shoulder, or the back, probably occasioned by sudden exposure to Cold Air: Calves kept in the House are never affected by it, but in open sheds, neglected and become dirty, so that the Calves lie out on the cold ground, they are very commonly struck by it, and often, in five hours after they first appear to loath their food and look heavy in the eye, they die of it. Yearling Calves, if very thriving, are often struck. As a preventive, Mr. Blaikie gives all his yearling Calves, in October, in March and April, each two ounces of Nitre, dissolved in water, at three separate periods, and bleeds them as many times, alternately.

Croxall: Thomas Prinsep, Esq. formerly lost many of his yearling Calves by the Black-leg, but by persevering in the following preventive course, has not lost any for some years past; from the beginning of March to the end of April, he bleeds his rearing Calves, regularly every fortnight, giving them each time a dose of Salt and Nitre, dissolved in human urine.

Foremarke: Mr. John Pearsall, formerly lost a great number of the best-conditioned of his rearing Calves, by the Black-leg, which is not a catching disorder, but seems to originate in each individual; appearing
sometimes

sometimes first on the back, but more commonly in the fore-quarter near the heart, causing a swelling of the skin, which under the hand feels like dead leaves, and rustles almost like dried and shrivelled parchment, when the hand is rubbed over it: he used to bleed in the Spring, and give Nitre and Camphor, or Camphor and Brandy, which lessened, but did not prevent the mortality; out of six or seven attempts at a cure, by means of rather cruel kinds, he succeeded in curing only one calf. For some years past, Mr. P. and numerous other Farmers, have employed Mr. James Horsley in preventing the disease, which he does, by a drink Medicine, given Monthly, from October to March; he charges 9*d*. each dose; and he uses a seton, or settering of a black Helebore sprig and grease, in the brisket of the animal.

Ilkeston: Mr. Samuel Cocker, has sometimes lost one-third of his rearing Calves, under the year old, by the Black-leg: in various attempts at a cure, during 25 years, he never succeeded but once, with a Cow-Calf, and that was, by scarifying the part where the mortification had begun, and putting small pieces of Saltpetre under the skin, a hot fire-shovel being held to the same to cauterise it; after which the part was bathed by cloths dipped in hot saltpetre brine: after two or three weeks the skin and mortified part came off, and by carefully dressing the wound to keep off the flies in summer, it healed, and she grew up to be a useful Cow.

Ingleby: Mr. Robert Charles Greaves, by way of preventing the Black-leg, bleeds his yearling Calves, and puts a rowel in their briskets: the disorder is said to arise from a quick transition, either from good to bad, or from bad to good keeping; no cure is known.

Mark-

Markeaton: Francis N. C. Mundy, Esq. formerly suffered much from the Black-leg, usually when his Calves were between eight and ten months old, but he did not observe, that the fattest of them were more subject to be attacked than the leaner ones; of late years he has engaged Mr. George Draper, to apply preventive medicines to his Calves, and has thus greatly lessened the mortality among them. In a conversation with Messrs. Drapers, they informed me, that they can pretty certainly prevent this disorder, but very rarely cure it; and that it prevails most in bleak and cold situations.

Norbrigs: Mr. Joseph Butler described to me, a disorder of his yearling and two-year old Heifers, when not in calf, under the name of Speed, which may not be the same exactly with the Black-leg, but more allied perhaps to Blood-water, which will be treated of below: of this disorder many of Mr. B.'s young Cows have died suddenly, and their blood has afterwards been found quite black: as a preventive, he bleeds about the third week in March and first week in April, which generally succeeds; but if the blood is still found black or dark coloured, at the second bleeding, he repeats it a fortnight afterwards.

Stanton in the Peak: Mr. Joseph Gilbert, agent to Bache Thornhill, Esq. stated, that the Black-leg was so prevalent there some years ago, that no Calves could be raised, on some Farms, before it was discovered, that the disease commences in a vein under the skin between the Claws before, whence it spreads in eight or twelve hours time, and produces an almost general mortification under the skin, and invariably kills the animal; unless, that immediately after the appearance of the disorder, the diseased vein is cut out and

the

the part scarified and hot fomented, which has saved some few Calves. Bleeding and roweling, are the preventives now used: it occurs most after cold and wet weather, and the Calves being brought suddenly into sheltered and warm situations, which sudden changes are therefore now avoided.

Waldley: Mr. Thomas Bowyer says, that the Black-leg is prevalent about there, and no cure is known for those Calves affected. On Hanson-grange Farm in Thorpe, this disease has very rarely occurred.

Blood-water, Bland-water, Bad-water, Foul-water, Red-water, Staling Blood, Bloody Flux, &c. is another serious disorder of the Neat Cattle of this County; and seems to affect Cattle of all ages.

At Blackwell, Mr. Joshua Lingard informed me, that tho' the Red-water has not been known there, that it has prevailed in Taddington, Priestcliff, Brushfield, and lately in Wormhill, near to there, tho' it is not supposed to be contagious: the cause of this disease seems to him to be local, and is unknown: a great deal of quackery has prevailed, in treating Cows under this disorder, but purgative medicines seem the most effective towards removing it.

Little Longsdon: Mr. Charles Greaves of Rowlee, informed me when at his house, that very few of the Beasts which graze in the Limestone Pastures N of Little Longsdon and about Wardlow-Hay, escape an attack of the Blood-water, tho' it has been said, that they never have it a second time. Sudden changes from heat to cold, and *vice versa*, are supposed to occasion it; a looseness attends its first attack, succeeded by obstinate costiveness. Small red *Ants*, and their eggs and nest, are there bruised, and mixed in a drink, for Cattle under this disorder.

Rowlee

Rowlee in the Woodlands of Hope: Mr. Charles Greaves' Cattle grazed in the Woodlands, are very subject to the Blood-water, about the time that they begin to mend in condition, in the Spring: those which were previously the poorest, being perhaps the most subject to it; a looseness precedes, the Cattle beginning to stale light red, deep red, or blackish Urine, and if their water is not *turned,* or become again of its natural colour in four or five days' time, the animal dies. At or before the turning of the water, obstinate costiveness ensues. Mr. Greaves prepares an infusion of Moorhawk Grass (   ), which grows in the Valleys in the Woodlands, a large handful of which is boiled for half an hour, in three pints or two quarts of skimmed Milk, and squeezed out; this tea is given warm, with the drenching-horn, and the animal is turned out to graze, if the weather is cool, but kept in the House and fed with cut Grass or Hay, if it be hot weather: in 24 hours after, if the water is not turned, the same dose is repeated, or instead thereof a pint of good red Port Wine is given, cold. Or, a pint of green Nettles, when chopped small, are mixed in about two quarts of cold sour thick Butter-milk, and given with a horn. Cattle-doctors and Cow-leaches, give different medicines of their own preparation.

Out of 24 of his Cattle which were affected, in the Spring of 1809, Mr. G. lost two, one a pretty good Bull, of four or five years old. It is maintained by many, that Beasts never have this disorder but once, and others say, that it generally attacks them in the first Spring after coming to graze in the shrubby Valleys of the Shale Woodlands; but these Mr. G. considers as doubtful positions.

Stanton in the Peak: Mr. Joseph Gilbert says, that

the Blood-water often occurs there, and is frequently cured by purgative medicines.

Waldley: Mr. Thomas Bowyer says, that Cows of all ages there are affected, occasionally, with the Blood-water; that a Medicine furnished by Mr. George Draper of Derby, at 3s. per dose, generally effects a cure, tho' sometimes it wants repeating.

In a conversation with Messrs. Drapers, they stated this disorder to prevail much on the Limestone-Shale lands in Shottle, Turnditch, Hazlewood-lane, &c., and also on the Red-Marl lands in Osmaston, Waldley, Doveridge, Needwood-Forest Staffordshire, &c. Cattle of different ages being so affected, soon after being pastured on these soils; which are supposed, by the diuretic quality of their herbage, to disease the kidneys of the Animals. This disease very rarely occurs on Hanson-grange Farm, on the 4th Limestone.

Mr. John Nuttall of Matlock, thinks that herbage in want of Sun, on steep banks facing the N, occasions this disease, and refers to Bonsal, Matlock, Snitterton, Winster, and some other places which have been mentioned above, as places where it prevails. Others maintain, that old Pastures, and particularly where Nut-hazel bushes and hedges prevail, communicate this disease, as observed, Vol. II. p. 91: it occurs at Cox-bench on Coal-Shales and Grits; and some there say, that their stall-fed Cows have been attacked by it.

In Staffordshire, Cudweed *(Gnaphalium germanicum)* is said to be good for Cattle labouring under this disorder: see Mr. W. Pitt's Report, page 222: see also Mr. J. Pilkington's "View of Derbyshire," Vol. 1. p. 453.

Scouring, Ger, Gerring, Running-out, Shooting-out, &c. is a disorder of Cattle, less prevalent, I believe,

lieve, than the above. At Ash Mr. Richard Harrison informed me, that there, as also at Quarndon in Leicestershire, Calves of three or four weeks old, are much subject to scour, and that Swede Turnips given to Cows, seem to occasion their sucking Calves to contract this disease (as mentioned page 21, Note), for which Mr. Henry Bowyer of Brailsford, Surgeon, prepares powders, which Mr. H. has found a sure remedy.

Hoon: Mr. Thomas Harrison informed me, that Calves, when not suckling or eating Hay, if suffered to graze the rich aftermaths on the Red Marl Meadows, are soon attacked with a husky cough, which ends in Scouring, of which many die. Mr. George Draper is said to have given an opinion, that Blood is produced by this rich food, more copiously than the lungs will bear, which occasions the Cough, and at length affects the Stomach. His treatment of this disorder is very successful, I understand, if commenced in time.

Stanton in the Peak: Mr. Joseph Gilbert mentioned that the early spring grass on the Limestone-Shale Pastures about there, frequently occasion the Cattle to Scour, when first turned upon them; and which in some instances produces also the Blood-water, as above-mentioned, and even rots the Calves in some instances; as I have been told.

While I was at Derby, and at Burton, I met with Mr. Thomas Bellamy, of Bath, who had come there, to call on the principal Dairy-Men and Breeders, and to deposit a quantity of his well-tried medicine for the Scouring of Cattle, with Mr. John Drury, the Printer in Derby, for sale; the six doses proper for a Beast at 21s., or at 12s. to the Subscribers to his Work on this subject.

Garget, Gargle, Longsough, &c. is found a troublesome disorder in the Bags or Dugs of some of the Milking Cows, in Darley Dale, in Longford, and some other places, for which a medicine sold by Mr. John Drury, printer in Derby, which is prepared by Mr. Leo, a Surgeon at Gayton in Norfolk, is said to be an efficacious remedy. More commonly, the Garget here, attacks Cows when dried, and put to graze, respecting which, my travelling Notes are as follows, viz.

Ash, Etwall, &c. ; on the strong Red Marl lands, grazing Cows are very subject to the Garget, as already observed, p. 42.

Blackwell: Mr. Joshua Lingard finds his feeding Cows, on the Limestone land, very subject to Garget; in order to cure them, he draws their bags twice a day, giving a quarter of a pound of Nitre, dissolved in cold spring water, rubs the part affected with Goose Grease, which, with bleeding, mostly succeeds, so that few die of this disorder. As a preventive, he bleeds and gives Nitre as above, in July and three weeks after: the disorder is rather local, and old Cows are most subject to it.

Brailsford: Mr. Edward S. Cox has desisted from fattening Cows, in his small Inclosures on the Red Marl, on account of the prevalence of the Garget, or humours in their Bags, when half fat; Cows which have given much Milk, and are six or seven years old, seem most subject to it; old Cows are not so liable to it; one in six of his fatting Cows, that had been Milkers, were affected by it, and some died. Severe bleeding, from the milk vein before the bag, on the side which is most affected, with cooling purgative drinks, have best succeeded in its cure.

Little Longsdon: Mr. James Longsdon, has his feeding Cows sometimes, in hot weather, affected by Garget,

Garget, and they sometimes lose the use of their limbs: bleeding, with two or three ounces of Nitre dissolved in cold water, relieves them; tho' sometimes they relapse, and are affected in a different Pap.

Giddiness: Mr. Joshua Lingard of Blackwell, stated, that Stirks or Young Cattle of two years old, are sometimes attacked by giddiness: that the late Mr. Robert Walker, a Farmer and Colt-cutter of Wormhill, cured many of this disease, by trepanning their sculls in a certain place, and extracting a bladder of water, from which operation very few died!

Foot-rot or Foul: I heard of some instances of proud flesh, growing between the claws of Beasts, and occasioning dreadful lameness, to which caustics were applied, as soon as discovered: it is not improbable, but some relation exists between this disorder and that at Stanton, mentioned by Mr. Gilbert, page 79.

Scab; was a disease which formerly prevailed on the Beasts in the Peak Hundreds; and it is said, that Mr. Francis Bagshaw's Pastures of early grass on Shale in Hazlebage in Hope, were in considerable repute for curing this disease; engendered only by poverty and previous neglect, in all probability.

Warts, or excrescences on the bellies of young Cows, seemed rather common at Little Longsdon.

Hoven, Risen-on or Swelled: the cause of this dangerous disorder of Cattle, has been spoken of in Vol. II. pages 158 and 164. A flexible smooth long cane, with a smooth knob tied over with greased Leather or Rag; or what is better, a flexible leathern tube, put down the animal's throat with care, so as to enter the stomach, afford instant relief in this dangerous disorder: a smooth piece of rather stiff rope, well greased, is used by some Farmers on these occasions.

Mr. Samuel Rowland of Mickleover, has an old Whalebone-whip, with a knob of leather stuffed with tow at its end, which is oiled before using.

I was glad to hear, that the barbarous and dangerous practice of stabbing Cows when hoven, to let the confined air out of their intestines, is now very rarely resorted to in Derbyshire: Mr. Rowland has had several Cows recover, after lingering a long time, from this treatment.

Belland: this is a species of poison, to which Cattle are subject, from grazing too long in the vicinity of the Lead Smelting Furnaces or Cupolas (as mentioned, Vol. I. p. 392), and sometimes from drinking the Water that has been used in the Buddling or dressing of Lead Ore, as mentioned, I. p. 377. The symptoms are, violent gripings and costiveness. At Bakewell the following was given to me as a receipt for relieving Cows or Horses under this disorder, viz. half a pound of soft soap, and a handful of hog's dung, well mixed in a quart of human Urine, and three gills of Buttermilk, to be given with a horn.

Slipping, Picking, or Casting their Calves, before the natural time of their birth, seems to have prevailed greatly at particular periods, in some districts: In Longford, from 1804 to 1807, this was the case, as Mr. Christopher Smith, the Bailiff to Edward Coke, Esq. informed me.

On Hanson-grange Farm, Mr. William Gould informed me, that for the last twenty years, seven or eight out of every twenty of his Cows, part Long-horn and part Short-horn, have slipped their Calves at half their time; the youngest and best Cows seem most subject to it; it was difficult, Mr. G. said, to account for the prevalence of this disorder, within the period above-mentioned

mentioned (which is not now increasing) as no particular system or change of management has taken place on the Farm, to which it could be referred: his Mares and Ewes are not in the same way apt to slip their young.

Freemartins, or barren Cows, which were twins at the same birth with a Bull calf, are sometimes frequent in particular neighbourhoods: this was said to be the case about Foremarke, where the fattening of one is mentioned at page 27.

Roarer, or Common Buller, a diseased or depraved habit of some Cows, has been mentioned at pages 25 and 76.

The *Cutters* of Calves, Colts, Pigs, &c. have been said, not to operate when the Moon is in certain *signs*, but I think that the persons who mentioned it, could not be speaking of modern practitioners: if it were not for the shameless effrontery of the Stationers' Company, in yearly giving so large a circulation to their astrological fooleries, in old Moore's and others, I believe, of their Almanacs, this egregious folly of our Grandmothers' days, would soon cease to be remembered.

I have not heard, what degree of success the *Cattle Life Insurance Company* had, in making insurances on the *Live Stock* of the Breeders, Dairymen, and Farmers of Derbyshire.

The *working* of Oxen has been spoken of in p. 69, which I mention in this place, because the " Plan" does so, as well as in the former place.

SECT. II.—SHEEP.

It seems to me probable, that Derbyshire was formerly stocked with four distinct breeds of Sheep, only the first of which remain to this day, in their original situation and numbers, viz. the *Woodland* Sheep, the breeding flocks of which, still extend over the Gritstone and Shale Moors in Yorkshire and Cheshire, which adjoin to the mountainous district in the north of this County, called the Woodlands (tho' now almost without Trees or Wood, as observed, Vol. I. p. 382), and whence this breed have been named: the large tract of land over which this breed is spread, having remained to the present time in its original and unimproved state, and the breed of Sheep without any attempts to alter or improve it, until that in 1810, His Grace of Devonshire introduced several Merino Rams into his Woodlands of Hope, with a view to improve the quantity and quality of his Tenants' Wool.

On the high Limestone district, which adjoins south upon that above-mentioned, and extends S W into Staffordshire, there was, while a state of Common generally prevailed, a second breed known by the name of the *Old Limestone* Sheep, but which, since the Inclosure of the Commons, have almost entirely given place to Dairy Cows, or to more useful varieties of Sheep. In all the southern parts of the County, where no natural distinction prevailed between its Soil and Climate, and those of Leicestershire adjoining, a third breed was common to both of these Counties, and was known here by the name of the *Old Leicester* Sheep, but which have now been almost universally crossed with,

with, or have given place to the *New Leicester* breed, and to others.

On the eastern side of the County, adjoining to Nottinghamshire, and to which it somewhat approaches in Soil and Climate, it seems probable, that the *Forest* Sheep, the fourth sort above alluded to, a good deal prevailed, until the New Leicester crosses, and others, have in a great measure taken their place.

My travelling Notes, on the breeds of Sheep, besides a great variety of mixed and uncertainly crossed animals, mention ten different *Breeds* of Sheep, and seven Crosses of these and others, viz.

1. Woodland.
2. Old Limestone.
3. Old Leicester.
4. Forest.
5. New Leicester.
6. South-down.
7. Ryeland.
8. Portland.
9. Merino.
10. Spanish.

11. Gritstone.
12. Old Limestone and New Leicester.
13. New Leicester and Northumberland.
14. New Leicester and South-down.
15. Woodland and Merino.
16. South-down and Merino.
17. Ryeland and Merino.

I shall proceed to notice these Breeds in their order as above, and to extract from my Memorandums, in the order of Persons' Names who keep each sort.

1. *Woodland* or Moorland Sheep, are rather a small and long-legged sort, of horned Sheep, whose wool is fine, except on the breech: in general these Sheep have white faces, but some have black specks on their

noses

noses and legs: it is customary with the Flock-masters, to cut the tails of their Ewes, but to leave those of their Rams and Wethers at full length.

There are no Walls or other fences between the Sheep Farms of this Woodland district, or even between the different Manors or Counties, but the whole lies open together; the divisions being in general along the water-head Ridges, or lines where the waters divide to the different Cloughs, Dales, or Valleys. The upper parts of which Valleys are often very rugged and narrow, and are entirely without habitations, or the Inclosures, which are found lower down in these Valleys, and where the Farm-houses are also situated; the Inclosures skirt round and indent by the Valleys, into this high, heathy, and barren tract.

Every Morning, the Sheep of each Farmer are found at a certain Gate leading out of the inclosed lands around the House, into the open Moors, at which Gate they are usually foddered in severe weather; and from this place the Flock is hunted, by means of the Shepherd's dogs, trained for the purpose, to the very extremity of the Farm, and something beyond, rather than short of the boundary lines of the next Farms; where, being left by the Shepherd, they soon begin to graze their way home, to the Lee or Gate above-mentioned, and this they quickly do, if the weather is stormy or bad; which the Shepherd or his Wife no sooner perceives, than a signal is given to the dog, who runs to the spot and hunts the Flock a great way up the Moors, even to the boundary of the Farm, often; and thus these poor Animals are, in bad weather, and in the winter season, except in snows, almost perpetually in motion during the day.

In order to guard against losing their Sheep, under this

this Gothic system, two *Shepherds' Societies* are established, one at Hayfield and another at Salter-brook House, as will be further mentioned in Sect. I, of Chap. XVIII. who publish the Names of all their Members, the Flock-owners, and *the Marks* by which each of their several Sheep are distinguished; and at stated periods, meetings are held, for the Shepherds to bring in, and mutually exchange their strayed Sheep.

The inclosed lands around the Farm-houses, are principally mown for Hay, and the After-grass is preserved for the Ewes in the Winter and Spring, but the Wethers and Rams are mostly foddered at the Lees or Moor Gates above-mentioned, or at other Walled Lees or Shelters, at a distance in the Moors, and where their dung is frequently raked or swept up, for spreading on the inclosed lands. The Lambs are usually sent out *to be wintered*, in the lower lands of Derbyshire, Yorkshire, and Cheshire, which surround this district; but besides the expense of driving, great loss and inconvenience is experienced, in their being often very badly kept and treated, and taking the *Rot*, when thus removed far from their owner's inspection, during the worst half of the year. A further extension of the Inclosures on to the sides of the hills, and a greater spirit for improving them (which Leases* to the Tenants,

---

* I have mentioned, Vol. II. p. 38, the very striking effects of Leases, on the improvement of just similar lands, in Macclesfield Forest in Cheshire, and cannot refrain from again endeavouring to call the attention of the Duke of Devonshire, Bernard Howard, Esq. and other Proprietors of these extensive Wastes (which are not commons) to their speedy and more effectual improvement. Wide belts of Planting and good high Walls, for separating the Farms, would by their shelter very much improve these lands: and it might be highly desirable, to inclose and cultivate some parts of the Bogs therein, with *Fiorin Grass*, as recommended, Vol. II. p. 202.

only

only can induce), and rearing the greatest possible quantities of winter provender, might enable the keeping of the Lambs at home, and much better providing for the Sheep also, than at present; and the flocks ought to be attended by Shepherd Boys, who never left them during the day.

Each Woodland Farmer uses his own Rams; Tup-letting is unknown among them, and but little care has been bestowed, I fear, for benefiting either wool or carcass, by selection in breeding. Mr. Charles Greaves of Rowlee, is the Duke of Devonshire's Agent for this part of his Estate, and a principal Flock-master, to whose name in the following list, of Woodland Sheep-breeders in Derbyshire, I must refer for further particulars, which follow, respecting their management: these *Woodland Sheep-breeders* are as follows, viz.

Bennet, John and Joseph, of Turnlee, in Glossop.
Booth, Christopher, of Wood Farm, in Rowlee in Hope.
Bowden, George, of Brook-house, in Car-Meadow.
—— Joseph, of Lane-head, in Car-Meadow.
Bower, William, of Chinley-fields, in Hayfield.
Bradbury, Edmund, of Kinder, in Glossop.
Bridge, Thomas, of Grimbo-car, in Darwent-chapel.
Brocklehurst, John, of Raworth, in Glossop.
Carrington, Robert, of Barber-Booth, in Hope.
Cresswell, Nicholas, of Ollerbrook, in Edale.
Dean, Ralph, of Grindsbrook, in Edale.
Eyre, George, of Alpert, near Rowlee.
———— of Upper House, in Kinder.
—— John, of Ronksley, in Darwent.
—— Joseph, of Upper Ashop, in Rowlee.
—— Rowland, of Goars, in Darwent.

Eyre,

Eyre, Thomas, of Alpert, near Rowlee.
Fielding, William, of Hurst, in Glossop.
Fox, Duce, of Marebottom, in Darwent.
——— Joseph, of Lane-head, in Darwent.
——— Thomas, of West-end, in Darwent.
Frost, Thomas, of Deep-clough, in Torside.
Furniss, Henry, of South-head, in Kinder.
Gee, Thomas, of Kinder.
Greaves, Charles, of Rowlee: this Gentleman stated, that his Woodland Ewes begin to take the *Tup* about the 5th of November, and about Old Lady-day they *Lamb*, the best conditioned Ewes being the earliest; has about one Twin Lamb in 40. He usually *washes* about the Longest Day, and *clips* in the last week in June: five fleeces usually go to the stone of 14 lb. and his *Wool* sold in 1808 at 17s. 6d. to 19s. 6d. average 18s. 6d. per stone. The *Lambs* are all clipt, and produce ¼ to ½ lb. of Wool each, which sold in 1808 at 13d. and 14d. per lb. for the Hatter's use. The long tails of the male Sheep are separately sheared, and the wool called Birling or Belting, sold at 6d. or 8d. per lb. for Carpet-making. The *Lambs*, wintered about Bolsover, cost 5s. to 6s. per head, from Michaelmas to Lady-day. The *Wethers*, kept till five or six years old, are sold lean in September and October, and fetch 21s. to 28s. average 24s. per head. The *Ewes* sold off at different ages from two to five years old, sell at 12s. to 20s., average 16s. per head.
Hadfield, John, of Deep-clough, in Torside.
——————— of Padfield, in Glossop.
——————— Joseph, of Lees-hall, in Simondley.
——————— Moses, of Simondley, in Glossop.

Hampson,

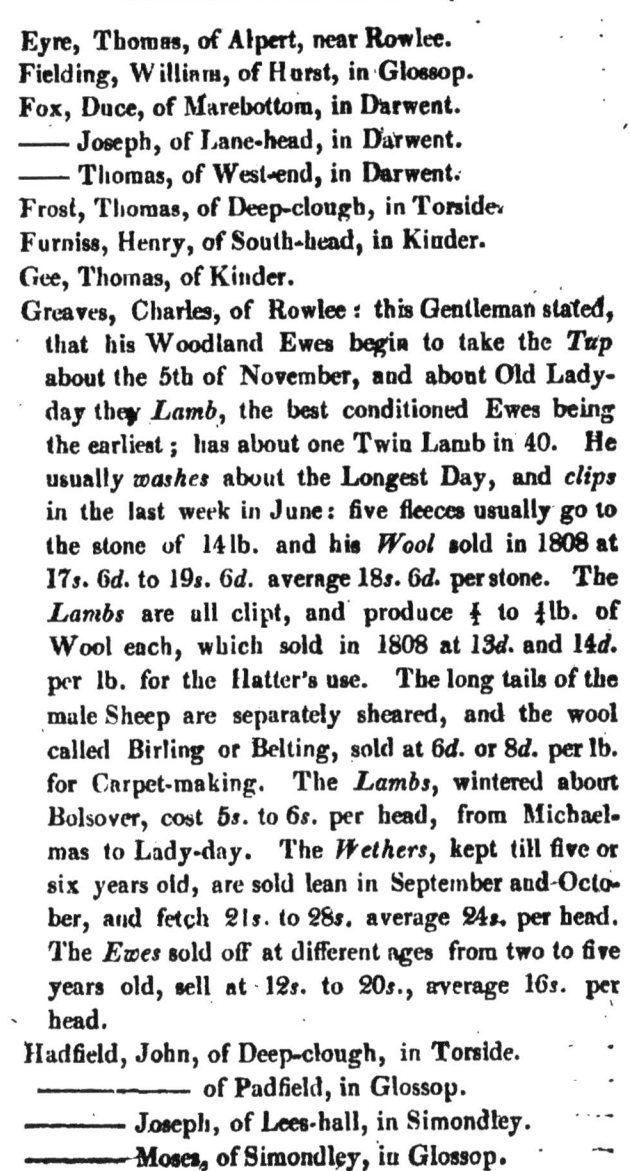

Hampson, John, of Whitfield, in Glossop.
Higginbottom, Robert, of Long-lee, in Car-meadow.
Hyde, Joseph, of Hollingworth-head, in Car-meadow.
Jowl, Joseph, of Nether Ashop, in Darwent.
Kinder, John, of Kinder, in Glossop.
———— Samuel, of Shire-Oaks, in Malcalf.
Longden, Benjamin, of Upper House, in Rowlee.
Marriot, Joshua, of Hayfield, in Glossop.
———— Thomas, of Shude-hill, in Hayfield.
Marshall, Thomas, of Tagsney or Egg's-nase, in Barber-Booth.
Middleton, Isaac, of Bridge-end, in Darwent.
———— Robert, of Two-thorn Field, in Darwent.
Newton, Joseph, of Torside, in Glossop.
Nield, James, of Chunall, in Glossop.
Ollerenshaw, Joseph, of Dingbank, in Darwent.
Pickford, Joseph, of Spray-house, in Car-Meadow.
———— Samuel, of Chunall, in Glossop.
Pursegrove, George, of Upper Booth, in Barber-Booth.
Roberts, Joshua, of Deep-clough, in Torside.
Robinson, James, of Pye-grove, in Glossop.
———— Robert, of Chunall, in Glossop.
Rose, Daniel, of Tinwood, in Darwent.
Rowbottom, John, of Lane-head, in Glossop.
Shepley, John, of Woodcock-road, in Glossop.
Shirt, George Kirk, of Upper Booth, in Barber-Booth.
Slack, Robert, of Little Hayfield, in Glossop.
Sykes, David, of Torside, in Glossop.
Taylor, William, of Ollerbrook, in Edale.
Thorp, George, of Darwent.
———— Jacob, of Darwent.
———— John, of Darwent.
———————— of Hancock, in Darwent.
———— William, of Ash-house, in Darwent.

Tomason,

Tomason, Ann, of Grainfoot, in Darwent.
Tymn, John, of Nether or Lady Booth.
Wagstaff, John, of Glossop.
Wain, Charles, of Darwent-Hall.
———— Thomas, of Birchin-lee, in Darwent.
———— William, of Hollin-clough, in Darwent.
Walker, Robert, of Fairholmes, in Darwent.
Webster, Elizabeth, of Ridge, in Darwent.
———— Thomas, of Bank-top, in Darwent.
Wilcockson, Thomas, of Wood Farm, in Rowlee.
Winterbottom, James, of Glossop.
————————— John, of Blackshaw, in Padfield.
Wood, John, of Glossop.

There are a few other Breeders of Woodland Sheep, in Abney, Bamford, Bank-hall, Hathersage, Rushop, and other places bordering on the Shale and Shale-grit Moor districts, in the northern parts of the County.

Aged Sheep of this breed, however poor, when bought in, are in great repute, after well fattening, to furnish *Mutton* for the tables of the higher classes. Sir Joseph Banks, Bart. annually, has a score of these sheep sent to be fattened in his Park at Revesby in Lincolnshire, and sometimes jocosely remarks to his guests, " Here is Derbyshire Bone and Lincolnshire Mutton."—F. N. C. Mundy, Esq. after shearing time in July, buys Woodland Sheep in the Peak, at about 25s. each, and fats them on Grass in Markeaton Park, for his own Table; they weigh 16 to 20lb. per quarter, and their flavour is excellent.—Lord Vernon, (the late) annually in September, bought in 60 Woodland Wethers, of 3 to 5 years old, and in 12 to 17 months feeding on his Farm at Sudbury, brought them to 14 or 20 lb. per quarter. When I saw this flock, it was much infected with the *foot-rot*.

2. *Old Limestone*, Derbyshire Limestone or Peak Sheep; these were large, heavy, bony, polled Sheep, with a thick skin and coarse wool, which for years did not fetch much above half the price of the Woodland Wool, already mentioned: and they were but little disposed to feed: they are rarely met with now, except, I believe, with such Peak Farmers as have paid little attention to the improvement of their Lands or Stock: Lord Waterpark is the only improving Farmer that I was told of, as keeping the old Limestone or Peak Sheep, and these, I think, were, more probably, 60 or 80 Woodland Sheep, purchased for fattening for his own Table, as above-mentioned.

3. *Old Leicester* Sheep; these large coarse-woolled Sheep are now rarely found in Derbyshire, without a mixture of other breeds, as has been mentioned, page 88. Mr. William Pitt has described this breed of Sheep in his Leicestershire Report, p. 245, and has mentioned Mr. Frisby of Waltham, near Melton Mowbray, N E, and Mr. Moses Miller of Kibworth, near Market-Harborough, N W, as breeders, and as sellers or letters, of improved Tups of this breed.

4. *Forest*, Sherwood or Nottingham Sheep, these are natives of the large gravelly tract, extending from Nottingham to Doncaster, called Sherwood Forest, on which vast numbers were annually bred and kept, previous to the Inclosures, planting and making of Parks, which has otherwise occupied a considerable portion

of

of this district; the numbers of them reared, are still, however, very considerable. They are a small short-legged, polled breed of Sheep (though some few of them have horns), with grey faces and legs, and fine Wool, of 13 to 18 fleeces to the Tod (of 28 lb.), and when fat, weigh 7 lb. to 9 lb. per quarter, as is stated by Mr. Robert Lowe, in the Notts. Report, p. 124. The Ewes of this breed have lately been thought a good cross for Merino Rams.

The Earl of Chesterfield, of Bradby Park, annually buys in a lot of four-shear Forest Wethers: they clip 5 lb. of Wool, which sells at half as much more per pound, as New Leicester Wool fetches, and are fatted to 18 lb. per quarter, for his Lordship's table, their *Mutton* being in much esteem. Samuel Oldknow, Esq. of Mellor, had in his flock 20 Forest Ewes, which I suppose to be of this or a similar breed, tho' he mentioned their coming from the Moors beyond Bolton in Lancashire.

Francis Robinson, of Melborne, and others of his neighbours, who have Watered Meadows, much too flat in their construction, as mentioned Vol. II. p. 476, annually buy in lots of Forest Ewes and their Lambs, and fat both off in the same Autumn, on account of the prevalence of the *Rot* among them.

5. *New Leicester*, Bakewell, or Dishley Sheep. Mr. William Pitt, in his Leicestershire Report, pages 248 to 268, has given several particulars of the early History of this famous breed of Sheep; which he considers to have been originally derived by Mr. Robert Bakewell of Dishley, from the Sheep of the Chalk Wolds or Downs, in the East Ridings of Yorkshire, crossed by the native old Leicesters, and perhaps also

by the large Lincolnshire Sheep, and subsequently further crossed with the Durham breed of Sheep; but to his Work I must refer, for the further details.

It seems agreed, almost generally in Derbyshire, that about 20 years ago, these Sheep had been *bred too fine;* they were too naked of Wool, and Flies, in consequence, attacked their heads and flanks; evils which the several Breeders and Tup-letters, have in different degrees, been striving of late years to remedy; and it seemed to me, in conversing with the practical Farmers of Derbyshire, that the Sheep of those Tup-letters were most in repute, who had attended most to *size and wool*, in the breeding of their Rams.

The following is a List of *Tup-letters*, or Professional Breeders of Sheep of the improved or *New Leicester* kind, not now or formerly residing in Derbyshire, but who have let their Rams, &c. into this County.

Astley, Richard, of Odstone Hall, near Market Bosworth N, Leicestershire.
Bakewell, Robert (the late), of Dishley Grange, near Loughborough N W, Leicestershire.
Bennet, John, of Hoton-hill, near Loughborough N E, Leicestershire.
Bettison, Jonas, of Holmpierpoint, near Nottingham, E.
Boultbee, William, of Sutton Bonnington, Notts, near Loughborough N E, Leicestershire.
Breedon, W. and H. of Ruddington, Notts, near Nottingham S (began 30 years ago to let).
Buckley, Michael, of Normanton (Notts), near Loughborough N, Leicestershire.
Buckley, Nicholas, of Normanton (Notts), near Loughborough N, Leicestershire.

Burgess,

Burgess, Robert, of Hugglescote, near Ashby-de-la-Zouch S E, Leicestershire.
Deverell, of Clifton, near Nottingham, S W.
Dyott, Richard, of Freeford-hall, near Litchfield S E, Staffordshire.
Farrow John (the late, now Daniel and J.), of Loughborough, Leicestershire.
Green, Valentine, of Normanton on the Heath, near Ashby-de-la-Zouch S, Leicestershire.
Harvey, of Blithfield, near Abbots Bromley W, Staffordshire.
Hill, Richard, of Thorpe-Constantine, near Tamworth N E, Staffordshire.
Knowles, Samuel, of Nailstone, near Market Bosworth N E, Leicestershire.
Maltby, Gilbert, of Hoveringham, near Nottingham, E.
Meek, Richard, of Dunstall, near Burton W, Staffordshire.—Sale October 1811.
Moore William, Senr. (the late) of Thorpe-Constantine, near Tamworth N E, Staffordshire.
Morris, William, of ditto.
Paget, Thomas, late of Ibstock, near Ashby-de-la-Zouch S (until 1793), now of Leicester.—See the account of his Sale in the Leicester Report, p. 251.
Sikes, Joseph, of Balderton, near Newark S E, Nottinghamshire.
Stone, John P. of Quorndon, near Loughborough S E, Leicestershire.
Stone, Samuel, of Knighton, near Leicester, S.
Stone, Thomas, of Barrow on Soar, near Loughborough, S E.
Stubbins, Nathaniel, of Holmpierpont, near Nottingham, E.

Talbot,

Talbot, Earl, of Ingestry, near Stafford E, (and *South-Downs*).

Tomalin, John, of Knightsthorpe, near Loughborough N W, Leicestershire.

Watkinson, George, of Woodhouse, near Loughborough S, Leicestershire.

White, John, of Cotes, near Loughborough E, Leicestershire.

Williams, Thomas, of Elford Park, near Tamworth N, Staffordshire.

A List of Derbyshire *Tup-letters*, or Breeders of Improved Sheep, of the *New Leicester* kind.

Bancroft, William, of Barrow, near Derby, S.

Bowyer, Thomas, of Waldley in Cubley, near Ashburne, S W.

Brown, Thomas Cave, Esq. of Stretton-le-Fields, near Ashby-de-la-Zouch S W, Leicester.

Cocker, Samuel, of Ilkeston Hall, near Derby, N E.

Cox, William, of Culland, in Brailsford, near Ashburne, S E.

Creswell (late Robert, Sen. now), Robert and Richard of Ravenstone, near Ashby-de-la-Zouch S E, Leicestershire.

Greaves, Robert Charles, of Ingleby Hall, near Derby, S.

Harvey, Thomas, of Hoon Hay, in Marston, near Derby, S W.

Hoskins, Abraham, Esq. of Newton Solney, near Burton on Trent E, Staffordshire.

Jowett, Thomas, of Draycot in Sawley, near Derby, S E.

Jowett, Thomas, and Robert, ditto.

Matthews,

Matthews, James, of Loscoe Farm in Repton, near Burton on Trent S, Staffordshire.

Moore, Thomas, of Lullington, near Burton on Trent S, Staffordshire.

Mousley, Benjamin, of Honey Hill in Chilcote, near Ashby-de-la-Zouch S W, Leicestershire.

Mundy, Francis N. C. Esq. of Markeaton (until 1808) near Derby, N W.

Needham, Robert, of Ashford, near Bakewell, N W.

Oakden, Philip, of Bentley-Hall in Longford, near Ashburne, S.

Prinsep, Thomas, Esq. of Croxall, near Burton on Trent S W, Staffordshire.

Smith, William, of Swarkestone Lowes, near Derby, S.

I shall now proceed to my Memorandums, made on the Flocks of the various Persons, in order, who breed or keep these *New Leicester* Sheep in Derbyshire, and by reference to *the origin* of their Flocks, and the persons in the above Lists, of whom they have *hired Tups*, endeavour to shew, how each flock *has been bred;* with such other particulars respecting their Wool, Carcasses, &c. as will, I hope, prove acceptable to Sheep Breeders.

Mr. John Bainbrigge of Hales-green, has 30 New Leicesters, from the stock of Abraham Hoskins, Esq.

Mr. William Bancroft of Barrow, has a flock of these Sheep, and lets Tups as above-mentioned.

Mr. John Blackwall of Blackwall, did keep thorough-bred New Leicesters, but found that they had *too little Wool* for his exposed situation; his Wethers clipt 7¼ lb. of Wool, and the shearlings when fat, 24 lb. per quarter of mutton.

Mr. Thomas Bowyer of Waldley (several years a pupil and assistant of Mr. Bakewell at Disbley), originally began, 24 years ago, from the stocks of Mr. John Farrow, Mr. Thomas Harvey, and Mr. Philip Oakden: he has now 20 breeding Ewes, and lets a few Tups, at 5 to 10 guineas each, or sells, at about the same prices: his Rams are kept in a state, fit to kill, and in general, if not let or sold, they are sent to the Butcher.

Rev. Joseph Bradshaw of Holbrook, hires his Tups, of Mr. R. C. Greaves; fat Wethers 23 or 24 lb. per quarter.

Mr. Thomas C. Brown of Stretton-le-Fields, lets Tups, as above.

Mr. Fletcher Bullivant (the late) of Stanton-Ward, kept 240 New Leicester Ewes, originally from Mr. R. C. Greaves, and hired Tups of Mr. Valentine Green; his Wethers at 22 months old, weighed 24 to 26 lb. per quarter.

Earl Chesterfield of Bradby Park, annually buys in 5 New Leicester Ewes, and hires a Tup of Mr. William Smith.

Mr. George Clay of Arleston, keeps 55 Ewes, originally from Mr. R. C. Greaves, and from whom he has hired Tups; his Wethers at 18 months, average 20 lb. per quarter.

Mr. Samuel Cocker of Ilkeston Hall, keeps 80 Ewes, hires Rams of Mr. Nathaniel Stubbins; clips 5 or 6 to the tod, of 28 lb.; his Wethers at 22 months old, weigh 22 lb. per quarter, and culled Ewes at 1 to 4 years old, 18 lb. per quarter. Mr. C. lets 5 or 6 Tups annually, to little Farmers in the neighbourhood, at two to six guineas each.

Edward Coke, Esq. of Longford, keeps a New Leicester flock, derived from Mr. Robert Bakewell, Mr. Nicholas Buckley, Mr. Stone, and Mr. Samuel Knowles; as his Bailiff Mr. Christopher Smith, informed

formed me, and who added, that Mr. C. thought he had *bred too high*, for size, wool, and hardiness.

Mr. Edward S. Cox of Brailsford; his sheep were derived from his Father's flock, Mr. William Cox.

Mr. William Cox of Culland: it is said, that more than fifty years ago, Mr. C. was the first who gave Mr. Robert Bakewell 20 guineas for the hire of a Tup; he lets Tups, as above.

Messrs. Robert and Richard Cresswell of Ravenstone (late their Father Mr. Robert C. Sen.); their flock was raised originally from those of Mr. Thomas Paget, and Mr. Valentine Green, and since, they have hired Tups of Mr. Robert Bakewell, Mr. Nathaniel Stubbins, Mr. John White, Mr. Thomas Stone, and Mr. John Stone; they let a good many Tups, at considerable prices, but the particulars seemed a secret.

Mr. Henry Dethick of Willington: this gentleman in 1797 won the Derby Society's Prize for two-shear New Leicester Wethers; the one which he exhibited, weighed 37¼ lb. per quarter.

The Duke of Devonshire at Chatsworth, has a New Leicester flock; formerly hired Tups of Mr. Thomas Harvey, but has for several years used his own Rams.

Sir Henry Fitzherbert, Bart. of Tissington, had a New Leicester flock, and when I was at this place, in 1808, Sir Henry was about to enlarge his farming establishment, and to enter with spirit into breeding, as I was informed.

Mr. Joseph Gould of Pilsbury, has a New Leicester flock, and hires Tups of Messrs. Creswell; clips his Lambs about Midsummer: his fleeces in the years 1803, 4 and 5, were 4 to 5¼ lb. each; average about 5 lb.: he fats a few two-year Wethers to 18 or 20lb.

per quarter, but more commonly sells them lean, or partly so, to make way for others.

Mr. William Gould of Hanson-grange, has about half his flock composed of New Leicesters.

Mr. Robert Charles Greaves of Ingleby: Mr. G.'s late Father, Mr. William G. prior to 1777, selected the original Ewes of this flock, and hired Tups of Mr. Valentine Green, and previously of the Father of that Gentleman: and Mr. G. has now a very closely bred flock; firmness of Mutton, in handling, a light Pelt, and general lightness of Offal, having been, as he stated, the objects principally kept in view; but several practical Farmers whom I met with, stated, that this flock was become rather too *deficient in size, wool, and hardiness* for their use. Mr. G.'s prices were stated to be from 10 to 30 guineas for the let of a Tup, except to his neighbours with small flocks, some of whom had them as low as two guineas.

Mr. Timothy Greenwood of Newhaven, has a New Leicester flock, and hires Rams at 8 or 10 guineas each, from Mr. Philip Oakden, and Mr. Harvey of Blithfield.

Mr. Richard Harrison of Ash (and previously his Father Mr. John H. now of Hoon), has a New Leicester flock; Rams hired of Mr. Samuel Knowles. Mr. H. lets Tups, as mentioned in the above list.

Mr. Thomas Harvey of Hoon Hay: he keeps 300 to 400 New Leicesters; has hired Tups of Mr. Thomas Paget, Mr. Nicholas Buckley, Mr. John Farrow, and Mr. Valentine Green, he lets a good many Tups at 5 to 40 guineas each.

Thomas Hassall, Esq. of Hortshorn, keeps 100 of these Sheep, and hires a Tup of Mr. Valentine Green: his Wethers at 18 months old, weigh 20 lb. per quarter.

Mr.

Mr. John Holland of Barton Lodge, has left off keeping these and other Sheep on his Farm, conceiving that when they ate the land close, it *rotted* them?

Abraham Hoskins, Esq. of Newton Solney : keeps 120 New Leicesters, originally from Mr. Valentine Green's stock, and of whom Tups were hired, as also of Messrs. Daniel and J. Farrow, Mr. Robert Creswell, Sen. Mr. Robert C. Greaves, and Mr. John Tomalin; Mr. H. is intending to get more *size and wool* in his flock: at 22 months old his Wethers average 24 lb. per quarter: he lets 8 or 10 Tups, at 5 to 20 guineas each, annually.

Mr. Thomas Jowett, Sen. of Draycot: his flock was raised from those of Mr. Jonas Bettison, Mr. John White, Mr. Farrer, and Mr. John Bennet: his fat Wethers at 18 months old, weigh 20 to 24 lb. per quarter; his Ewes at 3 to 6 shear, weigh 18 to 22 lb. per quarter; he annually sells off a part of his Theaves, and lets Tups at 5 to 20 guineas; the average price about 10 guineas.

Messrs. Thomas and Robert Jowett of Draycot: keep 200 New Leicesters, derived principally from the flock of Mr. John White: they let Tups at 5 to 30 guineas each, average about 15 guineas.

Mr. Robert Lea of Burrow-fields: his flock was originally from Mr. Robert Bakewell's, of whom Mr. Lea, more than 40 years ago, hired the Tup great A, at 30 guineas, being the first that Mr. B. let at so high a price: he has hired Tups also of Mr. Thomas Paget, Mr. Valentine Green, and Mr. Thomas Prinsep: to the Tups of the last breeder, he has also frequently sent Ewes to be served, at $2\frac{1}{4}$ guineas each. Mr. L.'s shear-hogs clip $5\frac{1}{4}$ lb. of Wool.

Mr.

Mr. Thomas Lea of Stapenhill: keeps 110 New Leicester Ewes, derived from the flock of Mr. John White: hires Tups of Mr. Robert Burgess, and saves no Tups of his own: his shearling Wethers at 18 months old, weigh 18 to 26lb., average 22lb. per quarter.

Mr. James Matthews of Loscoe Farm: has hired Tups of Mr. Robert Bakewell, and Mr. Robert C. Greaves; his Wethers at 19 months old weigh 18lb. per quarter; he lets a few Tups at 3 to 5 guineas each, in the neighbourhood.

Earl Moira of Donnington Park, keeps a New Leicester flock: see Mr. William Pitt's Leicestershire Report, pages 264 and 268.

Mr. Thomas Moore of Lullington: has a New Leicester flock, raised from his late Father's, Mr. William Moore; has hired Tups of his Father, and since of his brother William M. who succeeded him: he lets Tups, principally about Daventry in Northamptonshire, at 5 to 40 guineas each.

Mr. Benjamin Mousley of Honey-hill, keeps 52 breeding Ewes, originally from the flock of Mr. William Moore, Sen., has bought Tups of Mr. Abraham Heskins, and hired of Mr. Robert Creswell, Sen. Mr. William Moore, Mr. Thomas Williams, and Mr. Richard Hill; annually lets 2 or 3 Tups at 2 to 10 guineas each.

Francis N. C. Mundy, Esq. had a New Leicester flock, and let Tups, prior to his Sale in 1808.

Mr. Robert Needham of Ashford, has a New Leicester flock, and lets Tups, as mentioned in the List above.

Mr. Philip Oakden of Bentley-Hall: his flock was
derived

derived from those of Mr. John Farrow, and Mr. Robert Creswell; Mr. O. in 1808 let 50 Tups, few of them in Derbyshire, at 5 to 30 guineas each.

Samuel Oldknow, Esq. of Mellor: had 20 New Leicester Ewes: one of his Tups gained a Cup at the Manchester Society, in October 1808.

Mr. John Pearsall of Foremarke, keeps New Leicesters: some years ago they were bred much *too fine*, and bare of Wool, and were greatly complained of by his brother William P., Butcher of Repton, and others, as being very light behind; that they have been greatly improved of late in size and Wool, by Mr. William Smith, and some other Tup breeders of the County.

Messrs. John and Robert Porter, of Barrow: in 1809 these Gentlemen gained the prize at the Repton Society, for two-shear New Leicester Wethers, which weighed 47¼ lb. per quarter.

Thomas Prinsep, Esq. of Croxall: I was informed in 1809, that this Gentleman still continued to hire Tups from the first flocks in Leicestershire, and let none himself under 50 guineas; or, he took in Ewes to his Tups at 2½ guineas each: a three-shear Wether of his, killed a few years ago, was said to weigh 53 lb. per quarter.

The late Mr. Francis Radford of Little Eaton: kept New Leicesters, and gained a prize from the Derby Society in 1795, for a two-shear Wether, which weighed 43¾ lb per quarter, with an unusually small proportion of offal.

John Radford, Esq. of Smalley: keeps a large flock of New Leicesters on his Farm in Great Hucklow: they rest quiet, are fed on Seeds and Turnips; and, however *fat* they are made, he never finds their mutton

ton complained of on that account, by the Butchers who supply the manufacturing people.

Mr. Thomas Rowbotttom of Lee-Hill: keeps 40 or 50 New Leicester Ewes; hires Tups of Mr. Thomas Prinsep.

The Hon. John Simpson of Stoke-Hall; in the Leicester Journal of the 6th of October 1809, the following account appeared, of the Sale of 230 fat Wether Hogs from Mr. S.'s Babworth Farm, near Retford in Nottinghamshire, viz.

	£	s.	d.
Carcass of each on the average,	2	8	1
Wool, shorn when Lambs in July 1808, each	0	1	2
Wool, of the Hogs in June 1809, 6lb. on the average at 1s. 2d.	0	7	0
Average produce of each	£2	16	3

Mr. Thomas Broom Simpson, late of Repton Park, kept New Leicesters; his Wethers averaged about 22lb. per quarter, sold at Candlemas.

Mr. John Smith of Repton: keeps New Leicesters, and hires Tups of Mr. Robert C. Greaves: at 20 months old, his Wethers weigh 23lb. per quarter, or more.

Mr. William Smith of Swarkestone Lowes, and Foremarke-park: Mr. S.'s account of his flock was as follows, viz.: about 1790 he began collecting his Ewes from Mr. James Matthews, Mr. Robert C. Greaves, and from different flocks, in Leicestershire, and has since hired Tups of Mr. Valentine Green, and Mr. Robert C. Greaves. That several years ago, his flocks on his different Farms, having been kept separate, and one of them being much *closer bred* than the other

other, and from finer Tups, he found these high-bred sheep, grown mossy-woolled, bare-headed and bellied, their eyes staring and necks stooping (see Mr. Richard Parkinson's Rutland Report, p. 131); while his other flock, were of a good size, a fair fleece of wool, open-back't, and were equally or more disposed to early fattening, than the other; which last kind of sheep he found vastly more profitable than the high-bred ones, and that his customers for Tups, among the practical Farmers, invariably preferred these larger and better woolled sheep: which has induced him, of late years, to turn his principal attention in selecting and Tupping his Ewes, to these qualities in particular, and he has an increasing demand for his Tups, of which, he let 16 in 1809, at 10 guineas each on the average, besides taking in 200 Ewes to his Tups at half a guinea each. For 5 years out of the 6, previous to 1809, Mr. S. had won the first prize offered by the Derby Society for three Theaves of this Breed. Mr. Smith had an Ewe, which at 13 years old, brought two Lambs.

Lord Vernon (the late) of Sudbury, kept 45 New Leicester Ewes, hired Tups of Mr. Philip Oakden: the Wethers at 18 months old weighed 20 lb. per quarter; the culled Ewes fatted at 1 to 6 years old. Mr. Thomas Brain, his Lordship's Bailiff, had been attentive to *size and wool*, in selecting Ewes and Tups for breeding.

The late Mr. John Wall of Weston Underwood, had a New Leicester flock; hired Rams of Mr. William Cox and of Mr. Thomas Harvey.

Lord Waterpark of Doveridge, has 100 of these Ewes; hired Tups of Mr. F. N. C. Mundy and Mr. Thomas Prinsep.

Mr.

Mr. Henry Wayte of Milton, has a flock of New Leicester Sheep.

Mr. John Webb of Barton Lodge, has a New Leicester flock, and has hired Tups of different Leicestershire breeders: he clips 6 lb. per Sheep: his Wetbers, in the September after they are one-year old, weigh 21 or 22 lb. per quarter, and sell for 49s. each on the average.

Sir Robert Wilmot, Bart. of Chaddesden, had until lately a highly bred New Leicester flock; hired his Tups of Mr. Jonas Bettison, Mr. Robert C. Greaves, and Mr. Valentine Green: his present flock have more of *size, wool, and hardiness.*

6. *South-Down* Sheep: this valuable breed of fine-woolled Sheep, which are natives of the Chalk Downs in the neighbourhoods of Arundel, Brighton, and Lewes in Sussex, had found their way into Derbyshire, but not in the numbers that their merits demanded, I think, previous to my examining the County in 808 and 1809; but their numbers have since increased, I am informed, partly for the purpose of crossing with Merino Rams.

Mr. John Bainbrigge of Hales-green, had 12 Southdown Ewes, which he was going to put to a Merino Ram.

The Earl of Chesterfield, on his Farm in Bradby Park, had 20 Ewes of this breed and a Ram.

Mr. Edward Soresby Cox of Brailsford, commenced this breed in 1804, with small lots of Ewes from Lord Anson, Lord Talbot, William T. Coke, Esq. of Norfolk, and the Duke of Bedford: has hired Tups of Earl Talbot, of Ingestry near Stafford, E. Mr. Cox had

had for five years tried these Sheep against his New Leicesters (not perhaps of the very first rate, never having hired Tups, nor were his South-downs of the very prime kind), and, except in the greater prolificacy of the South-down Ewes, when considering Wool and Carcass, it was not possible to decide, which of these breeds were best adapted to his good Pastures, on the Red Marl.

Sir Henry Crewe, Bart. of Calke, has 280 South-downs; the Ewes originally purchased of Mr. John Ellman, Sen. of Glynd near Lewes SE, in Sussex, at 66s. a head, after being tupped: has since hired Tups of Lord Anson of Shugborough, near Stafford, ESE. They clip about 3 lb. of Wool: at 20 months old the fat Wethers weigh 20 lb. per quarter, and the two-shear Wethers 28 lb. per quarter. At the Show of the Derby Society in July 1810, Sir Henry exhibited a fat Wether of this breed, which, when killed, weighed 128 lb. and 18½ lb. of loose fat. Mr. William Smith, Sir Henry's Steward, informed me, that he thought these the best fine-woolled sort of Sheep in the County; and that on close-fed Paddocks, among a variety of Park Stock, they will do much better than New Leicesters, tho' not so on rich and less fully or variously stocked Pastures.

Francis Eyre, Esq. of Hassop, has a flock of South-downs.

Thomas Hassall, Esq. of Hartshorn, has 20 South-down Ewes, that came from Sir Henry Crewe's at Calke.

William Drury Lowe, Esq. of Locko-park, in 1809 had 100 South-down Sheep of different descriptions, and valued them much.

Earl Moira of Donnington Park, had 20 South-down Ewes

Ewes and a Ram, as mentioned in Mr. William Pitt's Leicestershire Report, pages 20 and 264.

Samuel Oldknow, Esq. of Mellor, in 1809 purchased of Francis Eyre, of Hassop, 10 South-down Ewes, intending to cross them with Merino Rams.

Mr. Walter Plimley of Styd-hall, had some South-down Sheep.

7. *Ryeland,* or Ross Sheep: these are a small fine-woolled breed; natives of a tract of Red Marl Country, in the south-east corner of Herefordshire, near to the Forest of Dean. At the time that I was on my Survey, in 1808 and 1809, I did not hear of any person having kept these Sheep in Derbyshire, except Samuel Oldknow, Esq. of Mellor, for crossing with his Merinos; but since, their numbers have increased, intended for the same purpose.

8. *Portland* Sheep: a small, horned, fine-woolled breed; are natives of the Island of this name in Dorsetshire.

The Earl of Chesterfield of Bradby Park, has some of this breed from Sir Henry Crewe, Bart. which are kept for supplying his own table with *Lamb* and *Mutton:* when properly kept, they will take the Tup at any time, and bring Lambs twice in the year: they clip about $2\frac{1}{4}$ lb. of Wool each, which sold in 1807 and 1808 at 3s. 3d. per pound: on their backs and sides it is very fine, but on their breech, tail, and legs, is rather coarse: at three or four years old they fat to about 12 lb. per quarter, and their Mutton is considered a great delicacy.

Sir Henry Crewe, Bart. of Calke, has some of these Sheep, whose ancestors were brought from Portland
about

about 40 years ago, by the late Sir Henry Harper, Bart.: their Wool, tho' fine, in part, sells for 6d. per pound less than South-down Wool.

9. *Merino*, or Travelling, very fine-woolled Spanish Sheep; were first imported within a few years past, from that Country, where very immense flocks of this breed of Sheep have almost from time immemorial been kept, and summered on the wild and open Mountains, and are driven thence to great distances in the Autumn, to their Winter Pastures in the more even country, and whence they return again in the Spring, to the Mountains, annually.

Notwithstanding the absolute importance of the Wool of these Merino Sheep, to our staple Manufacture of fine or broad Cloth, and the vast and increasing Sums annually sent out of the Country to purchase Merino Wool, an opinion so strongly prevailed, as to the soil and climate of Spain, and the *travelling life* of these Sheep, being essential to the production and preservation of their superfine Fleeces, that no attempts were, I believe, made to import these Sheep, and naturalize the growth of this Wool in the British Islands, until that His Majesty, on the advice of the Right Hon. Sir Joseph Banks, Bart. principally, undertook their importation, and the breeding and rearing of them in England; in which design Lord Somerville also very meritoriously embarked, and was ably seconded in the treating and management of them, when imported, by Dr. Parry, Mr. George Tollet, and others.

At length, after a long period of struggling against deep-rooted prejudices, as to the almost certain prospect of the debasement of their Wool, and as to the quantity and quality of their Mutton, the Merinos began to make their way among English Breeders. Mr. Benjamin Thompson, late of Red-Hill Farm near Nottingham, and Mr. Wooton Berkenshaw Thomas of Chesterfield, having early and seriously engaged in the breeding of these Sheep, they had began to attract attention, and to spread in this County, a short time before I commenced my Survey in 1807.

Since then, the very large importations of Spanish Sheep, which followed the advance of our Armies into the interior of that Country, and the consequent fall in their prices, and on account of the greater facility of procuring Merinos, their numbers have greatly increased in this County; so that Mr. W. B. Thomas, to whose zeal and kindness I am greatly indebted, by a correspondence purposely opened with all the known possessors of Merino Sheep in Derbyshire, was enabled to ascertain, for my present purpose, that in the last Month (March 1813), not less than 450 pure Merino Sheep and 1750 Merino-crossed Sheep, were kept in this County.

I shall now proceed with my Notes on the Sheep of this Breed, in the order of their Owners' Names, and shall introduce therewith and distinguish, such recent information as I have been favoured with from Mr. Thomas, as above.

Mr. John Bainbrigge of Hales-green, has some Merino Sheep.

The Rev. William Bagshaw of Stavely, had from Mr. Tollet, 10 Merino-crossed Ewes, and in 1808 sent the same to be tupped at Red-Hill Farm, by Mr. Thompson's Merino Ram, Don Felix.

Mr. John Bunting of Bunting-field in Ashover, has a few Merino Sheep.

The Earl of Chesterfield, of Bradby Park, has a Merino

rino flock, which in May 1810 received an addition, in a present from His Majesty, of part of the Spanish Sheep then lately imported.

Mr. Francis Hutchinson of Arleston has now, as Mr. W. B. Thomas informs me, 21 pure Nigrette Ewes and Rams: this Gentleman, I believe, partook of the Sheep distributed by His Majesty in June 1810.

Mr. Edward S. Cox of Brailsford, has a few Merino Sheep.

His Grace the Duke of Devonshire, of Chatsworth: in the year 1810 His Grace, thro' his Agent Mr. Thomas Knowlton, purchased at Colonel Downie's first Sale, 6 Merino Rams and 25 Ewes, intended partly, for distribution among his Tenants, the principal breeders of Woodland Sheep (see pages 88 and 92), and in order, that the Farmers' Ewes might be sent to the Chatsworth Rams, on the same truly liberal terms, as their Mares have long been, to the Stallions there kept.

Philip Gell, Esq. of Hopton: had in 1809, 45 aged Merino Ewes, 35 of their Lambs of two-years old, and 39 one-year old Lambs; they were bred from the King's flock, from which also his Rams were procured. Mr. Gell found these Sheep perfectly hardy; for in Lambing on Hopton-Moor, in a cold and rainy season, he lost not a single Ewe or Lamb; he found them, however, rather more subject to the *Foot-rot*, than other Sheep.

Mr. Charles Greaves of Rowlee: in 1810 three of the Merino Rams purchased by the Duke of Devonshire, as above mentioned, were placed in Mr. Greaves's care, for crossing some of the best of his own and his neighbours' Woodland Ewes, as mentioned page 88.

The Earl of Harrington, of Elvaston, has now, according

cording to the return to Mr. W. B. Thomas, 46 pure Paular Ewes and Rams.

Mr. Anthony Holmes, of Stanton in the Peak, in August 1809 advertised, to take in 15 Ewes at 3 guineas each, to a Paular Ram of his.

Joshua Jebb, Esq. of Walton-lodge, entered early into the breeding of Merinos; at Lord Somerville's Sale he purchased a valuable Ram, and in July 1810 was presented by His Majesty with two choice Ewes: his return lately made to Mr. W. B. Thomas, is 13 pure Nigrette Ewes and Rams, and 591 Merino-crossed Sheep, being the largest Merino flock in the County, I believe.

Mr. William Lovett of Boythorp, has some Merino Sheep.

William Drury Lowe, Esq. of Locko-park, in 1809 had 3 Merino Wethers from the Earl of Harrington's flock, which he was going to fatten: as a prelude, as I understand, to entering on the breeding of these Sheep.

Messrs. William and John Milnes of Ashover: in July 1810 these Gentlemen received a present of two Ewes and a valuable Ram, from His Majesty: their return to Mr. W. B. Thomas, mentioned their present Merino flock to consist of 10 pure Paular and Nigrette Ewes and Rams, and 30 Merino-crossed Sheep.

Samuel Oldknow, Esq. of Mellor, entered early into the Merino Breed, and purchased at the King's Sales; at Colonel Downie's Liverpool Sale, in September 1810, his principal purchases were made: the return of his present flock to Mr. W. B. Thomas, is 44 pure Paular Ewes and Rams, and 298 Merino-crossed Sheep; some of these I shall notice further on, page 126.

Mr. Walter Plimley of Styd-hall, has some Merino Sheep.

Mr. John Rayner, is stated in some account which I received, to occupy a Farm in one of the Peak Hundreds, and to keep Merino Sheep, but I don't happen to have noted more particulars.

Mr. Wooton Berkenshaw Thomas of Chesterfield, occupies farms in Boythorp, Brampton, and Barlow, on all of which he has Merino Sheep; the two latter Farms lying adjoining to the High Moors, in very exposed situations. The laudable zeal with which Mr. Thomas entered into the design of rearing a Merino flock, has been glanced at already, and cannot soon be forgotten by the Agriculturists of this County. Mr. T. was at first forcibly struck with the same circumstances, which Lord Somerville has since mentioned in his " Facts and Observations relative to Sheep and Wool, &c." published in 1809, viz. that " notwithstanding the great importance of *short-woolled* Sheep to the Nation, the *whole* attention both of Farmers and Breeders has, for these 20 years past, been absorbed, in *size and frame*, and carrying to a degree of perfection (hardly credible) the heavy *long-woolled* Sheep, such as Lincoln, Cotswould, Romney-Marsh and New Leicester; but more particularly the last; altho'," observes his Lordship, "every practical Man will admit, that one-half of the kingdom, at least, is by nature appropriated to the short-woolled breed:"—he might with safety admit much *more than half;* " for it at length appears, that our climate (from the most northern parts to the most southern) *can* grow Wool, of the finest possible quality." The Mutton of "the short-woolled Sheep, being close in the grain, consequently heavy in the scale and high-flavoured as to taste; the large-woolled Sheep more open and loose in the grain" of flesh.

Mr. T. acting on these persuasions, procured first, I believe,

I believe, from Mr. Tollet 20 of his Anglo-Merinos, and omitted no opportunity of increasing his flock from other quarters, particularly by the purchase of Rams and Ewes at the King's and Lord Somerville's Sales: in July 1810 he was presented with two fine Ewes by His Majesty.

In order to excite attention to the progress and advantages of breeding Merino Sheep, Mr. Thomas has for some years past, invited a large party of Agriculturists of the County to be annually present at his *Sheep-shearing*, and to whom he has been anxious to explain fully, every circumstance that could conduce towards forming a practical and safe judgment, on the merits of this breed of Sheep: and for this purpose, the live Animals in all their states, their Wool, their Mutton, and Cloth, both for Ladies' and Gentlemen's wear, manufactured from the Wool grown on his own Farms, were exhibited; and it may not be improper to state, that in Mr. T.'s Family, no other Habit or Broad Cloth, but this of his own growth, is worn, and which Cloths, many competent judges have declared to be equal in quality to the best that can be made from imported Spanish Piles.

Mr. Thomas's account of his Flock at present (March 1813), is 108 pure Escurial, Paular, and Nigrette Ewes and Rams, and 321 Merino-crossed Sheep, of various descriptions. In 1812, Mr. T. clipt 386 fleeces, which sold for 340*l*. 7*s*. (besides 22*l*. 5*s*. 6*d*. for Lamb's Wool) or near 17*s*. 8*d*. for the Wool of each Sheep, through his whole flock!

Mr. T. finds the pure, as well as all the crossed Sheep of this Breed (and he has tried most of the usual English crosses) to be perfectly hardy, not only as to *keep*, doing well on the High Moors, where, he says, that

large

large Lincolnshire and Leicestershire Sheep could not exist, much less be kept in store order, but likewise as to bearing *cold* and exposure. On his two Farms adjoining the High Moors, Mr. T. had more than 100 pure and crossed Lambs, dropped this season, which have all done remarkably well: which hardiness, he attributes, to the closeness and quantity of their Wool, and. says, that not a solitary instance has occurred in his crossing, where the Wool of the produce of native Ewes by pure Merino Rams, has not been *doubled in value per pound*, and also *very considerably increased in weight*.

In the autumn of 1812, and spring of 1813, Mr. Thomas sold to the same Butcher, near 30 of his fat Merino and crossed Wethers and aged Ewes, and who reports, that tho' the *Rot* prevailed in a degree which he never before remembered, yet all of Mr. T.'s sheep died sound, and, as he expresses it, turned out better than they handled while living: that on his shopboard, this *Mutton* went off readily at the best price of the day, and that some of his Customers gave it a decided preference, for its moderate size, fine grain, and *age*; because, the annual profit from the wool, had enabled the keeping of these sheep round, to advantage, till they were 3 or 4 shear.

Henry Bache Thornhill, Esq. of Stanton in the Peak, has now, according to the return to Mr. Thomas, 17 pure Merinos; the Ewes of the Paular and the Rams of the Nigrette breeds, and 100 Merino-crossed Sheep, of different descriptions.

Before I proceed with my Notes on the eight varieties or breeds of Sheep which remain yet to be mentioned, I beg here to present a few observations, by a Gentleman who has a good deal interested himself in the questions, which have been much agitated of late, as to the Individual and National Advantages, of breeding from *pure Merino* Ewes and Rams only, rather than by *crossing* the best of our native Ewes, (selected for carcass and fineness of Wool) with pure Merino Rams; who has furnished me with the following Table of Calculations, intended to shew, the comparative advantages of each practice: on the suppositions, that 1000 Native Ewes are selected at two years old, ready to drop their Lambs, got by Merino Rams, which are always supposed to be pure ones, *hired* for the purpose: that each Ewe at and after two years old, shall annually bring a Lamb, and one in every two of these to be Ram Lambs; all which last are supposed to be cut, and sold as fat Wethers, at three years old; and that the Ewes when 10 years old, are also fatted off: allowing 5 per cent., or 1 in 20, for casualties in each year, with Ewes and their progeny.

BREEDING OF PURE, OR CROSSED MERINO SHEEP? 121

Years of the Experiment.	Native Ewes.	Ewes of different Crosses at the year's end, for Tupping, viz.					Total of Ewes from 2 to 9 Years old.	Three-year fat Wethers to be sold.	Ten-year fat Ewes to be sold.	Total Stock of Ewes, Lambs, Wethers and Tegs, at each year's end.
		1st.	2d.	3d.	4th.	5th.				
1st	1000	—	—	—	—	—	1000	—	—	1900
2d	950	—	—	—	—	—	950	—	—	2707
3d	903	451	—	—	—	—	1354	428	—	3856
4th	858	856	—	—	—	—	1714	407	—	4885
5th	815	1220	204	—	—	—	2239	580	—	6381
6th	774	1366	760	—	—	—	2900	734	—	8265
7th	735	1674	1265	92	—	—	3766	960	—	10,732
8th	698	1974	1775	439	—	—	4886	1243	—	13,927
9th	663	2208	2427	1002	42	—	6342	1615	663	18,074
10th	—	2359	3191	1779	238	—	7567	2095	—	22,194
11th	—	2539	4019	2795	678	19	10,050	2718	299	28,641

By considering the last figure of every number in the Table as a decimal, and rejecting it, except when it exceeds 5, when the last figure but one is to be taken *one more*, this Table will answer for 100 breeding Ewes, either Native or Merino; thus, from 100 Native Ewes, as Woodland, South-down, Ryeland, &c. at the end of the 11th year, or 10 complete years distant from the commencement, 254 Tegs and Ewes of the *first* cross, 402 of the *second* cross, 279 of the *third* cross, 68 of the *fourth* cross, and 2 of the *fifth* cross, in all 1005 crossed Ewes, will be ready for the Tup; or, if Merino Ewes only were used, 1005 Tegs and Ewes, all of the pure breed, will be ready for the Tup.

It also appears, that in that eleventh year, 272 fat three-year old Wethers, and 30 ten-year old Ewes will go off to market; the total stock then remaining being 2864 Sheep of all kinds, except Tups, which are not included, but as many of them are to be hired, or used from another flock, as is necessary.

The necessity which thus appears, of persevering in crossing for 10 years, before breeding Ewes of the fifth cross, and then only 19 from 1000 Ewes at first used, can be obtained, has been much insisted on, by the advocates for the *pure blood*, without its being, I believe, before so clearly shewn, how fast, after this period in crossing is reached, the number of the fifth cross will multiply, as is evident from the increases of the first, second, third, and fourth crosses, respectively, in the Table, after they have once begun.

The improvement of the *Carcass*, seems also kept entirely out of view by the advocates of this side of the question. A doubt cannot, however, remain, but the public are much interested, in the increase of the
good-

good-carcassed and early-fattening kinds of Sheep, being as extensively as possible clothed with fine Wool; for which last object, tho' an important one, we cannot afford to sacrifice Mutton, to any considerable extent.

10. *Spanish*, spotted coarse-woolled Sheep: these are of a dark brown colour spotted with white, with thick white tails, and they a good deal resemble some, that I have seen among the fancy stock in Gentlemen's Parks, and were called Cape Sheep. Sir Robert Wilmot, Bart. of Chaddesden, had 6 Theaves and 3 Lambs of this breed, which came from Lord Middleton's at Wollaton in Nottinghamshire, and had put a New Leicester Ram to the Theaves.

11. *Gritstone* Sheep, an ancient, and formerly rather a common cross here, between the Woodland and the old Limestone Sheep, mentioned pages 89 and 96; they had black or grey faces, with coarsish wool, fetching, a great many years back, not much above half the price of the Woodland Wool: few, if any, besides the small and most unimproving Farmers, bordering on the Peak Hundreds, now keep these Sheep: of which I have not noted any instances.

12. *Old Limestone and New Leicester* Cross: since that fine, and rather too high-bred New Leicesters have been common, and have been found too delicate and tender, the above cross has been resorted to, by some persons, for increasing the size, quantity of Wool, and hardiness of their Sheep.

John Berrisford, Esq. of Osmaston Cottage, keeps these Sheep.

Mr.

Mr. William Gould of Hanson-grange, has about half his flock thus crossed.

Mr. Joshua Lingard of Blackwell, has a flock of these Sheep, and thinks, that they are hardly so profitable now, as before they lost size, by being too much crossed with New Leicesters: he has bought Tups at Leicester Fair, but never hired any, about one-third of the Ewes bring two Lambs, and on the average Mr. L. has raised 110 Lambs from 100 Ewes put to the Tups. They clip 6 to the Tod of 28 lb.: the Wethers are sold in September or October, when about 18 months old, to those who winter them on Turnips: they come to 16 or 18 lb. per quarter: he generally sells his Ewes at 4 years old, in October, with their Lambs, to those who fatten both of them next season.

Mr. Thomas Logan, late of Buxton, some years ago tried the Lambs of a score of the Old Limestone Ewes, tupped by a New Leicester Ram, against the progeny of a score of good New Leicester Ewes, with the same Ram, and represented the result, when the young sheep went to market, from similar keeping, to be greatly in favour of the crossed breed.

Mr. Robert Needham of Perry-Foot: has a flock of Limestone Sheep, 12 or 14 years crossed by New Leicester Tups, hired of Mr. Robert Needham of Ashford, and others.

Lord Waterpark of Doveridge: I am not sure, from my Notes, made from the information of Mr. Thomas Bowyer, whether his Lordship uses this cross or not, as hinted already, page 96.

13. *New Leicester and Northumberland Cross*: Lord Vernon (the late) of Sudbury, keeps 40 Northumberland

berland Ewes, and puts them to New Leicester Tups, hired of Mr. Philip Oakden: they lamb in the latter half of March and first half of April; the *Lambs* at 3 to 4 months old, are fatted to 7 to 11 lb. per quarter.

14. *New Leicester and South-Down* Cross: the Earl of Chesterfield, in Bradby Park, has a small lot of South-down Ewes, put to a New Leicester Ram; at 6 months old the *Lambs*, when fat, weigh 12 or 13 lb. per quarter.

15. *Merino and Woodland* Cross: the person first using this cross, whom I heard of, was Mr. James Buckley of Greenford in Saddleworth, Yorkshire, about 1807, and that he found considerable profit from the increased quantity and quality of his Wool.

Mr. Charles Greaves of Rowlee, and others of the Duke of Devonshire's Tenants in the Woodlands of Hope, have since 1810 adopted this cross, as mentioned pages 88 and 115. Several other persons in the County have also lately tried or adopted this cross.

16. *Merino and South-Down* Cross: Mr. John Bainbrigge of Hales-green, in 1809, began by putting 12 South-down Ewes to a Merino Tup.

Mr. Edward S. Cox of Brailsford, in 1809, put a Merino Tup, from the King's flock, to some South-down Ewes, and was intending to give this cross a fair trial. The number of persons who have adopted this cross, has considerably increased, since my travelling notes were taken.

17. *Merino and Ryeland* Cross: Samuel Oldknow, Esq. of Mellor, in 1809, stated, if I mistook not, that

he

he had 9 Rams of the 4th and 5th cross? (see p. 122) with pure Merino Tups from the King's flock, 8 Ewes ditto, and 36 Ewes of the 1st cross: he reared 26 Lambs per score, from old and young Ewes reckoned together, and had few casualties: Mr. S. seemed to approve much of this cross, and was in treaty with his Wool-stapler for this Wool, at 3s. 6d. per lb. washed on the Sheeps' backs. Since the above period, several others in the County have adopted this cross.

I have heard it mentioned, that a sort of Sheep from the Northward, called *Cumberland* Sheep, have sometimes been introduced, for crossing with the native Sheep of this County.

The *food* of Sheep, in Summer and Autumn, is principally grass, either of old pastures or of new Leys or Clovers; and in Winter and Spring, most commonly, Hay, or common or Swede Turnips*, as has been mentioned, Vol. II. p. 137 and 147, or Cabbages, Vol. II. p. 140; and more rarely, *kept-grass* or Rowen, as mentioned Vol. II. p. 183, or *Watered-grass*, Vol. II. p. 469, 471, 472, 479, &c. or Cole or Rape, Vol. II. p. 140, or Khol Rabie, II. 149, or Carrots, II. 151. *Oil-cake* is rarely if ever used here for Sheep, I believe.

Sheep-Walks, or rights of pasturage for Sheep over the Lands of others, are unknown here, as observed, Vol. II. p. 197. The prices at which Sheep are summered, in certain Ley or joist Pastures, are mentioned, Vol. II. p. 198.

It does not appear, that *Salt* is often administered to

---

* In feeding off his Turnips with Sheep, Mr. Joseph Gould of Pilsbury, is careful to plough in the *Sheep-dung* as soon as possible after the Turnips are eaten, having ascertained, that the frost and rain are very injurious to this manure, when exposed on the surface.

Sheep,

Sheep, for any purpose in this County; nor is *Folding* much if at all practised, I believe; Mr. William Greaves, Jun. of Bakewell, mentioned, that it was not all approved in the Peak Hundreds.

Cotting of Sheep, and providing of *Yards* expressly for their accommodation and shelter, seems rather less attended to in this County, than its importance demands, in such a climate as its hills present: independent of the additional quantities of excellent manure that would thereby be collected. *Sheep-lees,* consisting of high walls in form of a cross, or curving like an S, &c. are not uncommon on the Moors, intended to afford shelter from the driving wind and storm, from whatever quarter it may blow; to these Lees the Sheep retire, and are found by the Shepherds, in case of a sudden fall of Snow in their absence. On his Farm in Ashover, Mr. John Milnes has lately erected a neat circular *Sheep-house* at the angles or meetings of four Fields, having sheds and partitions all round it: the inside being a store-house of Winter provender for the Sheep.

In Plate I. facing page 128, I have, by permission of Sir Joseph Banks, Bart. copied a Drawing which he caused to be made, and engraven some time ago, for private circulation, in order to shew to Flock-masters a ready mode of erecting a temporary *Lambing-Fold,* by means of rough Poles and common fleak or fold Hurdles, on which straw Thatch was previously laid and bound, by means of tarred cord or withy bands: the construction of which building will, I trust, be sufficiently obvious from the Plate.

When Sheep are turned out into the Lanes and small Commons in this County, it is very common to chain or couple every two of them together, at about two

feet

feet distance, to prevent their straying far away, or leaping the Fences.

The occupiers of the Inclosures next to the large Moors and Commons, find it very difficult, by means of over-hanging Copings of large Stones to their tall *Wall-Fences*, to prevent the climbing of the Sheep over them; which they effect in a surprising manner, by running obliquely at the wall, and catching their feet on its inequalities, unless prevented by these over-hanging Copings, or by a row of short Bushes, laid projecting from the top of the Wall, and confined there by large Stones laid on them.

Fatting of Sheep: the kinds of food used for this purpose, have been mentioned and referred to, p. 126. In favour of the early and *excessive fatness of New Leicester Sheep* (or Lambs rather, as some have called them, at the age of sending to market), it has generally been urged, that such *Mutton* was relished, and even generally preferred by the Work-people in manufacturing districts, and which opinion I found to be supported by John Radford, Esq. of Smalley, as observed p. 107, by Mr. William Smith of Swarkestone Lowes, and by Mr. William Beardsley, Butcher, of Belper, who seem to agree in thinking, that this fat Meat is as much in request as ever: a somewhat different opinion was, however, expressed by F. N. C. Mundy, Esq. of Markeaton, who said, that such very fat Mutton has less ready sale now than formerly; Mr. George Clay of Arleston said, that the labourers and manufacturers have sickened of very fat Leicester Mutton, and such won't readily sell; Mr. William Pearsall, Butcher, of Repton, maintained also the same thing.

Mutton

Mutton Hams: In the excellent Farming establishment of the Earl of Chesterfield in Bradby Park, I was shown the process of salting and drying New Leicester Legs of Mutton for Hams, which are said to be much relished at his Lordship's Table, and that these kind of Hams are now very common in the neighbourhood of Bretby. In Section VI. of this Chapter, I shall mention the curing of common or Pork Hams, which exactly agrees with the process in this case.

Fat Lamb.—The Earl of Chesterfield has fat Lamb on his Table, at different seasons of the year, from his Portland Ewes, as mentioned page 112. A few Southdown Ewes are also annually tupped by a New Leicester Ram, for producing Lambs that early come to a good size and degree of fatness, as mentioned p. 125.

Lord Vernon, on his Farm in Sudbury, uses Northumberland Ewes with a New Leicester Tup, for producing Lamb for his Table, as mentioned p. 124.

Live and Dead Weights of Fat Sheep.—I have here again, as in the case of Neat Cattle, page 73, to lament the too little attention which has generally been paid, to ascertaining and recording of the *live weights* of fat animals, after proper fasting, and just before slaughtering, as a check upon, and to complete the detailed weighings of their several parts afterwards. From the records of the *Derby Agricultural Society*, I am enabled to present the live and dead weights, of the two-shear New Leicester Wethers, in two competitions for the Prizes offered by this Society, in the years 1794 and 1797, viz.

1794.	Live Weights.	Carcass.	Heads.	Loose Fat.	Skins.	Plucks.	Entrails.	Blood.	Loss*.
	lb.	lb.	lb.	lb.	lb.	lb.	lb.	lb.	lb.
Mr. Francis Radford, the late, of Little Eaton	175	113	4½	13	16	4¼	8½	6	9¾
Sir Robert Burdett, Bart. the late, of Foremarke	170	102	4½	11½	17¼	4½	16	6¼	7¼
Mr. William Cox of Brailsford, since of Culland	173	102	4¼	13½	17	4¾	16	7½	7½
Mr. Robert C. Greaves of Ingleby	165	95	4¼	13	19	4	15½	6	8¼
Sir Robert Wilmot, Bart. of Chaddesden	160	94	4	13	18	3¼	14½	5½	7¾
1797.									
Mr. Henry Dethick of Willington	225	150	—	16	17½	5¼	10½	6½	—
Mr. William Cox of Brailsford	169	112	—	12½	15½	6½	9½	5¼	—
Mr. Robert C. Greaves of Ingleby	190	126¾	—	13	14	4½	11	5¼	—
Mr. Francis Radford, the late, of Little Eaton	192	128	—	13	16	5	11	5¾	—
Mr. John Radford of Smalley	189	126	—	13	16½	5	14	5¼	—
Sir Robert Wilmot, Bart. of Chaddesden	152	101¼	—	13¼	14	4	9	5	—

* I have supplied the numbers in this column, such as make up the live-weights in column 1. On adding the sums of this and the preceding column together, and dividing by five, there results, however, 14¾ lb. very nearly, as the average weight of blood and loss, which far exceeds, and is more than double what the returns of the London Butchers to the Smithfield Club state, as the Blood and loss of considerably heavier Sheep of this breed, viz. 6¼ lb. on Sheep of 200 lb. live weight (as I find on an average of 15 of their Prize Wethers in different years), whereas these Sheep average alive but 168½ lb. nearly. Perhaps this may partly arise, from the entrails being weighed full in London, and empty at Derby, leaving their contents to go with the loss. It seems, however, that the blood at Derby, as stated above, 6¼ lb., making allowance for the size of the Sheep, must equal the blood and loss in London.

The average live weights in the first of these years is 168.6 lb., and in the last 186.2 lb.; whether this increase of 17.6 lb. per Sheep, is to be wholly ascribed to improvement in size and early ripeness, in these three years from the first establishment of the Society's exhibition; or whether earlier lambing, more forcing food, or later exhibition, had concurred, I did not happen to learn. Following Mr. John Layton Cooke's method (as ought to be done by all experimentalists and calculators on this subject), and supposing the *live weight to be units*, or 1.000, we find the average proportional weight of *Carcass* in the first year to be .600, and in the second year .667*; so that there appears an increase of near 6¼ per cent. in the proportion of meat to live weight, in these three years.

---

* In Mr. Cooke's "Tables for the use of Farmers and Graziers," page 32, he says, if the gross weight of a fasted fat Sheep be 1., the proportion that the carcass bears to the whole, will vary from .58 to .68. His 16th table gives the weights of Carcasses by inspection. These Tables are in a principal degree derived from the numerous documents on this head, preserved and published by the Smithfield Club, as mentioned, page 73.

Since the above was prepared for the press, Mr. L. Cooke has kindly furnished me with his unpublished Table, of the proportionate weights of the several parts, of ripe or perfectly fatted sheep, expressed in decimals of the live weight 1. (as in page 74 for Cattle), viz.

Carcass or Quarters	.680
Head	.030
Loose Fat	.070
Skin, or Pelt (without shearable Wool)	.055
Pluck (Heart, Liver and Lights)	.025
Entrails and contents	.085
Blood and loss, by evaporation in cooling	.055
	1.000

From this Table it would appear, that fat New Leicesters of 200 lb. live weight, ought each to have (200 × .055) = 11 lb. of Blood and loss, instead of 6¼ lb. as stated in the preceding Note, as the average of 15 Returns.

The proportionate average weight of loose *Fat* in the first year is .076, and in the latter .072, so that a decrease of near ¼ per cent. in the loose fat, calculated on the live weight, appears to accompany an increase of near 10¼ per cent. of that weight: which agrees well with a favourite point aimed at, by the late Mr. Robert Bakewell, that of laying on the fat outside the ribs, rather than within them, where it could neither be seen or felt.

The proportionate weight of *Skin* (and wool at the time of slaughtering) is in the first year .103, and in the latter only .084; shewing a *decrease* of near 2 per cent. on the live weight *in Skin and Wool*, while an *increase* of near 10¼ per cent. on that weight was making;—circumstances strongly corroborative, of the refinements in this respect having been carried too far, and having rendered these valuable animals tender, and less adapted to the uses of the practical Farmer, as has been mentioned in pages 98, 103, 104, 107, 109, 110.

From the records of the *Repton Sheep Society*, I am enabled to present the *dead weights* of the five two-shear New Leicester Wethers, kept on Grass and Turnips only, which were exhibited on the 7th of February 1809, viz.

	Carcasses.	Loose Fat.	Heads and Plucks.	Skins.	Entrails.	Blood.
	lb.	lb.	lb.	lb.	lb.	lb.
Messrs. John and Robert Porter, of Barrow	189	17¼	11½	29	12¼	7½
Mr. Henry Dethick, of Willington	164	16¼	10	21¼	12¼	7
Capt. John Smith, of Repton	188½	23¼	11½	20¼	14¼	7½
Mr. Wm. Smith, of Swarkestone Lowes	171	21¼	9	19	11¼	6¼
Mr. Henry Wayte, of Milton	167¼	22¼	10¼	25	12¼	7¼

Unfor-

Unfortunately the *live weights* are here omitted, but the average weight per Sheep, of all the particulars in the above 6 columns, being 249¾ lb., if we add 10¼ lb. for the average loss, we shall get 260 lb. for their live weight, which, considering what is said in the Note, in page 131, and the amount of this loss, by making proportions for it in different ways of comparing this Table and that of 1794, in the page referred to, we are probably not much wide of the truth; at any rate, the errors cannot be considerable in thus assuming the *unit*, for calculating in Mr. L. Cooke's method: whence the following observations arise, viz. 1st, As to *size*, we have here an increase of 38¼ per cent. on the gross average weight of the Sheep of 1794 at Derby, and of 28 per cent. on that of 1797. 2nd, The *Carcass*, here (176 lb.) is .677 of the live weight, which is equal very nearly to Mr. Cooke's highest proportion, for fully ripe Sheep, and exceeds the Carcasses of 1794 by near 7¼ per cent., and those of 1797 by 1 per cent. on the respective live weights. 3rd, The proportionate weight of *loose Fat* is here .078; shewing an increase of something less than ¼ per cent. on the live weight, on the internal fat of the Sheep of 1794, and more than ½ per cent. on those of 1797. 4th, As to *Skin* and Wool, they are .088 of the live weight, which exceeds the proportion in 1797 almost ⅐ per cent., but still falls short of that of 1794 by 1½ per cent., notwithstanding the exertions of the Members of this Society of late years, for giving *more wool and hardiness:* which last property, is probably more connected with the quantity and distribution of the Wool on the animal, than with the collective weight of its Skin and Wool.

In the preceding account of different breeds of Sheep, from pages 101 to 126, many particulars of the dead weights of mutton or quarters are mentioned, which need not be here repeated.

Before proceeding to speak of *Wool*, it will be right to mention a few particulars, respecting the Washing and Shearing of Sheep, and of the Chambers provided for Storing their Wool, while in the hands of the Growers; subjects, which have no distinct places assigned them in the " Plan" of Report which I follow herein.

The *Washing of Sheep* preparatory to Shearing them, is here generally performed in the open and stagnant parts of the Rivers and Brooks, in rather a slovenly manner, as it appeared to me: I saw it performing first in 1808, in the Dove River, on the 30th of May, in order that they might *shear early*, to prevent the *Maggots*, as I am told. In the smaller streams bordering on the High Moors and Woodlands, temporary walls of stone and earth are built across the stream, to make a dam called a *Sheep-wash*. In the lane N E of Long Duckmanton leading to Nunnery, I saw a permanent stone Sheep-wash, consisting of a walled Chamber 10 feet long and 6 wide, across the Brook, into which the water ran over a stout stone coping, and out over a similar one: at one side of the Brook and end of the Chamber, permanent Pens were erected, for holding the Sheep previous to Washing, and on the opposite side, a paved inclined gangway, led up out of the Chamber, for the Sheep to walk out, after being washed, by a Man who stands in the water, up to his middle, as is quite general here; and no provision seems in use, for the Washer to remain out of the Water, as in the Earl of Winchilsea's Sheep-wash, which Mr. Richard Parkinson has drawn and described in his Rutland Report, p. 182.

In

In Bedfordshire, a more simple and effective kind of *Sheep-wash* is not uncommon: which my Son Joseph has represented in Plate II. facing page 144. Two Dams are made across a Brook, the upper one A, being 1½ or 2 feet or more higher than the lower one B; thro' or over the upper dam, a trough or spout C, is laid, which discharges a constant and moderate stream of water into the Sheep-wash, which is between the two Dams. D, is a tall and narrow Tub, firmly fixed by straps or extra hoops round it, which are screwed to posts firmly fixed in the ground; which precautions are necessary, for confining down the Tub, when the Man is not in it, and to make it very firm and secure, so that the Washer may lunge against its edge, in exerting all the force of his arms, in washing and parting the Wool of the Sheep, and presenting every part of them successively under the stream of water which falls from the spout, as shewn in the Plate. A Pole E, fixed across the Sheep-wash an inch or less above the water's surface, prevents the Sheep which is washing, and another which is soaking, ready for the operation, from escaping beyond the reach of the Washer: who has no sooner finished a Sheep, than he sinks it under the pole E, and it escapes out of the water to a clean pasture field, by the inclined paved passage F: the Shepherds, or those in attendance, assisting and guiding the Sheep when necessary, in coming up out of the water, as also in throwing in a fresh Sheep to soak, as at H, from the Sheep-pen G, as often as one is finished by the Washer. The advantages of this jet of water on the Sheep while washing, are, that they can be more effectually cleansed, without that long soaking which is often injurious to the Sheep. When the Washing at any place is ended, the

the same tub and spout might be carried and used at other Sheep-washes, or put by in a dry place, until wanted next year.

The *Shearing* of Sheep, presented nothing to remark in my Notes, except that about Cowdale I saw Sheep very badly shorn, particularly a flock belonging to Mr. William Goodwin, as I was told, which certainly looked more like Rats having gnawed off their Wool, than a Man having shorn them.

In many instances, I saw in the Peak Hundreds, Sheep too long neglected to be shorn, until a considerable portion of the most valuable of their wool had peeled off, and was lost, and their fleeces had become broken, by which the value of the whole was depreciated.

The *Shearing of Lambs* is practised by Mr. Joseph Gould of Pilsbury, as mentioned page 103; by Messrs. William and John Milnes of Ashover, as will be seen by their Wool account, p. 139; by the Hon. John Simpson of Stoke Hall, as mentioned page 108; by Mr. Charles Greaves of Rowlee, p. 93; by Mr. W. B. Thomas of Chesterfield, p. 118, and by many others, whom I have omitted to note.

Wool Chambers.—In the description of the excellent new Farm Premises of the Earl of Chesterfield in Bradby Park, in Vol. II. p. 10, the situation of his Lordship's Wool Chamber, which is a very complete one, is pointed out, as over the Implement-house, No. 22 on the plan. In the same volume, page 209, the situation of Mr. Samuel Oldknow's neat Woolchamber, in connection with his Gardener's House, is mentioned. In most others of the new and considerable Farm Premises in the County, this necessary appendage

pendage is well contrived, and very neatly kept, for the preservation of the Wool while in the grower's possession: the Wool-staplers afterwards stow it in large Warehouses, partly loose in Bins and Chambers, and partly in Bags, ready to be sent off to the Manufacturers.

Sale of Wool.—Throughout Derbyshire, either the *Stone* of 14 lb., or the *Tod* of 2 stone or 28 lb. is in use, I believe, in the dealing for Wool: I heard of no *Wool Fairs* or Markets expressly for the sale of the Wools of this County, tho' the establishment of such would, I think, be very desirable and useful, if generally attended; but the Wool-staplers here, as is too common also in other Counties, go round to certain Farmers, of whom they usually buy, and thus bargain for the greater part of the Wool, at their own houses; or the Farmers call on the Wool-staplers at their houses, in secresy, and without that fair competition, which an open market would give, where the article in bulk, or ample samples, should be brought, for inspection and comparison.

Of Woodland Sheep, the quantities and prices of Mr. C. Greaves's Wool has been mentioned in p. 93; Forest Sheep belonging to the Earl of Chesterfield, page 97; New Leicesters, the Hon. John Simpson, page 108; Portland Sheep, Earl Chesterfield and Sir Henry Crewe, pages 112 and 113; Merinos and their crosses, Mr. W. B. Thomas, pages 118 and 119, and Samuel Oldknow, Esq. page 126; besides which, various notes on the quantities of Wool, from most of the different breeds of Sheep, will be found, from pages 88 to 126.

In June 1808, Lord Vernon's Bailiff sold the three previous

previous clips of his Lordship's Wool of several sorts, averaging about 4 lb. per fleece, at 29s. per Tod of 28 lb. altogether.

Messrs. William and John Milnes of Ashover, keep about 40 breeding Ewes, and annually shear about 90 Sheep and 50 Lambs, of the better kind of polled Sheep, selected from the mixed breeds of the neighbourhood, and improved by using New Leicester Tups; by which cross, tho' the Carcass has improved, they have found a great decrease in Wool therefrom. Their shearling Wethers, at 18 months old, usually weigh 18 lb. per quarter. These Gentlemen have favoured me with an extract from their Wool Account for 17 successive years, which may serve to record the progress, as to quantity and prices, of this important article.

SALES OF WOOL.

Sheep's Wool.

When sold.	To whom.	No. of Fleeces.	Weight.		Price per Tod.			Amount.			Average weight per Fleece.		Average value per Fleece.	
			Tod.	lb.	*l.*	*s.*	*d.*	*l.*	*s.*	*d.*	lb.	oz.	*s.*	*d.*
1792, April 26,	Mr. Dutton, the Clip of 1790 and 1791	164	32	6	0	19	8	31	13	6	5	8	3	10¾
July 11,	Mr. Gell, Wirksworth, 1792	83	17	3	1	4	0	20	10	6	5	12	4	11
1795, Sept. 10,	Mr. John Basset, 179?	95	20	—	1	0	0	20	0	0	5	14	4	2¼
1796, Aug. 13,	Mr. Dutton, 1794, 1795, and 1796	210	46	26	1	4	0	55	4	0	6	2	5	3
1799, Nov. 15,	Mr. Wm. Jefford, 1797, 1798, and 1799	212	47	22	1	1	0	50	6	6	6	5	4	9
1800, Aug. 12,	Ditto, 1800	67	11	4	1	2	0	12	19	4	4	15	3	10¾
1801, Aug.	Ditto, Green-house Farm, ditto	47	9	4	1	2	0	10	1	2	5	7	4	3¼
July 21,	Ditto, 1801	119	24	2	1	6	0	31	5	0	5	10	5	3
1802, Nov. 5,	Mr. Charles Johnson, 1802	126	24	15	1	11	6	38	12	10	5	7	6	2
1804, Aug. 17,	Mr. William Jefford, 1803	131	25	4	1	11	0	38	19	0	5	6	5	11
Ditto	Ditto, 1804	112	18	7	1	11	0	28	5	0	4	9	5	0
1805, Sept. 26,	Ditto, 1805	108	17	19	1.15	0	30	18	9	4	9	5	8
1806, Sept. 12,	Ditto, 1806	95	14	13	1	13	6	24	4	6	4	4	5	1

Lambs'

Lambs' Wool.

Year	To whom sold.	Number of Lambs shorn.	Weight.	Price in Pence.	Amount.			Average Weight.	
			lb.		l.	s.	d.	lb.	oz.
1791	Messrs. Smiths, of Bakewell	—	41	10	1	14	2	—	
1792	Ditto	46	78	11	3	11	6	1	11
1793	Ditto	41	63	8	2	2	0	1	8½
1794	Ditto	26	46	7	1	6	10	1	12
1795 & 6	Ditto	67	107¼	10	4	9	7	1	9
1797 & 8	Ditto	—	97	10	4	0	10	—	
1800	Ditto	43	59	9	2	4	3	1	6
1801	Ditto	48	59	9¼	2	6	8	1	4
1802	Ditto	66	78½	12	3	18	6	1	3
1803	Ditto	60	93	12	4	13	0	1	8½
1804	Ditto	49	57	12	2	17	0	1	2¼
1805	Ditto	41	43¼	13½	2	8	11	1	0½
1806	Ditto	69	62	14	3	12	4	0	14

Distempers of Sheep.—I shall commence my account of these, with the most prevalent, as well as most destructive, among the diseases of this useful Animal, in this as well as all other Counties, that I am acquainted with, viz.

The *Rot*, is here, as in most other districts, the subject of differences of opinions, as to its causes, and the success or the probability of any cure for it. Mr. W. B. Thomas of Chesterfield, seemed to lean to the opinion, that different breeds of Sheep are more or less subject to this disease, in mentioning (see page 119) the remarkable escape of his Merinos in the Autumn of 1812, when Sheep were rotted in so unusual a degree.

Mr. Charles Greaves of Rowlee remarked, that sound Sheep, when killed, have sometimes large Flukes, Hydatids, Liver-flukes, or Gourd-worms (*fasciola hepatica*) in their Livers; and that examining the

the corners of the eyes next the nose, to see whether the same be white, instead of red, its proper colour, is the only known method of discovering this disease, before it has proceeded so far, that the Sheep have become Powk't, Shockil'd or Bottled under their jaws, after which, few if any recover, from this terrible disorder. With respect to the cause of this malady, Mr. Greaves supposes it to be produced, in most wet Autumns: poverty, and eating soiled herbage, where Brooks and Rivers overflow, contributing to the effect. His Lambs often take the Rot while out at their Winter keeping, as observed page 91.

Mr. John Holland of Barton-fields, found the New Leicester Sheep, when they eat close, on his gravelly Pastures, on Red Marl, to contract the Rot, as observed page 105.

When I was enquiring into the circumstances of Roston Common (see Vol. II. p. 343), I was told, that thick and hard stocking thereon, is supposed to occasion this disease.

Mr. George Nuttall, Surveyor, late of Matlock, stated to me, that he had found the most prevalent opinion to be, that Sheep in a poor state, having access to luxuriant pasture, commonly contracted this disorder.

About Bakewell I heard an opinion, that the early grass produced on Shale Lands, in warm situations, occasioned this disorder in Sheep, in low condition.

Mr. Christopher Smith, Bailiff to Edward Coke, Esq. of Longford, informed me, that on the clayey gravels on Red Marl, around there, Sheep are often rotted after dry weather, when rain has fallen, and set the grass growing frim, which is supposed to occasion it. He mentioned, that some years ago Mr. Coke purchased a

lot

lot of Sheep from very poor land near there, part of which he sent to his Brother's in Norfolk, where they were fatted on poor land, and all died sound: the remainder of the lot he kept at home, on very good land, which never before or since had rotted Sheep, and yet many of these Sheep proved rotten.

The late Mr. Francis Bruckfield informed me, that prior to the Inclosure of Calow Moor, which was esteemed a very sound pasture for Sheep, though hard stocked, he had a right of Common there, but never exercised it, except for his Ewes, for 2 or 3 weeks, after taking away their Lambs, and in order to lower their Milk; that on part of the same hill (of Limestone Shale), and soil, but within his Alton Farm, a field that was a dry and good Sheep Pasture, well sheltered, in places, by the forms of the hills, and which was usually in these places, highly dunged by his Sheep lieing on them; that a considerable rain once happening before the weaning time, the grass sprung up in these places luxuriantly, and like a salad, as he expressed it, and his Sheep were rotted thereby, as he had reason to conclude, because they had never been on other land, but this sound field and the Moor, as above mentioned.

Mr. John Milnes of Ashover, conceives that wet lands in want of draining, most commonly communicate the Rot: Oaker Hill, of cold Limestone Shale, near Wensley, is remarked for the rotting of Sheep.

Mr. John Pearsall of Foremarke, thinks wet land the cause, and that the Rot seldom appears on dry pastures.

Mr. William Cox of Culland, stated, that his old Pastures, on Red Marl, rotted his Sheep, before he drained them, but not since.

In

In Ashford, the swampy land, on a base of calcareous Tufa, by the Wye River, is very apt to rot Sheep.

It has been mentioned, at page 97, and also at page 476 of Vol. II., that the flat and improperly formed *Water-meadows* in Melborne, uniformly rot the Sheep, fed on them through the Summer and Autumn: and an opinion prevails, I know, through most parts of England, that Sheep are in great danger, if not a certainty of being rotted, by *Autumn-feeding* on Watered-Meadows: it seems, therefore, very material, to again revert to the well-attested cases to the contrary of this, in Hartshorn, Ingleby, and Measham, pages 472 and 475 of Vol. II., and to urge again the importance of properly constructing and *raising*, Meadows when first made, so as to have a pretty steep declivity for the running water, *in every part of them;* and to suggest, that on all such, the experiment with a few Sheep annually, might with some confidence be tried, during the Autumn, until it can be determined, whether the flock can safely be admitted, to assist in eating up the grass, when the expected period of Autumn floods is approaching. It being very often, of great importance, to be able to use the first and all the subsequent of these floods, on the Meadows, wherever water is scarce, in particular.

That wet and boggy lands are not always liable to rot Sheep, I am enabled to state, from the example of the Bog at Prisley, in Bedfordshire, prior to the late Mr. Elkington's attempts to drain the same, mentioned at page 366 of Vol. II.; but where this circumstance, of Mr. Oliver's Sheep always escaping the Rot thereon, has been omitted to be mentioned, tho' it was among the principal reasons alledged by him to shew, that

Mr.

Mr. E.'s operations had injured the Farm, as there stated.

Dr. Darwin has treated of this disease in his "Phytologia," and considers watery food, as the cause, of enabling the Flukes to commit their ravages on the liver, and recommends salt and water, or iron-filings, or flour and salt, to be given to the diseased Sheep.

Mr. James Pilkington, in his View of Derbyshire, Vol. I. p. 470, has stated, that Misletoe (see p. 215, of my second Volume) eaten by Sheep, will prevent the Rot; see also his 376th and 525th pages. Its cure, he also mentions, Vol. I. p. 373, as having been effected by eating Elder Bark.

Mr. Benjamin Chambers of Hurst, in Tibshelf, adjoining to Nottinghamshire, is stated, in that Report, p. 129, to have known hundreds of Sheep affected by Rot, to have cured themselves, by grazing on (yellow or magnesian) Limestone land, not eaten too bare, "their livers healed again," he says.

Mr. John Pearsall of Foremarke, mentioned to me, that he had often cured, or at least stopt the progress of the disease, so that Sheep, after being much affected by Rot, got fat, by giving a tea-spoonful or two of Spirits of Turpentine in a small tea-cup full of Wine or Ale, shook up together in a phial; repeated every other day, until they gain strength, and the powk or bottle decreases.

In the "Pantologia," or General Dictionary, lately published, under the article *Rot*, a clear view of the opinions of Dr. Harrison, as to the cause and treatment of this terrible disorder, is given: and I am happy to have heard, that the ever-to-be-remembered and revered *Dr. Jenner*, to whom we owe Vaccination,

has

has for some time turned his attention to the scientific investigation of this disease, not less general, but even more fatal among these very useful animals, than the Small-Pox was among our own species.

Red Water, or *Resp*.—Mr. Joshua Lingard of Blackwell, mentioned this disease, as very prevalent among his Lambs, from six to 11 months old; a bloody fluid collecting, between the skin and carcass; some die, and no remedy is known: it is not a contagious disease, or ever happens with his Lambs, much after 18 months old.

Bache Thornhill, Esq. on his Shale and Clay Lands in Stanton in the Peak, has often lost Sheep by this disorder, and no cure is known. A dose of Onions and Tar, and bleeding at the end of the tail, repeated at intervals of 3 or 4 months, or change of Pasture, have been known to prevent this disease.

The Wood Anemone (*Anemone nemorosa*) has been thought to benefit Sheep under this disease: See Mr. J. Pilkington's View of Derbyshire, Vol. I. p. 416.

Brown Ger.—This is a disorder sometimes happening in Mr. Joshua Lingard's flock, at Blackwell, on Limestone; it is a sort of *scowering*, or discharge of brown fluid matter; it attacks Sheep of all ages, without any known cause, but seems not to be contagious: no cure or preventive is known, and few recover which are attacked.

Scowering, or the black scowering, in Sheep, differs from the above, in the colour of the discharge, and in the symptoms: Mr. Thomas Bellamy of Bath, who some time ago published, by subscription, a Work on his remedy for Scowering in Sheep or Cattle, continues to prepare his medicine, and has deposited a quantity of the same for sale, with Mr. Drury, printer, in Derby.

Turn, Turney-headed, Dunt, or Giddy: this is a disease in Sheep, not very uncommon, I believe. Mr. Joshua Lingard of Blackwell, mentioned, that in some years 6 or 8 of his Sheep die of this disease, at from 6 to 18 months old: no cause is known for this disorder, or cure or preventive, nor do they ever recover. It is said to be occasioned by a bladder of watery matter, containing an insect, which appears after death, on different parts of their Brain. Some in his neighbourhood have attempted to extract this bladder, but always killed the Sheep in the attempt.

Mr. Charles Greaves of Rowlee, said, that his Woodland Sheep, of all ages, are subject to the Turn or Giddiness. By taking hold of the horns of Sheep so affected, and feeling the skull with the thumb, a soft part is discovered therein, where the bone is very thin, and yields and springs up again under the thumb: and this is found to be occasioned by a bladder of clear watery matter, on the Brain, which seems to have corroded the skull. The wool is carefully sheared off this place on the head, and a hole clipt in the soft skull, about half an inch diameter, and the Sheep is then held up by its heels, with the head downwards, when often without, and generally with a little assistance of the finger, the bladder will fall out on the ground, without bursting; it is sometimes four inches long, and contains a tea-cup full of watery matter, in a very thin bag. A small bit of tar is gently rubbed on the wound, and a pitch plaister carefully put over the whole. Mr. Greaves has known half a dozen of his own sheep recover from this operation, and

and others of his neighbours! Some persons prick the skull and the bladder without cutting, but such Sheep don't recover.

Mr. Thomas Bowyer of Waldley, considers the Turn in Sheep, as probably an hereditary disease, for which no cure is there known, or attempted: they often lose their eye-sight; and yet Sheep thus affected, will generally become tolerably fat, in good keep, and their meat does not seem any the worse for it.

A very able Writer, in the article *Nervous System*, lately published in Dr. Rees' Cyclopedia, mentions, that hydatids on the Brain of Sheep, apparently the same as occasion this disease, generally cause the opposite limbs of the animal to be paralized, to the seat of the insect on the brain.

Rickets, Warfar, or Evil.—Mr. Charles Greaves of Rowlee, stated, that the Woodland Lambs are subject to this disease, one in every 20 or 30 of them are so affected, and not more than 1 in 10 of these recover: they lose the use of their limbs, in part or wholly; and if this happens to the legs, they seldom recover. Sometimes one leg only is affected, and a gathering takes place about the knee: the opening of this tumor sometimes gives relief.

Foot-Rot, or *Foul*.—This very troublesome disease has appeared to me, to be more common among the fine-woolled flocks than others: it is said to be contagious to Sheep walking over the same ground, with those affected by this disease. When the disorder first appears, it has been found, that washing their feet very carefully, and keeping them on clean dry straw for a time, will effect a cure: other Flock-masters pare and wash the diseased parts of their feet, and then pen them for six or seven hours, on a dry floor strewed 3 or 4 inches thick with Lime, fresh from the kiln, but not hot.

Philip Gell, Esq. of Hopton, found his Merinos very subject to this disorder, as observed page 115.

The late Lord Vernon's Woodland Sheep, kept at Sudbury, are much affected by this disorder, as observed page 96; when very bad, Butter of Antimony and Tar is applied, and the feet wrapped in coarse rags firmly bound on.

Mr. Samuel Cocker of Ilkeston-Hall, finds his New Leicesters sometimes subject to the foot-rot: he pares the hoof off, as far as it is hollow, and generally finds this sufficient, without other applications.

Scab, or *Shab*.—In some parts of the County, where the Shepherds are less attentive than they ought to be, this disorder too much prevails: Tobacco-water is the most common application for its cure, but some use Mercurial Ointments.

Maggots, or *Fly-blown*.—This dreadful evil, affecting Sheep, seems far less common or destructive in Derbyshire than I observed it to be in Worcestershire in 1808. In the Vale of the Dove, early shearing is resorted to, as observed page 134, for preventing the Sheep being *struck by the Fly*, as the operation is called, of depositing its eggs in the Wool, and whence the Maggots soon after proceed, which, if neglected, soon eat into the flesh, and miserably destroy the animal. It has been found, that in low woody places, the flies are most apt to strike flocks, and seldom on hills. Mr. John Milnes of Ashover, finds Goulard's Extract of Lead, diluted in 3 or 4 times as much water, very efficacious in destroying the Maggots, and

in

in promoting the healing of those or other wounds, to which Sheep are liable.

Ticks (*acarus reduvius*) are much too common for the comfort of the Sheep.

Bud.—Mr. Charles Greaves of Rowlee, informed me, that this is a disease affecting one in 30 or 40 of the Lambs in the Woodlands of Hope, and none of whom recover. It occurs in May, and seems occasioned by the Lambs eating the Buds and young shoots of the shrubs in the Valleys, to which the Wool lost by the Sheep, is adhering: this occasions them to swell and die, and in their stomachs, masses of Wool felted together, and mixed with the buds and leaves of Trees, are found.

Mr. Joshua Lingard of Blackwell, mentioned a disease, by which he annually loses from 10 to 20 Sheep, and is more destructive in the neighbourhood than any other perhaps, except the Red-water: it has no particular name, but some persons imagine, that Sheep thus affected are *poisoned*, by eating a red insect with their food. It occurs most commonly in bare pastures, which are thickly stocked; the Sheep die suddenly, and their skins turn black.

Goats, have no place assigned them in the "Plan" prescribed for my Report, I suppose, on account of the few of these animals that are now kept in England, compared with former times. It seems proper, however, in this place to mention, that tho' wild Goats were very common in the Peak Hundreds some ages ago, they are now nearly, if not quite extinct. Mr. Robert Arkwright of Lumford Cotton-mill, near Bakewell, is the last person whom I heard of, as keeping a herd of Goats, on the Steep Banks N of the Mill. Mr. Robert Needham of Ashford, had some also, on Fin-Cop Hill.

EGYPTIAN RAM—HORSES.

An Egyptian Ram, much resembling a Goat, is kept among the fancy Stock in Earl Moira's Park, which is partly in this County, as mentioned in the Leicestershire Report, p. 264.

SECT. III.—HORSES.

Derbyshire has long been famous, and ranked next after Leicestershire, of all the English Counties, I believe, for its stout, bony, clean-leg'd breed of work-horses, principally of a black colour. A great part of the Dairy Farmers keep brood Mares of this kind, for doing their Farm labour, in part, and annually rear some colts.

Mr. William Pitt, in his Leicestershire Report, pages 283 and 285, has given a short history of this improved breed of Horses, whence it appears, that Leicestershire had long enjoyed a good breed of Cart-horses, before the late Mr. Robert Bakewell began his celebrated improvements on animals: in the prosecution of this object, Mr. B. went to Flanders and Holland, and there selected some West Frieseland Mares, which he imported, and crossed them with Stallions selected in Leicestershire, as is supposed, and thence, after a perseverance of some years, he procured his celebrated stallions *Gee*, &c. &c. By means of the Dishley Stock, most of the Mares and Stallions of those who keep the latter to let have been improved, since the period above mentioned, and hence this is often called the Bakewell breed of Horses.

In order to enable my readers to judge of the breed or blood of the several Stallions and Mares which are kept by the most noted Derbyshire breeders, I have been

been careful, as in the case of Neat Stock and Sheep, to minute as many particulars as I could learn, respecting the history of each person's Stock, as to the persons from whom they originally purchased, and have hired Stallions, &c.

A List of such *Stallion-letters*, or Professional Breeders of Improved Work-Horses, as are not resident in Derbyshire, but have let their Stallions, &c. into this County.

Astley, Richard, of Odstone-Hall, near Market-Bosworth N, Leicestershire (Cart and Blood Horses).
Bakewell, Robert (the late), of Dishley Grange, near Loughborough N W, Leicestershire.
Challener, ——, of Nether Thorp in Shire Oaks, near Worksop N W, Nottinghamshire.
Grice, Joseph, of Blackfordby, near Ashby-de-la-Zouch N W, Leicestershire.
Hart, William, of Culloden Farm in Norton, near Ashby-de-la-Zouch S, Leicestershire.
Inge, ——, of Thorpe, near Hinckley E, Leicestershire.
Knowles, Samuel, of Nailstone, near Ashby-de-la-Zouch S E, Leicestershire.
Moira, Earl of, Donnington-Park, near Loughborough N W, Leicestershire.
Stevenson, Thomas, of Snareston, near Ashby-de-la-Zouch S, Leicestershire.
Wilds, ——, of Coton, near Market-Bosworth S W, Leicestershire.

A List of Derbyshire *Stallion-letters*, of the improved Cart-horse and Nag kinds.

Abbot, John, of Spondon, near Derby E, (2 Cart and a Blood Horse, Fairfield).
Arnold, John, of Radburne, near Derby, W.

Bancroft, John, of Synfin, near Derby, S.
Blunston, John, of Risley, near Derby, E.
Chesterfield, Earl of, Bradby-Park, near Burton-on-Trent E, (2 Hanoverians).
Clarke, Job, of Repton, near Burton N E, (Nag).
Cockayne, ———, of Walton, near Burton-on-Trent, S W.
Devonshire, Duke of, Chatsworth, near Bakewell E, (Nag).
Edge, ———, of Quarndon, near Derby, N W.
Hassall, Thomas, of Hartshorn, near Ashby-de-la-Zouch, N W.
Moore, Daniel and John, of Winshill, near Burton-on-Trent, E.
Morley, Joseph, of Draycot, near Derby, S E.
Orpe, William, of Birchwood-Moor, near Ashburne, S W.
Plimley, Walter, of Styd-Hall in Shirley, near Ashburne S, (Blood).
Robinson, John, of Weston-on-Trent, near Derby, S.
Shepherd, Thomas, of NewtonSolney, near Burton-on-Trent, E.
Shirt, Robert, of Beighton, near Sheffield S E, (Nag).
Sitwell, Sir Sitwell, Bart. (the late) Renishaw, of near Chesterfield N E, (Blood).
Smith, Edward, of Draycot, near Derby, S E.
Smith, James, of Aston in Sudbury, near Uttoxeter, E.
Smith, John, of Coton, near Burton-on-Trent S, (Cart and Nag).
Smith, John, of Sawley, near Derby, S E.
Smith, Joseph, of Lullington, near Burton-on-Trent, S.
Twig, Joseph, of the Common in Marston-Montgomery, near Ashburne, S W.
Ward, John, of Lullington, near Burton-on-Trent, S.

I shall

I shall now proceed to mention such particulars as my Notes furnish, respecting the breeding of Work-Horses, in the order of the Names of Persons, viz.

Mr. John Arnold, of Radburne, keeps two or three Stallions to let, as mentioned in the above List, of the heavy Leicester or Bakewell breed, one of them brown; which are used principally among his neighbours, who rear great numbers of Colts: the Geldings go off to the Southern Counties, the largest for Dray-Horses, &c. in London, and the middle-sized for Farmers' uses and the Cavalry: the Mare Colts are principally kept at home, for breeding.

Mr. Luke Ashby, of Eggington, keeps 4 or 5 breeding Mares, but no Stallion.

Mr. John Bancroft of Synfin, keeps 8 breeding Cart Mares; has hired or used the Stallions of Mr. Samuel Knowles, and Mr. John Robinson: he keeps a Stallion (see the above List) for his own and his neighbours' use: usually sells his Colts at 18 months old, at perhaps 25 guineas each on the average.

The Earl of Chesterfield, of Bradby-Park, has two Hanoverian Stallions, of the Nag kind, from which he breeds a few Colts annually, and they serve the Mares of his Tenants and the neighbourhood. His Lordship keeps also some black Cart Mares, and breeds therefrom for the use of his Farm.

Mr. Joseph Clarke, of Willesley, keeps 5 breeding Mares, of the black Cart kind: he has hired or used Stallions of Mr. Joseph Grice, Mr. William Hart (whose Stallions have long been in great repute in this part of Derbyshire), Mr. Thomas Stevenson, and Mr. Wilds. See the above Lists.

The Duke of Devonshire, in Chatsworth-Park, keeps 3 Stallions, named Arun, Morwick, and Plough-boy, for

for the free use of his Tenants and the Farmers of the district, making only a small compliment to the Groom, who has the care of them.

Mr. George Glossop, Mr. — Greaves, and others in and near Eckington, breed very useful heavy black Cart Horses.

Mr. Richard Harrison of Ash, keeps several black Cart Mares; hires Stallions of Mr. Samuel Knowles; in 1809 Mr. R. was offered, and refused, 150 guineas, for a 2¼ year old Cart Stallion, bred on his Farm.

Thomas Hassall, Esq. of Hartshorn, keeps some Cart Mares, and a Stallion, which serves the Mares of the neighbourhood, at 21s. each.

Mr. Thomas Jowett, Sen. of Draycot, keeps 3 black Cart Mares: uses his neighbour Mr. Edward Smith's brown Stallion, and Mr. Joseph Morley's black Stallion: he sells his Colts at 18 months old, at 25 guineas each, on the average.

Messrs. Thomas and Robert Jowett of Draycot, keeps 5 black Cart Mares, and one Hack Mare, for breeding: they use Mr. Edward Smith and Mr. Joseph Morley's Stallions, as above.

Mr. Joshua Lingard of Blackwell, keeps 2 or 3 common Cart Mares, works them on his Farm; winters them abroad with his Sheep; uses the Stallions of the neighbourhood, and sells the Colts at 2 years old. He also keeps some Nag Mares, and sends them to the Duke of Devonshire's Stallions; these Colts go off at 4 years old: his Colts fetch from 5 to 20 guineas each, according to circumstances; the Colts of the Cart kind are found the most profitable.

Mr. John Pearsall of Foremarke, keeps 6 or 7 breeding black Cart Mares, and hires Stallions.

Mr. James Radford of Eggington, keeps 4 Cart and
1 Nag

1 Nag Mares; uses the Stallions of Mr. Thomas Shepherd, Mr. Edge, and Mr. John Arnold: Mr. R. had in 1809 a remarkable fine Gelding, which he bought of Mr. John Massey of Hilton.

Mr. Thomas Shorthose of Eggington, keeps 4 or 5 Cart Mares.

Mr. Thomas Smedley of Eggington, keeps 4 breeding Cart Mares and a breeding Nag Mare: his stock was originally derived from the famous old blind Horse at Packington, and Mr. Thomas Shepherd's Horse *Peach*, to whom he yet sends Mares: he also uses the Stallions of Mr. Edge, and Mr. John Arnold (a brown Horse). Mr. S. sells his Colts at 18 months old, at 30*l.* each, on the average: he deals in Colts, bought up in the neighbourhood.

Training of Nags: Mr. Thomas Lea of Stapenhill, breeds no Colts, but annually buys in about 90 three-year old bay Nag Colts, which he keeps proper and experienced persons to break and train; and when rising 4 years old, sells them again at 50 to 90 guineas each.—This plan of employing steady and proper persons constantly, in the breaking and training of Saddle Horses, seems greatly preferable to the imperfect and improper treatment which this noble Animal too often receives, from the Servants and Boys employed on a Farm; whence many good Horses contract bad habits, which greatly lessens their values: the ready sale and high prices which Mr. Lea finds for his Horses, shews, that his plan is worthy of more general adoption.

Racing, or *Blood* Horses, are bred in the County by several persons, yet the late Sir Sitwell Sitwell was the only person whom I heard of, who had any considerable establishment for training these Horses for the Course.

Course. The Breeders of these Horses whom I heard mentioned, were, Mr. John Abbot of Spondon, Mr. Joseph Butler of Killamarsh Forge, Mr. George Glossop, and Mr. Greaves of Eckington, Sir Thomas Windsor Hunloke, Bart. of Wingerworth, the Earl of Moira of Donnington-Park, Mr. Walter Plimley of Styd-hall, the late Sacheverel Chandos Pole, Esq. of Radburne, and the late Sir Sitwell Sitwell, Bart. of Renishaw.

Mr. Joseph Butler of Killamarsh Forge, is understood to keep, the most methodical and exact general Studbooks, with accounts of the Pedigrees and Matches of all the noted running Horses in Britain, superior perhaps to those of any person, who does not wholly devote their time to this subject.

At Over-Haddon there is a large Pasture Field, devoted to the Summering of *Stallions*, in which 15 or 16 grown ones are seen together, at some times: see Vol. II. p. 189.

At Ashby-de-la-Zouch in Leicestershire, on the Monday before the 25th of March annually, a large *Stallion Show* is held; and where the breeders of Colts usually assemble, to inspect them, and fix on those they mean to hire or use, in the ensuing season.

Considerable *Horse Fairs* are held at Ashburne, and at Burton-on-Trent, and Ashby-de-la-Zouch in Leicestershire, near to this County. Mr. Reed Denham, jun. of Calow near Chesterfield E, was mentioned to me, as the most considerable Horse-dealer resident in the County.

Work.—The Team labour of this County is now very generally performed by Horses; a good deal of it by Brood Mares, as has been mentioned page 150. The use of Oxen and Heifers, seems nearly confined to the

the Noblemen and Gentry, and a few experimental Farmers, as will be inferred from the accounts that I have given of Ox Labour at page 69: and they appear to have instituted or carried on, no comparative experiments, of sufficient extent or duration, to be likely to convince their neighbours the Farmers, of the superior benefit of the practice, here at least. In a district like this, where the rearing of Horses of a superior breed, is extensively practised, and a ready sale is found for all that can be produced, at constantly increasing prices, and where a considerable share of work is performed by the Brood Mares, Oxen seem to have far less chance of being generally used for Labour, than in some other Counties, differently circumstanced.

Food.—Horses in this County are generally *grazed* in the Summer, on natural and artificial Grasses, and during the Winter season, are fed on *Hay* of the natural Meadows, or of Clover and other cultivated Grasses. The practice of Mr. Joseph Butler at Norbrigs, with regard to *Sainfoin Hay*, cut for his Colliery Horses, has been mentioned Vol. II. p. 165. He finds it thus go half as far again as loose Hay; Horses can't, on account of mouldy patches, pull out and waste the Hay under their feet: they also eat the Hay-seeds, which are very nourishing.

Oats is the grain principally given to the Working Horses. The *Bran* of Wheat is not so plentiful here as in the Southern Counties, where wheaten Bread only is used by the Inhabitants, and in some districts it is hardly to be procured for the use of Horses, when disordered, or to mix with their Corn. Mr. John
Aveson,

Aveson of the old Mill in Glossop, grinds down the Shudes, Shillings, or Husks, from his Oat-meal Mill (see Vol II. p. 186 and 457), as a substitute for Bran, for his Horses: a practice worthy of imitation, I think, wherever Oat-meal is extensively manufactured.

Mr. Joseph Gould, never gives Hay to his Farm Horses, but unthrashed Oats and Straw.

Some instances of feeding Horses with *Carrots*, are mentioned Vol. II. p. 151 and 210; Mr. Robert Burrows, in his Communication to the Board, has stated 70 lb. per day to be his allowance of Carrots to each Working Horse.

The practice of Mr. Thomas Bowyer, in giving *Swede Turnips* to his Horses, has been mentioned Vol. II. p. 148.

Soiling, or giving of cut *Green Food* to Horses, is not near so extensively practised as it ought to be. That very able improver, the late Joseph Wilkes, Esq. of Measham, who kept a great many Horses on his Colliery, and other works, always kept them within doors, and as much as possible fed them with cut *Grass and Edish* or Aftermath, as is mentioned by Mr. William Pitt, in his Leicestershire Report, pages 191, 192, and 290.

Among the many interesting and very useful Agricultural Communications in the "Farmer's Journal," published Weekly, I have noticed several good hints on this subject, particularly in a Letter from Lancashire, published 26th April, 1813: *Red Clover* is there an article most highly commended, for soiling Horses and Cows, with a portion of Winter Tares, to supply them between the first and second cutting of Clover: in this way, in 1811, 6 Horses and 22 Cows were,

after

after the 21st of May, very completely summered, on only 14 A. 3 R. 6 P. of green crops!

The practices of the late Mr. Francis Bruckfield and several others, in soiling with *Winter Tares*, has been mentioned Vol. II. p. 134. Mr. George Clay of Arleston, begins about the middle of May to cut his Stubble Tares for his Horses, and contrives the quantity so, that the land shall be cleared (for fallowing and a late crop of Turnips) about the end of June, when the Clover Edishes, shall be fit for turning the Horses into them.

The practices of several persons in soiling with *Spring Tares*, is mentioned Vol. II. p. 134; and of several others with *Lucern*, Vol. II. p. 166.

In one instance I heard of *crushed Gorse* or Whin, being used for soiling Horses, as mentioned Vol. II. p. 356. In Anglesea, I have lately observed this to be a common food for animals.

In *Littering* or Bedding of his Horses in the Stable, Mr. John Aveson of the old Mill in Glossop, uses the Shudes, Shillings, or Husks of the Oats; and thus saves Straw, and makes much valuable Manure, from an article which is generally, at the Oat-meal Mills, either blown away into the stream by the fanners, or set on fire in order to get rid of it, when accumulated in a heap.

Distempers of Horses.—The benefits of the patriotic exertions made about 20 years ago, for reforming our National Veterinary Practice, have extended themselves to Derbyshire, and is very visible, in the increased knowledge and respectability of the persons who now practice as *Veterinary Surgeons*, Cattle Doctors, Farriers, &c.

Messrs.

Messrs. Thomas and George Draper, Veterinary Surgeons, of Allsaints in Derby, and of Castle-Donnington in Leicestershire, are extensively employed, and much relied on, by the owners of Cattle of all kinds, for their skill and success in treating them, under distempers or accidents. In a conversation with these Gentlemen, I learnt, that an *Inflammation of the Lungs* is here the most prevalent disease of Horses, occasioned, as they believe, by the frequent and great changes of temperature, to which they are exposed. It has been calculated, that 1 in 27 of Husbandry Horses, are lost annually by casualties.

One of the Horses which I rode during this Survey, and had been lent for the purpose from the Stables of Sir Joseph Banks at Overton, as one of the means which the worthy President of the Royal Society generously supplied, towards my undertakings (which are alluded to at the beginning of my Preface to Vol. I.) was suddenly seized in September 1809, with a swelling in the Throat, which seemed in danger of causing Suffocation: on consulting Messrs. Drapers on this case, they mentioned, that a similar disease was then very prevalent among the Horses of the County, many of which they had soon cured, and so it happened with my Nag, which I left in their care for 8 or 10 days.

Henry Bowyer of Brailsford, Surgeon, applies himself to the disorders of Horses and other Animals: his powders for scowering Calves, have been mentioned page 83.

Mr. James Horsley of Melborne, Cattle-doctor, has been noticed, for his preventive medicine for the Black-leg in Calves, at page 76.

Mr. William Robinson of King's-Newton, Farrier and

and Cattle Doctor, is much resorted to. Under an open shed in his yard, he exhibits a complete Skeleton of a Horse, by way of making his profession known.

When Horses have had the misfortune to break the skin of their knee or other parts, the growth of the hair over the healed wound, is said, in this County, to be promoted, by rubbing it with the swarf or coomb from a Bell-axle, or other brass centre of machinery.

SECT. IV.—ASSES.

The number of Asses kept in this County is rather considerable, but from being principally in the possession of the lower class of Carriers, of Coals from the Pits in the vicinity of the Towns, for supplying the poor People, the Carriers of Pottery Wares, &c. and other things which are hawked about the Country, little or no attention has been paid to the improvement of the breed of these useful Animals, either as to size or any other particular.

Earl Moira of Donnington-Park, has two fine Goza Stallion Asses, of 14 hands high; which are the only selected or improved Males of this species, which I heard of in the County: his Lordship employs several Asses for carrying off Turnips, Cabbages, &c. to the Stock in his Pastures, and other like uses of the Farm, as mentioned by Mr. William Pitt in his Leicestershire Report, page 293.

Mr. Joseph Butler used Asses, on the Iron Rail-way from the Ironstone and Coal-pits to Wingerworth Furnace. In his underground Coal-works at Lings and Norbrigs, he also uses Asses, for dragging the Corves of Coal from the Banks or faces of work, to the bottom.

of the drawing Shaft, as described in Vol. I. p. 345. Mr. Butler has found these Animals capable of enduring the Choak-damp or Carbonic Gas in the Pits (see Vol. I. p. 535) better than either Horses or Men; and in cases of accidental choking of the works, the Asses have been found alive and little injured, three or four days after the Men had died, in consequence of the want of air, and two days after Horses had perished from the same cause. But Asses are principally preferred in those works, where the height or head-way of the Gates is insufficient for Horses. At Riddings, and other Collieries, where Asses are used underground, I have been much amused, at seeing these Animals drawn up on the Saturday Evening, in order to graze on the surface until Monday Morning; which they seem from habit to expect, and when let loose, testify their joy, by the most frantic braying, running and kicking up. The contrast is striking, on seeing them driven to the Pit-head, and having the sling-chains wrapped round them, and expecting to be caught up, and suspended high in the air, before they are let down the shaft, to their labour.

The Coal-masters and their Agents who employ Asses in their works, should carefully see, that the Boys who drive them, do not, as is far too common, use a thick stick, and always strike them in the same place on the rump, until a shocking sore is thereby occasioned, as I have several times shuddered to witness. And the conduct of a brutal Girl, employed in driving a poor Ass, that dragged the Corves from the top of the Pit to the Stack, at Ballyfield Colliery in Hansworth in Yorkshire, which she incessantly cudgelled with a truncheon two inches or more diameter, excites horror, even now by its recollection.

When

When turned out to graze in the Lanes, a stout stick of some length is often suspended by a small chain or a cord, round the necks of Asses, to prevent their creeping through hedges, or leaping ditches or walls, which this mode of yoking seems as effectually to prevent, as the fastening of their fore legs together, or loggering them, as is commonly practised in other places, and cannot but injure their paces and limbs, when in use.

SECT. V.—MULES.

This mixed breed of Animals, between the Ass and the Horse, is very little known in Derbyshire, I believe, except on its southern and western borders. At Donnington-Park, Earl Moira has two Spanish Ass Stallions, as mentioned in the last Section, which cover Mares either of the Ass or Horse kind at 23s. 6d. each. The Mules of this breed are of good size, and are used both for Work and the Saddle. Mr. Dawson, the Steward to Earl Moira, prefers a Mule for his own riding, as Mr. Pitt informs us in his Leicestershire Report, page 294.

The Duke of Devonshire employs several Mules, tho' the number is now far less than formerly, in carrying the roasted Copper Ore from his Ecton Mine to Whiston in Staffordshire (see Vol. I. p. 353), and bringing Coals for the Lead-smelting and Copper-roasting works at Ecton; and others in fetching Coals from Hazles-cross to Whiston, for the above purposes, and for supplying the Copper-smelting works there with fuel: this last distance is 3 miles: each Mule carries 3 cwt., in a pack on its back, and they make two journeys a day. The Mules are the property of persons who carry for hire.

SECT. VI.—HOGS.

Derbyshire being a considerable Dairy County, large numbers of Hogs are kept in the County, which yet can boast of no particular or characteristic Breed, tho' some persons have distinguished the very excellent sort which are commonly found on the larger Farms, by the name of Derbyshire Pigs: others have called them the Burton, and the Tamworth Breed, I believe. Mr. Brown in the Original Report says, that they would be called Berkshires, in the Southern Counties of England.

These Pigs are often white, sometimes spotted with black and white: the quantity of hair on them is small, and it often curls over the whole surface, of these thin-hair'd Pigs. Their legs are rather short, and so are their heads and ears, broad in the back, rather light in the bone, and they shew a strong disposition to fatness, even on indifferent keep. Experience proves, that a valuable breed of Pigs can sooner be acquired, by attention to breeding, than of any other of our domestic Animals; and it is not less evident, that they almost as rapidly decline, by close breeding, and without continual crossing. In this and several other Counties I have met with Farmers, who at one period had a breed of these Animals, that they not only found extremely profitable in the ordinary course of their business, but which were sought after by others, in a pretty extensive circle, to breed from; and yet, under

the

the same treatment which would have preserved and even improved the breed of Neat Cattle or Sheep, the valuable properties of their Pigs decreased by insensible degrees, until they were no longer sought after by other breeders, or found profitable to be retained by themselves: and I have even found some, whose Pigs were a second time degenerating.

For the reasons above stated, professional *Boar-letters* are almost unknown here. A few Noblemen and Gentlemen keep *Boars*, which are used by their Tenants and Neighbours.

Earl Chesterfield keeps on his Bradby Farm, a Derbyshire Boar, for the use of the neighbourhood: he keeps also a Suffolk Boar.

Earl Moira keeps in Donnington-Park a German and a Berkshire Boar, partly for the use of his Tenants, &c. I believe.

Mr. Thomas Moore of Lullington, keeps a handsome white Boar, which serves the Sows of his neighbours, at 7s. 6d. each.

It is on the above accounts also, that I have not been able to class, or treat of the different breeds of these Animals separately, as I had wished. After going through thro' my Notes in alphabetical order, on the most prevailing of the improved kinds, as above, I can only place after them, a few, whose owners' gave them particular denominations.

John Berrisford, Esq. of Osmaston Cottage, has a very fine breed of deep-sided Pigs.

Mr. Thomas Bowyer of Waldley, keeps a valuable sort of Pigs: has fatted several on boiled Swede Turnips, mixed with Barley-meal.

Earl Chesterfield of Bradby-Park, has remarkably fine Derbyshire Pigs, black and white, thin curled hair,

hair, with erect or prick ears and short noses, broad backs and deep sides. In a store state, kept on Whey and refuse Vegetables, they appear almost fit for Bacon; they are fatted at 9 months old to 10 or 12 score, and at 15 months to about 28 score.

Mr. Samuel Cocker of Ilkeston Hall, has a good breed of Pigs; at 18 months old they weigh 18 to 24 stone (14 lb.)

Edward Coke, Esq. of Longford, has an excellent breed of Pigs.

Mr. Edward S. Cox of Brailsford, had a sort of Pigs, with broad backs, short ears and noses, and whose constant fatness, on store keep, were greatly admired; but the system now pursued on his Farm, furnishes no food for this species of Stock.

Mr. William Cox of Culland, has a handsome sort of white, thin-hair'd, deep-sided Pigs, with small heads, and little bone.

Mr. Thomas Jowett, Sen. of Draycot, has a good sort of thin-hair'd short-ear'd Pigs, reared from the Stock of Mr. Michael Buckley of Normanton, Notts. near Loughborough N E, Leicestershire.

Francis N. C. Mundy, Esq. of Markeaton, has long been famous, for a fine white breed of Pigs.

Mr. Walter Plimley of Styd-hall, has a fine large breed of white Pigs, which he fattens to 30 score.

Thomas Prinsep, Esq. of Croxall, has been famed for his Pig Stock.

Mr. William Smith of Swarkestone Lowes, has a black and white, thin-hair'd breed of Pigs, which are kept on Whey, and at 12 months old fetch 5*l.* each, on the average.

Mr. George Toplis of Brassington, cures Bacon on a large scale (as will be mentioned further on): in May he he usually buys in his Pigs, at from 2 to 12 months old, of the thick-back short-eared sort; keeps them on Whey and Oat-meal, and boiled Potatoes, until the Autumn. Mr. Toplis lays out about 1000*l.* annually in Pigs; he don't chuse large ones, as requiring more in proportion, to fatten them: he prefers Barrow Pigs to open Sows, as usually having more fat inside, and proving less red in the flesh: but Spayed Sows, if cut when young, are rather superior to Barrow Pigs. In some years Mr. T. kills 180 Pigs, whose sides weigh from 11 to 20 score, or 14 score on the average.

Earl Vernon (the late) of Sudbury, kept great numbers of Pigs of a small mixed breed, black and white, and rather short-eared: they were hardly kept during the first 16 or 17 months, but being very disposed to feed, in 6 weeks after, they arrived at 8 to 11 score.

The late Mr. John Wall of Weston Underwood, kept several fine deep-sided Pigs: which he fed on Whey (see page 54), aftered being soured in the Swill-tub, where it was preserved.

Mr. Robert Lea of Burrow-fields, has a useful sort of Pigs, a cross between the Berkshire and Hampshire breeds: he keeps them on Whey, and finishes their fattening with Barley-meal and Beans. At 12 to 18 months old they come to 14 to 22 score.

Mr. William Greaves, Jun. of Bakewell, keeps several large long-eared Hertfordshire Pigs, and others smaller and prick-eared, and of the China breed. About Christmas he kills the larger kind at 12 to 14 months old, for Bacon.

Earl Chesterfield, keeps on his Bradby Farm, an Otaheitean Sow, which he crosses with a Suffolk Boar; the Pigs are very handsome, and at 15 months come to 17 or 18 score. Their Pork, Bacon, and Hams, are much

much better relished at his Lordship's Table, than those from any other breed.

Sir Henry Crewe, Bart. has rather a singular, wild kind of Pigs, yet small and handsome, the flavour of whose Pork is highly relished.

Abraham Hoskins, Esq. of Newton Solney, in 1808 gave Swede Turnips to his Store Pigs, with excellent effect.

Mr. Thomas Moore of Lullington, has a fine breed of Pigs, which are called Ginger, in that neighbourhood, bred from the common slouch-eared Sows of the County, and a wild Boar belonging to Mr. Elliot. They are kept on Whey, and at 14 or 15 months old weigh 15 to 17 score; he annually kills 16 or 17 such, for family use.

Mr. William Garman of Persal Pits, has a useful breed of sandy, black-spotted Pigs, allied, I believe, to the above.

Roasting Pigs.—This delicious food is to be procured in considerable plenty and perfection, in all the Markets of the County. The Earl of Chesterfield, from his Otaheite Sow, supplies his table with delicious Sucking Pigs of a fortnight old.

Food.—Many Agriculturists seem to be aware, of the great benefits that would result, from cultivating *Green Crops* of Clover, Lucern, Chicory, Grass, &c. near to the Piggery, and feeding the same off the ground, in small portions at once, by Pigs. The difficulties attending this practice, have been found, in the propensity of Pigs to root or break the surface of the ground and destroy the plants, particularly such as have large roots, which seem indeed the most natural food of the Hog; and in the strength of their noses,

and

and the propensity which they have, to lift up hurdles or other temporary fences, set to confine them to limited portions of food, unless the same were made heavy and quite unwieldy. Now the first of these evils, and a great one it is, in most districts of England, and for which ringing them is but an uncertain preventive, is easily remedied, as is practised in Cheshire (see Mr. Henry Holland's Report, page 292), viz. by shaving off the gristly or horny projection of the snout, thro' which the ring is usually passed; and by which their future *rooting* is effectually prevented. Several years ago, some fat Pigs exhibited at Lord Somerville's Spring Show in Barbican, had been thus treated, when very young, by a very sharp razor; which operation seemed at the time to give little pain, and the wound, as was certified, was quickly healed, and by which simple means, the propensity to turn up the ground seemed effectually removed, without any evil in the taking of their food, being experienced.

The difficulty of confining Pigs, seems effectually got over by the Highlanders, as I noticed last Summer (1812), in Sutherland, and others of the Northern Counties of Scotland, where a Pig is seen *tethered by the neck*, eating Grass, and such refuse of the Garden as may be thrown to it, in the front of nearly every Hut or Cottage. *Ropes* were there used, often made of Heather or Heath, and sometimes of the long and flexible fibres torn from the Fir-trees found buried, and partially decayed, in the Peat Mosses, where they dig their Peats. It has struck me, since seeing of the above, that light chains of sufficient length, having plenty of swivels in them, and iron or chain collars, might be contrived, and advantageously substituted for tethering Ropes; and that instead of a Stake,

which

which requires driving, and more difficult operations frequently, for drawing it up again, a Screw-pin should be used, attached to each chain, having a square head, to be turned by a winch-handle: the screw to be cut in a particular manner, contrived by Mr. Robert Salmon, for fixing down his harmless *Man-Trap*, which is described and drawn in the 27th Volume of the Transactions of the Society of Arts, p. 183. And perhaps such an apparatus, and mode of flitting or tethering, would be found greatly preferable for larger animals, than tieing them by the leg, as is at present practised.

The giving of *green food* to Pigs, except the refuse of the Garden, is at present far less practised in Derbyshire than it ought to be: I have no Notes on the subject. Mr. Thomas Bowyer has used *Swede Turnips*, as mentioned Vol. II. p. 148; and III. p. 165; and Abraham Hoskins, Esq. page 168. Several persons have used *Potatoes*, as mentioned Vol. II. p. 155, and III. p. 167.

The principal liquid food of Pigs in this County, is *Whey* from the Cheese Pan, mixed with the Slop or Pot-liquor, &c. from the Cooking operations: and very little *Skimmed-Milk* is given them, comparatively, on account of the practice of making Butter here, from Whey instead of Milk Cream, as mentioned pages 43 and 65. *Grains* are not used for this purpose, I believe, beyond the produce of the Farmer's own brewings. I saw or heard here of no instance of the use of *Oil-cake Tea*, or Jelly, for Pigs, such as I saw used in Lincolnshire; a quart of dried and pounded Cake, being boiled in three quarts of Water, for forming a gelatinous liquid, of which young Pigs are very fond.

The

The most common application of Corn to the feeding of Pigs here, is *Barley-meal*, or Oat-meal, see page 165. Some *Pease* are used, and *Beans*, see page 167. The Earl of Chesterfield applies *Buckwheat* in this way, as observed Vol. II. p. 135.

Boiling of Swede Turnips has been practised, see page 165; and Potatoes, page 167.

I could hear of no instance of *Steaming*, or giving *warm* food to Pigs, in this County.

An excellent Paper on the management of *Pigs*, by the Rev. J. T. Hamilton, will be found in the " Letters and Papers" of the Bath Society, Vol. XII. p. 128.

Sties.—In the Plan which is given of the Earl of Chesterfield's new Farm Premises in Bradby-Park, Vol. II. p. 10, the situation and arrangements of his Lordship's excellent Piggery will be seen, Nos. 1 and 2. Sir Thomas Windsor Hunloke, Bart. in the improvements made, on taking Lydgate Farm in Wingerworth into his own occupation, in 1810, paid great attention to the Piggery: and the like may be said, respecting most of the modernly improved Farm Premises, mentioned in Vol. II. pp. 9 and 11.

Stone *Troughs* are in very general use for feeding Pigs, and often are circular, with a lump left in the middle while excavating them, for throwing the Whey or wash to the outsides, so that master Pigs cannot prevent the others from feeding, by getting into the trough, &c. see Vol. I. p. 433. For pouring Whey, &c. into the Trough, fixed within the Pig-stie, a stone hopper is often seen, worked into the wall, into which the liquid food is emptied, on the outside of the

Stie: sometimes this hopper is within the common kitchen.

Mr. John Aveson of the Old Mill in Glossop, applies part of large quantities of Shudes or Shillings of Oats produced at his Oat-meal Mill, to the *littering* and bedding of his Pigs in their Sties, and by that means produces a good deal of Manure from an article too commonly wasted.

Mr. Edward Bradbury of Corder-Clough Farm in Glossop, fetches Peat from the Moors, and strews it from time to time in his Pig-yard, to suck up the urine, and thus he procures a rich Manure.

Live and Dead Weights of Fat Pigs.—I regret to mention, that few records of this kind appear to be kept in the County, though so important, for forming and maturing the judgment of all those, who sell their fat animals alive, and are furnished with a weighing machine : I met with no document of this kind in the County, respecting Pigs.

Mr. Layton Cooke, in his excellent set of Agricultural Tables, which have been quoted in the Notes at pages 74 and 131, has, from the documents collected and published by the Smithfield Club, and from various others, calculated at page 33, that a well-fed Pig of the standard weight 1.00 when alive and fasted, ready for killing, will have its sides, head, feet, and flae collectively, of the weight from .75 to .85*.

Mr.

---

* Since the above was written, Mr. L. Cooke has very kindly furnished me with his unpublished Table, of the proportionate weights of the several parts of ripe or perfectly fatted Hogs, expressed in decimals of the live weight 1. (as in page 74 and 131, for Cattle and Sheep), viz.

Carçass

Mr. Thomas Brown mentions a calculation, p. 26 of the Original Report, that a profit is made annually on Hogs, of from 20s. to 25s. for each Cow, in the Dairies of this County.

Bacon.—The greater part of the Pork killed in Derbyshire, is either used fresh or made into Bacon, Pickled-pork not being in use for the Farm Servants, about Bakewell or elsewhere in the County, that I heard of.

Brassington is famed for its Bacon, about 400 fat Hogs being now killed there annually, and cured for sale*, and some years ago the numbers were greater: this decrease in the trade, is supposed to arise from the decrease of the Mining in the Peak Hundreds, and of Manufactures of late, and perhaps owing to Bacon being less preferred as a summer food, than formerly.

Mr. George Toplis, whose practice in buying and fattening his Pigs, has been mentioned page 167, commences killing in August, and continues till Christmas, and sometimes much longer. The hair is taken off his Pigs by scalding. When the carcass is cold,

Carcass or sides .7200		
Flae .... .0747		.8500
Head .... .0500		
Feet .... .0053		
Fat, Crow, and Caul		.0558
Pluck		.0125
Entrails and contents		.0214
Blood		.0180
Hoof, Hair, and loss by evaporation in cooling		.0423
		1.0000

* At Clonmell, in Tipperary in Ireland, it is said, that 1900 hogs are annually killed and cured for Bacon.

and

and cut in two, the bones are taken out, all but the small ones within the sides, and the Hams are cut off: the Spare-ribs and Offals are principally sold at Derby. On the next day after killing, 1 lb. or more of Saltpetre in fine powder, to each Hog, is rubbed well all over the sides or flitches; they are then salted, each side in a shallow oval wooden Tub, called a Kimnel, 6 feet long and 3 feet wide, in which they lay from 3 to 14 days, according to the heat of the weather, but are never immersed in brine. After this the flitches are hung up to dry, in a room, over a Baker's oven, or at 10 or 12 feet distant from a fire in a room of the house, with their edges towards the fire, which should be steadily kept up during the day, but not so intense as to sweat or rancify the surface. In 2 to 5 weeks time, according to circumstances, the flitches will be dry and ready for sale, but as the markets seldom demand it at this time, the flitches are piled up in a close dry room, with dry salt strewed between them, packed as close as possible, and generally the whole is closely covered by cloths, to exclude the air more effectually from the Bacon, until wanted for market. The principal sales take place in June, July, and August: in 1808 the wholesale price was $8\frac{1}{4}d.$ and $9d.$ per lb.: it is a ready-money trade. Mr. Toplis has warerooms in Derby and in Sheffield, and keeps a retail stall on the Friday at Derby, at which places his Bacon will generally sell for a halfpenny per pound more than his neighbours', on account of his superior skill and care, in curing and preserving it. Mr. Joseph Charlton of Brassington, is another considerable Bacon-man.

In Earl Chesterfield's Farming Establishment in Bradby-Park, they are very curious in curing of Bacon,

con, and his Lordship has a conical Building or *Stove*, on purpose for drying of the flitches and smoaking of them, by a fire lighted on the ground in the centre, and supplied with Saw-dust and other slow-burning articles.

Hams.—Mr. George Toplis of Brassington, (whose practice in curing Bacon has been described above), first rubs over each Ham, a quarter of a pound of Saltpetre, in fine powder, and two or three hours afterwards, dips them in fresh strong salt Brine, to dissolve the Salt-petre. During the next 6 to 10 days, they are laid immersed in the Brine, in wooden Salting Tubs, and are moved every day, to make them touch the bottom, and each other in fresh places. They are then taken out of the Brine, and laid singly on boards, with the rine downwards, and a little salt is strewed on them from time to time: the best Hams are thus cured, between Michaelmas and Christmas, and they lay thus on boards without turning until about January, when they are hung up and dried, just as flitches of Bacon are, as above, except that the Hams are hung rather nearer the Fire for a few days. After hanging 2 or 3 months, each Ham is put into a strong brown Paper Bag, whose bottom has been sometime previously painted, to prevent its softening if the Hams should give, and become moist, and thus they are hung up to the cieling of a dry room, until sold, which is principally in the Summer season. The Inns in Derby take the largest, averaging about 14 lb. each, at the retail price of Bacon at the time, which in 1808 was 10*d.* per lb.

The person who cures Hams for Earl Chesterfield at Bradby Farm, makes a curious pickle of Salt-petre,

Sal-

Sal-prunella, various Herbs, &c. which is kept in a leaden Pickling-cistern, over which the wooden Salting-trough is placed, wherein the Hams are salted and drained, by means of small holes in its bottom, for letting the dissolved salt drop into the cistern. The same pickle is used for 10 years together, being only boiled and well skimmed once a year, except when fresh ingredients need adding to the pickle. I was much at a loss to comprehend the propriety, of using such very old pickle, and especially the keeping of it in a leaden cistern: and, cannot but recommend to his Lordship, the propriety of submitting a bottle of this old pickle to some able Chemist, for analysis, fearing that it contains a most deadly poison in solution, in quantities sufficient to undermine the health of persons, who frequently eat of Hams cured therein.

Distempers of Hogs.—This very useful animal, seems here, alike free from diseases or casualties (after the Sow has pigged a day or two), as in every other district that I am acquainted with; so much so, that I never heard a Derbyshire Farmer, name or allude to a disease of his Pigs, that I recollect. This is one among the most valuable of their properties.

SECT. VII.—RABBITS.

I DID not observe any regular Rabbit-Warren remaining in the County: the only instances that approached to these, in the number of Rabbits, were on a rough and rocky piece of land S of Griffe-Hill, near to Hopton, and in the northern part of Sudbury-Park: near which last situation, the culture of Wheat was

obliged

obliged to be abandoned, on account of the depredations of the Rabbits, as Mr. Thomas Brain, his Lordship's Bailiff, informed me: the Oaks also in the
young Plantations were quite destroyed by them, as
observed Vol. II. p. 244.

In the Hedge-rows and small steep or rocky Woods,
these Animals still too much abound, in various parts
of the Country, and commit considerable depredations on the Crops of the Farmers, and on the young
Plantations of their Landlords. I have little hesitation in saying, that the sooner *wild* Rabbits are extir-
minated the better, since due encouragement given to
the Cottagers, in rearing *tame* ones for sale, as in
Buckinghamshire, might (as Mr. Richard Parkinson
has observed at p. 144 of his Rutlandshire Report)
fully supply their places on our tables.

As to the soil of *Warrens*, being unfit for other purposes, as has often been alledged; in many situations,
Marling might entirely alter their nature, as has so
happily been proved in the north-east part of Norfolk:
and even where this is impracticable, useful Plantations would certainly grow, wherever Rabbits can pe-
netrate and find hiding places: and these accessions of
Plantation, might enable the clearing of as many
acres of *the best* of the Woodland soils, in order to
meet our pressing want, of *more food* for the People,
as is further shewn in Vol. II. pages 226, 236, and 261.

SECT. VIII.—POULTRY.

In the Plan, which is given of the excellent Farming Establishment of Earl Chesterfield in Bradby-Park,

Vol. II. p. 10, the arrangements will be seen, of as complete a Poultry Establishment, as perhaps exists in the kingdom. The *Roosting-house* is well contrived, with covered places for the Ducks and Geese, under the Fowls, and the whole is constantly kept strewed with fresh saw-dust. The *Sitting-house*, and which serves also for *laying*, is furnished with flues, to preserve an equal temperature in Frosts. In the *Feeding-houses*, the fronts, partitions, and floors of the Pens are all of lattice-work, which readily take out, in order to wash them thoroughly; shallow drawers with fresh Saw-dust, pass under each Pen, to catch the dung. The fatting Poultry are fed twice a-day, and after each the food is taken away, and the day-light excluded, for them to rest and sleep.

Turkies.—This kind of Poultry, is thought by the housewives in many Counties, to be more difficult to rear, on account of their tenderness, than either Geese, Fowls, or Ducks: but which must, I think, be a mistake, since they seem rather common in the Peak Hundreds of this County. On passing Clod-hall, a wretched small Farm, E of Baslow, near the top of the High Moors, in the Spring of 1808, I was quite surprised to see the large number of fine young Turkies, that were running around the house, apparently unaffected by the bleakness of the situation.

Mr. Edward S. Cox of Brailsford, has a breed of very large reddish brown American Turkies, and the same are now considerably disseminated in the neighbourhood, pure, as well as crossed with the common Turkey. Mr. Cox has two Brothers, one of whom, William, has long resided in America, and there keeps this sort of Turkey, which being noticed and admired

by

by Mr. Roger Cox, when there on a visit to his Brother, several years ago, he brought over some of them. They roost upon trees, or high parts of the buildings: many of the Cocks weigh near 20 lb. when fat, and are thought to be well flavoured; the hens of this breed are small, in proportion to the Cocks.

Geese.—The numbers of this species of Poultry has decreased, since the inclosure of the large Commons, but many Cottagers still keep them, in the wide Lanes and small pieces of Common, that are left adjoining to them: in Pilsley in Bakewell, I noticed a good many thus kept. When Geese are turned out, it is common in this County, to sling a stick about 2 feet long before the breasts of the old ones, balanced on a loop of string round their necks, which is found as effectual, in preventing their trespassing on the Farmer, by creeping Hedges, &c. as the same mode of hampering the Asses, Sheep, Pigs, &c. and is not cruel, like the thrusting of a long feather through the nostrils of Geese and Ducks, as some thoughtless Persons do. Mr. James Pilkington, in his "View of Derbyshire," Vol. I. p. 334, says, that Geese are benefited by eating the seeds of the *Festuca Fluitans*, or Water Fescue Grass, that they find in wet ditches and brooks.

Fowls.—The Farmers' and Cottagers' Wives are as attentive in this County, as in most others which I have visited, to rear or provide as many Hens, as the offal of the Barn or Cottage door will supply, and even to purchase food for them, in frequent instances: Barley, whole or ground, being most commonly preferred for this purpose. I observed nothing to note, as to the treatment or the breed of Fowls, except that

Plesley,

Plesley, on the borders of Nottinghamshire, has long been famous for its fine breed of black Fowls. Around Wingerworth, considerable numbers of *Game Fowls* are kept, for the truly idle and disgraceful purpose of *Cocking!;* a few are kept in Killamarsh and other places, and formerly they must, I suppose, have been more common than at present, in Tansley, before their Cock-pit was converted to a Methodist Meeting-house!

When pursuing my Survey in the less frequented parts of this County, and its vicinity, I very frequently dined upon *Eggs*, which I found in general to be large and good; and observed, that the Landlady of my Inn, often kept them tied tight up in a strong net, which was hooked up to the cieling, by one of its meshes. This method they adopt, for the purpose of giving free access to the air, and in order to see and be reminded, of *turning* their stores of Eggs, into a fresh position each day: this being the main essential in preserving Eggs, whose yolks subside slowly, when left unmoved, and come at length to touch the shells on the lower side, when rottenness almost immediately commences.

Ducks.—In the Poultry Establishment of the Earl of Chesterfield in Bradby-Park, I saw a very fine and large breed, called *Rouen* or *Rowen Ducks*, which are greatly esteemed at the Table, for their flavour. I saw nothing to note on the breed or management of the common Ducks of the County.

Where waters much impregnated with *Lime*, usually called Petrifying Springs, break out, and deposit Tufa (see Vol. I. p. 457), as at Alport in Yolgrave, &c. it is said, that the *Eggs* of Ducks and Geese, which frequent these waters, are so much thickened

ened thereby, that hatching becomes difficult, and is even prevented, unless these hard and thick shells are properly cracked, just before the hatching time, in order to facilitate the escape of the young one.

Wild Ducks and other Birds and Animals, Fish, &c. which are not the objects of the Farmer to raise, will be noticed at the beginning of Chapter XVI.

SECT. IX.—PIGEONS.

It seems probable, that *Dove-cotes* or Pigeon-houses were formerly much more common in this County, than at present. I have observed them only in the following places, viz. Brampton, Breaston, Burrow-fields, Catton, Chaddesden, Draycot, Dronfield, Great Wilne, Milton, Newton Solney, Pilsbury, Repton, Sawley, Stapenhill, Willington, and Willsthorpe.

I confess myself one of those, who never could see the advantages of this species of Farm Stock. During the ten years that I was Agent for an extensive Estate, whereon all Repairs were done by the Landlord, I never repaired a single Dove-house, but have pulled down a great many; principally, on account of the great damage that large flights of Pigeons do, to the Ridges of Thatched and Tiled *Roofs;* to the latter, by picking out the mortar, which they seem very fond of doing, especially when it is tender and decomposing*.

That

---

* The propensity of Pigeons for any thing *salt*, is well known; and it is the Salts, produced by decomposition, and that exude or appear

That they consume and destroy a great deal of good Corn, cannot be denied: and as to the arguments in their favour, founded on their use in picking up the *Seeds of Weeds* in the Stubbles, during the Autumn, if it be true to the extent urged, how much better is the practice of Mr. John Blackwall of Blackwall, see Vol. II. p. 124, in *paring* his Stubbles, causing all such seeds to vegetate on the surface, or to be fermented in his Dung-yard, and so destroyed, instead of being buried deep in the soil, in a state fit for a future vegetation, whenever they may be turned up; nearly the whole of the shed Seeds, being thus buried, in early *Autumn ploughing*, and no inconsiderable portion of them, it is to be feared, in Spring ploughing of stubbles, or whole-lands, as such have been called of late. For it is to be observed, that most, if not all of our troublesome *Seed Weeds*, owe this property, to the power which their seeds possess, of laying almost any length of time in the earth, ready to vegetate, whenever brought within the proper influence of the air, light and moisture, in soil sufficiently pulverised.

There is little analogy, in the force of the arguments in favour of Rooks and other Birds, who live principally on Grubs and Insects (which we know, perhaps, no other mode of destroying, and still less can we prevent their propagation on our lands), and those in favour of Pigeons or other Birds, who pick

---

appear as an effervescence on old walls, or on the faces of some Cliffs or old stone Quarries, that occasion them to be often frequented by flights of Pigeons, as is not uncommon in the yellow Lime district, on the east side of the County, and as I particularly noticed on this Rock in Yorkshire, at Thorpe Arch N of the River, at Bramham S W of the Town, Fairburn W of the Town, &c.

up

up the Seeds, which our slovenly carelessness, in imperfectly-conducted *fallowing*, and the neglect of efficient *weeding*, can alone scatter on our lands, and those of our neighbours, in the case of winged seeds which blow with the wind. See Mr. John Wright's Experiments and excellent Remarks on this subject, in the Agricultural Magazine for April 1813, p. 261.

A few couples of Tame Pigeons are kept, about the premises of most of the Gentlemen, and more opulent Farmers and Tradesmen of the County, but which are seldom observed to stray into the fields.

SECT. X.—BEES.

The number of Hives or Stocks of this very useful Insect, kept in the County, seem as considerable as in many others, particularly about Darley in the Dale, Heath, Lullington, &c. In Heath, a person, whose name I have not happened to note, makes a good many Bee-boxes, glazed Hives, &c. for sale, among the curious in the management of their Apiaries.

Wasps and some other Insects, Reptiles and other small Animals, will be further noticed in Sect. 8, of Chap. XVII.

CHAP. XV.

RURAL ECONOMY.

SECT. I.—LABOUR.

Hired Servants.—The time and mode of hiring Farm Servants, seems to vary in different parts of the County; in several places, Old Martinmas, the 23d of November, was mentioned as the usual time of change. About Bakewell they are usually hired for 12 months, but not at any particular season of the year. About Bretby it is customary, at Michaelmas, to hire only for 51 weeks, for avoiding Settlements; hence an idle and dissipated week, occurs yearly at this season, and numerous disputes arise.

Earl Chesterfield pointed out to me, that the public Statutes for hiring, occasioned strangers without characters to be hired; and that after Michaelmas, numerous Servants ran away from their places.

Mr. William Cox of Culland pays 15 or 16 guineas a year to the Servants kept in his House.

Labourers.—The progressive prices of Farm Labour within a few years past, will in some measure appear, from stating the general results of enquiries or experience at three different dates, and which general particulars can also be compared with the results of my more particular enquiries which follow.

Mr. Thomas Brown, in the Original Report, written

in 1794, says, " Labourers from Candlemas to Martinmas, 1s. 4d. to 1s. 6d. per day: hours of work from six to six. From Martinmas to Candlemas, 1s. 2d. per day, working from daylight till dark.

" Near market-towns the wages are from 1s. 6d. to 2s. per day: between Derby and Ashborn, from Lady-day to Midsummer, the wages are from 1s. 6d. to 2s. per day; from Midsummer to Michaelmas, from 2s. to 2s. 6d. per day; and from Michaelmas to Lady-day, 1s. 4d. to 1s. 8d. In the harvest month, beer is always added to the above wages; and the hours of working, early and late, as the dews will admit. Reaping Wheat, 8s. per acre, Oats, 5s. per acre; Mowing Barley 2s., Grass 2s. 6d.; Thrashing Wheat, 2s. 8d. per quarter, Barley 1s. 8d., Oats 1s.

" It is not unusual for a Farmer to find the Labourer his Board; where this is done the hours of working are the same as above stated; and the price of Labour is, from Lady-day to Midsummer, 5s. per week; from Midsummer to Michaelmas, 8s. (always 9s. in Harvest); from Michaelmas to Lady-day, 4s.

" *Hired Servants.*—The best Man-servant from ten to twelve guineas a-year; a woman from four to five guineas."

A Gentleman who made some enquiries on this subject in the County in 1805, found the price of Labour not to vary much, in different parts of the County: in Winter the average about 1s. 8¼d., and in Summer about 3s. 2¼d. per day; Harvest being included in the latter, at about 2s. 6d. per day, with Board and Beer. Thrashing Wheat 6d., Barley 4d., and Oats 3½d. per bushel. Mowing Grass 3s. Reaping Wheat 12s. per acre. Hoeing of Turnips the first time 6s., and the

second

second time 3s. per acre. Carpenters and Masons, 21s. per week.

Mr. George Nuttall, Surveyor, late of Matlock, in the Spring of 1808 gave me the following, as the result of his experience and enquiries, in various parts of the County, viz. Farmers seldom give their Labourers more than 6s. per week from the end of Corn Harvest till the following Lady-day, and 7s. per week for the remainder of the year, with their Board (including Sundays) all the year round: 12s. per week is usually paid from Michaelmas to Lady-day, and 15s. for the other half year, or even 18s. in some cases, where the Master does not Board his Men.

The results of my enquiries have been as follows, viz.

At Alton in Wirksworth, the late Mr. Francis Bruckfield, mentioned his own and neighbours' practice to be, from Michaelmas to Lady-day to pay their Labourers 12s., from Lady-day to Midsummer 14s., and from Midsummer to Michaelmas 12s. per week, and their Board during the last quarter.

At Bakewell, Mr. William Greaves, Jun. stated, that the Farm Work there, is principally done by Day-labourers, whose pay was 1s. 6d. per day, until the Inclosures commenced, about 1805; since then, 2s. to 2s. 6d. per day is given.

At Barton Lodge, Mr. John Webb gives his Labourers 13s. per week and small Beer, and for 14 weeks (including Hay and Corn Harvest) 3s. per day, Small Beer, and about a quart of Ale per day.

At Bretby, the Earl of Chesterfield gives 12s. per week from Michaelmas to Lady-day; in the other half year the Labour is principally performed by the piece.

At Burrow-fields, Mr. Robert Lea employs 10 regular

lar Day-labourers; whom he supplies with Corn, Wheat, and Barley, mixed in equal proportions, at 5s. per bushel of 36 quarts, however the prices at Market may fluctuate. From Michaelmas to Lady-day he pays them 8s. per week: during the other half of the year they principally work by the piece; but if in Hay-time or Harvest they work by the day, they have 1s. 4d. and their Victuals and small Beer and Ale. When they work by the Job, they usually earn 15s. in the Winter, and 20s. per week in Summer.

At Buxton, Mr. George Wood mentioned, that 9s. to 10s. 6d. per week, with Board and Lodging, is the usual price of Labour there: for a single day 2s. 6d., or 3s. without Victuals.

At Chatsworth, the Duke of Devonshire's Labourers have 1s. 6d. per day the year round, with the Summer keep of a *Cow* in the Park, and Grass-land sufficient to mow for its Winter support, at a merely nominal Rent.

At Croxall, Thomas Prinsep, Esq. pays his Labourers 9s. per week from Michaelmas to Lady-day: in the other half of the year they earn about double that sum, at Job-work.

At Culland, Mr. William Cox gives his regular Labourers 4s. to 5s. in the Winter, and 6s. to 7s. per week in the Summer; finds them a *Cottage* and good Garden, at 20s. Rent; fetches their Coals; keeps them a *Cow* at 50s., including Baulks to cut Hay enough for her during the Winter (the whole keeping worth about 4l.) These Cows are as well kept as his own, and some of their owners sell 2 cwt. or more of Cheese, from their Cow: they each keep a Pig. The prices without these privileges is, in the Winter, 8s. to 10s. per week, and Board in the House; and 12s. to 15s. in

Hay-

Hay-time and Harvest, with 3 quarts of Small Beer and 1 of Ale daily.

At Foremarke, Mr. John Pearsall, from Michaelmas to Lady-day, pays 10s. per week and Small Beer.

At Ingleby, Mr. Charles Greaves pays 1s. 10d. per day and Small Beer, and 2s. 6d. per day (of 12 hours) and for extra hours, during three months of Hay-time and Harvest, with Small Beer and Ale.

At Markeaton, Francis N. C. Mundy, Esq. pays 2s. 6d. per day, from Michaelmas to Lady-day, and during the other half year they earn near double this at Job-work.

At Newhaven, Mr. Timothy Greenwood, to two regular Labourers who occupy *Cottages* on his Farm, Rent-free, he pays 7s. and 8s. per week; fetches their Coals; supplies them with Oatmeal at 1s. per peck; keeps a *Cow* for each of them, and finds ground, prepared, for planting their Potatoes; besides which, they have the Plucks of the Sheep killed for the use of his Inn.

At Over Haddon, Mr. Isaac Bennet, Jun. pays his Labourers 7s. per week and their Victuals, and during the Hay and Corn Harvest, 2s. per day and their Victuals and Small Beer.

At Pilsbury, Mr. Joseph Gould, from the conclusion of Harvest till Lady-day, pays his Labourers 2s. 2d. per day; thence till the beginning of Hay-time 2s. 4d. per day; and during Hay Harvest 2s. per day and Board and Lodging. The cutting of his Corn is done by his Yearly Servants, and by Labourers employed to reap by the Thrave, of 24 Sheaves (each a yard in circumference); a Kiver or Shock is half a Thrave.

What relates to the Wages of persons employed in

in the Manufactories, will be found in Sect. 8, of Chap. XVI.

Piece-work, or *Job-work*.—The full advantages of this mode of letting of work, to Master and Man, and to the Country, are seldom appreciated as they ought to be. I practised it very extensively, in the late Duke of Bedford's Rural works under my direction; principally thro' the able assistance of my Brother Benjamin, now Surveyor of the Whitechapel and Essex Roads; but the late Mr. John Billingsley, of Ashwick Grove in Somersetshire, carried this system further than any other Agriculturist that I have conversed with; for besides the several works which are enumerated below, in alphabetic order, and are frequently *let* in this County, his Ploughing, Harrowing, Rolling, Sowing, Turning of Corn when cut, Hay-making, &c. &c. were all done by the Acre, and from which he found great advantages, even where his own Oxen and Horses were used, by the takers of the work.

Cutting of Hedges, Ditching, &c.: an instance of this is mentioned Vol. II. p. 88.

Dibbling of Beans, see Vol. II. p. 99.

Draining, see prices of this work Vol. II. pages 385 and 386.

Hoeing of Corn, see Vol. II. p. 99; of Turnips, see Vol. II. p. 99, 137, and 147.

Mowing of Grass, see Vol. II. p. 183.—At Foremarke Park, Mr. William Smith pays' for Mowing Barley and Oats, each 2s. per acre, with 1 gallon of Small Beer and 3 pints of Ale, per Man, per day.

Planting-out, Cabbages, Trees, &c.; of these I have noted no prices.

Reaping of Wheat, see Vol. II. p. 122.—At Buxton,

ton, Mr. Thomas Logan employed Women to reap his Oats, and paid 3d. per Thrave of 24 Sheaves.

Spreading of Dung, Lime, &c.: see Vol. II. p. 435.

Stone-digging, or Quarrying, see Vol. II. p. 435.— At Doles Quarry in Plesley, the Surveyor of the Road, after paying for unbareing the blue Stone, pays 10d. per cubic Yard stackt, for getting it.

Taking-up Potatoes: no prices have been mentioned.

Thatching: ditto.

Thrashing of Corn by the Mill for Hire, has been mentioned Vol. II. p. 124; by the Flail my Notes are, of *Wheat*: at Barton Lodge, Mr. John Webb pays 6d. to 7d. per bushel; at Burrow-field, Mr. Robert Lea pays 7½d.; at Chatsworth, the Duke of Devonshire pays 6d.; at Foremarke, Mr. John Pearsall, 6d. and Small Beer; at Foremarke Park, Mr. William Smith, 6d. to 8d. per bushel (35 quarts) and Small Beer.

Thrashing of *Barley:* at Barton, Mr. John Webb says, they pay 3d. to 4d. per bushel; at Burrow-fields, 4d. per bushel; at Chatsworth, 3¼d. per bushel; at Foremarke, 3d. and Small Beer; at Foremarke Park, 3¼d. per bushel (35 quarts) and Small Beer.

Thrashing of *Oats:* at Barton Lodge, 1½d. to 2d. per bushel; at Chatsworth 1½d.; at Foremarke Park, 2d. per bushel (35 quarts) and Small Beer.

Thrashing of *Beans:* at Barton Lodge, 2¼d. to 3d. per bushel.

Trenching, or Digging; a case of this kind is mentioned, Vol. II. p. 347.

Walling, or building dry stone Fence-Walls; the prices of this work will be found at page 85 of Vol. II.

Whether we regard dispatch, economy, perfection of Rural Works, or the bettering of the condition of the

the Labourers therein, nothing will contribute so much to all these, as a general system of *letting works*, at fair and truly apportioned prices, according to the degrees of labour and skill required, in each kind of work. Few persons have doubted that dispatch and economy are attainable by this method, but those who have indolently or improperly gone about the letting of their Labour, have uniformly complained, of its being slovenly done, and of the proneness of the Men to cheat, when so employed. These last, are to be expected in all modes of employment, and can only be counteracted, or made to disappear, by *competent knowledge and due vigilance* in the Employer, or his Agents and Foremen, who ought to study and understand the time and degree of exertion and skill, as well as the best methods, in all their minutiæ, of performing all the various works, that they have to let.

At first sight, these might seem to be very difficult and unattainable qualifications in Farmers, Bailiffs, or Foremen, but it is nevertheless certain, that a proper system and perseverance, will soon overcome these difficulties. One of the first requisites is, the keeping of accurate and methodical day-accounts, of all Men employed; and, on the measuring up and calculating of every job of work, to register how much has been earned per day, and never to attempt abatement of the amount, should this even greatly exceed the ordinary day's pay of the Country; but let this *experience gained*, operate, in fixing the price of the next job of the same work, in order to lessen the earnings by degrees, of fully competent and industrious Men, to $1\frac{1}{4}$ to $2\frac{1}{4}$ times the ordinary wages when working by the day.

Select the men into small gangs, according to their abilities

abilities and industry, and always set the best gang about any new kind of work, or one whose prices want regulating, and encourage these by liberal prices at first, and gradually lower them, and by degrees introduce the other gangs to work with or near to them at the same kind of work. On the discovery of any material slights of, or deceptions in the work, at the time of measuring it, more than their proportionate values should be deducted for them, and a separate job made to one of the best gangs or men, for completing or altering it: by which means shame is made to operate, with loss of earnings, in favour of greater skill, attention and honesty, in future. When the necessity occurs, of employing even the best Men *by the day*, let the periods be as short as possible, and the prices considerably below job earnings, and contrive, by the offer of a desirable job to follow, to make it their interest and wish, to dispatch the work that is necessary to be done by the day, in order to get again to piece-work. The men being thus induced, to study and contrive the readiest and best methods of performing every part of their labour, and of expending their time, the work will unquestionably be better done, than by the thoughtless drones, who usually work *by the day*. And that these are the true methods of bettering the condition of the Labourers, Mr. Malthus has ably shewn in theory, and all those who have adopted and persevered in them, have seen the same in practice.

Rewards are offered by the Agricultural Society at Derby, as by most others in the kingdom, for long and meritorious *hired or day Service*, but seldom for having performed the greatest quantities of *job work*, or earned the most money by such, at fair prices;

which

which last seem to me, far more proper objects of their commendation and encouragement; since it cannot but be evident, that their rewards for *long services*, are most commonly bestowed on those easy and listless characters, who have preserved their places through personal attachment, or charitable feelings in their Masters or Mistresses, rather than by efficient services, such as benefit the Country. And as to their Rewards for not having applied for parochial aid, under trying circumstances, what are these but libels on the *Poor Laws?*, which cannot indeed be too much libelled or decryed, for the frightful train of evils which they are bringing on Society; but the attacks upon them should be addressed to another class, those who have the power of reforming them, or at least of desisting from almost every year arming them, with some new and deadly sting, against the vitals of Society.

It seems to follow indisputably, I think, from the documents which Mr. Thomas Batchelor has collected, in his Bedfordshire Report, and from his calculations therein, that the pay of Agricultural Labourers is now considerably less than it ought to be, or indeed can be, without occasional assistance from and *dependence on the Parish**: and surely no plan can be so proper and beneficial in its effects, for removing this evil, as enabling those able to work, to do more labour and earn more money.

For these reasons it has appeared to me, as very

---

* At the beginning of the present Century it was calculated, taking the labourer's wages at 2s. 6d. per day, that he must work 4½ times as many days to earn *the same quantities of food*, as from 3 to 5 centuries back he could, when his daily wages was 4d. to 2d. per day! Part of this was doubtless occasioned, by the many *idle* Saints' Days which the Church of Rome imposed on the people at the earlier periods.

desirable, that our Agricultural Institutions would offer Premiums, honorary and pecuniary, for Essays on the subject of *letting and conducting Rural Works*, accompanied by correct and ample details of *prices*, and *time* occupied, on all the various works, drawn from considerable experiments, and computed on the standard *measures* and *weights* of the Country; with the daily *earnings* in each case, and the ordinary day's *wages* at the time.

Such documents, collected from all the most judicious and attentive to these subjects, would now or at any future time, prove generally useful, towards commencing and establishing the change, so much desired in Agriculture, that of substituting piece-work for day-work.

Cottages, attached to Farms.—Instances of this kind, at Culland and at New Haven, have been mentioned, pages 187 and 188: there are doubtless many others, and it would be well for the Country, if such were far more numerous than at present. In Vol. II. p. 21, a favourable, and I trust a just account, has been given, of the state of the Derbyshire Cottages in general: and in Sect. 8, of the next Chapter, something will be said on the letting and attaching of *Cottages* to Manufactories.

Cottagers keeping Cows, and renting Land.— However desirable it may be, that Farmers should keep a Cow, on reasonable terms, for the most deserving of their stated Labourers, residing on or near their Farms, as above; yet the impolicy of suffering *Commons* to remain uninclosed and in a state of total neglect, like that of Hollington and others, Vol. II. p. 343, for the sake of Cottagers' Cows, is strikingly apparent,

and

and so would the favourite schemes of some persons, of allotting them one or more Pastures in Common. Under the baneful influence of the Poor-laws, but few of the Cottagers will be found provident enough to purchase a Cow, and fewer still perhaps, who would long keep one, that was given or provided for them by others: see Vol. II. p. 76. The less of business of his own, that a labourer for others has, which should cause him at any time to leave his regular employ, and be his own master, the better for his habits, and his Family and the Country. A *Garden*, as large as he can cultivate at over-hours, and his Family and his Pig can consume the produce of, has, after much observation, seemed to me the proper extent, of a Labourer's occupancy of Land. *Milk* the Farmers might, and ought in general to furnish to the labouring Poor, and at a cheap rate to their own Labourers, without their keeping Cows: see pages 30, and 40. After I had been at Cromford, I heard of a Cow Society having existed there, I believe among the Cottagers, but the precise nature and objects of the same, I am unacquainted with.

SECT. II.—PRICE OF PROVISIONS.

THIS County seems well supplied with Markets, and every article of food to be procurable, at or near the ordinary prices of the surrounding districts: my very slow progress thro' the County and its environs, precluding the opportunity of ascertaining the prices of provisions in all the different parts of it, at or near to the *same time*, so as to state the average results for any precise period, and the frequent fluctuations or advances

vances which have taken place in prices, then deterred me, from crowding my Notes with detached particulars, which it was unlikely I could turn to any useful account. What I have to remark further hereon, will come more properly under the head " Food and Mode of Living," in Sect. 11, of the next Chapter.

SECT. III.—FUEL.

PIT *Coal* is the fuel of far the greater part of the Inhabitants of the County, the prices of which are moderate, in most parts, as may be inferred from the number and distribution of the Coal-Pits therein, given in Vol. I.. p. 188, and from the Canals, Rail-ways, and Turnpike-roads, made for facilitating their conveyance to almost every part, which will be mentioned in Sections 1, 2, and 3, of the next Chapter.

Mr. Arthur Young has recorded, that in 1693 the price of Coals in Derby Town, to the consumer, was $3\frac{1}{2}d.$ per cwt. The modes of selling and prices at present, in several instances, are mentioned, Vol. I. p. 340 and 341.

Wood is very little used for Fuel in the County, as observed Vol. II. p. 222; less so now, than before the Canals and new Roads, enabled the cheaper conveyance of Coals, to places distant from the Pits.

Peat is also a great deal less used now as fuel, than formerly, from the same causes: the districts within which Peat Bogs are principally found, in the northern parts of the County, are particularly described in Vol. I. p. 309, and the names of the principal Mosses or thicker fields of Peat, are mentioned in Vol. II. p. 348.

p. 348. It may be proper further to mention in this place, that at Crookstone Peat-pits, which used formerly to supply Hope, almost entirely with fuel, altho' 5 miles distant, over very uneven and bad Roads, the face of peat is about 6 feet high, which is dug in Turves or Peats 12 inches long or high (cut a little inclining), and 6 inches broad and 2 inches thick: these, when first cut, being very soft and tender, are laid flat on the heap of refuse Peat which has been dug over, and after being partially dried and hardened, are set up on end, against each other, until thoroughly dry, when they are carried home on Pack-horses, or in Carts.

Candles are principally used for procuring Light, and far less use is made of *Lamp-Oil* in the County, than would be desirable.

Gas-Lights had been successfully introduced in the private House of Joseph Strutt, Esq. in St. Peter's, Derby; and for lighting the Factory of the Butterley Company near Ripley, before I finished my Survey; at which last place, the necessary apparatus for making and burning the Gas, were preparing for general Sale, and by this time I hope, that many Houses and Factories are lighted, without the consumption of either Tallow or Oil. It has been calculated, that 6 lb. of good Coal will produce about 18 cubic feet of purified and cooled Gas, and that the same is about equal to a pound of Tallow Candles, in illuminating effect; and that the *Coke*, of the best quality, which remains, will measure about $1\frac{1}{4}$ times as much as the Coal did, that was used in the Retort.

CHAP. XVI.

POLITICAL ECONOMY:

CIRCUMSTANCES DEPENDANT ON LEGISLATIVE AUTHORITY.

UNDER the general head of this Chapter, it may be right to refer back, to the *Mineral Laws* of the Peak Hundreds, treated of Vol. I. p. 356; *Bridges* paying Tolls, Vol. II. p. 22; *Tithes*, p. 29; *Poor's-Rates* and other Parochial Taxes, p. 32; and *Inclosures*, p. 71. Another important subject of Legislative Authority, of great concern to the Farmer, viz. *Game*, has not been noticed in the " Plan" prescribed for our Reports, and accordingly it appears omitted in the Reports on the adjacent Counties, except in the York West-Riding Report, where at p. 279, the *Game-Laws* are recommended to be altered, for improving the condition of the Farmer; and in the Warwick Report, p. 166, where a Section for Game is made, after that for Rabbits, but I have preferred introducing here, the operations of the Laws and Customs respecting

GAME.

I shall first speak of BIRDS.—In this County, all the usual kinds of feathered Game, or Birds that are protected by the Laws, or by the Game-keepers, &c. of the Land-owners, for their sport or use, are found, in as great plenty and perfection, as in most others, and

some

some that are almost unknown in the Southern Counties.

Black Game, which frequent the Heathy Moors, are less plentiful than formerly, and considering the very worthless nature of the herbage amongst which they feed, the *Ericas,* (see Vol. I. 305, and II. 356, &c.), the sooner they disappear altogether, the better for the community. These and Grouse, constitute the *Moor-game,* on the East Moor and the High-Peak Moors, &c. the shooting of which commences on the 12th of August.

Dotterels: some of these are found in Foolow, Peak-Forest, Taddington, &c.

Ducks, wild, frequent several of the large pieces of Water in the List, in Vol. I. p. 497; the large Mill-dams at Mellor, Vol. II. p. 261, protect considerable numbers of these Birds, and so do the larger Rivers.

Geese, wild, are only found occasionally to frequent the larger Rivers.

Grouse: these, with Black Cocks, constitute the Moor-game, as above mentioned: formerly, some black Grouse were shot on Eggington Heath, it is said.

Partridges, are the most common feathered Game, and the numbers of which in some places, do considerable injury to the Farmer's Crops.

Pheasants: these are bred in considerable numbers, in the Woods in Wingerworth, where Buck-wheat is sown for them (see Vol. II. p. 135), and in other places. In Bradby Park the berries on a grove of Elder Trees are reserved for them, as mentioned Vol. II. p. 216.

Pigeons, of the Dove-cote, have been spoken of, as rather detrimental than otherwise, at page 181;

Wood-Pigeons are found to attack Swede Turnips, in some situations, see Vol. II. p. 148.

Reeves and Ruffs: Mr. Pilkington, in his " View of Derbyshire," Vol. I. p. 493, says, that these Birds were formerly found on Synfin Fen, Vol. II. p. 350; those which are now fattened in the Poultry-Houses in Bradby Park and others, are purchased in the Fens of the Eastern Counties, I believe.

Snipes: as draining has advanced, these Birds have considerably decreased, and those remaining might well be spared.

Woodcocks: the Woods in several parts of the County harbour this Game, in its season, in considerable numbers, I believe.

The ANIMALS which contribute to the pleasures of the Sportsmen and Gentry are,

Deer, which are principally of the Fallow or spotted kind; but few of the large Red Deer are now kept in the County, and the cruel sport, of turning out these Animals, on purpose to *hunt* and torture them, is here quite laid aside. I noticed *Deer-Parks* in Alderwasley, Alfreton, Barlow (very small), Bretby or Bradby, Calke, Chatsworth, Drakelow, Hardwick, Kedleston, Locko, Norton, Sudbury, Sutton in Scarsdale, and Wingerworth. Large tracts of Land, now disparked and cultivated, are found in Codnor, Crich (Chace), Haddon, Hulland (Ward), Denby (Kidsley P.), Morley (near Heage), Peak-Forest, Shottle, &c.

In the Earl of Chesterfield's Park at Bradby, a regular account is kept of the Deer, and of the Hay supplied to them in the Winter season; these, in the eight years preceding 1809, were 450 head, consuming 38 ton

—38 ton 5¼ cwt. of Hay, worth 197*l.* yearly, on the average; that is, each Deer consumed about 190 lb. of Hay in the Winter, worth 8*s*. 9*d.* on the average. During Snows, the Deer in the Park are also supplied with a portion of Cabbages, and if such are much frozen, they are chopped in pieces, see Vol. II. p. 144. A certain number of the Males are cut each year, or made Haviers, as soon as dropped by the Doe; in doing this, only the lower half of the testicles are cut off, by a sharp pair of scissars (instead of drawing the whole out, as with Lambs, after the purse has been opened). These Haviers, head, or put up their horns like Bucks, and they grow to the size of 24 lb. per Haunch, but do not in October, thicken in the neck, and fret, as Bucks do, but are then in order to kill, before the Doe-Venison comes in, near Christmas.

Considerable parts of Chatsworth Park (on the Shale-grit Rocks) are covered with Fern, which might easily be removed, and kind Herbage produced in its place, by Liming, as observed Vol. II. p. 437, but for an apprehension which was entertained, that the Deer would be injured, and made riggle-backt, by this improved herbage! After I had visited Sudbury Park, I could not but suspect, that something more than ignorance in the Park-keeper, had dictated these apprehensions.

The late Lord Vernon's Deer-Park at Sudbury, is near 500 Acres, and has an elegant Deer-House in its centre; about 1000 head of Deer were kept; only one-third of the Fawns from which, were raised annually: I found the greater part of this large Park covered with Rushes, Thistles, Docks, Nettles, &c. to a most shameful degree; and was told as the reason, that the late Henry Hall, the Park-keeper, would not suffer the

Bailiff

Bailiff to remove one of these disgraceful intruders, pretending, that the shade and cover they afforded, was useful to his young Fawns.

Foxes.—The number of these useless Vermin is not considerable, considering the facilities which they have, of securely hiding themselves, and breeding in holes in the rocky Woods, and steep Banks, in most of the northern parts of the County; *Fox-hunting* being here little practised, the Farmers are at liberty, and are encouraged, to kill them whenever they can, which in many districts of England, they dare not do.

Subscription Packs of *Hounds* (Harriers I believe) are kept at Derby, at Locko-Park, and at Sudbury. Richard Arkwright, Esq. keeps Harriers at Willersley, and Mr. John Milnes has a few in Ashover.

Hares.—The Farmers in Wingerworth, Stanton, and some other places, experience at times great injury from the numbers of Hares that are preserved; tho', on the whole, I think, that the Derbyshire Gentlemen are less rigorous respecting their Game, than those of several other Counties. The mischiefs occasioned by them to Plantations*, have been instanced in Vol. II. p. 244; and to Crops of Barley, Sainfoin, and Swede Turnips, in pages 126, 165, 145, and 148. The Rev. Mr. Groves is said to have shot a Hare some years ago, near Bakewell, which weighed $13\frac{1}{4}$ lb.

Rabbits.—These mischievous little Animals, are much too common in the Hedges and Banks, in some parts of the County, as observed page 177, where the injuries that they do to Plantations* have been mentioned.

---

* In the last Autumn (1813), the Government *Plantations* of young Oaks, in the Forest of Dean in Gloucestershire, and in the New Forest in Hampshire, were observed to be overrun with field *Mice*, in a very unusual

tioned. Wheat Crops often suffer greatly from them; so have Oats, Sainfoin, and Swede Turnips, in the instances mentioned in Vol. II. pages 244, 256, 165, and 145, and in numerous others.

FISH constitute the third description of Game to be here mentioned, and with which, the Rivers, Brooks, Canals, Reservoirs, Meers, and Ponds, in Derbyshire are, in different degrees, pretty well stored, viz. with Barbel, Carp, Chub, Craw-fish, Dace, Eels, Grayling, Gudgeons, Perch, Pike, Roach, Salmon, Samlets, Tench, Trout, &c.

Some account of the Rivers of the County has been given, Vol. I. p. 468 to 487; and in page 377, the injury has been mentioned, which the Fish in the Derwent receive, from the thick water, unnecessarily let down by the Buddlers; because they ought to retain this water in Reservoirs, until it has deposited all its sediment, consisting of Clay, slime of Lead Ore, &c. (Vol. II. p. 447).

A List of all the most considerable Reservoirs, Mill

---

unusual degree; and these, as the Winter approached, began to attack and strip the bark from the young Plants, near to the ground, in a very alarming manner. The Officers who had charge of these Plantations, at length succeeded in destroying great part of these small intruders, by digging numerous shallow ditches, with smooth and upright or overhanging sides and ends; into which the Mice fell in the night, and were thus taken and destroyed.

Since learning the above particulars, it has occurred to me, that the barking of young Trees *near to the ground*, may sometimes have been charged on *Hares* and *Rabbits*, as above, when more minute observations would have shown, that the mischief had been occasioned by field *Mice*. I now think it very probable, that the almost entire destruction of numerous young Plants in Hate Wood in Ashover, belonging to Sir Joseph Banks, was thus effected, and not by Rabbits, as I concluded when I saw them, in 1810 and 1811.

Pools,

Pools, and *Fish-Ponds*, in the County, is given, Vol. I. p. 497.

The Fish Ponds of John Berrisford, Esq. at Osmaston Cottage in Shirley, seem well managed; the uppermost of them is used as a mud-trap, as mentioned Vol. II. p. 448; his pond of ten acres was at first stocked with *Carp* and *Tench*, but was found to breed *Perch* to such an excess, that some *Pike* were put into it, in 1809, in order to keep them under; which doubtless they will do, if it be correct, as asserted by Mr. Joseph Butler and others, that a Pike eats his own weight of other Fish, every six weeks.

Certain Ponds in Sir Thomas Windsor Hunloke's Park, in Wingerworth, are appropriated to the feeding of *castrated* male *Carp* and *Tench*, which are found very superior in size and flavour, to other fish: the late Sir Windsor H. saw this practised in Italy, many years ago, and had one of his Servants, who was with him, instructed in performing the operation; which is less difficult or dangerous, than might be supposed, and in consequence of which, not more than one in fourteen or fifteen of the fish die: see the article FISH in Dr. Rees' Cyclopædia.

Francis N. C. Mundy, Esq. some years ago, made a large Lake on the Morledge Rivulet, in his Park at Markeaton; for about the three first years, the Fish of all sorts grew much faster than they have since done: a *Carp* of one pound, which had lost an eye, and was otherwise marked, increased one pound in weight the first year; when taken for the table, at four or five years old, he weighed seven pounds.

The reason of the above fact appears to me, to be, that the Lake had more remains of the vegetation, and land Insects, &c. on its bottom and sides, in the first year

year that it was used, and was besides less fully stocked, than afterwards. See Mr. William Jessop's excellent suggestion, as to sowing vegetables on the sides of Reservoirs, when their waters are low, in "Communications" to the Board, Vol. I. p. 178; and M. De Luc's "Geological Travels in the North of Europe," Vol. I. p. 314; and in France, &c. Vol. II. p. 89.

Mud now accumulates in the bottom of the Markeaton Lake, and the *Pike* caught therein, are, in consequence, now kept on a gravelly bottom and rapid stream, at the tail of the Mill, between grates placed for that purpose, for the space of four or five weeks, before they are taken up for the Table, by which their condition and flavour is entirely changed, and their flesh becomes firmer, than that of Pike fresh taken from the Lake.

Samuel Oldknow, Esq. of Mellor, is possessed of the Fishery of the *Combs-brook* Reservoir, of 45 acres, belonging to the Peak Forest Canal Company, situated 1¼ m. WSW of Chapel-en-le-Frith. Here, by means of Tickets, which himself and some Friends at Buxton and elsewhere circulate, the Company from that place and others, are allowed to *angle* at their pleasure, and take away whatever Fish they catch (exceeding an ounce in weight) on paying to Mr. O.'s Agent, who lives in a Cottage on the spot, 6d. per pound of Fish.

Messrs. Strutts, in their very capital Weirs and Flood-gates at Belper's Bridge, see Vol. II. p. 398, have constructed a very complete *pass* for the *Salmon* in going up the Derwent to spawn, which prevents the necessity of their leaping the Weir, and a *trap* for taking them as they come down again, after spawning; and the same at their works at Milford.

The

The *Trout* in the Dove, the Derwent, the Wye, the Lathkil, and others of the Rivers and Rivulets of this County, are very fine, and during their season, most of the Inn-keepers procure and dress them in a good style, for their Guests; they usually run from 1 to 2¼ lb. weight; and such would probably be more plentiful, but for the *Groupers*, a kind of Poachers, who in dry and hot weather, wade the Rivulets and Brooks, and gently feel for, and take these Fish with their hands, when asleep under the ledges of the Rocks, Roots of Trees, &c. A Trout which was caught at Chatsworth, and was marked and let go again in the Derwent, when it weighed 2 lb., was taken again four years afterwards, and then weighed 11¼ lb.!

Improved Communication by Means of Turnpike-Roads, Rail-Ways, and Canals.

The great national importance of the subjects to be principally treated of in the three first Sections of this Chapter, and the great value of its *Improved Communication*, by means of Turnpike-roads, Rail-ways, and Canals, to the County of Derby in particular, has induced me to present a *Map*, facing page 193, (of the same scale and size as those of *Ridges of Hills* facing page 1, and of *Strata** facing page 97, in Vol. I.), which should

---

* Into one or other of these three Maps, I was very desirous of introducing the *Rivers*, Brooks, and minuter streams of water, in or near to the County, but found, that it could not be done, without confusing them. In Vol. I. p. 469, I have referred to Mr. Brown's Map in the Original quarto Report, for the Rivers and principal Brooks, but the minuter ramifications of these, which in Geographical, Geological, and other points of view, are important, have not any where yet been published. Mr. Aaron Arrowsmith, of Soho Square, will soon supply this defect,

should shew the whole of these, in connection with those in the adjoining Districts, of the seven adjacent Counties.

Such a connected view of the great and meritorious exertions of the Inhabitants, for opening facilities of communication, amongst themselves and with the kingdom at large, besides serving to mark the progress of these grand Improvements of the condition of the Country, may perhaps stimulate some other British districts to similar exertions, and suggest some hints, for improving the legislative provisions, under which they are conducted, the Roads in particular.

In order to shew more readily the connection of this Map with other Maps, and render it more useful, I have placed a series of Roman and Italic Capitals and small Roman Letters round its margin, beginning at the top left-hand corner, and proceeding to the right; and shall proceed now to explain the several lines which pass off the Map, viz.

County Boundaries (see Vol. I. p. 2), *that pass off the Borders of the Map,* p. 193.

B Between Lancashire and the West Riding of Yorkshire.

---

defect, regarding Derbyshire and its environs, by the publication of his grand 18 sheet Map of England, and his four sheet reduction of the same Map.

The probable period of *publishing* my intended large *Map and Mineral History* of the District, being now more distant, than when I wrote the Preface to my first Volume (see Phil. Mag. Vol. XLII. p. 55, Note, and p. 246); in order that the Public may not lose the benefit of this part of my labour; I have permitted Mr. Arrowsmith to copy the Turnpike-Roads, and the Rivers, Brooks, and Rills, from my Manuscript Map, into those which he is about to publish, as above-mentioned.

M Be-

M Between the West Riding of Yorkshire and Nottinghamshire, at the Point where Lincolnshire joints these two Counties.
Z Between Nottinghamshire and Leicestershire.
I, Between Leicestershire and Warwickshire.
V Between Warwickshire and Staffordshire.
k Between Staffordshire and Cheshire.
p Between Cheshire and Lancashire.

Turnpike Roads, see Section 1, of this Chapter.

C
C'
E } To Huddersfield, and Halifax.
F
G

H
J } To Wakefield.

K To Ferry-Bridge;—the *Great North Road*, for York, Edinburgh, &c.
L' To Thorne and Snaith.
O To Gainsborough.

P
Q' } To East Retford
Q,

R
S } To Newark.

T
U } To Southwell, and Newark.

W To Bingham and Grantham.

Y
B } To Melton-Mowbray.

C To Uppingham.
D To Market-Harborough, and Northampton.
E To Welford and Northampton.
G To Lutterworth.
H To High-cross and Coventry.

I To

$\left.\begin{array}{l}I\\J\end{array}\right\}$ To Hinckley, and Nuneaton.

M To Nuneaton, Hinckley, Lutterworth, and Daventry.

O To Coleshill, and Birmingham.

P To Upper Whitacre and Coventry.

$\left.\begin{array}{l}R\\S\end{array}\right\}$ To Coleshill and Warwick.

$\left.\begin{array}{l}T\\U\end{array}\right\}$ To Sutton-Coldfield, and Birmingham.

$\left.\begin{array}{l}X\\Y\end{array}\right\}$ To Walsall.

a To Wolverhampton.

$\left.\begin{array}{l}a'\\b\end{array}\right\}$ To Newport and Shrewsbury.

c To Eccleshall and Drayton.

d To Nantwich.

$\left.\begin{array}{l}e\\f\end{array}\right\}$ To Drayton and Shrewsbury.

g To Whitchurch.

j To Nantwich and Chester.

l To Sandbach, Middlewich, Northwich, and Knutsford.

n To Cheadle (Chesh.) and Manchester.

o To Knutsford.

r:—From Manchester, Roads proceed direct for Congleton, Altringham, Warrington (and Liverpool), Wigan, Bolton, Bury (and Haslinden), and Rochdale.

The *Mail-Coach Routes*, that cross this district of Map, p. 193, are as follows, viz.

PK,
London and Edinburgh Mail, through Hoddesdon, Ware, Royston, Caxton, Huntingdon, Alconbury-hill, Stilton, Stamford, Grantham, Newark, Tuxford, East Retford, Bawtry, Doncaster, Ferry-Bridge, Tadcaster, York, Easingwold, Thirsk, Northallerton, Darlington, Durham, Newcastle-on-Tyne, Morpeth, Alnwick, Belford, Berwick, Dunbar, &c.

London and Glasgow Mail, thro' Barnet, Hatfield, Welyn, Baldock, Biggleswade, Eaton-Socon, Alconbury-hill (and thence, as above, to) Ferry-Bridge, Abberford, Wetherby, Borough-Bridge, Catterick-Bridge, Greta-Bridge, Brough, Appleby, Penrith, Carlisle, Longtown, &c.

YH, *London and Leeds* Mail, thro' Barnet, Hatfield, Welyn, Hitchin, Bedford, Higham-Ferrers, Kettering, Rockingham, Uppingham, Oakham, Melton-Mowbray, Nottingham, Mansfield, Chesterfield, Dronfield, Sheffield, Barnsley, and Wakefield, to Leeds.

Dr, *London and Carlisle* Mail, thro' Barnet, St. Alban's, Dunstable, Hockliffe, Woburn, Newport-Pagnel, Northampton, Market-Harborough, Leicester, Loughborough, Derby, Ashburne, Leek, Macclesfield, Bullock-Smithy, Stockport, Manchester, Bolton, Chorley, Preston, Garstang, Lancaster, Burton, Kendal, and Penrith, to Carlisle.

Mc, *London, Chester, and Holyhead* Mail, thro' Barnet, St. Alban's, Dunstable, Hockliffe, Woburn,

Woburn, Newport-Pagnel, Northampton, Welford, Lutterworth, Hinckley, Atherstone, Tamworth, Litchfield, Rudgley, Wolsey-bridge, Stafford, Eccleshall, Ashley-heath, Wore, Nantwich, Chester, Hawarden, Holywell, St. Asaph, Abergeley, Aberconwy, Aber, Bangor, and Llangefni, to Holyhead.

Rn, *London and Liverpool* Mail, thro' Barnet, St. Alban's, Dunstable, Hockliffe, Fenny-Stratford, Stoney-Stratford, Towceter, Daventry, Dunchurch, Coventry, Stone-Bridge, Coleshill, Tamworth, Litchfield, Rudgley, Wolsey-Bridge, Stone, Newcastle-under-Line, Congleton, Holme-Chapel, Knutsford, Warrington, and Prescot, to Liverpool.

U, *Birmingham and Sheffield* Mail, thro' Sutton-Coldfield, Litchfield, Burton-on-Trent, Derby, Ripley, Alfreton, Chesterfield, and Dronfield, to Sheffield; from hence also a Coach goes forwards with the Letters, to Rotherham and Doncaster.

an, *Birmingham and Manchester* Mail, thro' Walsall, Wolverhampton, Stafford, Stone, Newcastle-under-Line, Congleton, Holme-Chapel, and Cheadle (Cheshire), to Manchester.

r, *Manchester and Sheffield* Mail, thro' Stockport, Bullock-Smithy, Buxton, Tideswell, and Stoney-Middleton, to Sheffield.

From Derby to Nottingham, a Coach carries the Letters daily; besides which, there are several Horse Posts for the conveyance of Letters to and between the different Post-offices in the County of Derby and its environs: a List of the Derbyshire Post-offices, will be found in Sect. 1, of this Chapter.

The *Canals* and Navigations (see Section 3) leaving the district of Map, p. 193, conduct as follows, viz.

A, The *Rochdale* Canal (whose western commencement at q is mentioned hereafter)—To Rochdale, Todmerden, Sowerby-Bridge near Halifax, whence, by means of the *Calder and Hebble, Ayre and Calder,* York *Ouse,* and the *Humber* Navigations, it connects with the Port of Hull and the German Ocean; and also with the six following Canals and Navigations, and several others N and NW of this line, as I have particularly described in the article *Canal,* in Dr. Rees' " New Cyclopædia," written in 1805.

D, The *Huddersfield* Canal—To Marsden and Huddersfield; whence, by means of *Ramsden*'s Canal, the *Calder and Hebble, Ayre and Calder* Navigations, &c. it connects and supplies Water-carriage as above.

I, The *Barnsley* Canal—To Wakefield; whence, by means of the *Ayre and Calder* Navigation, &c. its connects as above.

L, The *Don* Navigation—To Thorne and Goole-Bridge; whence by means of the York *Ouse,* &c. and by its cut to the *Ayre and Calder* near Snaith, and by the *Stainforth and Keadby* Canal to the *Trent* River, it connects as above.

N, The *Idle* Navigation—To West-Stockwith; whence, by means of the *Trent, Humber,* and York *Ouse* Rivers, &c. it connects as above.

Q, The *Chesterfield* Canal—To East-Retford and West-Stockwith; whence, by means of the *Trent, Stainforth* and *Keadby, Humber,* and York *Ouse,* &c. it connects as above.

V, The

V, The *Trent* River—To Newark (by its back stream, called Newark-Dyke), Gainsborough, West-Stockwith, Keadby, and Burton (upon Strather), to Trent-fall; whence, by the *Humber* and York *Ouse*, &c. it connects as above, and with the German Ocean.

By the junctions beyond the limits of the Map, of the last-mentioned seven lines of Navigations, the most extensive water communications are formed, from the northern and western parts of Derbyshire and its environs, with Leeds, Bradford, Skipton, Colne, Burnley, Blackburn, Tadcaster, Cawood, Selby, York, Borough-Bridge, Ripon, Topcliffe, Sheriff-Hutton Bridge, New-Malton, Yedingham-Bridge, Market-Weighton, Howden, *Hull*, Beverley, Great-Driffield, Barton, Brigg, Caistor, Lincoln (by the *Foss-dyke* Canal), Tattershall, Horncastle, Sleaford, *Boston*, Wainfleet, Spalding, Wisbeach, Crowland, Market-deeping, Stamford, *Lynn*, &c. &c.

X, The *Grantham* Canal—To Bingham, Bottesford, and Grantham.

A, The *Leicester and Melton-Mowbray* Navigation—To Melton-Mowbray; where it connects with the *Oakham* Canal, to that town.

F, The *Leicestershire and Northamptonshire Union* Canal—To Gumley-Wharf (near Market-Harborough, to which there is a Cut); from whence, by means of the *Grand Union* Canal (now very nearly finished, April 1814), it will connect with the *Grand-Junction* Canal in Norton, near Daventry, and with the five following Canals, and several others to the SE, S and SW from Derbyshire.

K, The

K, The *Ashby-de-la-Zouch* Canal—To near Hinckley, and Marston-Bridge near Nuneaton; whence, by means of the *Coventry*, *Oxford*, and *Grand Junction* Canals, &c. it connects as above.

N, The *Coventry* Canal (main line)*—To Nuneaton and Longford; whence, by means of the *Oxford* and *Grand Junction* Canals, &c. it connects as above.

Q, The *Birmingham and Fazeley* Canal—To Digbeth near Birmingham; whence, by means of the *Warwick and Birmingham*, *Warwick and Napton*, *Oxford*, *Grand Junction* Canals, &c. it connects as above.

W, The *Wyrley and Essington* Canal—Near to Walsall and Wolverhampton; whence, by means of the old *Birmingham*, *Birmingham and Fazeley*, *Warwick and Birmingham* Canals, &c. it connects as above.

Z, The *Staffordshire and Worcestershire* Canal, to Penkridge and Aldersley (or Autherley) near Wolverhampton; whence by means of the old *Birmingham*, *Birmingham and Fazeley* Canals, &c. it connects as above: and with the *Severn* River and Bristol Channel.

By the various junctions, beyond the limits of the Map, made by the last mentioned six lines of Navigations, the most extensive water communications are formed from the southern and eastern parts of Derbyshire and its environs, with Northampton, Wellingborough (near), Thrapston, Oundle, Peterborough, Wisbeach,

---

* The detached part of the ("grand trunk") line, belonging to this Company, between Whittington-Brook and Fradley-heath, will be mentioned in Sect. 3.

Lynn

CANAL, &c. LINES, AT LIMITS OF THE MAP. 215

Lynn (see page 213); Downham, Brandon, Thetford, Mildenhall, Bury St. Edmund's, Ely, Cambridge, St. Ives, Huntingdon, St. Neots, Bedford, Biggleswade, Daventry (near), Stoney-Stratford, Buckingham, Fenny-Stratford, Leighton-Buzzard, Tring, Wendover, Great Berkhampstead, Hemel-Hempstead (near), Watford (near), Rickmansworth, Uxbridge, Brentford, Paddington and *London*; Coventry, Rugby (near), Southam, Warwick, Banbury, Oxford, Abingdon, Wallingford, Reading, Newbury, Henley, Great Marlow, Maidenhead, Windsor, Staines, Chertsey, Kingston, Brentford and London; Stourbridge, Kidderminster, Bewdley, Bridgenorth, Shrewsbury, Worcester, Droitwich, Upton, Tewksbury, Pershore, Evesham, Stratford on Avon, Gloucester, Newnham Stroud, Berkley, *Bristol*, Bath, &c. &c.

h, *Newcastle* (under-Line) *Junction* Canal, which (with Sir N. B. *Gresley's* Canal) extends to Partridge-nest, and other collieries.

m, The *Trent and Mersey*, or Grand Trunk Canal —To Sanbach (near), Middlewich, and Northwich, to Preston-brook; whence by means of the Duke of *Bridgewater's* Canal, the *Mersey and Irwell* Navigation, &c. it connects as follows.

q, The *Rochdale* Canal (whose northern exit from the Map at A has been mentioned, p. 212)—To Knot, Mill or Castlekey on the S W. side of Manchester Town; whence by means of *Bridgewater's* Canal, and the *Mersey and Irwell* Navigation, &c. it connects as above.

s, The *Mersey and Irwell* Navigation—To Warrington (near), Frodsham, Liverpool, &c. connecting

necting with *Bridgewater's* Canal near Manchester?, and at Runcorn-Gap, with the *Manchester, Bolton, and Bury,* the *Sankey,* and *Ellesmere* Canals, the *Weaver* Navigation, and with the Irish Channel near *Liverpool.*

By the various junctions, beyond the limit of the Map on this side, made by the three last mentioned lines of Navigation, most extensive water communications are formed, from the western and middle parts of Derbyshire, with Chester, Nantwich, Prescot (near), Newton (near), Wigan, Chorley, Bolton, Bury, &c. &c.

The Canals, Navigations, and *projects for such,* which have been made public, that fall within the limits of the Map, will be separately noticed in alphabetical order, and many particulars respecting each will be given, in Section 3, of this Chapter.

SECT. I.—ROADS.

To many persons, the *Roman Roads* and Stations, are objects of considerable interest: I shall therefore follow the example of the Leicestershire Reporter and some others, in shortly mentioning those which are said to traverse this county.

The famous Watling-street has part of its route shewn across the Map facing page 193, from *M* in a WNW direction, through a place called Chesterfield, about 2¼ m. SSW of Litchfield in Staffordshire: a Roman Station or Town probably existed at or near to this Chesterfield, from whence a Road proceeded NW, through Street-hay, Wichmor-Bridges, Branston, near to Burton (on its NW side, and where perhaps

haps there was a Station) and entered Derbyshire at Monks Bridge; passed near to Littleover and to Derby, on their N.W sides, to another Station at Little Chester, about ¼ NNE of Derby.

From Little Chester, where the Romans had a wooden bridge over the Derwent (Pilk. View II. 197), a Roman Road is supposed to have passed nearly SE, through Burrowash and Draycot, and thence towards Nottingham.

Near to Parwich there appears to have been a Roman Station, and some have supposed that a Road led from Little Chester, nearly NW to this Station, although its exact route is unknown, I believe: from near Parwich, the continuation of this Road is visibly seen to Buxton, whose Hot Baths are said to have been in repute at that early period, and about 2¼ m. N of which, there is said to have been a Roman Camp, on Combes Moss Hill, half a mile SSE of Bank-hall: from Buxton this Road proceeded forwards, and quitting the County a little above Whaley-bridge, passed through Disley and Stockport (where probably there was a Station) to Manchester.

From Stockport (or from Manchester) it is supposed that a Road passed eastward to a Station at Melandra Castle, just within the borders of Derbyshire, three-quarters of a mile N of Gamesley in Glossop; and which road, there are reasons to suppose, proceeded E, through Glossop and up the Dale E of it, then turning near SE, along what is called Doctor Gate (Vol. I. p. 49), a deep cut and paved Horse Road, across the Peat Moss, on the Grand Ridge between Glossop and Hope Parishes, from whence this Road descended by Ladyclough into Ashop-dale, and

thence

thence SSE to the Roman Station, at Brough*, about three-quarters of a mile SE of Hope.

From this Station at Brough, it is said that there was a Roman Road, by Small-dale (in Hope), near Peak Forest on the S, and through Small-dale, to Buxton, mentioned above.

Near to Castleton there is said to have been a Roman Camp, perhaps on Mam Tor Hill; (I. p. 45) and some have supposed, that there was another on the E side of Hathersage Church. It has also been conjectured, that a Road led from the Brough Station to another on the south side of the Town of Chesterfield, but its precise track is little if at all known.

From the Station near Chesterfield, a Roman Road can plainly be traced southward to another Station at Stretton (as is supposed), and thence forwards to Little Chester; already mentioned, a Roman Camp having been traced on Pentrich Common, near to this last Road.

From Burton on Trent, a Roman Road *(via devana,* see Mr. Pitt's Leicestershire Report, p. 307), proceeded nearly in the track of the present Road, across the south-east corner of Derbyshire to Ashby-de-la-Zouch, and thence forwards to Leicester.

Some persons have mentioned, that a Roman Road proceeded from Repton (perhaps its southern end?) SW to Edingale on the confines of Staffordshire, and with this Road some have connected the ancient Camp,

---

* About the year 1796, a gold coin of Cæsar Augustus (of about 18s. value) was found by a ploughman, in Horstead field, 300 yards W of the village, which the Tenant told me, is now in possession of his Landlord, the Rev. Henry Case Morewood, of Alfreton-Park.

or earth-work near Bretby Church, but with little probability I think.

Lastly, a Roman Road is supposed to have proceeded from Nottingham, northward passed on the E side of Skegby, and entered this County at Newbold-Mill*, three quarters m. WSW of Plesley, and proceeding through Clown, to have left the County again on the NE of Knitaker in Barlborough, and proceeded through Harthill, &c. in Yorkshire.

The principle which seemed most to govern, in the setting out of most of the very ancient Roads above mentioned, and others of nearly the same period, throughout Britain, seems to have been, the proceeding in *as straight a line as possible* between one Station and another, ascending and crossing the Hills, rather than deviate from the right line to avoid them, because Horses principally, and few if any Carriages were then, I believe, used; and the elevated points in these Roads, served to command the country which the Roman soldiers then held in military subjection.

At the same time it is observable, that the straighter *ridges of dry and rocky land*, are very much followed by the lines of Roman Roads; probably, because they were found most clear of wood, and the most passable in their first incursions through the country;

---

* This is a singular spot, owing to *two* Counties, *five* Parishes, and *seven* Roads all meeting here, viz. Nottinghamshire and Derbyshire: Mansfield, Skegby, Teversall, Alt-Hucknal, and Plesley Parishes; and the Roads to Mansfield (two ways), Skegby (Roman), Teversall, Alt-Hucknal, Stoney-Houghton (Roman), and Plesley.

In two other instances, six roads meet in one point, in this district, viz. at Lane-head NE of Tideswell (before inclosure), and at the place called Six-hands (from the hand-post erected in the centre, Vol. I, p. 53), on Needwood Forest, 1 m. W of Hanbury, Staffordshire.

and that the military ways thus marked out by Nature, and found partially clear, gave rise to the forming of various Stations and Towns upon those lines; and that afterwards, these Roads were further improved, and straightened, and other lines of Road in directions less adapted by Nature for the passage of Men and Horses, were cleared and formed, and paved with great labour, through low and clayey districts, and where many streams of water required to be crossed, for connecting these original Stations.

The situations of shallow and rocky places in the beds of the larger Rivers, where *fords* naturally existed, or could easily be made, seem also much to have influenced the laying out of the Roman Roads; and that at these fordable points of Rivers, great numbers of their Stations and Towns were fixed.

In after times, the pursuing of straight lines for any considerable lengths of Road, began to be rather less attended to; and the following of *dry and sound ground*, between Town and Town, chusing the fordable points of Rivers and Brooks, and avoiding streams of water as much as possible, became the leading principles, in chusing tracks across the woods, wastes, and commons, which then almost universally prevailed. In still more modern times, when some wheel Carriages began to be used, and the Inhabitants began from necessity to repair and attend a little to their Roads, such Roads became most attended to and used, where stone or gravel were most plentiful and accessible, for repairing them; and many of the disused and worst of the original Roads, by degrees got taken into and stopt, by the partial inclosures which progressively took place; so that at length, the Roads that remained to Travellers, were very crooked, hilly, and

PRINCIPLES OF SETTING OUT ANCIENT ROADS. 221

and uneven, and in further process of time, were rendered very narrow* in numerous instances, by the progress of inclosures on their sides, and by their wearing into deep hollow ways, in many instances, of sandy or soft, rocky or lamellar strata, to which, or to mixed alluvial soils, such *hollow roads* are peculiar, and are never found on strong clay strata†.

Nearly in the above state the art and practice of Road-making and mending continued, when *Turnpikes* first began to be erected in England, and the management of particular Roads to be placed, by special Acts of Parliament, under the management and controul of Trustees appointed for the purpose.

The first Turnpike Act that had reference to Derbyshire, was for repairing and improving the Road from the Bridge over the Trent at Shardlow, thro' Derby to Brassington, situated on the southern slope of the Peak Limestone Hills; the reasons alledged for this first Derbyshire Turnpike Road, *terminating* at so small and obscure a Town as Brassington, were, that the Traveller towards the north, having by means of this improved Road, been helped over the low and deep lands of the County, and landed on the rocky districts, might find his way therein, without further assistance, to Buxton, Tideswell, Castleton, Stoney-Middleton, Ashford, Bakewell, Winster, Matlock, Wirksworth, Hartington, Longnor, &c.

The first nine or ten miles of this Road was already

---

* Instances of very *narrow* fenced lanes are found, in Elmton, in Overton in Ashover, &c.

† Illustrations of these facts may be seen, in the *Hollow Roads*, S of Ashburne in Gravel Rock, S of Heath in 12th Grit, at the Holloways (NW of Crich) in Shale Grit and first Grit Rocks, in Roston in laminated Red Marl, in Smithsby in Grit Rock, &c.

so nearly straight and level, as to need or admit of very little improvement, besides widening it in places, and levelling and covering it with gravel, which was every where near at hand. But in the remaining 12 or 13 miles, where no direct or straight Road previously existed, the summit or ridge between the Vales of the Morledge and the Schoo Rivers required to be passed (see the Map of Ridges, Vol. I. p. 1), and the Gravel and Shale Limestone for repairing this part of the Road, were distributed on the tops of hills, with deep intervening Marl or Shale Lands in the Vales.

In this part, therefore, acting on the principles that formerly had prevailed in Road-making, as mentioned above, the Projectors adopted a line for their Road, neither the most direct, or the nearest to a level, that might have been chosen, but turned aside out of the Morledge Vale, to mount the gravelly hills near Bullhurst, and after following their unequal heights to near Cross o' th' Hands in Turnditch Liberty, turned then to the left, and crossing uneven ground to the top of another high gravelly Hill, on the N of Hulland Ward; from which hill, a course as direct as could be, by the old Lanes, was taken, right across the Schoo Valley (instead of obliquely descending into and crossing it), in order to mount a considerable Shale Limestone eminence at the NW end of Hognaston, and soon decend again from the same; which, with other minor defects in the construction of this Road (notwithstanding that a Turnpike Road was some years after, carried forwards to Buxton, as shewn in the Map facing page 193) have occasioned this part of this first of the Derbyshire Turnpike Roads, to be nearly abandoned, except by the People of the Neighbourhood,

and

and to be almost unknown to Travellers, who in general go by Ashburne, and join the line of Road above mentioned, near to Newhaven, although such is considerably further about, and not free from very long hills, to be ascended and descended.

I have been thus particular in describing this early Turnpike Road, in order to account for the vast number of lines of Turnpike Roads that are seen in the Map, often running by the side of, and intersecting each other, without apparent meaning; which has, in a considerable degree, arisen, from the very injudicious and hilly lines, which were chosen for the first Turnpike Roads, which have successively been succeeded by others, constructed somewhat on better principles, yet all of which Roads, except a very few among the latest, are still defective, and many of them will continue to be replaced in whole or in parts, by better chosen and more level lines of Road: but still *debts* remain, on all these old lines of Road, for the paying of some Interest on which, from the local traffic, or for preventing evasions of the Tolls on the new lines, the Toll-bars are in most instances kept up, on these very neglected and bad lines of Road, and which circumstance is daily increasing into a dreadful nuisance, to the occasional Traveller on these old Roads, but more especially to many persons locally situated, who are obliged to use and pay heavy Tolls, on Roads, whose state of neglect is a disgrace to the Country.

Some of the *old Roads* which have been abandoned, nearly, as to repairs, since improved parts of these Roads have been opened for avoiding them, are, Buxton (1 *m.* N over Within-hill) to Shallcross; Chapel-en-le-Frith (¼ *m.* W over the end of Roshop Edge), to Sparrow-pit Gate; Chapel-en-le-Frith

(over

(over Eccles-hill) |to Whaley-bridge; Disley (over Jackson's Edge) to Hoo-lane, and Disley (over Whaley Moor) to Whaley-bridge (Cheshire); Dronfield (over Cole Aston and Norton Hills) to Heeley-mill (nearly); Duffield (over Chevin and Alderwasley Hills) to Wirksworth; Great Hucklow (over Sir William Hill) to Grindleford-bridge; Higham (over a long succession of Hills) to Duffield-bridge; Newbold-common (1¼ m. N of Chesterfield, over Whittington-hill) to near Unston; Quarndon to Little Ireton (over New Inn-hill*); Sandy-brook (over Spen-lane and other Hills) to near Alsop; Thornsett (over Chatham-hill) to Marple, Cheshire, &c.

The early projectors and makers of the Turnpike Roads in this County (as in most others), too closely imitated the defective system on which Roads had been previously set out; not only in unnecessarily ascending hills, where more level lines might have been chosen, but in descending directly into, and thus crossing valleys at right angles, instead of the more oblique and easy descents which might in most instances be had.

Many tremendously sudden and *steep Hills* in the public Roads, have resulted, and long been the terror of Travellers, and too many of which yet remain; my Notes hereon are as follows, viz.: Ashburne ¼ m. S,

---

* This, if I am rightly informed, was an alteration rather for the worse, as to Hills and straightness, in order to remove the Road out of the Valley thro' Kedleston Park: this part of the Road is, however, well constructed (except that Trees are planted too near it), and if instead of descending again into the Morledge Valley, to pass through Weston Underwood (see page 222), a higher line had been continued to the E, with a gradual rise to near Bullhurst, there might have been less to complain of, than au present.

now altered: Amber-lane 1 m. WNW of Ashover, very steep; Birkin-lane E side, long; Calow-mill ¼ m. E of Hathersage, altered; Curbar* ¼ m. ESE; Great Rowsley ¼ m. E; Hill-top 1 m. WSW of Temple Normanton; Kirk-dale 1 m. SW of Ashford; Knouchley ½ m. E of Stoney Middleton (6 inches in a yard); Long-hill 2 m. NW of Buxton, altered (the old road rose 6¼, the new one only $1\tfrac{1}{16}$ inches per yard); Monksdale* 1¼ m. W. of Tideswell; Over Haddon ½ m. SE; Walton (SH) 1 m. SW, long: Winnets ¼ m. W of Castleton, lately altered; &c.

Here I beg to make mention of some *new lines of Road*, which, besides avoiding numerous Hills, would, I think, materially benefit the County, by greatly adding to the convenience of those persons travelling thro' it, and to the pleasure of such as visit this fine district to contemplate its natural beauties. I will begin with the course of the Derwent, in which noble Valley, a short line of almost level Road seems wanting, to which there appear to me no obstacles, between Little Eaton, and the E end of Duffield-bridge, by which means the new Road to Wirksworth, up the Ecclesburn Valley, would be rendered very complete, and be a proper counterpart to the excellently contrived Road up the Bootle Valley to Ripley and Alfreton. At present, after descending so very regularly and pleasantly from Wirksworth to Duffield, the satisfaction

---

* The tremendous descents into this Valley, and others scarcely less formidable, in the Road between Tideswell and Buxton, might have been avoided, and a very good line of road adopted, about 55 years ago, by passing through the village of Wheston, by Dale-head and Small-dale, but for the opposition of a Mr. Robert Freeman, who then resided at Wheston, and *did not like a Turnpike Road thro' his village!* Egregious folly this, very common in the last age.

which a Traveller has experienced, from contemplating the happy change, in avoiding the monstrous hilly Road by Cross o' th' Hands (see page 222), is considerably abated, on having, after all, unnecessarily, to mount the high hill, a branch from the same range, near to Allestry, in his way to Derby.

A good Road exists at present up the Vale of the Derwent, from Duffield to Belper-bridge, from whence it would be very important to the public, that a good Road should be made up the W side of the River to Cromford, nearly along the line of a private Carriage-way, between these places, belonging to Messrs. Strutts, Charles Hurt, and Richard Arkwright, Esqrs. whose public spirit and liberal views, would, I think, induce them readily to concur in making this public Road, provided the scheme was taken up in earnest by the Country, of making a good Road up the entire course of the Derwent, or near it, as I am about to mention.

From Cromford to Matlock-bridge, there is already a pretty good Road; beyond this, several Hills want avoiding, by keeping nearer to the River, in the way to Great Rowsley. On account of Chatsworth-park occupying both sides of the River, and in order not to miss the Town of Bakewell, the Road leaves the Vale of the Derwent (for that of the Wye), and does not join the same again until near Calver. From this place to Shuts-Dint Bank, 1¼ SSE of Darwent Chapel, there is a Road up the Vale, in want only of a few local alterations for avoiding hills, by keeping nearer to the River.

From Shuts-Dint Bank to near Holden-house, the Road should follow the Derwent Vale, and proceed thence, still up the same Vale, or up the Westend Vale,

Vale, which ever presented the easiest line for ascending the Grand Ridge; descending from the same, and crossing Longdendale (and the Sheffield and Manchester Road) so as to join the Road leading to Huddersfield, on the NE side of Woodhead in Cheshire.

The upper part of this line of Road, could not fail of proving highly important, towards improving the extensive Woodlands of Hope, which remain yet so nearly in a state of nature, see Vol. I. p. 236.

An improved line of Road is very much wanted between Tideswell and Buxton, as mentioned in the Note on page 225.

From Buxton to Ashford, a public Road is now completed, I believe, following in great part, the romantic *Vale of the Wye*, but independent in part, of the intended private or pleasure Road, through all the deep recesses of this very surprising Vale, that I have mentioned in Vol. I. p. 72. From Ashford, this new line is continued forwards by a private Road across Birchill's Farm, to join and cross the Bakewell and Baslow Road, and thence by a new public Road, through Pilsbury to Edensor Inn (near to Chatsworth House): whence there is a private Road thro' the Park, to Chatsworth lower Bridge, and thence a Parish Road, in pretty good repair, completes this line to Great Rowsley Bridge, by which, company passing from Buxton to Matlock, may make Chatsworth in their way, instead of Bakewell.

If on this last-mentioned line of Road, an Inn were built opposite to the mouth of Great-rocks Dale, and a Road therefrom were conducted up the same, and thro' the Black-hole Lime Quarries at its head, to join the Buxton and Castleton Roads to Chapel-en-le-Frith, the same would form the least elevated passage across the Grand Ridge* in the way to Manchester, Liverpool, &c. that is to be met with any where in this part of the Country, but with the disadvantage of missing Buxton.

A great improvement of the very hilly Road between Chesterfield and Matlock, has been talked of, by passing up the Hipper Vale, and the Holy-moor Brook Vale, almost to its source, crossing the Amber Vale near to its head (in or near the line of the Darley Road), then turning to the S, and passing down Lumsdale to Matlock.

Such observations as have occurred to me, respecting the parochial and private Roads, will be mentioned further on, after I have spoken more particularly of Turnpike Roads, in compliance with the order prescribed in the " Plan" for these Reports.

Turnpike Roads.—The period at which the passing of Acts of Parliament for the raising of Tolls from the passengers, on particular districts of Road, that were intended to be improved, or in part new made, first became general, forms an important era in the history of British Improvements; and in process of time, these separate Roads have grown into a connected and complicated system, very essential to the local and to the

---

* In the year 1789, Mr. John Nuttall ascertained by a careful levelling, that the summit over which the Turnpike Road passes 1¼ m. NW of Buxton, is elevated 1198 feet above the Canal-bason at the E end of Cromford, and 1250 feet above the Mersey River at Stockport Bridge. On each side of Gun-Moor Hill, 2¼ m. N of Leek in Staffordshire (Vol. I. p. 8), remarkably low places occur on the Grand Ridge, through the westernmost of which, the Manchester Road passes, and in which inosculation of the opposite Valleys, the Rudyard Vale Reservoir is situated, and Turner's Pool in the other, on the Ridge! See Vol. I. pages 498 and 499.

national

national prosperity, as the Map of these Roads in and around Derbyshire, facing page 193, will strikingly exemplify.

At the period alluded to, the profession of *Civil Engineer* (see the Preface to *Smeaton's Reports*) had not arisen; nor were sufficiently scientific or competent persons in the habit of being employed or consulted, in contriving, setting out, and executing the new or improved Roads, which were to become Turnpike: the Legislature were at this period unprovided with any materials or documents, from whence to judge, whether the most eligible and best lines were chosen for the Roads, proposed for their sanction, such as their late excellent regulations or *Standing Orders*, are calculated to produce, by requiring accurate *Maps and Sections* of all proposed alterations of Roads, to be presented, made by competent Engineers or professional Road-makers, such as had long been required from the proposers of new Canals or Rail-ways, applying for Acts of Parliament.

In this absence of essential information, the Committees of Parliament could do no more, in numerous instances, than receive and decide between the evidence of persons, wholly incompetent to the business, who were brought forwards by the parties, to serve their own local interests and views, in supporting, opposing, or varying the proposed new Turnpike Roads: and hence have resulted, the many badly contrived, and as badly executed lines of Road, which have already been alluded to in pages 221 and 222, and which were since obliged to be superseded by others, set out and constructed on better principles.

In conversing with two or three sensible men in Derbyshire, on the great waste of Money which has hereby been occasioned, they remarked, that a good to the Country has resulted therefrom, which ought not to be overlooked, viz. that if the first Turnpike Roads had been conducted thro' the Valleys, as at present, the hilly and rocky, and often barren districts, over which the first Road-makers contrived so absurdly to mount, would have remained yet, and perhaps for long periods to come, without practicable Carriage Roads, which are so essential to their Agricultural Improvement, and which these Roads, imperfect as they are, have very beneficially supplied, in many districts.

The spread of Canals and Rail-ways, contrived and executed on correct and scientific principles, and conducted under professional Men, competent to the overcoming of the many difficulties which Nature, and Art also, in some instances, opposes to such improved modes of communication, can hardly fail in time, in any district, of having a beneficial influence on the principles and practice of *Road-making*, and on the description of persons who are employed and entrusted with their management.

Accordingly we find, that in South Wales, where Canals, Rail-ways, and similar works, connected with its Coal and Iron Mines, have multiplied in an unusual degree, within a few years past, that there the example has been set, of forming an " Association for the improvement of Roads," of the principal Inhabitants, with the Duke of Beaufort at their head, who, at a Meeting held on the 23d of May, 1805, appointed a Committee, under whom, Mr. Evan Hopkin, Civil Engineer, was employed, to make a perambulatory Survey of the considerable lines of Roads from Gloucester, through Ross, Monmouth, and Brecon,

con, to Caermarthen, and from Gloucester, through Newnham, to Chepstow; Mr. H. being " directed to point out the most obvious defects, and also what improvements can be made, along these lines of Roads."

The Report of Mr. H. was divided into several distinct parts, corresponding with the different Turnpike Trusts, and copies of these several parts were made, and laid before the respective bodies of Trustees. In the abstract of Mr. Hopkin's Report, that has been printed, he mentions nine instances where the Turnpike Roads had steep parts in them, that rose for considerable distances, five inches in the progressive yard, eleven instances of six inches rise, and three instances of seven inch rises! All which Hills he has found capable, except one, of being reduced under three inches rise per yard, viz. in seven instances the rise might be reduced to two inches in the yard, two others to one inch and a half, two others to one inch each, and in eight instances be rendered level, or the hills entirely avoided, and these often for very considerable distances.

Important improvements might be suggested and effected, in most districts in England, in consequence of similar Surveys, with estimates of the cost of making the necessary alterations of the public Roads, made by skilful and competent Engineers, who should be free and unbiassed by those local prejudices and interests, which, not less than incompetency, in the contrivers and executors of the present Roads, have contributed to their being in so many instances, less perfect and useful to the public, than they ought to be.

The new Roads lately made in the northern and western

western parts of Scotland, by order of the Parliamentary Commissioners for Highland Roads and Bridges, set out and executed under the direction of Mr. *Thomas Telford*, the Engineer, and competent assistants, are without doubt the best contrived, for avoiding Hills or steep descents, and the most perfectly constructed, of any similar lengths of Road (in districts as unfavourable for travelling) in Britain, or perhaps any where else.

Around Disley, for the space of four or five miles, in Cheshire and in Derbyshire, hints have been taken from the Canals and Rail-ways in the neighbourhood, and through the employment of persons competent to such works, greater improvements have there been made in the Roads (as is mentioned in page 224), than in almost any district that I am acquainted with in England; and which improvements, Gentlemen who are travelling, or are making a stay at Buxton, and possess influence in the direction or management of Road concerns in other districts, would do well to notice particularly.

When *Turnpike Acts* were first applied for, the idea seemed to prevail, that it was merely necessary to widen, straighten, and substantially repair the Roads, and continue the Tolls thereon, only so long as the principal Sum borrowed for such repairs, and interest thereon, could be paid off; that then the Toll-gates might be pulled down again, and such Roads would thenceforwards need no other than the ordinary attention of the Parish Surveyors, and the Statute-duty: the popular prejudices that existed at [the time, against the establishment of Tolls, on Roads that had previously been free (however bad and difficult it may have been to pass on them) seem also to have concurred,

with

with the mistaken idea above-mentioned, in occasioning the first Acts, granting powers to particular Trustees, to borrow Money and collect Tolls for the repayment thereof, to be merely *temporary Acts*, for 21 years at furthest.

Soon after this period, a set of sanguine projectors came forwards, to assure the Public and the Legislature, that if heavy carriages were made to move on *rollers*, or on *very broad-wheels*, such would do no damage to the Roads, and ought consequently to be wholly or in great part relieved from the Tolls; and some of these projectors went even so far, after Parliament had listened to their claims of partial exemptions for Broad Wheels, as to maintain, that the advantages of these heavy broad-wheel or rolling carriages, would soon be so striking, in highly improving the Roads (instead of injuring them, as they have effectually done*), that Parliament might soon, safely, compel the few persons, carrying large weights, who had not done so of their own accord, to adopt very broad wheels, and thus, according to the visions of these schemers, Roads would become and remain universally good, and the obnoxious Tolls and Tollgates, might again disappear from the Country: and thus, what I have long viewed as the two fundamental errors in British policy with regard to Roads, viz. *temporary Acts* relating to them, and *Broad-wheel exemptions* from Tolls, gained a firm footing, and have ever since kept their place, in spite of all experience, and evidence of their mischievous effects.

When the 21 years granted to Trustees have expired,

---

* See Holt's Lancashire Report, p. 192, and York West Riding Report, p. 217, &c. &c.

it has, I believe, almost without any exceptions, happened, that a Petition has come before Parliament, stating, that *Debts yet remained, which the parties must entirely lose* without a renewal, and that the Roads could not be repaired or upheld, without a renewal of their term; which granted, a second, a third, &c. application, has periodically followed, precisely to the same effect, and yet the system has gone on; until now, owing to the immense number and ramifications of the distinct Turnpike Roads, a vast sum, drawn *from the public in Tolls*, and which ought sacredly to have been applied to the repair of the Roads and discharging the debts, originally contracted in improving them, is expended periodically, in *Fees* to Officers of the Houses of Parliament, *pay* and travelling *expenses*, &c. of Attorneys, Surveyors, and Witnesses, assembled from the most distant parts of the kingdom, and detained for weeks together in London, on numerous occasions.

In the last Session of Parliament, 53 Geo. III. *fifty-five* distinct Road Acts were thus obtained (whose titles are recorded in the Monthly Magazine, Vol. 37, p. 154) most of them for extending terms, I believe, and yet only two of these 55 Acts were for any parts of the Roads within the square of Map facing page 193, viz. from Leeds to Sheffield, entering it at H, and from Market-Harborough to Loughborough, entering at D; whence some idea may be formed, of the great annual number of Road Acts that are obtained, for the whole of England, Wales, and Scotland.

Owing to the necessity of the frequent advance of Toll, during the rapid decrease of the value of money in the last 20 or 30 years, and of making improvements, and raising more money, &c. before the expiration

piration of the existing terms, I think it can hardly be reckoned, that the terms have existed much above 15 years on an average, and have scarcely cost the Trustees less than 500*l.* each Act: and that this very injurious and partial tax on the British Public, has probably exceeded 27,000*l.* in the past year. See the Durham Report, p. 360.

When *Weighing Engines* had been introduced, and were approved by the Legislature, a remission of this *renewal tax*, to the extent of adding *five years to the term*, to all such Trustees as would erect a Weighing-engine or Machine, was agreed on: a circumstance strongly marking the sense of the Legislature itself, on the character and operation of its *temporary Acts*, for authorising *permanent Roads* to be made for the general accommodation and benefit, *at the expense of individuals*, but whose claims to reimbursement, or even to interest on their Money, *were to cease* with the term granted, except on paying anew, this most impolitic tax.

The Board of Agriculture, and Proprietors of Lands in general, will, I hope, ere long, take the general subject of *Turnpike Roads*, into their serious consideration, and endeavour to obtain a general Act, that shall establish an entire new system in these respects; or at least, that Acts should be obtained, consolidating and regulating the Turnpike concerns of each district of country, into which Britain might be divided, or even each separate County, on one uniform plan.

The establishment of some plan, whether by the formation of *a Board of Roads and Bridges*, or otherwise, by which the *Accounts* of the several sets of Trustees should be collected, compared, and *audited*, by persons conversant with expenditures of

this

this nature, and with the peculiar facts, as to lengths of Road, and the levels thereof, nature, distance, and costs of materials, ordinary prices of labour, traffic, &c. &c. within each Trust, from previous Surveys, and documents made for and lodged in such general Office by competent Road Engineers, would, I think, be attended with much good to the country.

Such a general and consistent auditing of Road Accounts annually, that should not merely stop with the *fact*, of the Trustees *having expended so much money*[*], but enquire fully into the fairness and *propriety* of such expenditure, might furnish relief to numerous Parishes which now suffer severely, from the prejudices and mismanagements of the Trustees and Officers of Turnpike Roads, which pass through their Parishes, perhaps only across an outside of, or along the border thereof, so as to yield little benefit to the Inhabitants, as a substitute for Parish Roads; which Trustees nevertheless claim *half* of the Statute-duty, and Compensation and Composition Money in lieu thereof, either separately or among them, where there is more than one Road, although in numerous instances, there are six, eight, or ten times as great lengths of useful and necessary *Parish Roads* to be maintained out of *the other moiety* of Statute-duty, &c.

I beg here to mention a case relative to this subject, that occurred within my own knowledge in a Midland County, conceiving that many similar ones exist, in various parts of England. Thirteen Miles of a prin-

---

[*] The accounts of Turnpike Trustees are at present *without any kind of audit* or controul, I believe; and the Parish Surveyors' Accounts seldom undergo further examination, than as to the fact of expenditure! the propriety or judiciousness of which, is little thought of by the Vestry, and the Justices merely swear the Surveyor thereto.

cipal line of Turnpike Road, under a set of Trustees, the *acting ones* principally Clergymen, passed for the length of two miles through a Parish, of which I was Surveyor of its Roads, the deep, sandy and excessive bad state of which part of the public Road, had for ages been proverbial, throughout England and further; the half of the Statute-duty, &c. of this Parish (consisting of 4000 acres of pretty good land) had been regularly called for, by the Turnpike Surveyors, and a productive Toll-gate stood in the Parish, yet the public Road grew progressively worse and worse, under the antiquated and absurd system pursued for its repair; until at length, the Post-office Surveyor caused *the Parish* to be indicted : the Trustees, when applied to by the Parish Surveyors, very coolly replied, that it was *the duty of the Parish* to put the Road into a thorough good state, and until that was done, the Indictment neither could be removed, *nor should they do any thing* at this part of the Road, until this was done!

On investigating the circumstances, it appeared, that, although a *sand* Ridge was crossed for about a Mile in length, by the Road in question, beyond the limits of this Parish (and not indeed in the same County), that only two short distances of about 150 yards in length, each, within the Parish, were upon Sand strata, but on the contrary, the remainder was upon a most tenaceous *alluvial Clay;* and that the proverbial depth, and heaviness of the mire in winter, and loose sand in summer, upon the Road of this Parish, through nearly its whole length, had entirely resulted from bringing, for ages past, the very *soft sand-stone,* collected on the ridge above-mentioned, at great expense of digging and cartage, for repairing

the

the Road: but which was no sooner spread thereon, than the first heavy carriage that passed over it, crushed the whole to sand, which at first, if wet weather ensued, made a loamy tenaceous mire, and afterward, when washed by the rains, and dried, an intolerably loose and heavy sand.

It might have been expected, that where so very improper a material as this soft sand-stone, had been carried six or seven miles, into Parishes still further from the Sand Ridge, for repairing this Turnpike Road, that hard *gravel* was unknown in the district: the case, however, was far otherwise, for several of the valleys that crossed this Road, were known to have beds of flinty and quartz gravel under them, and under other parts, although a good deal mixed with loam and clay, and requiring a very careful and effectual sifting: and it further appeared, that the principal Proprietor of the Estate, who had long lamented the state of the Road, and the ineffectual methods pursued for its repair, had some years previously directed his Agent, to commence a system of *employing Labourers in the winter* season, when work was scarce, to dig, and thoroughly sift and cleanse, and stack this gravel; levelling and returning the top-soil on to the pits, as they proceeded, according to a regular plan to be laid down and well enforced: it also appeared, that great quantities of excellent gravel had been thus prepared, in moderate sized regular stacks, or square heaps three feet high, near to the Road, and such were offered to the Turnpike Surveyor and Trustees, at the mere cost of digging it, (upon the Labourer's bills, who worked by measure), without any charge for the gravel, loss of crops, or damage in fetching it away: but which offer they declined accepting,

cepting, alledging, that their Surveyor ought to be allowed to dig or procure Materials *himself*, where and how he thought fit;—after laying many Months, the gravel so prepared, was used on some private Roads; and thus the humane, and patriotic intentions of the Proprietor were entirely frustrated.

The *Indictment* having thrown the Road on the hands of the Parish, as above-mentioned, they embraced immediately the renewed proposal of the Proprietor, as to furnishing gravel, and commenced a substantial covering of the Road in question, with cleanly sifted gravel, divested of all its larger and softer stones near the top, by carefully raking the surface, at the time of spreading, and for a time afterwards, in frequently levelling the ruts formed therein, and throwing or carrying forwards all the large, flat, or soft pieces of stones, into the foundation, where the wheels of Carriages should not operate, to crush or disturb them from their places, as flat or irregular shaped stones, are so apt to be, on the surface of a Road.

While this was doing, it happened that the Trustees applied to Parliament for a renewal of their Act, but against which, the Parish in question petitioned; alledging the injudicious expenditure of former Tolls, the evils that the Public had in consequence suffered, and the heavy expenses that the Parish were then bearing, for effectually repairing the Road, through the same mismanagement; but which repairs, a recurrence of the old system, as was to be expected, would soon cover with mire or loose sand, as formerly. A compromise in consequence took place, and met the ready concurrence of the Legislature, by which it was enacted, that the Trustees or their Surveyor, should thenceforth have no right to interfere in the

manage-

management of the Road through the Parish above alluded to: but after deducting their Toll-collector's and Clerk's salaries, from the Tolls taken on the whole of their Road, they should, half yearly, pay over two-thirteenths of the remainder, to the Parish Surveyors.

The consequence was, that a very improved and substantial Road was completed through the Parish; the Public have ever since enjoyed the benefit of a hard and good Road, in place of deep Mire or Sands, and the Parish have since been greatly relieved from expenses, and their other Roads have much improved, in consequence of this arrangement. But I have been sorry, tho' not surprised, at seeing, when lately travelling this Road, that 14 or 15 years experience since, has only yet in a small degree taught these Trustees, or their Surveyor, the propriety of abandoning soft sand-stone in repairing their Road, and using only the hard and cleanly sifted gravel, that is fully within their power, in ample quantities.

A Board of Roads, such as I have suggested, might be entrusted to remedy such want of skill or propriety as I have instanced, or any other important failure, in the execution of the trusts or duties committed to their charge; perhaps by presenting the Parties and misconduct, to the Quarter Sessions of the County, and if not attended to there, or amendment ensue, then to one of the Courts in Westminster.

Another duty of such a Board, where *Indictments* had fallen on particular Parishes, who had not previously failed in doing their assigned proportion of Statute-duty to the Indicted Road, should be, to relieve such Parishes, by obliging the Trustees to repair such Road, in case it appeared to the Board that
they

they had, on the whole, sufficient Funds, (as the general Accounts in their possession would often enable them to judge of, without a special Survey) but which Funds they were in the habit of applying unequally or partially in the different Parishes (an evil that I know frequently exists): or if, on the contrary, a real want of funds appeared to authorise, perhaps with the sanction of the Quarter Sessions, the increase of the Tolls, and borrowing of further sums of Money, on the credit thereof, sufficient for effectually repairing the Road, without partially and very unfairly burthening the Parish, as at present.

It has occurred to me, from a long and careful attention to Roads, in all situations, and I know numbers of intelligent Travellers and Road Surveyors, Mr. Beatson, Mr. Hornblower, &c. &c. who have made the same observation, viz. that nothing is more essential to the goodness and permanence of a Road, than *causing the wheels of Carriages continually to change their places on the Road,* by which alone, *Ruts* thereon can be avoided, and a smooth surface obtained and preserved; this is remarkably exemplified, at the meetings or turnings of Roads in most situations, notwithstanding the *grinding action* of the wheels thereon (that Mr. Cummings and others so greatly magnify in conical wheels, &c.) while turning, and on the slopes of hills, of considerable steepness, where the horses, in order to ease the ascent or descent, endeavour to cross continually from one side of the road to the other: as also in parts of Roads, that are often full of carriages going different ways and paces, and are consequently obliged often to turn out and change their tracks, as in the immediate environs of great Towns.

In the Act passed in 1808, for the new Turnpike Road between Ashover and Tupton in this County, Mr. Joseph Butler of Killamarsh, introduced Clauses, which upon the principles above mentioned, granted an exemption of tolls to the extent of $27\frac{1}{2}$ per cent. on the average, on loaded carriages of different descriptions, provided their wheels ran clear of the usual tracks of such Carriages*, by having wheels, making a track of less width than 4 feet 5 inches outside, or a greater width than 5 feet 8 inches inside: with a further exemption of $24\frac{1}{2}$ per cent., or 52 per cent. on the average, on these carriages of *new widths of tracks*, below the Tolls on ordinary carriages, provided that at the same time, the rims of their wheels and their axle-trees and bushes, are made perfectly *cylindrical*.

The propriety of the latter exemption, to such an extent, I do not pretend to understand, but the former one, if acted on, would, I feel confident, have a very beneficial operation on the Roads, in preventing Ruts, and the consequent disturbing of the materials, and unequal wear.

Owing to the separate, and sometimes the opposing interests of different sets of Trustees, it not unfrequently happens, that particular Towns and places are oppressed, or greatly inconvenienced, by the numbers and situations of the *Toll-bars*, erected around or near them : Matlock furnishes an instance of this kind, in five directions, Toll-bars are placed very near to

---

* The Warwickshire Reporter, on the contrary, p. 172, wishes all wheels reduced to *one width and distance*, in order that they may pass on Rail-ways, by the sides of the present Roads; without being aware, that scarcely any of our lines of Public Roads are, or could be, on account of levels, adapted for Rail-ways.

this Town, one hilly and indifferent Lane, only remaining open, towards Willersley, by which the Inhabitants can stir abroad, or take the necessary exercise, without paying Tolls, as great in most instances as would be required for travelling many miles on the same Roads: in the environs of very large places like London, these closely encircling Toll-bars, seem less objectionable than at smaller places; although Liverpool (see Holt's Lancashire Report, p. 189) and some other places, furnish instances of the very opposite kind, by all the Roads around the Towns remaining open to unreasonable distances, to the injury of the more distant parts, that are, in consequence, locally burthened by the situations of the Gates, and amounts of Tolls collected thereat. These and many similar evils, arising out of the present incongruities of our Turnpike System, might, and seem alone capable of being remedied, through the general information and enlarged views of the subject, that a Board of Roads might take, and either be authorised to act upon, or to present to the County Sessions for approval.

At many of the *Toll-bars* in Derbyshire, I observed the neglect, of writing up the date and title of the *Act*, and *Table of Tolls* thereby authorised to be taken: and at some (Hague Bar in particular, near New Mills) neither the printed Act was kept, nor could (or would) the Keeper tell, the name and residence of the *Clerk* to the Trustees, &c. Information was alike wanted, to be posted up, in most instances, as to what other Gates (if any) the payment there demanded, would *free*, nor did the Tickets express this, as ought always to be the case: the importance of these notifications, as checks on imposition, made me glad to observe, in the general Bill for the regulation of the Roads, that was discussed in the House of Commons, in May 1810, that more effectual Clauses for enforcing these and other similar and useful regulations, were provided: but, unfortunately, the regulation of Stage-coaches, so as to draw further taxes from the Travellers thereby, having passed, in a separate Act, these proposed improvements and regulations of our Road Laws in general, fell to the ground, like numerous former ones.

Weighing-Engines are very common on the Roads in the County, and as far as I could learn, are pretty fairly and impartially used in general, for the suppression of over-loaded carriages, which otherwise, by *crushing the materials*, would so materially injure the Roads; it has appeared to me, however, that the excessive and prohibitory character of the present *fines* for over-weight, might be very advantageously changed, for easier and progressive *additional Tolls* at each gate, according to the number of wheels, and the degree of over-weight, including also the consideration of the breadth of wheels, but in a very subordinate degree to what is done at present, so as to make it the obvious interest of the Carrier, to carry less weights on each carriage or pair of wheels: or rather, that the Tolls ought to increase with the weight carried on the pair of wheels, in much more than a simple ratio, so as greatly to discourage the larger weights, on whatever breadth of wheels they may be pretended to be carried: since experience has amply taught, that wheels of greater breadth than 6 or 7 inches, or nine at the most, are only kept in use in any part of England, by pretended and fallacious compliances with the Law, and solely *for claiming the exemptions,*

exemptions, that have so injudiciously and improperly been continued, to these enormous *stone-crushing machines*, as the most favoured carriages, in these respects, ought to be called, instead of rolling waggons, &c.

I have already noticed in Vol. II. p. 64, that the Road Weighing-Engines are, in this County, often made subservient to the trade of the Inhabitants, by ascertaining the weights of Hay, Straw, Manure, Coals, &c. that are sold; and I would add in this place, that it would be well perhaps, if Trustees of Roads were empowered, by a general Act, to take moderate and fixed rates, for weighing Goods at their Engine; in which case it might answer their purpose, and in some cases add considerably to their funds, to erect their Engines, in or near the Towns, instead of attaching them to their Toll-gates (which on many Roads is now entirely laid aside on other accounts), where they might be freely resorted to by the Inhabitants for ascertaining large weights, on paying the stated tolls for so doing, to the Engine-keeper employed by the Trustees.

The Mile-stones, through considerable parts of the County, are too much neglected, and I do not remember, that on any Road, there appeared a system, of annually, or every two or three years at most, repairing and re-painting the figures thereon, as ought generally to be the case, even where the paint is not liable to scale from them, as is mentioned, Vol. I. p. 428; this periodical and *frequent attention* to these very useful appendages to the Turnpike Roads, might perhaps deter or discover some of those idle and disorderly-persons, who now so shamefully deface the Milestones, by their wanton and mischievous attacks on them, and on another, often still more essential appenda pendage to the Roads, the Way-posts, or Finger-boards, which to the disgrace of many parts of the County, are entirely defaced, even in parts, where, from the great numbers of crossing and branching Roads, these dumb directors of the Traveller are of the utmost importance; yet wherein, mutilated posts and parts of boards only, are seen, and scarcely a single inscription remains legible, from the peltings of the idle vagabonds above alluded to.

About the year 1798, the Justices assembled at their Quarter Sessions, made an order for the erection of Way-boards in all necessary places, throughout the County, and which seems, in several districts, to have been pretty fully complied with; but about Langley and other places to the NE of Derby, scarcely one of these remained, in 1808. At the Christmas Sessions of that year, a fresh Order was made by the Justices, and which, I hope, has ere this been obeyed by the Parishes, and effectual steps taken, to suppress or punish severely, the disgraceful mischief above-mentioned. Mr. James Swainson's new Direction-boards, with *open* iron letters, for rendering them visible or tangible in the night, which will be described in Vol. 31, p. 240, of the " Transactions of the Society of Arts," might be very useful in some particular places.

At different points on the beautiful private Road leading from the end of Bonsal Dale to Hopton Hall, the residence of Philip Gell, Esq. the Way-boards are inscribed " *Via Gellia:*" and near the S end of Shirley, I saw a board, at the entrance of a Bridle-way, on which was painted " *Equus via Longford.*"

On the Turnpike Roads of the County, there are many excellent *Inns*, some of them built at great expense to the parties, and proving of proportionate accommodation

commodation to the public: on this head, my Notes refer to Bakewell (Rutland Arms), Buxton (Eagle and Child, &c.), Derby (King's Arms), Edensor, Hathersage (Ordnance), Kedleston (New Inn), Matlock Bath (Hotel, &c.), New Haven, Pebley Lane, NE of Barlborough, Smithy-houses, &c.

I would take the present opportunity of mentioning a great inconvenience, that Travellers who have occasion to leave the main Roads, like myself, while engaged on this Survey, suffer, from the operation of the *Spirit and Wine Licenses*, as at present managed. I am told, that Maltsters, Common Brewers, Tobacco Manufacturers, and many others, are allowed a License from the Excise Offices at the beginning of the year, but do not pay for the same until the end of it, and then its cost or amount is proportioned to the business that has been done under it: but the Licenses to Victuallers, that I am alluding to, must be paid for at the beginning of the year, according to the amount that the Landlord may be *rated to the Poor*, the lowest cost of a Spirit License for a year, being 94s. (and the highest 142s.) The consequences of which are, that nearly all the small Inns on the bye-roads, now take only Ale Licenses; and, as owing to the inferiority of most of their taps of this beverage, few Travellers now stop to dine or refresh with them, very few of them, in consequence, keep now, either Corn or Hay, or even a Stable for baiting or sheltering a Horse, in many cases:—for weeks together, I have suffered these inconveniences, almost daily, after the trouble of riding perhaps a mile or two, in search of, what was reported to be *an Inn;* and which had indeed been such, until the mischievous operation of the Spirit and Wine Licenses, as above-mentioned.

Some persons have injudiciously applauded these prohibitory Licenses, as lessening the practice of dram-drinking, but the practical effects of them are quite otherwise, since we have only to enter any considerable Town, and there we shall find, every one of its numerous public-houses with a Spirit License, and with a *trade in Gin*, fully sufficient to enable them to pay for it, although ten or more such dram-shops may sometimes be counted in a single street, and into which a Traveller rarely, and in some instances scarcely ever enters, except on Fair or Market days; the detached Inns and better sort of public houses in the very smaller Villages, were never the resort of the Gin-drinking manufacturers and labourers, whom it is wished to discountenance, and the only operation of the Licenses I am condemning, is, to inconvenience the country and injure the Revenue, for the Exciseman and Supervisor have yet to attend and survey the Ale casks of most of these ruined Inns, at as great an expense to the public, as might keep the usual checks, on their sale of Liquors, which *Travellers*, in the winter season in particular, cannot do without.

The several *Mail-Coaches* that traverse this district, have been mentioned already (pp. 210, 211); but it remains, that I should join the intelligent Reporter on the West Riding of Yorkshire, p. 217, in mentioning, the partial and oppressive operation on many Road Trusts, of the total exemption which these Coaches have enjoyed from Tolls: this evil was lately removed in Scotland, by the 53d Geo. III. chap. 62, which charged additional postage on Letters there, for enabling the Post-office to increase the allowances to Mail-coach Proprietors; and who now in consequence pay

pay tolls, equally with other Stage Coaches, and I hope that ere long, the same thing may take place in England, where many Roads, around London in particular, suffer from these exemptions, in an equal or greater degree, from the number of Mails travelling the same Roads, than was the case around the Scottish Metropolis. In travelling across the Lothians from Edinburgh, in the Autumn of 1812, I was forcibly struck, by the very enormous tolls and restrictions imposed on the Stage Coaches travelling the Dunbar Road, and was told, that the Mail-Coach exemptions, had in part been made the plea, for introducing these.

The *Post Towns* in the County, the hours of arrival and departure of the Letters from *London* daily, and the postage of a single Letter from London at present (March 1814), are as follows, viz.

Alfreton, arrives at 4 Afternoon, departs at 7 Morning, 10d.
Ashburne, ... 5 After. .... 8 Morn. .. 10d.
Bakewell, ... 9 After. .... 5 Morn. .. 10d.
Burton on Trent, .. 5 After. .... 11 Morn. .. 10d.
Buxton, ... 8 Morn. (of another day), 3 After. .. 11d.
Chapel-en-le-Frith, 11 Morn. (of another day), 9 After. .. 11d.
Chesterfield, ... 7 After. .... 5 Morn. .. 10d.
Derby (All Saints), 2 After. .... 10 Morn. .. 10d.
Glossop*, ... 10 Morn. (of another day), 12 After.
Matlock-Bridge, .. 8 After. .... 7 Morn. .. 10d.
Stoney Middleton, 9 Morn. (of another day), 11 Morn. .. 11d.
Tideswell, ... 2 After. (of another day), 9 Morn. .. 10d.
Wirksworth, ... 7 After. .... 8 Morn. .. 10d.

From several of these Post Offices, there are walking Postmen, dispatched to some of the adjacent

---

* This I find, is not a regular Post-office, where Letters are stamped and charged.

Towns,

Towns, either daily, or two or three times, &c. per week; as from Chesterfield to Ashover, daily in summer, and four times per week in winter, &c.

Materials for making Roads are very plentiful and good, of hard Limestone, in most parts of the Peak Hundreds or north-western districts of the County: over most of the southern half of the County, quartz Gravel is extensively scattered, and supplies a pretty good road material; but in the Coal districts, on the eastern and northern parts of the County, fit materials for the purpose are very scarce and expensive to procure. Except in the streets of the larger Towns, there are no paved Roads in this County, as is common in Cheshire and Lancashire. I shall speak of the several kind of Materials in order, viz.

1. *Gravel.*—In the Map of Strata and Soils, facing page 97 of Vol. I. the light brown colour (1), shews all the most considerable tracts and patches of quartz Gravel, above-mentioned, and at page 134, a considerable number of smaller detached patches, of Gravel of different kinds, are mentioned and described. In applying this useful material to the Roads, I observed in general, two principal defects;—*first*, the Gravel was rarely sufficiently clean sifted; and *second*, the larger smooth round stones therein, were not *broken*, as they ought to be, as well as all the large, long, or irregular stones, which in that state can never wedge and fix down along with the others, as I have mentioned in a Letter, which the House of Commons did me the honour to print, in the second Report of the Committee on Broad Wheels, 18th July, 1806, and 8th March, 1808.

In crossing Mansfield Forest from Newark, in the Autumn

Autumn of 1807, I observed, about two miles from Mansfield, a considerable length of new Road that had been formed, at considerable expense, entirely of smooth round quartz pebbles, from the size of pullets to geese eggs; yet the difficulty and almost impossibility of travelling on these, was so apparent, that no teams had ventured on them in the course of several Months, but the Carters chose rather, to fly off to new tracks in the deep sand and loose pebbles of the Forest, for avoiding this loose pebble Road, which for ages to come might remain thus unused, unless covered first by the blowing sand of the Forest, or that all the largest and smoothest of the Pebbles were broken with hammers, into several pieces, so that their new surfaces might wedge together and fasten; in which case, no materials in Britain would make a better Road.

During the next Autumn, in crossing a noted Hill, (for its loose and bad Road, formerly) called the " Lickey," between Broomsgrove and Birmingham, I had the satisfaction of seeing a very hard, smooth, and fast Road, produced from the pebbles there, just similar to those on Mansfield Forest, but having been well broken at the time of laying them on the Road, as above recommended: it is surprising that this very simple method of making good gravel Roads, has not been more generally thought of and practised.

Francis N. C. Mundy, Esq. of Markeaton, has a large Gravel-pit NW of the village, where Labourers are employed in the winter season, in digging and sifting large stacks of Gravel, levelling the rubbish, and regularly returning the top-soil on to it, as the work proceeds; by which operation, the land is improved greatly, and this Gravel he sells to the Surveyors of the Turnpike and other Roads, at 1s. per cart-load of twenty

twenty bushels. I heartily wish that this practice of Mr. Mundy, and the Proprietor in another County, which has been mentioned, p. 238, was far more generally followed by Gentlemen, who might thus do a vast deal of good, by employing the Poor when short of Work, and confer on the country and themselves, the benefit of good Roads, often without the least loss, but some benefit to their Estates, and might even draw a small profit from the Gravel, if they thought fit.

This mode of *preparing Gravel*, with due attention to reinstating and improving the Land, might always obviate the necessity, of letting Road Surveyors, having no interest in preserving the Land, into private Estates, in search of materials, where often, thro' carelessness or ignorance, they commit terrible damage.

I had the opportunity 3 or 4 years ago, of seeing a good deal of the evil I am speaking of, when my younger Brother (Benjamin F.) was first elected Surveyor of the Whitechapel and Essex Roads, near which several valuable fields had been left in a most worthless state, through the mismanagement and neglect of former diggers of Gravel; the consequence of which was, that every possible resistance was given to the entry by the Surveyor, for this necessary purpose of procuring materials; but now, since the waste pieces above-mentioned have been reinstated, as far as was practicable, and an entire new system of proceeding adopted, by which the land is materially benefited by the operation, offers are voluntarily made to the Trustees, of liberty to dig, on account of the price per load that they pay to the Land-owners*. The general

---

* In April 1814, the prices paid near to the Whitechapel and Bow Road, for uncallowing, digging, clean sifting, stacking, and loading the

general importance of this subject, will, I hope, prove my excuse for having dwelt so long upon it: and even yet I have to express a hope, that Lords of Manors will unite with the persons having right of Herbage on their *Commons,* in putting a stop to the almost wanton and shameful, and sometimes irremediable damage that is now done to the Waste Lands, by the persons digging materials for the Roads.

2. *Limestone.*—In Vol. I. p. 408, a List of many of the Quarries of this stone in the County, is given, situated, principally, on the skirts of the calcareous districts, but within which districts, there are also numerous other smaller Quarries, whence this material is procured for the Roads: the Peak Limestones are in general very hard, and good for Roads, as already mentioned; but the yellow or magnesian Limestone of Scarsdale Hundred, is an indifferent Road material, grinding easily to a gritty mire in wet weather, and becoming loose in Summer, as may be noticed on the Nottingham and Chesterfield Road, between Mansfield and Glapwell, at the latter place in particular: and yet, on the brow of the hill here, a blue and very hard Limestone might be got, see Vol. I. p. 157, which it would be worth while to mine for, and to

---

the Gravel, and filling it into Carts, and levelling the siftings, and uncallowing in a uniform manner, for a *ton* or 23 cubic feet, being half a cart *load,* from 10½d. to 15d. according to the depth or thickness of top earth and gravel; average about 12d. per ton. The Trustees pay to the Land Owner and Occupier, together, 3d. per Ton, and level the pits: and the Cartage on to the Road costs from 6d. to 15d. per Ton; a Cart, 3 Horses, and Man, usually earning 20s. to 22s. per day of 8 hours, at this work.

carry

carry it considerable distances each way, for covering this important Road, when broken sufficiently small.

The important object of causing hard Limestone and other Quarry stones to be *broken sufficiently small* for the Roads, has been accomplished in some parts of this County, by the late Mr. William Gauntley, Mr. John Milnes, and others, by means of *guages* or iron rings, generally of $3\frac{1}{4}$ inches diameter inside for Grit-stones, for foundations, and others $2\frac{1}{4}$ inches diameter for Limestone, Crowstone, &c. for finishing or repairing; the labourers employed to break the stones, are each furnished with these Rings, and when the stones are brought on to the Road, they are subject to a small fine or deduction from their wages, as a farthing each, for every stone the Surveyor or his Assistant can find, that will not pass through the proper Ring. This subject will be further pursued presently, when I come to speak of Expenses of Road-making.

3. *Rider* stones and other hard vein-stuff (Vol. I. p. 248), from Westedge Mire in Ashover, and some others, have been used with good effect on the Turnpike Roads.

4. *Chert*, found in nodules, or irregular masses in the Limestone, Vol. I. p. 415, is collected in considerable quantities in the Crich Quarries, and sent off by the Rail-way and Canal, for the repairs of Roads to the eastward. Since in many places, where the Roads are hilly in particular, the Chert, when broken small, might be very advantageously mixed with the Limestone, in repairing the Roads, for correcting the

liability

liability to slip or slide on such limestone Roads, I shall here give a list of the places where I have observed *Chert* in considerable quantities imbedded in the Limestone; viz. Alport in Yolgrave, Ashover, Bakewell N, Bonsal NW and E, Bradwell, Buxton E, Castleton, Crich, Cromford, Dove-hole, Haddon Hall ¼ m. NW, Lanehead N in Tideswell, Little Longsdon NW, Matlock Bridge S, Middleton Dale, Over Haddon Pastures, Pindale, Snitterton S, Willersley NW, Wirksworth NW, &c.

5. *Crowstone.*—In Vol. I. p. 180, a list of places is given, where this very curious and valuable road material has been dug. It should be broken small, otherwise, if laid on in large lumps, as was formerly done in several parts of the Road between Chesterfield and Ashover and others, its hardness occasions intolerable choaks and unevennesses in the Road: it would well answer to extend the quarries of this stone, and to carry it much greater distances for repairing the public Roads, than at present.

6. *Cank.*—Lists of the places where this hard and useful stone for the Roads has been procured, will be found in Vol. I. pages 229 and 440: it requires, like Crowstone, to be broken small, for covering or repairing Roads.

7. *Gritstone.*—In Vol. I. p. 416, a copious list of Gritstone Quarries has been given: the stone from most of these, when broken into pieces of proper size, may be advantageously used in the foundations of Roads, provided the same is always kept covered

with

with a proper thickness of harder and tougher stones, broken smaller, as Cank, Crowstone, Chert, Limestone, or Quartz Pebbles, &c. for sustaining the immediate crush and wear of the wheels of Carriages and shoes of Horses, &c. But the ordinary gritstones of the district, are entirely unfit for covering or repairing the public Roads, because heavy carriages soon crush them into loose sand, or what is worse, into a loamy mire in wet weather, owing to the copious argillaceous cement, by which the grains of Quartz are united, in the greater part of the gritstones belonging to the Coal-measures.

8. *Burnt Stone.*—In order to remedy in a degree, the evil experienced, from using the argillaceous gritstones above mentioned (and in Vol. I. p. 161), when used on the Roads, great expenses are incurred on the eastern side of this County, in quarrying gritstone, and then stacking it again pretty closely, with layers of small coals between, and firing it: mostly these clamps of stone are made in a corner or end of the Quarry, where they are suffered to burn out, in the course of many days, and to grow cool; but in a few instances, a sort of rude Kiln has been built, for thus burning stone for the Roads. I have noted the following places in or near to the County, where I saw *stone burning,* viz. Barlborough (Marston-moor), Beighton, Beighton-Mill, Yorks., Brecks, Yorks., Brimington, Dronfield, Fackley-Lane, Notts., Handsworth, Yorks., Heath, Hollinwood-common, Intake in Handsworth, Yorks. Killamarsh, Mosborough NW, Pentrich, Staveley, Tupton, Unston, Woodseats in Norton, &c.

It

It is a common complaint, that the stone is not sufficiently burned, that the Labourers purloin the Coals furnished for the purpose, &c.; but the fact is, as observed in the Report on York West Riding, p. 213, (where this practice greatly prevails), that these burnt stones are utterly insufficient for the public Roads, and can endure but a very short time thereon, before they are crushed into mire and sand: and on all the principal Roads, like that between Mansfield and Leeds, &c. the practice ought speedily to be laid aside, in favour of greater exertions to search for, and quarry and bring the Cank stones, which are to be met with in this Coal district, although not very certainly or regularly, as mentioned in my 1st Volume, p. 440, in exploring fully the ranges of the several Crowstone strata, and bringing that valuable material on to the Roads; or if these cannot be got, in procuring, by means of the Canals, hard Limestone, or good cleanly-sifted quartz Gravel, for covering and repairing these important Roads.

8. *Bricks*, or *Burnt Clay*.—In Vol. I. p. 456, and Vol. II. p. 395, I have mentioned some instances of preparing these for the Roads; and according to Mr. Whitehurst's " Inquiry," 1st edit. p. 163, Clay that had been accidentally burned by a coal-seam beneath it, on Heanor Common, was formerly used for the Roads: the refuse of Potteries are used on Newbold Common, in Dunston, Whittington, &c. but all these seem of insufficient hardness for the main Roads, as observed above.

9. *Iron Slag*, or *Furnace Cinders*, as well as smaller quantities

quantities of *Lead Slag*, or *Cupola Cinders*, are used on the Roads, in the vicinities of the several Furnaces and Cupolas of the County, mentioned Vol. I. pages 396 and 382; but the very brittle nature of these slags, in general, cause them to be almost immediately crushed by heavy carriages, to small pieces, and ultimately to sand or dust, making the Roads very black and filthy; and were it not for a property that this crushed slag mostly posssses, of setting or adhering together again, in a slight degree, by the oxidation of the iron, I believe, this material would be, comparatively, of little use: to me it has seemed, that its value was greatly over-rated in many instances. At Somercotes Furnace, I observed the Slags to be accumulated in great quantities, and understood, that even at the very low price per load charged for them, the Surveyors of the adjacent Roads, were little inclined to take them away.

In particular instances, I have seen Roads wastefully formed or mended with Coals (I. 186), and in other instances, of striated *Gypsum* (I. 150), particularly at Coton, near to the River Dove.

Mr. Robert Lowe in the Nottingham Report, p. 135, recommends the using of kid or brush *wood* for the foundations of Roads, and when at Stapleford, on the borders of that County, I heard of the same having been practised, from thence over Bramcote Hill, towards Nottingham: to me, however, it seemed, that the *wood*, a perishable article, had in this case been something worse than thrown away, in a situation where Quartz Gravel in any quantities might be had, both for the foundation and finishing of the Road, if properly sifted, and the smooth stones broken, as has been mentioned already.

Expense

Expense of Roads.—I wish it had been in my power to have presented here, such an account of the several Turnpike Roads in Derbyshire, as Mr. John Bailey has given, of 18 of the Turnpike Roads in the County of Durham (Rep. p. 269), embracing 431 miles of Road, on which the total of tolls collected, is 15,898*l.* per annum, or near 36*l.* 18*s.* per Mile, on the average: and he particularises 13 of these Roads, amounting to 285 miles, whereon, after paying Interest and Salaries, the sum of 9,984*l.* remains applicable to their repairs, annually, or 35*l.* per mile on the average, but varying from 10*l.* to 100*l.* per Mile on the different Roads.

In some instances, in Derbyshire, respectable persons have contracted for doing the repairs on particular Roads, or parts of them: thus in 1808, Mr. John Milnes, of Ashover, undertook about 9 miles of Road in that parish, under the Trustees, for 68*l.* per annum, and the Statute-duty apportioned thereto, being $\frac{1}{4}$, $\frac{1}{6}$, and $\frac{1}{15}$ of the whole duty in different Townships, in which the Roads lay, and for enforcing of which he was appointed Surveyor of these Roads. Mr. George Nuttall, of Matlock, had in like manner undertaken 7¼ miles of Road, under the Chesterfield Trust, for 63*l.* and Mr. William Stubbin, of Chesterfield, 3½ miles of the same Road, for 30*l.* per annum.

The Road across Sheffield-Park (a short distance beyond the bounds of this County) to the Intake Toll-bar, is maintained by the Inhabitants of Sheffield, at the cost of 450*l.* per mile per annum! as I was assured; it is repaired with Ganister or Crowstone, of good quality, but laid on *in such large lumps*, and is in consequence so very rough, that the loaded Coal Carts are so violently jolted thereon, in descending the hill, as soon to crush even these hard stones, 3 or 4 coats of which are worn out in one year, in some places. If all the large lumps on this Road were picked up, and a thickness of 6 or 7 inches of *well broken Crowstone*, laid at once upon it, the same would settle down firmly and smooth, and the stone would thus wear five times as long as at present.

The mode of managing the Repairs of the Turnpike Road from Chesterfield to Tideswell, with its several branches, by Mr. William Gauntley, jun. of Bakewell, seems to me worthy of being recorded:—in a printed Notice, in 1808, this Road was divided into 18 lots, each terminating with the Townships or other known points, and measuring from 1½ to 4 miles each lot; the particular *quarries* were named against each lot, from whence the same was to be repaired, and whether any *statute-duty*, and from what Townships were assigned to it: these notices were circulated for the purpose of *letting the Repairs*, on the following conditions, viz. " The Stone for the Repairs of the Roads in the several Lots, shall be of the best quality, which the respective quarries and places therein mentioned will afford, and shall be got *and broken there:* and afterwards carried and laid on the side of the Roads, in single and equal loads, on such parts as the Surveyors shall direct. The quantity or number of *Tons* of stone so laid on the side of the Road, shall be ascertained by the Surveyor, who is to count the number of loads laid down, and to measure any one of them, with a two or four peck measure, and each load of stone laid down, shall be considered to
contain

contain the same quantity as the load measured. Four pecks or one bushel of broken stone, strickle measure, shall be taken for 1 cwt. or 20 bushels one *ton*. The Surveyor shall also examine every one of the said loads of stones, with a circular *guage* of 2½ inches diameter, and if there shall be found one or more stones which will not pass the said guage, in the load examined, each load of stones laid down, shall be considered to contain the same number of stones not sufficiently broken, and a deduction of one farthing for every stone which will not pass the said guage, shall be made from the charge, for the stone to be accounted for. The contracts to commence from the 25th of March next, and to continue in force until the 25th of December, 1808. Persons desirous of undertaking the repairs of the said Roads, in any one or more of the above Lots, may deliver to Mr. William Gauntley, junior, of Bakewell, in writing, sealed, their proposals of *prices per Ton, for getting, breaking, carrying, and spreading* the stone, on the said Roads, from each quarry separately."

In the 6th column of the Table, facing p. 34 of Vol. II. mention is made, of part of the expenses of the Parish Roads in each of the several Hundreds of the County, being collected with the Poors' Rates: I am unable to state, or even guess, at the amount or proportion so included; the Highway Accounts here, are generally kept separate from the Overseer's accounts, as they are in other Counties.

The following are the Names of Persons whom I noted as having distinguished themselves in setting out and undertaking improved Roads in this County, viz.

Mr. Joseph Butler of Killamarsh, Mr. William Gauntley, sen. (the late) of Bakewell, Mr. John Johnson of Union Lodge, Ashby Wolds, Leic., Mr. Thomas Marriot of Marple, Cheshire, Mr. John Milnes of Ashover, Mr. George Nuttall, late of Matlock, Mr. James Walls of Heanor, Joseph Wilkes, Esq. (the late) of Measham, &c.

Statute-duty.—The intelligent Surveyor of York West Riding, p. 216, and Mr. G. Holt, in " Communications to the Board," Vol. I. p. 186 and 189, and others, have recommended, that the Law in this respect should be entirely altered, in which opinion I heartily join :—at the time that these Laws were passed, the Legislature was far from being well informed on the principles of repairing and maintaining Roads, as I have endeavoured to shew at page 232, and it seems to have been assumed at that period, that *Teams*, and *Men to fill them*, were almost alone necessary in repairing Roads; the *duty* was therefore levied in kind, and the Surveyor is left *without Money*, for digging, sifting, and breaking Materials, and forming and preparing the Road for their reception, and (when brought there by the Teams) for spreading and attending to the same, constantly, or at short intervals, as ought to be the case; and notwithstanding, that the value of *manual labour* necessary for these purposes, and sometimes of purchasing materials, exceeds several times the value of *Team-work* connected therewith, except in particular cases, where the materials are very distant.

The Bill which was discussed in Parliament in 1810, as already mentioned, proposed to give greater discretion to Surveyors, for requiring *compositions* instead of of Team-work, than they have at present, but still the rate or amount of this composition was left liable, as always has been the case, to be fixed greatly below the value of the Team-work, remitted; and most of the evils of this complicated system would also remain. Surveyors who properly contrive their business, have never any difficulty now in hiring carriage, to almost any extent, at fair prices, as seems not to have been the case when the statute for this duty was passed; and at present, there seems no more propriety in authorising the Surveyor to require Teams in kind for his Road, than there would be, in authorising the Overseer to require Corn in kind, for his Work-house, or any other thing necessary for the Poor under his care; but which money, placed at his disposal, can so much better and conveniently purchase.

Parish, or *Township Roads.*—Considering the absurdity of the system, under which *Surveyors* of Parish Roads are generally chosen, or taken by a kind of rotation among persons almost equally unfit for the important duties of this office (see Comm. Board of Agriculture, I. p. 186), the Bye-Roads through a great portion of this County, are well managed, and are generally in a good state; but in the following places, some very deep miry and *bad Roads* were met with in the course of my Survey, viz. Blackwell (SH), Bolsover W, Bradley, Codnor, Dalbury, Dronfield W, Heanor, Ilkeston, Makeney NE, Mugginton S, Newton in Blackwell, Radburne, Shirebrook, Shirland, Stanley, Staveley SSE, Synfin (see Vol. II. p. 350), Tibshelf, Trusley, and West-hallam. In a few places, on the contrary, naked floors of Rock are seen in places of the Roads, as on the W side of Heathcote, and S side of Monyash, &c. : in more numerous cases, shelves of Rocks cross the Lanes, as in Alton and Overton SW, in Ashover, &c. and prove rather serious impediments to the Traveller.

So are the *Bye-sets* or hard ridges and gutters, made obliquely across the descents of hills in great numbers, both on the Parish and Turnpike Roads, for turning the Rain waters off the Roads into the ditches; the suddenness, height, and number of these on some Roads, are an intolerable nuisance; and very often, they are unnecessarily had recourse to, for carrying the Water across the Road, instead of either continuing a ditch along each side of the Road, or conveying the water under the Road, in a culvert or covered drain.

Ploddings, or raised paved foot-paths or causeways across the Roads, in the Villages and Towns, and in some other situations, are very great impediments to travellers, and ought to be removed, unless the Surveyor would pay more than ordinary attention, to often covering the Road with good hard and well-broken materials, on each side of these crossings, where there is always an extra wear, by the carriage wheels descending with a jolt, from this ridge across the Road. *Step-stones* standing up at intervals across some few Roads, liable at times to be flooded, are also dangerous nuisances.

Large stones laying loose on the descents of the hills, which the Carters have used for scoiting or *scotching stones* or checks, to stop the wheels, whilst the horses have rested in the middle of the Road, are far too common throughout the northern part of the County; sometimes the adjoining Fence *Walls* are pulled down by the Carters, to obtain the Stones; but not unfrequently, they

they are taken from the heaps of *large* stones, improperly laying by the road side, which have been provided for its repair. In the Bill of 1810, which I have several times had occasion to refer to, penalties were provided, for suppressing these improper and dangerous practices of the Carters, especially if the Farmer's Walls were destroyed for such purpose.

It might be well also to add further, a Penalty on Road Surveyors, for suffering the *breaking of stones* upon or within 20 or 30 feet of the centre of any public Road : and also, for suffering *heaps of stones*, or materials of any kind, to remain during a night, or more than two or three hours (without being effectually spread, or watched) within 15 or 20 feet of the centre of any Road. In Whittington and several other places, I have seen this nuisance of heaps of stone, and persons breaking them, by violent blows of large hammers, within 6 or 8 feet of the centre of the Road : I have more than once ridden Horses, who having previously suffered blows in their eyes or faces, by fragments of stone dispersed by the stone-breakers, could not be made to pass or approach them, without flying off the Road in the utmost fright and terror; very frequently the men are obliged to desist and leave their work, or they would injure the horses or persons passing, break the Carriage windows, &c. The breaking of stones could be best and cheapest done in the quarry, so that the Carts could afterwards be filled by shovels, or whiskets and coal rakes (see Vol. I. p. 367, and II. 262), as is very common in the northern part of the County, and when shot down by the Roads in reserve, the heaps might be neatly dressed up into cones, by the same means, as intended in Mr. Gauntley's ley's contracts, that have been mentioned, p. 260; a very great wear and destruction of the Carts, by throwing large stones into them by hand, and by the jolting of such in travelling, would also be avoided, by this beneficial change of system.

The very *large lumps* of Milstone Grit, limestone, Cank, Crowstone, &c. that were absurdly laid on the Roads in the past age, remain yet therein in many instances, as dangerous stumbling blocks, even on the Turnpike Roads, that are little used, see page 223, and on some others that are still important ones, as on the ESE side of Curbar, and some others: the old Roads from Duffield, up each side of the Derwent, are notoriously bad in this respect, on the E side in particular: this evil ought to be remedied every where, by picking up these great stones, and filling their places more than full, with small-broken and very hard stones.

In several instances, the Roads on sandy, soft rocky or lamellar strata, or on alluvial mixtures, are by long use and neglect, worn very deep and narrow, so as to present considerable difficulties to Travellers; several of these *hollow Roads* are particularised in a Note on page 221.

Owing to the small worth of *Brush Wood*, in most parts of this County, the Hedges by the Roads are often neglected to be cut, and it is too common for the Farmers, when cutting their hedges, and when falling or priming Trees by the Road sides, to throw out and leave the cuttings in the same, which in hollow and narrow Lanes, I have often found it difficult to pass, and such must in the night be very dangerous. I noted this nuisance in Brampton, Plesley, Wadshelf, Walton, Wessington, &c.

Another

Another less frequent practice, of converting parts of the Roads to *Dung-yards*, has been mentioned, Vol. II. p. 455; which improper practice forms a sort of contrast to another, almost general throughout the County and its environs, that of the Cottager's Children, and Women and old Men in some cases, frequently perambulating certain lengths of the public Roads, that each has assigned to him or herself, and most carefully picking up every piece of *horse-dung* that falls, into whiskets that they carry on their heads, by which practice, filth on the Roads is avoided, and much valuable manure collected, for sale to the Farmers, or for their Gardens; each one has a hole slightly sunk by the Road side, wherein he stores his dung, and very often, they cover it in part or wholly by flat stones, to prevent loss by evaporation or the treading of Cattle, &c. On some of the few Commons that remain, the Shepherds and others collect up the sheep and horse-dung therefrom, and store it for sale, in the same manner, see p. 91 of this Volume.

Thistles, and noxious weeds of different kinds, are too often suffered to infest the sides of the Roads, as observed, Vol. II. p. 193: and not unfrequently, Cattle are turned loose in the Lanes and Roads, and Asses and *Horses*, which last, from their propensity to kick at other Horses passing them, are very dangerous nuisances, as my legs, as well as those of my Horse, have several times been near experiencing.

Although, having pointed out so many defects and neglects, in the system pursued in Derbyshire, with regard to the important article of *Roads*, I would not be understood as thinking worse of it than of other Counties in this respect, but the reverse; because, after paying a good deal of attention to this subject in

most

most parts of England, I think few of the Counties excel Derbyshire as to its Roads, when the circumstances are duly considered. The Turnpike Roads, with a few exceptions, are good, and the Parish Roads seem fast improving.

In nearly all the modern Inclosures mentioned, Vol. II. p. 71, the Roads have been set out and formed on good principles; in several instances the Ditches have been ordered to be made inside the fields, with proper culverts under the hedges (see II. p. 85), for making the Roads equally or more safe to Travellers, with a less width of Land* lost to the community, for this purpose; and on which subject, I ought to mention, with approbation, the spirited conduct of Mr. John Nuttall, as a Commissioner of Inclosures, in resisting the claims of Trustees, or rather perhaps of their Lawyer Clerks, to have the Turnpike Roads left 60 feet wide, as formerly they might claim, instead of 40 feet wide, to which, since the 44 Geo. III. the Commissioners on new Inclosures were empowered to limit them, although few have done so; the latter width being fully sufficient for any Road (but in the immediate vicinity of a large Town), under a good and proper system of management, and with the reverse of these, the 100 yards wide, to which an ancient Statute laid claim for the Public Roads, I believe, would prove insufficient for Pits and spoil, and opportunities for avoiding the sloughs, and other nuisances of the Roads.

Since I can remember, the Legislature generally

---

* The new part of the Turnpike Road between Tupton and Ashover, was formed on these principles by Mr. John Milnes in 1808; so were the Roads on Ashby Wolds in Leicestershire, see II. p. 85.

required

required in Inclosure Acts, that the mere Parish Roads should be left 60 feet wide, although when further and better informed on the subject, they have progressively thought 50, 40, and even 30 feet, fully sufficient for this purpose, and I know several instances of this monstrous *waste of good land*, having been made, throughout large Parishes; in Hulland Ward in this County it is to be seen; the occupation Roads in Over Haddon seemed also very wastefully too wide.

In a future revision of the Road Laws, I think it would be highly proper, to authorise the reducing of all unnecessary widths of roads, with the consent of the Lord of the Manor, or even without it, by an order of two or more Justices, on a tender to the Lord, of a portion, perhaps half, the full value (ascertained by a Jury), of the Lands so to be taken into the adjoining Fields, by the owners thereof: but with the express condition, that the old Fence should be entirely destroyed, and the new Land, laid into the adjoining Field (except a Garden was made, or House built), and that no trees should be planted, or ever suffered to grow, in such new Fences next the Road, or in the old one, if suffered to remain, on any pretence whatever. This would check and prevent the too common and increasing practice, of Lords of the Manor inclosing these strips of waste next the public Roads, and *planting them*, by which means, the Roads, even the most public ones, in numerous instances, throughout England, are becoming choaked and incumbered with Trees, in a shameful degree: indeed a stop ought to be put to these *Road Belts*, as they may properly be called, and which might without

out any injustice be done, I think, because, the Laws invested the soil of the unused parts of the Roads, in the Lords of the Manors, for the public benefit, and not to enable them to commit or occasion nuisances thereon, more than other persons.

In many parts of Derbyshire, the laudable practice prevails, in the vicinity of Gentlemen's Seats, and of large Towns in particular, of *clipping the Hedges*, low and neat, see Vol. II. p. 87, and in some instances of pruning up the Trees therein, to such a length of clear stem, that they cease to incommode the Road in any material degree, see Vol. II. p. 259. Happening to observe in November, 1808, the Hedges of Mr. Daniel Hopkinson in Heath, to be cutting, by means of a scythe-blade fixed straight in a handle (Vol. II. p. 87), I conversed with the Men, and found, that same was doing *by order of the Surveyor* of the Road; I heartily wish that this practice was more general.

Complaints have justly been made, that considerable lengths of the Parish Roads in the southern parts of this County are not fenced off, but pass through, and not unfrequently go across the Fields, and that the *Gates* thereon are very inconvenient to Travellers, as also on the Bridle, and more private ways, leading to the detached Farm Houses, &c. In Vol. II. p. 92, and 93, I have mentioned the pains that some persons take, to render the Gates on Roads complete and convenient; and on the contrary, to the improper *strut* there mentioned, I ought to add, the very common and improper practice in some districts, of rearing large stones against the Gates, on Bridle and private Roads, in order to keep them shut, instead of providing

ing proper fastenings to them. In some parts it is common, for one of the bars to pass through the head of the Gate, and act as a fixed latch, requiring the whole Gate to rise or be lifted, in passing this latch into its catch, or out of it; this is a very inconvenient kind of fastening. To the northward of Buxton, it is common to find a rising catch for receiving the top bar of the Gate, which are convenient enough, but the lower part of the Gate is too much left free, to the action of the wind, or sheep, or pigs pushing against it, by this fastening at top of the gate: and the most simple and best fastening that I saw in the County I think, was a sliding latch, suspended by a short length of chain from the Bar above it, at about two-thirds of the whole height of the Gate.

Farm-ways:—I saw nothing particularly to note, on this head of the Board's inquiries in its printed " Plan:" but it may be proper in this place to mention, that access is had to the Gentlemen's Seats, in several instances, by considerable lengths of *private Roads* well laid out and kept, so as to do credit to the County; the Notes that I made on this subject, are as follows, viz. Barlborough Hall to Pebley Lane, Cornelius H. Rodes, Esq.; Belper (by Alderwasley, through the fine Meadows and Woods, by the Derwent), to Cromford-Bridge, Messrs. G. B. Strutt, C. Hurt, and R. Arkwright, Esqrs. see page 226; Buxton Baths, a circle of private Ride for the Company, through Mill and Sherbrook Dales, and on the W and NW of the Town, also down the romantic Vale of Wye to Ashford, the Duke of Devonshire, see page 227; Calke Hall to Ticknall, Sir Henry Crewe, Bart.; Chatsworth

worth House, across the East-Moor (with clumps of plantation adjoining it), and through Wingerworth to Hardwick House, formerly, but now in great part disused, except between Heath and Hardwick House; also from Chatsworth House to the Turnpike Road E of Baslow; and from Edensor Inn to Chatsworth lower Bridge, the Duke of Devonshire, see p. 227; Hopton Hall to Bonsal Dale, a beautiful Road called "*via Gellia*," Philip Gell, Esq. see p. 246; Locko Hall to Chaddesden, William D. Lowe, Esq.; Overton Hall to Slack, Sir Joseph Banks, Bart.; Radburne Hall to Mackworth, the late Sacheverel C. Pole, Esq.; Shipley to Heanor, Edward M. Mundy, Esq., see Vol. I. p. 456; Sutton Hall to Temple-Normanton, Clement Kinnersley, Esq.; &c.

Foot-paths.—A very commendable spirit prevails, throughout most of the Coal and Shale districts of the County, for providing very solidly paved Paths by the sides of the Road, from two to three feet wide, that are used by persons on foot and on horseback: these Paths are mostly laid with thick flat paviers of Milstone Grit, which is excellently adapted to this purpose, for wear, and as not apt to occasion the feet of Horses and Men to slip, as Limestone is dangerously liable to do: but owing to the constant use of these Paths by Horses, where in many cases the same is quite unnecessary, these Paths are too commonly worn quite hollow in the middle, holding puddles in wet weather, and being extremely unpleasant to walk on, the feet constantly tilting towards each other, unless by straddling, they are placed on the edges; these edges are in some places frequently dressed down,

and

and the path made flat again, by the rough tool of the Mason.

In the vicinity of Ashburne, Derby, and a few other places, wide and convenient gravelled Paths are provided adjoining the public Roads, but separated therefrom by neat white painted Rails, which are extremely pleasant and useful to the Inhabitants, especially if the adjoining Hedges are clipt and kept low, as near Ashburne. Between Belper and Milford, the Foot-paths are very good, and are protected from the Road by very stout stone posts, that are constantly kept whitened, for rendering them more conspicuous, particularly in the night.

There seems to have been rather important omissions in the Inclosure Acts, in several parts of Derbyshire, in not empowering and requiring the Commissioners to revise, alter, and lessen the number of Foot-paths across the Lands to be inclosed, and others adjoining, as much as was practicable or proper; for want of which, in many places, a large portion of the Fields have Paths diagonally across them, and many Fields have several such, to the great injury and damage of the Farmer.

While making a particular Mineral Survey for Sir Joseph Banks, of the Parish of Ashover, and its environs (see Vol. II. p. ix. and Phil. Mag. Vol. 42, p. 55, note), I laid down all the *Foot-paths* in my large Map, from which the nature and great extent of the evil I am complaining of, very strikingly appears: and in particular I noticed, in November, 1812, that by the side of the Foot-path, across a Field of valuable large Cabbages, belonging to Mr. Roger Wall, at North-edge (II. 141), some idle villain, had with his knife,

knife, deeply slit, and crossed the tops of a large portion of all the best Cabbages growing near this Path! In the Road Bill of 1810, to which I have often before adverted, a Clause was inserted, with a view to provide for the stopping up of useless Foot-paths in Fields near to Roads, but the same appears to me, not to go far enough, to meet the important evil that I have been complaining of.

Pack-Horses, were in general use in the northern parts of this County, until the last age, see Vol. I. p. 380, and some still continue to traverse its most northern parts, each muzzled, to prevent their stopping to graze by the Road sides, as I saw in Hathersage, and some other places N of Buxton: a good many *Asses* are used to carry burthens, as observed p. 161 herein, but they are seldom used to draw Carriages of any kind, except in and about the Coal Pits.

Waggons and *Carts* have been treated on in Vol. II. p. 58 and 59. In Vol. I. p. 380, some particulars of the *prices of Carriage* of Ore are given; to which it may be proper to add, that about Chesterfield in 1808, the hire of a one-horse Cart and Man, per day, was 5s.; of a two-horse Cart and Man, 8s.; of a three-horse Cart and Man, 10s. 6d.; and of a four-horse Cart and Man, 12s. per day: it being understood, pretty generally, that not more than two-thirds of the cartage would be performed by such *day-hired* Carts, as the owners thereof would perform therewith, if the same work was *let* to them, by bargain or job.

The carriage of Coals for the supply of Stanage Cupola, was at that period performed by a yearly contract, at 4s. 9d. per ton; the pits being $3\frac{1}{4}$ miles distant, the draught nearly all up hill, and a Toll of of 4¼d. per horse in summer, and 6¼d. in winter, each time of passing, laden or unladen, was payable by the Carter.

By the Acts of 3d Will. III. and 21st Geo. II. the Justices of this County have been used, at their Easter Sessions yearly, to assess and order, the *rates of Carriage* to be taken by common Carriers, to and from Derby and other Towns; in 1808, their orders from Lady-day to Michaelmas, were as follows, viz. Derby or Ashburne to or from London, 6s. per cwt.; Bakewell, Chesterfield, or Wirksworth, to or from London, 6s. 6d.; Tideswell, to or from London, 7s.; Buxton or Chapel-en-le-Frith, to or from London, 7s. 6d.; Derby, to or from Northampton, 3s.; and Derby to or from Leicester, 1s. 6d. From Michaelmas to Lady-day as follows, viz. Derby or Ashburne, to or from London, 7s. 6d.; Bakewell, Chesterfield, Wirksworth or Tideswell, to or from London, 8s.; Buxton or Chapel-en-le-Frith, to or from London, 8s. 6d.; and Derby, to or from Northampton and Leicester, 3s. and 1s. 6d. as above.

The "Plan" for this Report, reserves until this part of the present Section, the consideration of the different *forms of Roads*, which otherwise I should earlier have mentioned; these are as follows, viz.

Concave Roads.—The late Joseph Wilkes, Esq. of Measham, was a strenuous advocate for making Roads lowest in the middle, or concave, and in "Communications to the Board," Vol. I. he states, p. 200, that about 1763, such a hollow Road was made from Burton to Measham, and another from Burton to Derby: that others such Roads, from Sawley Ferry through Ashby-de-la-Zouch to Tamworth, from Bosworth to Measham, from Hinckley to Measham, from Atherstone to Measham, from Hinckley, through Melborne, to Derby, and part of the Road from Ashby, through Burton, to Tutbury, all which were originally convex, had, principally through his own exertions, I believe, been altered to concave ones: but on the contrary, the concave road at first made from Cavendish Bridge to Derby, had since been altered : at page 133, it is also said, that such Roads were in use near Dishley in Leicestershire, on the recommendation of the late Mr. Robert Bakewell.

In the parish of Measham, I observed, that these Concave Roads remained yet in use; but I found them in a very indifferent state, and as illustrating the absurdity of the principle on which they were formed and maintained, rather than any thing else: and such must I conceive have been the experience, on most or all of the other parts of the Roads above mentioned, wherein the hollow in the middle, transversely considered, has been filled up, although in many instances the waves or inequalities in length, still continue, in places, as will be further mentioned below. In Mr. William Pitt's Leicestershire Report, p. 309, a cut and description is given of the Roads formed on this plan on the Inclosure of Ashby Wolds. In April 1798, four small pages were printed, for recommending these Concave Roads for general adoption, which the President of the Board put into my hands; but I have been unable to discover a sufficiently valid reason that these pages contain, in favour of such a system.

Convex Roads: this form, so universally approved and acted on, almost throughout Britain, is undoubtedly the best one that a Road can have, if confined to 1 *inch rise in a yard* of breadth on each side; with

such

such a moderate degree of convexity, it has never appeared to me, in practice, that abutments for the materials, or sides of earth, of any form, are wanting to keep the materials together, on which so much has been said and written, and done, in numerous places: *the whole width* of the Road, from ditch to ditch (unless these are very improperly too distant) ought in every case to be *formed* to the above degree of convexity; the earth made as uniformly solid as is practicable, and a sufficient thickness of proper materials laid on at once, the largest and flattest stones at the bottom, but none such near the top, as I have mentioned in the Letter refefred to in page 250, and having a width proportionate to the traffic and importance of the Road. At High-moor, on the E side of Killamarsh, 1 observed a very commendable piece of industry, by Mr. John Hancock, in levelling to the proper slope, and covering with top-soil, a very uneven and useless piece of the unnecessary width of the Turnpike Road, before and on each sides of his house, by its side.

Flat Roads.—There have not been wanting, among very sensible and able Men, strenuous advocates, exclusively, form almost every possible form and manner of making and managing Roads: and the reasons appear to me to have been, that each one of such having himself produced good Roads, or seen them result from the practice of others around him, and though it might have been owing to local peculiarities perhaps, of materials, situation, traffic, or care and attention of the parties, more than to any thing in the *system* pursued, that such good Roads have resulted, yet such have too hastily concluded, that the *form* or the *system* pursued, &c. alone contributed to the effect

fect produced, and without that extended research and depth of thought, which is necessary to be exercised on so very complicated and difficult a subject, as the general principles and practice of Road-making embrace, they have become the unqualified advocate of the particular system adopted.

In this County, several intelligent Men, the late Mr. William Gauntley, sen. of Bakewell, and Mr. John Milnes of Ashover, and others, had adopted the idea, and maintained to me, that *perfect flatness*, in the breadth of a Road, was primarily, and almost alone sufficient, to ensure its being good!

Mr. Gauntley, in order to ensure the object of levelness across the Road, contrived what he called a *string level*, which consisted of a piece of box wood 11 inches long, 1¼ broad, and 1¼ deep, into the top of which a spirit-level tube was deeply sunk, and to the top, at each end of this level, several yards of strong whipcord was fastened. In using this Instrument, a Labourer was placed on each side of the Road, having the cord in his hand, which they pulled very tight and steadily against each other, and thereby made the bubble assume the middle of the tube, or either end, according as the two ends of the string were held level, or one higher than the other.

Mr. G. described to me some of his principles applicable to a Turnpike Road, as follows, viz. take out the soil 6 yards wide and 1 foot deep, along the middle of the line, and fill the space level full, with hard stone, broken so that it will pass through a three-inch Ring; and for repairing such afterwards, in flat level courses, the stones should be broken to pass a 2½-inch Ring: proper bye-sets must be made: if the road have an inclination lengthways of ½ or

1 inch

1 inch in a progressive yard, it will wear longer on that account, but no hill in a Road ought to remain with a greater ascent than 2 inches in a yard.

Mr. John Milnes, in constructing the new part of the Turnpike Road between Stubbing-edge and Briton-wood Nook in Ashover (see p. 242), formed a trench 15 feet wide and 1 deep, but not in the middle of the intended Road, 4 feet wide of level earth remaining, or being formed on one side, and 11 feet wide on the other, extending to the small banks on which the Hedges were planted (intended to be kept low) and the Ditches were made beyond this, next the fields: the lower nine inches of this trench was filled with Gritstone of the nearest Quarries, broken to pass through a $3\frac{1}{4}$-inch Ring, on which 5 inches thick of good Limestone or Crowstone, broken to pass through a $2\frac{1}{4}$ Ring, was laid, thus allowing 2 inches for the settling and crushing down of the materials before the whole width became flat. The side of 4 feet wide, was intended for a foot-path, and that of 11 feet for a summer Road, and on to which sides, it was intended evenly to spread the future shovelings of the stoned part, for gradually raising and improving them.

Waving or undulating Roads: It has already been mentioned, in speaking of Concave Roads, that the adoption of that form in level situations, often required the Road to be formed into waves or undulations, as to their lengths, for collecting the Rain-waters running along the middle of the Road, to furrows or low places across it, at from 50 to 100 or more yards distant from each other, where it might be conveyed off the Road, but where too often, through neglect, it stood in pools of water and mire, to be forded through by the Traveller; I have also mentioned, that in many places these have since had the hollows in the middle filled up, and are now flat and wavy Roads, and some such were formed originally, I believe, without being concave, while the folly prevailed, of thus expending the funds of the Roads.

I noted the remains of these wavy Roads, as being visible yet, S of Ashburne, in the disused part of the Road, W of Barton—turning near Burton on Trent, Staff., S of Derby, S of Rocester, Staff., E of Sawley, W of Spondon, &c. Near Rocester, this wavy part of the Road is very narrow and disagreeable to travel; near Spondon the waves are sudden, and the numerous furrows between them, give continual shocks to Carriages in passing; and on the whole I must say, that I saw no reason while on this Survey, to recommend this, or any other of the deviations from the usual form, of slightly convex and straight Roads.

Application of Water.—Mr. Ellis appears as an advocate for washing or irrigating the Public Roads, in the " Communications to the Board of Agriculture," Vol. 1. p. 207; and so does Mr. Jessop, p. 182 of the same Volume; and p. 160 of the 3d Report on Roads, printed by order of the House of Commons, 19th June, 1809: I have more than once been referred to the Road between Ripley and Little Eaton, as an example of a Road managed on this plan, and whereon the benefit of watering might be seen: candour requires me, however, to state, that great as the merit of this line of Road is, for avoiding unnecessary ascents, I saw nothing to commend in the state in which it was kept, while I was on my Survey: very early in two succeeding Winters, before frosts of the least consequence had occurred, I travelled this Road

by

by the Mail Coach, at which times it was miserably deep, loose and bad for several considerable distances together, so much so, that the Coachmen and Guards were continually swearing on account of its state, and threatening to report it at the Post-office: having at other times noticed this Road to be in, far from a good state, owing to the softness and insufficiency of the materials used in its repair, I could not avoid remarking, and will mention here, that on the hills to the east of, and near to the worst parts of this Road, near to the NE end of Horsley, and E of Kilburne, there are Gravel patches (Vol. I. p. 138), whence plenty of excellent quartz Gravel might be obtained for this Road, if adequate pains were taken to dig and clean sift the same, and to break all the larger smooth pebbles into 3 or 4 pieces with Hammers: it seems to me probable also, that on the range of hill which is situated something farther off to the westward, the bassets or rakes of Crowstone might be discovered or traced, and wrought, which were formerly dug at Bargate and Openwood-gate, and of which I saw numerous blocks in Killis Farm, see Vol. I. p. 180.

With the judicious use of *hard* and proper materials, such as are here recommended, I never saw the least use, or the desire by any one, for "applying Water" in their aid; at the same time, that all the various attempts, some of them rather expensive ones, which I have seen in various parts of England, for rendering improper or dirty materials effective, by such aid, have uniformly and utterly failed, and ere long been discontinued.

Connected with the subject of Roads, two other subjects remain to be mentioned, in addition to what is said on Bridges, Vol. II. p. 22, viz.

Fords:

Fords: Of these I noted as follows, viz. Alvaston NE, 2 formerly (across the Derwent), Ambaston (Derwent), Barrow S (Trent), Bredsall W (Derwent), Duffield NE, formerly (Derwent), Eaton W, in Doveridge (Dove), Ingleby, formerly (Trent), King's Mills SW, in Weston (Trent), Little Wilne (Derwent), Marston S (Dove), Newton-Solney (Trent and Dove), Ratcliff, Notts, (Soar), Rocester E, Staff. (Dove), Sudbury SW (Dove), Willington SE (Trent), Winchill NW (Trent), &c. A great part of these Fords were very dangerous to use in time of floods, and whereby many lives have been lost: they are now much less used than formerly, and that seldom, but by the persons of their immediate neighbourhoods.

Ferrys: In or near to this County, there are Boats kept for conveying Passengers, and Horses, and Carriages across the Trent River, viz. Barton, Notts, Drakelow, Thrumpton, Notts, Twyford, Walton, Weston-cliff, Wilden (formerly), and Willington.

For managing the Ferry above the Town of Willington, a strong chain is stretched across the River, by a block of pulleys: on the Boat a strong frame is erected, which carries a roller that acts always on the Chain, and prevents the Boat being borne down by the stream: the Boat has a square stage with wheels in its front, for the conveniency of getting on board, and is decked over within 1 inch of the gunnel, for carrying Carriages, Horses, and Cattle thereon. I was informed, that John Pearsall, Esq. is the owner of this Ferry, and is entitled to take 1*d.* for a Foot Passenger, 2*d.* for a Horse, 1*s.* for a Gig, 6*d.* for a one-horse Cart, 1*s.* for a Cart and Horses, and 2*s.* 6*d.* for a four-wheeled Carriage. Twyford Ferry Boat, is almost

almost similarly constructed and managed, and the Tolls nearly the same as above, but is less used. It belongs to Sir Henry Crewe, Bart.

SECT. II.—IRON RAIL-WAYS.

WITHIN the County of Derby, or near it, I believe, there is not any Public Rail-way under Act of Parliament, separate and distinct from the Canals, as there are in several parts of England; but with the exception of a few inconsiderable and private ones, constructed for the accommodation of particular Coal Works, &c. on the lands of their Owners and others, by private agreements, all the other Rail-ways of this district are appendages or *branches to the Canals*, were constructed under the powers of their Acts, and continue subject to the regulations therein contained; it would not be advisable, therefore, for me to give any details respecting the several Rail-ways of the district, in this Section; but I shall reserve the same for one general alphabetical Account of the Canals and their branches and appendages, that fall within the district of Map facing page 193, to be given in the next Section; including also therein, the accounts of the few Rail-ways that I have noticed, that do not connect with the Canals or River Navigations.

When I undertook in the year 1805, to write the article *Canal*, for Dr. Rees' "New Cyclopædia," Vol. VI. Part I. (of which an abridgment occupies nearly one half), I first perceived, that the manner in which our Canals and Rail-ways are blended together, precluded an entire separation of the accounts of them, as above men-

mentioned, and accordingly I have treated pretty fully in that article, on the principles of surveying, setting out, and executing Rail-ways as well as Canals, with all the necessary Inclined-planes, Deep-cuttings, Embankments, Road-arches, &c. so many of which are common to both of these branches of *the system of improved communication ;* and to these I annexed accounts, arranged alphabetically on one uniform plan, of about 120 Navigation and Rail-way concerns in the United Kingdoms, great part of them under Acts of Parliament, the dates of which Acts in most instances are enumerated, and to which article I beg to refer the Reader: yet, as in the interval since the above-mentioned accounts of these important Establishments were written, for the work above quoted, several of the Canals, &c. then in hand, have been completed, and others have been projected, which it will be desirable to record, I shall in the next Section, extract what concerns these Establishments in and near Derbyshire, supplying the corrections and additions that I am now enabled to make, in consequence of the laborious research that I have made in the district, preparatory to this Report; and I beg to submit these accounts as specimens, in part, of the revisals, corrections, and enlargements, that at some future period, I hope to make (or that my Sons will, in consequence of the reservation for this purpose that I made with the Cyclopædia Proprietors), of the complete List of the British Engineery Works of these important descriptions, to be given to the Public in a separate Work; towards accomplishing which, I have not omitted, and mean to omit no opportunities of collecting authentic information, either in Notes preserved, of my inspections of the works of

this

this Nature carrying on or completed, in all the various districts where my Mineral researches and employments lead me, or by Maps, Sections, Reports, &c. both printed and manuscript, with which the kindness of my friends frequently furnish me, as well as with particulars and hints on the theory, practical construction, and uses or misapplications of these various Works, and the great improvements of which they seem yet susceptible.

Mr. John Bailey, in the Durham Report, p. 380, points out an evil existing in that and the adjoining County, enormously affecting the Inhabitants of London, and almost all the south-eastern parts of the Kingdom, which are dependant on these Counties for their Coals, viz. the want of *Public Canals or Railways*, such as are common in almost every other Mineral district, which the owners of any Estates or Mines may freely use, and make rail-way branches to connect therewith, within certain limited distances, on paying moderate and fixed Tolls or *Tonnage-rates* per mile, to the Company who were empowered by an Act of Parliament, to take the lands necessary for making or laying such Canal or Rail-way, and pay for the purchase of same in fee, in case of dispute, only such sums as a Jury should assess them at, and as is also the case, quite generally, with respect to new Turnpike Roads; for want of this very equitable and proper plan, on a matter of so much importance to the public, as the carriage of Coals, to the places of shipment or consumption, the most enormous exactions are said to be made, in many instances, in the districts of the *Tyne* and the *Wear*, under the name of " way-leaves," for which, annual Rents are demanded, by the owners of the Estates, laying near to

the

the Rivers or Shipping Staiths, and by others again behind these, over a district several miles wide, across which, the greater part of the Coals now brought to market, have to be carried on Rail-ways, made by the owners of such Mines, at great expenses, and who yet have annually to pay for these *way-leaves;* often exceeding in amount the profits that are drawn by them and their Landlords, from the works now going!

I have even heard, that there are Land-owners whose Coals have long ago been worked out, who now draw larger incomes from the workers of the Pits beyond them, or rather out of the pockets of a part of the British Public, in a partial and improper Tax, than they could realize while their own Coals were in full work, and with even less damage or deduction from their Agricultural Rents or profits, than was then sustained; and in some instances, probably, Coals are now reserved from the Market under such Estates, nearer to the shipping than the present Pits.

To a useful little work printed in 1807, called the " Picture of Newcastle," a Map is affixed, which shews the many and great lengths of Rail-ways for Coals, which are subject to these exactions, at the wills of the Land-owners, except in such parts as pass thro' the same properties as the respective Coal-mines, but which are too inconsiderable a part of the whole, to preclude the conclusion, that similar advantages for way-leaves are not there obtained, either in the Rent from the Coal-workers, or in the prices that such Land-owners sell their Coals at the Staiths; but the Reader must not expect to find in the local History above referred to, even the mention of " way-leaves" (p. 178), or of some other very profitable, unusual, and concerted proceedings by the owners, workers, and shippers of

Coals

Coals on these Rivers, which Mr. Robert Eddington, in his "Essay on the Coal Trade," printed in 1803, and of which a very enlarged edition has lately been published, has explained to the Public.

In Derbyshire, the various lines of Canals, through most of the Coal districts (shewn by red lines in the Map facing page 193), from most of which, there are *powers provided in their several Acts to make Branches not exceeding certain lengths*, preclude the necessity of way-leaves, with high rents, or prohibitory oppositions from other Proprietors, to the conveying of Coals or other Minerals in the way to their market, in any material instances that I heard of: and even in the *Cromford* (and I believe in the *Erewash* Acts also), *the lengths of Rail-way branches that may be made from the line of the Canal, is unlimited*, or the *distance* is not defined, as often most improperly has been done, beyond which the powers of the Act shall be useless, for its most important intended purpose, that of facilitating the opening, and cheaply working of Mines of Coal and other useful Minerals:—on inquiry, I did not hear of a single evil that had resulted from the wider latitude thus given for Rail-way branches, but saw several instances of its very beneficial operation, so that I think similar principles may safely be adopted, in framing the Acts that are so very much wanted for the vicinity of Newcastle, as Mr. Bailey has stated, and in all other situations, except where previously existing Canal or Rail-way concerns should need protection, against unnecessary interference with their proper districts; the powers of making branches ought, by a general or special Act to be extended, as now might safely be done, when the wants and circumstances of the country are so much better

under-

understood, than they were at the time of passing many of the Canal and Rail-way Acts. In the *Chesterfield* Act, for instance, no branches are provided for, but common Roads may be made to the Canal, with consent of the Commissioners, not exceeding one mile long; but these must be *public Roads*, and *no* *Tolls* taken thereon.

Some few of the earlier Rail-ways laid in or near this County, were of wood; there were such *wooden Rail-ways* formerly to Greasley Colliery, Notts, Measham, Pinxton lower pit (which alone remained in use, at the time of my Survey), and Shipley, from the old wharf above Newmanleys Mill, disused and removed in 1796.

In the use of these wooden Rail-ways, the flanch or projecting rib for keeping the Waggon on the Rail-way, was on the wheel: but now, the flanches of iron Rail-ways are almost universally cast on the bars, and the wheels are plain, by which they are fitted for being occasionally drawn off the Rails on common Roads: I have heard it said, that the earliest use of these flanched Rails above ground (for they were first introduced in the underground Gates of Mines, it is said) was on the S of Wingerworth Furnace, leading to the Ironstone Pits, by Mr. Joseph Butler, about the year 1788.

I observed, however, three instances in the district, of flanched wheels being used on Iron Rails; viz. on a Rail-way branch of the *Ashby-de-la-Zouch* Canal, from Ilot Wharf in Measham, constructed about 1799, wherein pulley wheels ran on metal ribs, cast on the bars; another was on a separate Rail-way near *Congleton*, that will be mentioned in the next Section, whereon the bars were oval or egg-shaped, according

to

to Mr. Benjamin Wyatt's plan, see " Repertory," N. S. Vol. III. 285, and XIX. 15, and the other on the *Leicester* Navigation, Charnwood Forest branch.

There are, in the district I am speaking of, two instances of very considerable *inclined planes* for Railway waggons, viz. on the E side of Chapel-en-le-Frith, belonging to the *Peak Forest* Canal, and on the N and NW sides of Whiston in Staffordshire, belonging to the Caldon branch of the *Trent and Mersey* Canal, and in other situations; see the next Section.

In conversing with Mr. Michael Walker, Engineer and Coal-master of Eastwood (now of Falcon Lodge in Sutton Coldfield, Staff.) in 1808, on the *cost* of laying Rail-ways in that neighbourhood, he mentioned, that stone Sleepers, 2 in a yard, would cost 6d. each; 2 Rail (70lb.) 7s. 6d.; forming the ground 14d. per yard run; laying down the Rails and covering the Road 10d. per yard run, using only refuse sleck *Coals*, instead of rubble stone, in covering such Roads (see Vol. I. p. 186), and that such Roads have usually cost from 8s. to 15s. per yard run; on the average about 10s. See the estimate for laying the Rail-ways belonging to the *Ashby-de-la-Zouch* Canal, in p. 301.

In the second Volume of " Communications to the Board," the late Joseph Wilkes, Esq. has mentioned some experiments, in part made on the Rail-way branch of the *Nottingham* Canal to his Old Brinsley Colliery, for ascertaining the quantity of work done by a horse daily, on different Rail-ways; see also the article *Canal*, referred to in p. 283. On the first Wingerworth Rail-way, mentioned p. 288, Mr. Butler found two Asses fully able to do the work, that had previously

viously occupied four horses constantly, with Carts, besides an extra 3-horse Cart, hired one day weekly, to assist them.

In a great many of the Coal Pits mentioned, Vol. I. p. 188, Iron Rail-ways are laid along the counter-head, or Working-gate (I. 342), for conveying the Trams or Corves of Coals to the bottom of the drawing Shaft; but in Alton and Simondley Collieries, the Springs of Water (I. 502 and 505), proved so corrosive and destructive to the iron, that the Rail-ways were soon destroyed. At Thatch-marsh Colliery (I. 212 and 330) and at Ecton Copper-Mine, Staff. (I. 258 and 329), there are Tunnels of considerable lengths, driven into the hills, in which Rail-ways are used, for bringing out the Coals and Ore. The Thatch-marsh Tunnel has lately been driven by the Duke of Devonshire, for the better supply of Buxton with Coals, under the superintendance of Mr. George Dickens, his Colliery Agent.

SECT. III.—CANALS.

To enter in this place into general arguments on the utility or advantages of Canals, and their appendant Rail-ways, &c. either to the districts through which they pass, or to the Nation at large, seems perfectly unnecessary; all the portentous evils, and the forebodings of certain writers, and declaimers in the last age, of mischiefs to arise from the spread of Canals over the Country, have proved utterly groundless, and on a review of the actual results, from the multiplied and great undertakings of this kind, which do so much honour to our age and country, nothing appears
therein

therein to regret, but the want of more adequate profits or remuneration to the Individuals, who have so laudably embarked their Monies, in constructing many of these works.

In conversing with many able Commercial Men on this subject, I have met with several, who have argued, from the very low rates of interest or profits that particular lines of Canals and Rail-ways make to their Proprietors, and the average of all of them falling, perhaps, considerably below the usual profits, or interest on capital employed in commercial or manufacturing Speculations, that therefore this unproductive part of the national capital, as they called it, was thrown away, and respecting a further part of it also, doubts might be entertained, whether it was beneficially applied for the Public, because it proves to be in a small degree so to the parties, who invested this Capital.

In judging, however, of the utility of Canals, &c. it appears to me very improper, to refer the whole to a question of private interest to the parties: some few concerns of this kind have undoubtedly been entered on, and the number of them multiplied and extended, rather beyond the present necessities of the districts where they are situated, and which, consequently, in a calm and judicious view of the subject, offered no good prospects of return to the adventurers; but in a much larger proportion of the unproductive concerns, the necessity and utility of them were apparent, and the prospects of the adventurers were good, had not the opposing interests of the Park, and Mill, and Landowners, and of previously existing Navigations, Railways, Roads, &c. interposed in the Legislature, such difficulties, and increased expenses of execution, and in other instances such diminished Tolls, exemptions therefrom,

therefrom, or compensations to other concerns, &c. &c. as destroyed these prospects.

In many instances, also, the Act was obtained and began to be executed, with the best grounded expectations of success, but which have been frustrated by the want of knowledge, skill, and of integrity also in some cases, it is to be feared, of the parties who were entrusted to direct and execute the Works, and not unfrequently, in such cases, and in others, where the estimates of expense were made far too low, the money has been expended, and the Works have stopped, before the whole line was completed, or so much of its parts, as were essential to the expected Trade thereon: in this last case, the arguments above-mentioned must be admitted to apply; but in every other case, that is, wherever Canals or Rail-ways have been executed and remain *in use*, no doubt can be entertained, that the immediate district and the Public are not benefited thereby, however small the dividends or interest paid to the Share-holders may prove. And in several instances, it is even apparent, that the lowness of the Tolls that can be legally taken, and their amount so little exceeding the unavoidable expenses of management, and interest on Mortgages, perhaps, is the cause of the misfortunes of many of the original adventurers, while others among them, who are also Traders or Carriers on the Canal, &c. (as is very common) are in reality, owing to the lowness of the Tolls, reaping greater advantages from such concerns, than are usually made on others, that are apparently much more prosperous.

Mr. Joseph Butler of Killamarsh, remarked to me, that considerable benefit would accrue to the Inhabitants of many districts, that are supplied with Coals

by

by Canals from the Pits, as well as cause an increased Sale to Coal-masters, and of tonnage dues to the Canal Companies, if the latter would permit Coals to be carried during the Summer or Autumn seasons, and stacked in proper yards and sheds, in reserve, at the considerable Towns, Wharfs, and other chief places of consumption, and take *Bonds* for the Tonnage thereon, until they are disposed of, in the ensuing Winter, when frost may interrupt the Canal; by which means also, the distant fetching of Coals in Carts and Waggons during bad weather, to the destruction of the Roads, would in many instances be prevented.

In cases where Canals or Rail-ways have been made, almost exclusively for the conveyance of Coals from any particular district, where deep levels or soughs (Vol. I. p. 828) cannot be driven, so as to drain the several Collieries cheaply, and independently of each other, but Steam-engines must be resorted to, for pumping the water; it has sometimes happened, as on the *Leicester* Navigation, that will be mentioned further on, that disputes arise between the Coal-owners or their Lessees, as to the drawing of water, which occasion the shutting up of a considerable part, or the whole of the Coal-works, by which the Canal or Rail-way concern may be greatly injured, or wholly ruined. In order to guard against the pernicious effects of such disputes, it will be well for the Canal Company and the Country, and for the Coal-masters themselves, in many instances, if, as in the   Canal Act, 38 Geo. III. the powers of the Act, extended to the settling of disputes about pump-water, owing to the Engine of one party clearing the works of adjoining parties, in part or wholly from Water, by means of a Jury, to be impannelled to view, hear evidence, and determine the compensation or rate of the Engine expenses, that shall be borne by those whose water may be drawn as above-mentioned, so that the works may not stop, owing to such disputes.

In the Map facing page 193, all the several lines of Canals and Navigations are shewn by Red Lines, the Rail-ways by red dotted Lines, and the schemes or projects for such works, and the disused River Navigations are distinguished by blue dotted Lines; where these several lines pass beyond or off the border of the Map, Letters, as A, D, I, L, &c. are placed, referring to concise descriptions in pages 212 to 216, of the connections they form with the other Inland Navigations, &c. of the Kingdom.

I will proceed now to give, as was proposed at page 284, an alphabetical account of all the separate Canal, Navigation, and Rail-way concerns, that fall within the square of Map, preserving as much as possible the same method of arrangement, as in my general account of *Canals*, &c. that has been referred to as above: and always beginning the description, at the lowest end of the line of a Canal or Rail-way.

ADELPHI *Canal*:—Is a small private one, constructed about the year 1799, as an appendage to Mr. Ebenezer Smith's new Iron Furnace, on the west side of Long Duckmanton Village; its general direction is nearly N, by a bending course of about half a mile in length, to near Inkersall; it is rather considerably elevated: its objects are to convey goods, in small Boats, from the works to the Road from Duckmanton to Stavely, in their way to the *Chesterfield* Canal near the latter place, at such times as the intermediate and circuitous Road from the N end of this Canal to the Furnace, is in a bad state, which not unfrequently happens; this Canal is intended also to act as a Reservoir of condensing

densing Water for the several steam-engines employed at the Works, and it is almost entirely supplied by the water lifted from the Coal-mines: the Boats used on it carry only about 1¼ tons of goods.

ANKERBOLD *and* LINGS *Rail-way*:—A private one, constructed by Mr. Joseph Butler, as an appendage to his Lings Colliery, on the N side of North Winfield Town: its general direction is nearly E, by a crooked course of near 1¼ mile, its eastern end being rather considerably elevated: its object is to bring down Coke, burnt at Lings Colliery, on its way to Wingerworth Iron Furnace, and Killamarsh Forge, as will be further mentioned below: it commences at a Crane, in the Road by the small village of Ankerbold, and terminates at the several Pits and Coke-hearths at Lings Colliery, see Vol. I. p. 203.

The bars of this Rail-way are 4 feet long, and weigh 32 lb. laid at 20 inches apart, nailed down to wooden bearers, across the Road: the bodies of the Trams are large Boxes, made to hold a Ton of Cokes, and are contrived to lift off the carriages and low wheels of the Tram, by means of the Crane already mentioned, which the Men work, by means of a Winch-handle: while the body full of Cokes remains suspended from this Crane, a pair of wheels, axle and shafts of the common height and construction for Roads, and drawn by one horse, is backed under the Crane, and the body is lowered down, and becomes fixed on the axle, by means of steady pins, and catches thereon; and by this means the Cokes are part of them conveyed on the Roads, to the Bridge-loft-door of Wingerworth Furnace, for its supply, and the remainder of them are conveyed to the Wharf at the head of the *Chesterfield Canal*, N of that Town, where a similar Crane hoists the body off the Carriage-wheels, and lowers it down into a Barge, which, when filled with these large Boxes of Cokes, proceeds down the Canal to Killamarsh Wharf, and there another Crane hoists the bodies and their contained Coke out of the Barge, and successively places them on similar large wheels, drawn by one horse, which conveys them to Mr. Butler's Iron Forge (see I. p. 404) at a small distance off, and returns with the empty bodies to the Crane, by which they are hoisted off and placed again in the Barge, and when all of them have been so emptied, the Barge returns with them to Chesterfield Wharf, where these empty bodies are again hoisted, one by one, and placed on the wheels, from whence a loaded one has just been taken, and which is let down into the Barge in its place, and so on, until the Barge is again loaded, as before, in places of the empty Boxes, which go back in succession to the Ankerbold Crane, and are then changed for full ones, as already described. By means of these same Bodies on the Road Wheels, the Charcoal, burnt in the Wingerworth Woods for Mr. Butler (see II. p. 235), is collected and brought to the Wharf at Chesterfield, and is conveyed thence by the Barge, and on other Wheels, to his forge at Killamarsh, as already described.

Mr. Butler remarked to me concerning Rail-ways, that using several light Waggons (instead of one heavy one) is essential to economy in weight of Iron and first cost in laying a Rail-way, in the repair of broken Bars, and also in draught, because the chief friction and cause of labour to the horse, arises from the wheels grinding

grinding against the flanch or rib on the Bar, and this is greatly increased by heavy loads. It is nevertheless true, that a large share of the friction and labour of drawing Trams, originates with the imperfect form and workmanship of the Axle-trees, as has been proved since Mr. Collinge's patent Axle-trees (made in Westminster Road, Lambeth), have been applied to the Trams on the Rail-ways to the famous Pipe-Clay Pits near Corfe Castle, in Purbeck in Dorsetshire, and others in the South of England; by which also there is a great saving made, in the cost of incessantly greasing the wheels; and in the great durability of these new Axle-trees, and their Wheels.

ASHBY-DE-LA-ZOUCH *Canal*: Act, 34 Geo. III.— The general direction of this Canal, and its Rail-way extension, is nearly N, through a serpentine course of 36¼ miles, in the Counties of Warwick, Leicester, and Derby, entering the Map (facing p. 193) at *K*, about 3¼ m. from Hinckley. It is said to be elevated 289 feet above the high-water level in the *Trent* River at Gainsborough. It commences near and almost upon the *Grand Ridge*, or waterhead line of our Island (see I. p. 4), on its eastern side, and its line is tunneled through the *south Mease* and the *north Mease* Ridges, branching therefrom (see I. p. 14 and 13), and being each somewhat higher than the Grand Ridge near Bedworth.

Its principal objects are, the carrying of Limestone from Ticknall and Clouds-hill, and Coals from different Collieries in the Ashby-de-la-Zouch Coal-field, for the supply of the towns on the line, and the country southward. Coventry, which is not many miles from the S end of this Canal, is the 31st town in point of of Population* in Britain, appearing by the census of 1811, to have 17,923 Inhabitants: Hinckley, near to the line, is the 122d, and has 6,058 persons: Ashby-de-la-Zouch is the 284th, with 3,141 persons: Measham the 578th, with 1,525 persons: Nuneaton the 159th, with 4,947 persons; and Market-Bosworth is also another considerable town, on or near to the line of this Canal.

The commencement of this Canal is at Marston-Bridge near Bedworth, in the *Coventry* Canal; and at Willesley Wharf 1 *m*. N of Measham the Rail-way commences, and proceeds to the Lime-works on the E side of Ticknall. From Willesley Wharf, a branch of this Canal $3\frac{1}{4}$ miles long, proceeds westward, past Donisthorp Colliery, and Warren-hill Furnace and Colliery on Ashby Wolds (see I. p. 213), to near Union Farm on the Wolds: from the Tunnel-house near Old-Park, 1 *m*. N of Ashby, a Rail-way branch proceeds $3\frac{1}{4}$ *m*. north-eastward, past Lount Collieries to Clouds-hill Lime-works: and from Ilot Wharf on the

---

* It appeared to me important at the time of compiling the general account of *Canals*, &c. for the Cyclopædia, as mentioned, page 283, to shew the connection in each instance, that the same have, with *the populousness of the districts*, into which and for whose use the several Canals, &c. were made, by mentioning the names of the largest of the *Towns* (such as hold *Markets* in particular) on and near to the line of each Canal, &c. with the *number of Persons* (when above 5000), and the *numerical Order* in which they stood in a list of 120 such Towns, that had been prepared and printed from the Population returns of 1801. The list used on that occasion, has appeared since to be very defective, and in 1811 a new and more perfect enumeration of the People having taken place, I intend in Sect. xi. of this Chapter, to present a new list of 709 Towns, which I have prepared, after considerable pains in searching through the 511 closely printed folio pages of the Population Returns, which were ordered to be printed by the House of Commons on the 2d of July, 1812.

E side

E side of Measham there is another Rail-way branch northward, of about 1000 yards long, past Measham-field Colliery to the Coal-yard near Jewsbury Farm, by the Hinckley Road, see page 288.

From Marston-Bridge to Willesley Wharf the length is 28¼ m. without any locks, the branch to Union Farm 3⅞ m. is also without Locks, and the whole forms, with the summit pound of the *Coventry* Canal and its branches, and the adjoining pound of the *Oxford* Canal, the longest *level piece of artificial Water* in Great Britain, or perhaps in Europe, being *seventy-five miles and three quarters* in length! and being at the same time more singular, from its crossing the Grand Ridge without a Tunnel. From Willesley Wharf to the entrance of the Tunnel, near four miles, the Rail-way is laid double, with a moderate rise from the Canal; through the Tunnel 600 yards of single Rail-way is level: from the Tunnel-house at the NE end of the Tunnel, to Ticknall Quarries, 4⅞ miles of single Rail-way, and does not differ much from level; and from the Tunnel-house to Clouds-hill Quarry, the single Rail-way has not much inclination, I believe: the Measham-field branch has a small rise from the Canal.

This is a wide and deep Canal, adapted for 60 Ton Barges, but it connects at its southern end, only with narrow Canals, adapted for 20 to 24 Ton Boats, and of course such only will be used upon it, unless the *Coventry* and *Oxford* Companies should widen their Canals, between this and the *Grand Junction* Canal, as they agreed to do. At Willesley Wharf, I saw Boats constructed of wrought Iron plates rivetted together, used in the carriage of Limestone.

The Rail-way bars are flanched, three feet long, weigh 36lb. and are spiked down on to blocks of stone,

stone, of about 1¼ cubic feet each, by means of an Oak plug, inserted into a hole drilled in the stone.

Under the Town of Snareston the Canal passes thro' a Tunnel 250 yards long, and near to Old Park the Rail-way passes through a Tunnel 600 yards long, arched with Bricks, nine feet high and seven feet wide. At each end of these Tunnels are Deep-cuttings, others, where the Rail-way passes under the Turnpike Road S of Ashby, &c.

At Shakerston in Leicestershire, the Canal passes on an Aqueduct Bridge over the Sence River, and near Ilot Wharf on another, of one stone arch, over the Mease River, with Embankments at each end; under the westernmost of which, the late Joseph Wilkes's Irrigation sunk Culvert passes, see Vol. II. p. 474; half a mile N of Willesley, the Rail-way passes over a considerable Embankment, wherein Irrigation Culverts had been constructed, but improperly, see II. p. 480. On the Ticknall line, there is a considerable Embankment in Woodward's Farm; and on the Clouds-hill branch, another in Lount Valley.

On Union Farm, between Boothorpe and Overseal, Leicestershire, a Reservoir of 36 acres has been constructed (see Vol. I. p. 498, II. 468, and Leicester. Rep. p. 84) for supplying this Canal, and from the surplus water of which, Mr. John Johnson hopes to irrigate 50 acres of Land, in its way to the Canal branch, which terminates below this Reservoir.

From near the time of the passing the Act, until about May 1805, the works on this Canal and its extension and branches were in hand. The Company were authorised to raise 200,000*l*. in 100*l*. shares, of which it is said (in the Leicester. Rep. p. 315), that the Earl of Moira took more than 80 of these shares, besides expending upwards of 30,000*l*. in erecting an

Iron

Iron Furnace and works on its banks, at Warren-hill, on Ashby Wolds, as already mentioned: it is also said in the pages referred to, that the works executed as above described, have cost 180,000*l.* and upwards, and that in 1809, no dividend thereon had been made: the 12 miles of Rail-way having cost 30,000*l.* owing to the Tunnel and Embankments, &c. thereon.

I have a particular estimate now before me, made by the late Mr. Benjamin Outram, for executing the whole of this distance, with double lines of Rails, except 600 yards through the Tunnel, (instead of near eight miles of the distance having single Rails, as at present); a summary of which estimate is as follows, viz.

	£	s.	d.
Ground-work, to cut through high grounds, and embank hollow places, ...	3398	5	4
Driving and lining the Tunnel, ...	2100	0	0
Road-Bridges and Water-Culverts, ...	767	10	0
Forming the Road, six yards wide and 23,160 yards in length, ...	1101	8	0
Fencing the whole length on both sides, with Walls or Quick Hedges, except through Woods and the Tunnel; Gates, Stiles, &c.	1850	0	0
Rails, Blocks, Gravel or small stones (spread six inches thick), and all workmanship, in laying and completing 22,560 yards of double, and 600 yards of single Railway, with all the necessary crossing and turning places, and Road-crossings, &c.	23,267	0	0
	£32,484	3	4

The

The Rails between Willesley Wharf and the Tunnel, to weigh 38 lb. per yard, and on the remainder of the lines 30 lb. per yard: the Contractor to replace the Rails that break through use or wear, in the first three years.

At the time of obtaining the Act, but a part of these Rail-ways were intended, and the scheme then was, to have ascended by the Canal, nearly N from Warren-hill Furnace, through a rise of 140 feet in 1¼ miles, by means of Locks, to a Lock on the summit level near to Boothorpe, where a large Steam-engine was to be erected, for pumping up water from the Union Reservoir, by means of a culvert therefrom.

From the Boothorpe Lock, the summit level was to pass thro' Blackfordby, skirt the *Ashby and Burton Ridge* (I. 12), turn its point at the W end of Ashby-de-la-Zouch Town, and proceed thence to, and through a Tunnel of 700 yards long, in or near the place of the present one, a distance of 4¼ miles: from the NE end of the Tunnel the Canal was to lock down 84 feet in a quarter of a mile, to the junction of the Clouds-hill branch: from hence the line was to be a level Rail-way of 3½ miles to Ticknall Lime Quarries: the Clouds-hill Rail-way branch was to be level, from the point above-mentioned; and a branch northward of ⅞ of a mile was to fall 28 feet, to Dimins-dale or Calke Lime Quarries, and into Stanton-park Quarries adjoining, I believe, if desired by the Earl of Ferrers. From Boothorpe Lock, a branch of the summit level of the Canal was to proceed north-westward 2¾ m. to Swadlingcote Collieries; but the great difficulties and expense that would have attended these parts of the Canal, with Locks, caused them to be laid aside, as well as the Calke or Stanton and Swadlingcote branches, and the

Works

Works that have been described, to be adopted instead of other parts. At Ilot Wharf in Measham, there are Lime-kilns, &c. (see Vol. II. p. 428) and several others S of this.

In the Act it was stipulated, that there should be public Wharfs at Green-hills, near Sutton-Cheney, and on Ashby Wolds, Leicest. At Ilot, Measham, and SW of Willesley, there are Wharfs for landing Goods, and in Ashby-de-la-Zouch Town, a Yard and Warehouses for Goods brought by the Rail-way. Sir George Beaumont, as the Owner of Collieries in Cole-Orton, to which a Rail-way extension had been previously made from the *Leicester* Navigation, stipulated in this Act for a compensation from the Company, who were authorised to purchase certain quantities of his Coals for such purpose. The *Leicester* Navigation Company stipulated also for indemnification, in case this Canal or its Rail-way extensions were pushed within certain distances of their Works, within which limits they might demand *2s. 6d.* per Ton on all Coals carried thereon thro' Blackfordby. The *Coventry* Canal Company were also entitled by this Act, to take 5d. per Ton for all Coals, and a few other articles which might pass from this Canal or any of its branches, on to the *Coventry*, the *Oxford*, or the *Grand Junction* Canals, or from either of those three Canals to this Canal, or its branches; and for enforcing this last compensation, the *Coventry* Company were authorised to erect Toll-houses and Bars, and station their own Collectors on any parts of the works of this Company, that they might think proper.

The rates of Tonnage allowed to be taken on this Canal and its branches, vary from 2d. to ¼d. per ton per Mile on different kinds of articles, and some even are allowed

allowed to pass toll free, as may be seen at length in Phillips's 4to. "History of Inland Navigation," Appendix, p. 128.

In June 1796, Mr. Robert Whitworth surveyed and projected the proposed *Commercial* Canal, which was proposed to join the Swadlingcote branch of this Canal (as then intended, I believe), with the *Trent* River Navigation, the wide part of the *Trent and Mersey* Canal near to it, and at other points of that Canal also, &c. On the failure of this scheme, for enabling *River Boats* of 40 to 60 Tons burthen to pass between London, Hull, Chester, Liverpool, Manchester, &c. by the Route across our Map, from *K* to i, and a branch to V (the *Trent*), in the January following, it was proposed to extend this Canal (whether from its Union Farm or its Swadlingcote intended branches, I am unacquainted) to the *Trent* River, and to the *Trent and Mersey* Canal at Shapnall, W of Burton.

It has already been noticed, that the branches of this and the *Leicester* Navigation, do or were intended nearly to approach each other, in the vicinity of Cole-Orton and Clouds-hill; at the latter place they were expected to approach the then intended *Breedon* Rail-way; and a proposed branch on the other side, closely approached the proposed *Swadlingcote and Newhall* Rail-way; see these several accounts as they follow herein.

The late Joseph Wilkes, Esq. was Treasurer and chief actor in the concerns of this Company, while its Works were proceeding. The Agent now resident at Willesley Wharf, is Mr. John Crosley.

Ashover and Chesterfield proposed Canal.—In the
year

year 1802, Mr. John Nuttall made a Survey of the line for an intended Canal, from the Chesterfield Canal on the north side of that Town, by a very bowing course to the south, of 13·82 miles, nearly SW in general direction, to Amber-lane Bridge in Ashover, on the Turnpike Road from Chesterfield to Matlock, as shewn by blue dots on the Map, facing p. 193. The south-western end of this line is considerably elevated, having crossed that very considerable branch from the Grand Ridge, called the *East Derwent* Ridge (Vol. I. p. 12), without any Tunnel, or even a deep-cutting being necessary, and skirted in like manner round the end of another considerable Ridge branching therefrom to near Ford, which I have named the *Fabric* Ridge, in the Survey recently made for Sir Joseph Banks, as mentioned herein, p. 273.

The general objects of this Canal were, the carrying up of Coals to the Lime-works and Lead-mines, and Inhabitants of Ashover, and the bringing down the excellent Limestone that there abounds (Vol. II. p. 416), for the supply of the Country to the E and NE: in a part of its course, it occupied nearly the same line as the formerly, proposed, *Chesterfield and Swarkestone* Canal, and as the present *Wingerworth and Woodthorp* Rail-way, both which will be mentioned further on; and it approached within no great distance of the West end of the *Ankerbold and Lings* Rail-way, which has been mentioned, p. 295.

From the Wharf at Chesterfield to Woodthorp Hall, on the Inosculation or saddle between the waters of the Rother and the Amber Rivers, the distance is 5·67 miles, and the rise 219¼ feet: from thence to the Amber River below the Wier of Fall-Mill at Mill town in Ashover, 6·73 miles is level, and from hence to Amber-lane Bridge 1·42 miles, the rise is 77¼ feet; making the whole rise from the *Chesterfield* Canal to this place, 297 feet. This line was revised by Mr. John Rennie the Engineer, who seemed, I believe, to think, that the Amber in dry seasons was insufficient to supply the lockage on this proposed Canal, or the trade to pay for their expenses, and advised the substitution of a Rail-way, with an abridgement of 2·3 miles of the level length, by a Tunnel of half a mile long, thro' the *Fabric* Ridge above-mentioned, on the SW of Stretton Hall, in North Winfield Parish.

The people of Chesterfield, and its Canal Proprietors, are said, not to have seen their true interest, and that of the Country, in the great trade in Peak Lime, which this Canal or Rail-way might have brought to their Town, and to the various Wharfs on the line as far as Killamarsh, and even much further to the Eastward, it might have been presumed, owing to the very superior quality and high repute, with the Farmers, of the Ashover Lime, over that from the Magnesian Rock, which this Canal crosses between Pecks Mill and Shire Oaks in Yorks. and Notts., see Vol. II. p. 408.

BARNSLEY *Canal:*—Act, 33 Geo. III. The general direction of this Canal is S, for about 10 miles, in the West Riding of Yorkshire; entering the Map (facing p. 193) at I, near to Carleton, after having crossed the *North Don* Ridge* by a Tunnel near

---

* Vol. I. p. 13.—I observe now an error in the Map, Vol. I, p. 1, in continuing the *green* colour for this Ridge, eastward to near Rotherham, instead of its leaving the top of the map at about 2° 44' long. in order to include the *Dearne* River, which is a tributary of the Don; and from about 3 m. WNW of Penistone, towards the S: this Ridge ought to have another colour, and be named the *South Dearne* Ridge.

Roystone,

Roystone, I believe: its southern end and western branches are considerably elevated above the Sea: its general objects are, the export at both its extremities, of Coals, Iron, Grinding, Building and Paving Stones, the import of Limestone, Deals, &c. and forming a communication between the great manufacturing districts of Sheffield and Rotherham (by means of the *Don* Navigation, and the *Dearne and Dove* Canal) with Wakefield, Huddersfield, Halifax, Manchester, Liverpool, Leeds, Bradford, &c. &c. by means of the *Ayre and Calder* Navigation, and the Canals connecting therewith. Wakefield is the 76th Town in the order of Population, with 8,593 persons, and Barnsley the 158th, with 5,014 persons.

This Canal commences in the Calder River, near the south-western termination of the *Ayre and Calder* Navigation, $\frac{1}{3}m$. below Wakefield Bridge, and terminates in the *Dearne and Dove* Canal, at Eymingswood $\frac{1}{2}m$. E of Barnsley: from which last point a branch proceeds SW, up the Vale of the Dearne, to Haigh Bridge, SW of Woolley, and from which branch near Bargh Bridge, another proceeds SW to Barnby Bridge, E of Cawthorne: a Rail-way extension of this last branch $1\frac{1}{2}$ mile, is provided for in the Act, to Silkstone, and another Rail-way branch of 1 m. to Barnsley Town.

From the *Ayre and Calder*, in about $2\frac{1}{4}$ miles to Over-Walton, the rise is $120\frac{1}{7}$ feet, by means of 16 Locks; three of which were directed by the Act to be placed near Agbridge, near together, so that a Steam-Engine (that must burn its own smoke) might, by means of a culvert from the lower pound, connected with the Calder, force or lift the Water into the pound above them, that is necessary for their lockage, and the remaining 13 locks to be in like manner placed near together, near Over-Walton, and a Culvert and Steam-Engine provided, for forcing or lifting the Water again into the upper pound, that is let down therefrom in the use of these Locks. From Over-Walton to the *Dearne and Dove* Canal, about 7¼ miles is level, and the same level continues also along the branch to near Bargh Bridge; whence to Haigh Bridge there is a rise of 7 locks, which are directed to be placed together, with an Engine for returning their Water (so that none of the sources of the Dearne River may be diverted into the Calder); and the Barnby Bridge branch rises from the long level by 4 locks, whose Water an Engine is to return, as above mentioned.

In width and depth and Locks, this Canal is adapted for the same Boats as navigate the Calder River. It was stated in the " Monthly Magazine," Vol. 31, p. 404, that in 1810, the late Mr. Pinkerton, a famous Canal Contractor, had a lawsuit with this Company, as to the sufficiency of an Argillaceous Grit Rock, as the *lining* to part of the Tunnel, that he executed for them. Near the southern end of the line, at Eymings-wood, this Canal crosses the Dearne River on a very tall Aqueduct of 7 or 8 arches: Mr. William Jessop, Mr. William Wright, and Mr. Goll, were employed on this Canal as Engineers, and on the 8th of June, 1799, it was completed and opened for Trade.

It is provided in the Act, that if hereafter the *Calder and Hebble* Company should make a branch from their Navigation to connect with the Haigh Bridge branch of this Canal, then all Rail-ways or stone Roads that may have been laid under the authority of this Act, to the northward of Bargh Bridge (or Mill),

Mill), are to be disused and removed. The rates of Tonnage allowed to be taken on this Canal and its branches, on different articles, are various: some not to exceed 6d. to 4d. per Ton for the whole length of the Canal, and many others 3d. to ¼d. per Ton per Mile, all which may be seen in Phillips's 4to. "History of Inland Navigation," Appendix, p. 40 to 43, as well as the exemptions from Toll, and the Rates of Wharfage, &c.

The Company were authorised to raise 97,000l. in 100l. shares, which have not yet been productive of much profit to the holders of them. On the NE side of Barnsley, very convenient Wharfs and Warehouses are erected.

Baslow and Brimington proposed Canal.—In October 1810, when the *High-Peak Junction* proposed Canal was in agitation, Mr. John Gratton, jun. of Wingerworth, surveyed two different lines across the East Moor, of which this is the northernmost, for trying the practicability of a Junction between the *Chesterfield* Canal and the proposed one, that should open a water communication from the north-eastern Coal district of Derbyshire, and the north of Nottinghamshire, &c. with the Mining district of the Peak Hundreds, and with Cheshire and Lancashire, &c.: and although there may be small probability of either of these being adopted and executed, I think it will be useful to record the *levels* that have been taken on this occasion.

The general direction of this line is nearly west, by a crooked course towards the western end, of 12¼ miles, crossing the *East Derwent* Ridge by a long Tunnel, at a very considerable elevation. From the Chesterfield Canal above Wildens Mill (on the S side of Whittington Brook) to crossing of the Sheffield Turnpike Road, 1 mile, a rise of 6 feet : thence to crossing the Road from Baslow to Dronfield (on the N side of the Brook, crossed 1¼ m. below), 2¾ miles, a rise of 115¼ feet : thence to the Road at the S end of Millthorpe village, 1½ miles, a rise of 102 feet: thence to near the junction of the Brook that comes down from Bank Village, ¼ mile, a rise of 12 feet (the elevation of 235½ feet here, above Wildens Mill, being the same as the summit level of the southernmost line); thence up the N side of Smeathley Wood Brook, and across its N branch to the boundary of the Wood and Common S, 1¼ mile, a rise of 137 feet, to the summit level; which here was to enter the proposed Tunnel of 2 miles, nearly in a SW direction, into Barbrook Dale, SE of the Brook, near the old Smelting-mill: thence down the Valley, ¼ of a mile with a fall of 137 feet, to 300 yards below the Cupola or Lead Furnace, to the SE end of an intended Aqueduct across Barbrook (at 21 feet above it), where the southern, or *Baslow and Chesterfield* line, next to be described, joins this : from this point a level was traced, passing ⅛ m. NW of Baslow Town, and thence up the W side of the Derwent 2½ miles, passing between it and Curbar, and near to Cliff-House (Mr. Girdom's), to join the *Peak-Forest Junction* line, at the E end of the proposed Stoke Aqueduct thereon, S of Froggatt.

The levelling across the top of the proposed Tunnel, was as follows, viz. from Smeathley Wood to the summit or *East Derwent* Ridge, ¼ mile, a rise of 397 feet, making this point 769½ feet above Wildens Mill, and 534 above the proposed Stoke Aqueduct: thence in 1 mile, a fall of 199 feet, to the Turnpike Road top of Barbrook Bridge; whence in ¼ mile the fall is 198 to the S W end of the proposed Tunnel, as above mentioned.

Baslow and Chesterfield proposed Canal:—Surveyed in 1810 by Mr. John Gratton, jun. as mentioned in the last article, describing a line north of the present one, with similar objects in view. The *levels* taken on this occasion, commenced in the summit pound of the *Chesterfield* Canal, NNE of that town, (1¼ m. from the former point of departure at Wildens Mill, and 11 feet higher), viz. to the W side of the Chesterfield and Sheffield Road, about ¼ of a mile, rise 28 feet: thence to the W side of the Chesterfield and Newbold Road, ¼ mile, rise 18 feet: thence to the top of the *North Hipper* Ridge, N of the Windmill, ¼ mile, rise 35 feet: thence pursuing a level, bending to the north, to intersect and cross the Liniker Brook (or northern branch of the Hipper) on the N of Lounsley-green, 1 mile: thence up the S side of the Brook, to 40 yards above the Bridle Road between Brampton and Liniker, 1¼ miles, a rise of 143¼ feet (the elevation here being 235¼ feet above Wildens Mill, as at page 310): from this point, a Tunnel of 4 miles long nearly W, would emerge again at the bottom of the Ravine in Humbersley Dale, about ¼ of a mile below the Robin Hood Public-house: thence to the other, or *Baslow and Brimington* line at the proposed Barbrook Aqueduct, ¼ a mile, is level; and thence to the proposed *High-Peak Junction,* 2¼ miles, also level, as mentioned before.

In order to shew the depth of this proposed Tunnel under the Ridge, and how it may be shortened by equal lockage at each of its ends, it will be proper to mention, that from near the Liniker Bridle Road to the Brook at the W end of Berley Wood, is about ¼ of a mile, and a rise of 163 feet; thence to the Brook at its leaving the Common ⅛ of a mile, with a rise of 178½ feet; thence to the knowl of 6th Grit, S of Three Birches Quarry, ¼ of a mile, a rise of 74½ feet; thence to Grange-bar Road on the *East-Derwent* Ridge, ¼ of a mile, is a rise of 91½ feet, making this middle point of the Tunnel 743 feet above Wildens Mill, and 507½ feet above the proposed Stoke Aqueduct. From this Ridge to the knowl of 4th Rock, north of the Road-crossings, ⅛ of a mile, is a fall of 85 feet: thence to the top of the sudden hill in the old part of the Road, 1 mile, a fall of 81½ feet: thence to the meeting of the Roads at foot of this hill, ¼ of a mile, a fall of 108 feet: thence to the Road in front of the Robin-Hood, ⅛ of a mile, a fall of 12 feet; thence to the Road on the Bridge over Humbersley Brook, ⅜ of a mile, a fall of 37 feet; thence to the bottom of the Ravine, ⅛ of a mile, a fall of 184 feet. The distances given on this and the last line were not furnished by Mr. Gratton, but measured on my Map, and may not, some of them, be quite correct, although sufficiently near for this purpose.

Belper proposed Canal.—In September 1801, notices were given of an intended application to Parliament, for an Act for making a Canal with Railways, &c. from the *Cromford* Canal, at the S end of the Bull-bridge Aqueduct, to Black-brook Bridge, ¾ *m.* W of Belper (which is the 129th Town, with 5,778 Inhabitants): it was intended to cross the Derwent on an Aqueduct near Toad-moor Bridge, I believe, and to proceed by the Colliery, in the inosculating or meeting Valleys, near Belper Lane-end, instead of pursuing the Banks of the Derwent, through part

of

of Belper Town: but I am unacquainted with further particulars.

Belper and Morley-park *Rail-way.*—Since I finished this part of my Survey, I have heard, that a Rail-way has been laid in the Valley, from Belper to Morley-Park Collieries, and thence to Denby-hall Colliery (which also has a Rail-way to the *Derby* Canal) a distance of near 4 miles, for better supplying that large and increasing Town with Coals; but I have been disappointed of the levels and other particulars, that I hoped to have received, in time for this account.

Birmingham and Fazeley *Canal:*—Acts 23rd, 24th, 25th, and 34th Geo. III.—The second of these Acts is for uniting this Company with the old *Birmingham* Canal Company; and the last but one, for uniting under this Company, about $5\frac{1}{4}$ miles in length of the line, that they had purchased (of the *Trent and Mersey* Company) which was comprised and made under the powers of the original Act for the *Coventry* Canal, 8th Geo. III. The general direction of this Canal is SSW, by a bending course towards the E of $20\frac{1}{4}$ miles, in the counties of Stafford and Warwick, quitting the Map (facing p. 193) at Q; its southern end is considerably elevated above the Sea, and not much below the level of the *Grand Ridge* in the vicinity of Birmingham, to the SW.

The great objects of this Canal are, the export of the manufactured Goods of Birmingham towards Hull, and towards Liverpool, in part, and the import of raw materials of various sorts, Grain, &c. for the supply of that immense Town and its Neighbourhood: its northern $5\frac{1}{4}$ miles between Fazeley and Whittington

Brook,

Brook, forming also part of the grand line of communication between London, Liverpool and Manchester, &c. crossing our Map from N to m, as mentioned, pages 214 and 215. Birmingham, at its southern termination, is the 6th British Town, with a population of 80,753 persons; Litchfield the 157th, with 5,022 persons; Tamworth the 304th, with 2,991 persons; Sutton Coldfield the 314th, with 2,959 persons; and Coleshill the 552nd, with 1,639 persons, are also not far from the line.

This Canal commences in the detached part of the *Coventry* Canal at Whittington Brook, very near to the junction thereof with the *Wyrley and Essington* Canal, and terminates in the old *Birmingham* Canal at Farmer's Bridge, on the W side of Birmingham Town, very near to the junction thereof with the *Worcester and Birmingham* Canal: it connects with the *Coventry* main line at Fazeley, with the *Warwick and Birmingham* Canal at Digbeth, on the SE side of Birmingham, by means of the branch from this Canal, which leaves the line at the NE end of the Town, and skirts thro' its lower parts, to the Digbeth Basin and Wharfs.

From the detached part of the *Coventry* Canal at Whittington Brook, to the junction of its main line at Fazeley, and thence on the west side of the Tame River to near Middleton Hall, $8\frac{1}{4}$ miles is level; thence to the Aqueduct at Salford, $9\frac{1}{4}$ miles, has a rise of 90 feet by 14 Locks; thence to the commencement of the Digbeth branch, near $1\frac{1}{2}$ miles, has a rise of about 71 feet, by 11 Locks: thence through the NW side of the Town of Birmingham to the old *Birmingham* Canal at Farmer's Bridge, about $1\frac{1}{4}$ mile, has a rise of about 85 feet by 13 Locks: the Digbeth branch about $1\frac{1}{4}$ mile through the

the E side of the Town, has a fall of 40 feet by 6 Locks to the *Warwick and Birmingham* Canal. The width of this Canal is about 30 feet, and its depth 4½ feet. The Locks are 70 feet long and 7 feet wide in the clear, passing Boats with about 22 tons of lading. The Salford Aqueduct has 7 arches, each 18 feet span over the Tame River, and on the NE of Middleton Park, there are other smaller Aqueduct arches, over streams falling into the Tame. Parts of the Digbeth branch are arched or tunnelled over in the Town, and so is a part of the line, where the Sutton Road crosses it. The principal supplies of water for this Canal, are from the Reservoirs and Engines on the old *Birmingham* Canal.

The *Trent and Mersey* Company having by agreement with the *Coventry* Company (see that Canal) and this Company, completed the part between Whittington Brook and Fazeley; in October 1789, it was purchased of them, as mentioned, p. 313, and on the 12th of July, 1790, the Salford Aqueduct and other works were completed, and the whole was opened for Trade. The sums of Money to be raised for this Canal, were not all distinguished in the Acts, from what were intended for the extension and improvement of the old *Birmingham* Canal, now made one concern with this: the amount of each share was at first 170*l.* but the Act 24 Geo. III. limited them to 500 in number, and of course varied their amount.

The rates of Tonnage allowed to be taken on this Canal, are rather complicated, as will appear on consulting Mr. John Cary's " Inland Navigation," large 4to. pages 40 to 44, and pages 75 to 77. By the *Warwick and Birmingham* Act (33 Geo. III.), certain duties are secured to this Company, on Goods

passing

passing from or to that Canal, which may be seen in *Cary*, p. 44. It is provided, that the tonnage taken per mile on Coals, is here to be the same, as on the *Coventry* and *Oxford* Canals.

Breedon proposed Rail-way:—About the year 1793 it was in contemplation to make a Rail-way, and a Canal in part, I believe, between the *Trent* River near King's-Newton, and the Quarries of Magnesian Lime at Breedon, and I believe to those at Clouds-hill also (see Vol. II. p. 419 and 421), a distance of about $3\frac{1}{2}$ miles, nearly S, in Derbyshire and Leicestershire: by which this would have nearly or quite connected with the Rail-way branches of the *Ashby-de-la-Zouch* Canal, and with the proposed one from the *Leicester* Navigation.

In the *Derby* Canal Act, of 33 Geo. III. that company undertook, in case of this Rail-way, &c. being executed, to make a Canal about $\frac{1}{4}$ of a mile long, from the bank of the *Trent* opposite to its termination, northward, with the necessary Locks for connecting with the *Trent and Mersey* Canal near Weston Cliff: intended for giving a freer circulation of this Lime in the southern parts of Derbyshire; and in expectation also, of the shutting up of the Navigation on this part of the *Trent*; see that River herein.

BRIDGEWATER's *Canal*.—Acts, 32 and 33 Geo. II., and 2nd and 35th of Geo. III. with Clauses relating to it in the *Trent and Mersey* Act, 6 Geo. III. and others.—I have introduced the mention of this very important and early of the British Canals in this place (although it falls wholly beyond the limits of my Survey and Map, to the West of Manchester), in order to mention, that the same terminates, and the

Rochdale

Rochdale Canal commences, at the SW end of Manchester Town, (the 3rd on the List, with 98,573 persons), near to q in the Map. In Mr. H. Holland's Cheshire Report, p. 308, some account of this Canal has been given, and in my article *Canal*, in Dr. Rees' Cyclopædia, many other particulars may be found.

CHESTERFIELD *Canal:*—Act, 11 Geo. III.—The general direction of this Canal is nearly SW, by a crooked course of near 45 miles in length, in the Counties of Nottingham, York, and Derby, entering the Map (facing p. 193) at Q: the western part is rather considerably elevated above the sea, crossing a branch from the *South Idle* Ridge by a short Tunnel, and the *East Rother* Ridge, by a considerable Tunnel. Its principal objects are the export of Coals, Lead, Cast Iron, Limestone, Freestone, Pottery-wares, &c.; and the import of Limestone, Grain, Deals, Bar-iron, &c.

Chesterfield, the 183rd Town, with 4,476 persons; Worksop the 224th, with 3,702 persons; and East Retford the 463rd, with 2,030 persons; are situated on this line: and Dronfield the 640th, with 1,343 persons; Gainsborough the 147th, with 5,172 persons; and Blyth and Bawtry, considerable Towns, are also not far from this Canal. It commences in the *Trent* River, near its junction with the *Idle* River, at West Stockwith, 3¼ miles from Gainsborough, and terminates at Chesterfield Town.

Notwithstanding the want of powers to make public *branches* from this Canal to Mines and Works near it, mentioned, p. 288, several private ones have been made on the Proprietors' own Lands, or by consent of their neighbours, viz. from near High-house, a cut

W to Lady-lee Lime Freestone Quarry and Wharf (see I. 411 and 421); and another from Branchcliffe Grange S, to Shire-Oaks Lime Quarries: from Renishaw Furnace, a Rail-way NE to their Ironstone and Coal-Pits at Spinkhill: a cut S to Norbrigs Wharf, and a Rail-way thence E, to Norbrigs Colliery (see I. 205); on the SE side of Staveley 2 Rail-ways S to Inkersall Collieries; at Hollinwood-common, a Tunnel for small Boats SSW into that Colliery: near Brimington, a Rail-way branch N, to Glass-house Common Colliery, and the Glass-works.

The fine light yellow Freestone, from the upper Magnesian Rock at Roch Abbey (I. 420), which used to be brought in considerable quantities to the south-eastern Counties, is carted about 9 miles S, I believe, to Worksop Wharf: the very fine white, and almost crystalline Freestone of Steetley Quarry (I. 421), from the same Rock, is carted about $\frac{1}{4}m$. SE to Lady-lee Wharf: near Shire Oaks there are two considerable Brick Works by the Canal: Pecks Mill, or Dog-kennel Limestone Quarry (I. 411 and 434) in South Anston, is situated close on the N side of the Canal, and Burley Gritstone Quarry NE of Renishaw, is close on the E side of the Canal, so that the blocks of stone are hoisted therefrom into the Boats: the Gib or Derrick of this *Crane* was steadied by three long tarred Ropes from its top, to as many strong posts in the ground, by which any framing of wood-work was avoided.

From the tide-way in the *Trent* to Drake-hole Wharf, about 6¾ miles, has a rise of 27¼ feet; thence to East Retford Wharf, 8¼ miles, a rise of seven feet: thence to Babworth, 2¼ miles, a rise of 31¼ feet: thence to Worksop, 7¼ miles, a rise of 28 feet; thence to Pecks Mill, 5¼ miles, a rise of 156, to the summit

level,

level, 250 feet above the *Trent*. From Pecks Mill to the E end of the Tunnel, ¼ a mile, is level with the Tunnel, which extends two miles to Norwood: thence to Gander-lane, ½ a mile, is a fall of 100 feet: thence to the Norbrigs branch, 6¼ miles, is level: thence to Hollinwood-common, two miles, is also on the same level: thence to Wildens Mill (and the proposed junction of the *Baslow and W.* Canal) two miles, is a rise of 29 feet, and thence to the Basin at the NE end of Chesterfield, 1¼ miles, is a rise of 11 feet by one Lock; the Canal in this distance having crossed and connected with the Rother River.

The Lady-lee branch is about ¼ a mile long, and level: the Shire Oaks branch is about ¼ of a mile, and level: the Spinkhill Rail-way is about one mile in length, and rises considerably above the Canal: the Norbrigs cut is near ¼ of a mile, and level; the Rail-way therefrom, near 1 mile, rises considerably at its eastern end: the eastern Inkersall Rail-way is more than ¾ a mile, with a considerable rise; and the western one, near 1 mile, passing the " Common-spot" Coke-yard, is also much elevated above the Canal.

The Hollinwood-common *Tunnel* is 1¼ miles long, not connecting with the Canal, but kept one foot lower, by means of a Culvert under the Canal: the whole of this length, except the first 800 yards, is driven in the " Deep-end or Squires" Coal-seam, which it is used for draining, and also for working the same, and two other seams near it, all of good caking Coals, that go into Notts and Lincolnshire; its southern end is about 80 yards beneath the surface. The height of this Tunnel is 6 feet, its width 5¼ feet, and the depth of water therein, 2 feet: on which Boats are used,

used, 21 feet long, and 3½ feet wide, that hold seven Corves, weighing together 20 to 22 cwt. When these Tunnel Boats arrive at the side of the Canal, a Crane is used to hoist up these Boxes, and empty their contents into a Canal Boat: this curious Colliery (see I. 201 and 330) belongs to the Duke of Devonshire, and is wrought on his own account, under the direction of Mr. George Dickens, of Staveley, his Colliery Agent.

The Glass-house Common Rail-way is near 2 miles long, and considerably elevated at its northern end.

The first part of this Canal, from the *Trent* to East Retford, is constructed for large Boats of 50 or 60 Tons burthen: above this, the width is 26 to 28 feet, and depth of water 4 to 5 feet. The chain of 18 Locks, between Shire-Oak and Sand-hill Close, are numbered from 38 to 21; these were at first deemed a great curiosity, and obtained the name of the " Giant's Staircase," with many of the country people. Close to the W end of the Tunnel, there are four Locks (No. 19 to 15), formed by only five Gates: lower down at the Norwood Tonnage-house, three other Gates make two Locks, and below these seven other Gates make six Locks, the tail-gate of one Lock, answering also as the head-gate of the next adjoining Lock.

The Boats used in the upper part, are 70 feet long and 7 feet wide, carrying 20 to 22 Tons each: In 1794, when my friend Mr. William Smith (see I. p. 108) visited this Canal, in a considerable Tour he made for such purposes, such Boats were reported to cost when new, from 90*l*. to 100*l*. each: the Boat-owners then usually paying their Bargemen by the Ton of Goods, conveyed certain distances, instead of weekly wages.

The

The great *Tunnel* in Wales, and other Townships in Yorks. perforates Coal-measures in its whole length, and near to its western end, the Wales Coal-seam, ¼ of a yard thick, was discovered, and worked therefrom (Vol. I. p. 213) on its N side, at about 17 yards beneath the surface: a provision having been made in the Act, that in case of any such discovery, or in cutting the Canal in other parts, that sufficient Gates and Soughs might be driven from the Tunnel or Canal, for working and draining such Coal-seams, provided proper settling places or sumps were made, on such Soughs, and kept often cleansed, for preventing mud and silt from being driven into the Canal: but the Company were authorised, to demand to purchase the Coals from the Land-owners, under their Canal, in any parts, to preserve the same from being dug, to endanger the Canal. Near the middle of the Tunnel, 68 yards in length of it, is driven in a Gritstone Rock, which supports the same without being Brickt, as it is in the other parts, 12 feet high, and 9¼ feet wide inside. The great zig-zag Fault appears to me to cross this Tunnel, and considerably to elevate the measures on its western side (Vol. I. p. 168), but perhaps its place of crossing the Tunnel may be further from the eastern end of it, than I had at first supposed, as observed in my Note, on p. 410, Vol. II.

At the eastern end there is a considerable deep-cutting in Gritstone, Shale, &c. which was wheeled up on to the end of the Tunnel, and on its northern side, so as effectually to turn the Brook to the north side of the spoil-banks. This Tunnel was begun in November 1771, and finished on the 9th of May, 1795. Thro' a low part of the branch from the Red Marl Ridge, between Gringley-hill and Scawthorp, near Bawtry, there

there is a wide Tunnel of 153 yards long, at a place called the Drake's-hole. On the N of Staveley, the Canal is deep cut, through the west Dolee Ridge.

Aqueduct Bridges, and considerable Embankments, occur on this line; over the Dolee NE of Staveley; over the Brook at Renishaw Furnace: in Killamarsh Town, with a Road-arch under the Canal: on the E of Worksop, and at Shire-Oaks, there are also Aqueducts over the White-water, or northern branch of the Idle River.

On the west side of Woodhall, there are three Reservoirs, containing 13¼ acres of Water, which, by means of a feeder on the E side of Norwood, is conveyed into the summit level at the W end of the Tunnel: and on the SE side of Woodhall, there are three other Reservoirs below Pebley Mill, containing together 44 acres, the Water from which, is let in at the E end of the Tunnel: above this is a very large Reservoir, which, though made at the expense of the Company, as I have heard, owing to their neglecting, or being unable to stipulate for the purchase of Pebley Mill, it has been usual, on the approach of summer or dry seasons, for the Miller to draw it down to the level of his ancient Dam, by which it is rendered nearly useless to the Company, and the Trade of the country.

The above is not the only instance in which the Mill-Owners of this district, sought, not merely security or indemnity, but great and unfair *advantages*, at the expense of this concern, for in the Act it is stipulated, that a stop (lock) should be made at the Norwood end of the Tunnel, for returning all the Water into the Rother River, collected from its sources, meaning, I suppose, that the water from the Reservoirs, to be made by the Company NW of Harthill, should all (with whatever

whatever overplus could be had thro' the Tunnel from the other Reservoirs above mentioned) be returned into the Rother, not merely at its former point, but that the same should be conveyed backwards through 8¼ miles of the new Canal (locking westward against the course of the River), in order there to be discharged for the benefit of a Mill, with whatever additions the Company might obtain from Springs in the way, or the more direct supplies from Brooks, and from the Dolee River (where now they have a cut or feeder of 1½ mile long, for bringing part of its water into the Norbrigs branch): it being stipulated in the Act, that the Weir out of the Canal into Staveley Forge Dam (just above Hollinwood-common) should be made *four inches lower than any other Weir* between Chesterfield and Norwood; and although there is not any provision made, for compensation to the Company for this new advantage of all their lockage water, to a particular Mill property, there is added to the above, that if this Forge and Corn-mill is injured, *satisfaction for the loss of water*, is to be made by the Company; and also, that no new Mills are to be erected or supplied from the Canal.

The very mischievous effects of these restrictions, are apparent on the execution of this Canal, which on this account has a level pound of eight miles, between Hollinwood-common and Belk Lane in Killamarsh, in all the northern part of this course, skirting the high rocky points of land, with high intervening Embankments, loosing its water, and soaking and injuring the lands below: at the same time, *precluding Rail-way branches* from it, to the extensive Coal Fields and Ironstone Rakes on the opposite side of the Rother, and rendering a junction between this Canal and the

Don Navigation, so much more difficult, than it would have been, if the natural course of things had been pursued, and this Canal had locked down regularly as the valley falls from Norbrigs, almost to the Gannow Valley (*there* have discharged its water, instead of into Staveley Forge Dam) and have regularly locked up the same to Norwood. And after all, it seems, that the business was só managed by Mr. Brindley, that the Company have not escaped a claim of compensation for water to this Mill, which is yet annually paid, as I was told. It was to these, and many similar instances, that have occurred to me, in investigating Canals, of undue preference that the claims of Mill-owners appear to have met in the Houses of Parliament, that I alluded in p. 291, see also Vol. II. p. 489.

Mr. *James Brindley* projected this Canal, and directed its execution, until his death in September 1772, when his Brother-in-law, Mr. *Hugh Henshall*, succeeded to its management, and completed the whole in 1776. Mr. Joseph Gratton, jun. of Chesterfield, is the present Agent of the Company.

The Tonnage to be taken, is not to exceed, for Lime 1*d.* per Ton per mile, and 1½*d.* per Ton per mile on Coals, Lead, Timber, Stone, and all other goods, except Manures (but not Lime) for the lands of any person whose Estate has been cut by the Canal, in the Parishes through which it passes, which are to pay only an ½*d.* per Ton per mile. The Tonnage Rates must be the same throughout every part of the line, except that Coals delivered into Vessels on the *Trent* may be charged lower, than for the supply of places on the line.

Hay and Corn in the Straw, not sold, but going to be

be stack't, and materials for the repair of Roads (not Turnpike) in Parishes through which the Canal passes, may be navigated Toll-free for five miles, or Manures for the lands of the persons whose Estates have been cut within such Parishes, provided no Lock is passed, except when the water flows waste thereat, and having given six hours previous notice to the nearest Toll-collector, of such intention to pass Locks without Toll. The Ton to be 20×112lb. and one-sixth of a Mile, and one-fourth of a Ton to be taken into calculations of Tonnage.

Goods not to remain more than 24 hours on the Company's Wharfs without paying Wharfage; for the next six days, $3d.$ per Ton may be charged. Tolls are not to be liable to assessments for Taxes, but only the Land occupied by the Canal.

The Company were authorised to raise 100,000$l.$ in 100$l.$ Shares, and an additional 50,000$l.$ on Interest or Mortgage of the Tolls, or in new Shares if necessary: but the Works were not to commence until all the Shares were subscribed for, and five per cent. Interest to be paid, while the Works were in hand. The Works are said to have cost in all 160,000$l.$ including the expenses of a Survey and application for an Act, in the year before this passed, which expenses this Act directed to be paid by the Company. At first the Shares of this concern were much depreciated, and sold below par for a long time: from 1805 to 1810, the Dividends were 6$l.$ per share annually, and I believe they are not since much altered.

At East Retford and Chesterfield, there are extensive Wharfs and large Warehouses; and others at West Stockwith, Worksop, Drake's-hole, Killamarsh, Norbrigs, &c. At Killamarsh, and at Pecks Mill in South

Anston

Anston Wharfs, there are Lime-Kilns, &c. see Vol. II. p. 425.

When conversing with Mr. Joseph Butler, of Killamarsh, on the concerns of this Canal, he pointed out to me, as an injudicious and improper system, injurious to the Traders and the Country, that on the approach of dry weather, almost annually, orders are given, that Boats passing the Norwood Tunnel, must decrease their lading from 22 to 16 Tons, and their consequent draught of Water; under the mistaken idea, that thereby the Reservoirs will the better hold out!; although it is evident, that in so short a summit pound as 2¼ miles, and 2 miles of which is walled and arched over, and where, consequently, the extra soakage into the banks and evaporation would be so small, if the water were kept quite as high, or even higher than usual, that the contrary course, that of carrying only full loads, or somewhat larger than usual, would be advisable.

And further, that three-fourths of the Boats that go Eastward, *return empty* through the Tunnel, and yet the Company don't make provisions for and enforce the *waiting for turns*, as their Act empowers them, but on the contrary, this is rendered impossible, by the orders given on these occasions, that Boats shall only pass the summit east and west, each day alternately!; which in the loss of time in Horses and Men, and on capital in Boats and Goods, is a serious evil: aggravated, as he said, by the sub-agents of the Company being Stable-keepers, and deriving profit from providing the keep, in some instances, and from the Dung made by these unemployed Horses, and the Men cannot be kept from tippling in the adjacent Public-houses; the Boats during these delays being exposed

posed to extra pilfering of Goods, Coals in particular: all the other Boats and the Public-houses being supplied from those that are loaded on these occasions, either in barter for liquor and victuals, or Horse-Corn: and the Coal-master thus has to pay again in Coals, for the Corn that he has served out to his Boat-men, or Money advanced to them for purchasing it for their Horses.

Public-houses by the sides of Canals, especially in unfrequented places, Mr. B. deems useless, and a great nuisance, because every Boat has its Cabin, which the Men ought never to leave, when unemployed, to expose their lading to pillage, or their Boat to injury and mischief: and if Ale-house-keepers more distant from the Canal, suffer Boat-men to assemble and remain tippling in their Houses, the evil is even increased; for which reasons he has thought it an act of public duty, to cause informations to be laid, and the Acts enforced, against Publicans for this conduct, at Gander-lane, Killamarsh, Norbrigs, &c. I hope that his conduct in this respect, will be followed by others, in every district.

About the year 1771, the *Chesterfield and Swarkestone* Canal was in contemplation, intended to connect with this Canal at Chesterfield: and in 1810 the idea of this junction was renewed, and extended, under the name of the *North-eastern* Canal, intended either to join this Canal at Norbrigs or Chesterfield, and also near to Killamarsh. In 1802, the *Askover and Chesterfield* Canal or Rail-way was proposed to join this Canal at Chesterfield; as was also the *Baslow and Chesterfield*, or the *Baslow and Brimington* at Wildens Mill, in 1810, as already mentioned.

Chesterfield and Swarkestone proposed Canal.—
The late Mr. James Brindley, about the year 1771,
made a Survey, and proposed a Canal from the *Trent
and Mersey* Canal at Swarkestone, to the *Chesterfield*
Canal at the latter place: its direction being nearly N
for about 29 Miles: the first part of this line from
Swarkestone, passing Derby (the 45th Town, with
13,043 Inhabitants) to Little Eaton, very nearly
corresponded, I believe, with this part of the *Derby*
Canal, as since executed, see the Map facing p. 193;
from thence past Belper (the 129th Town, with 5,778
persons) to Bull-bridge (in the latter part, nearly along
the since proposed *Belper* Canal line); which line is
represented, nearly, by blue dots: thence for about a
mile, the track of the *Cromford* Canal as at present,
was nearly followed, to Pentrich-lane, and crossed:
thence to a summit near Woodthorp Hall, the line
is shewn nearly by blue dots; as it is also thence to
the *Chesterfield* Canal; in part of this course, nearly
along the *Wingerworth and Woodthorp* Rail-way
track, and nearly along the track proposed to be followed by the *Ashover and Chesterfield* Canal: and
more recently, the *North-eastern* Canal was proposed
to occupy the northern part of the line above described, or nearly so, between Pentrich-lane and
Chesterfield, which is the 183rd Town, with 4,476
persons.

All the objects that were originally in view in this
proposition, and others more extensively important to
the Country, might now be accomplished, in consequence of the completion since, of the *Barnsley,
Dearne and Dove, Don, Chesterfield, Cromford,
Erewash, Loughborough, Leicester, Leicestershire and Northamptonshire Union, Grand Union,*
and

and *Grand Junction* Canals and Navigations, if the proposed *North-eastern* Canal were carried into effect, for supplying the two chasms between the *Don* and the *Chesterfield*, and between the *Chesterfield* and the *Cromford* Canals; when the trade of the Country would enjoy a water communication between London and Wakefield, whose distance by the Roads is 184 miles (crossing our Map from *F* to I), not greatly exceeding this nearest distance, and opening at each end and along each side of its course, the most extensive communications, with almost all the great Towns in the interior of England, and with several of its sea-ports on the different Coasts.

Commercial proposed Canal.—In the year 1796, Mr. Robert Whitworth surveyed the Country between the head of the *Chester* Canal at Nantwich, and the northern end of the *Ashby-de-la-Zouch* Canal (then in progress), in consequence of the *Grand Junction* Canal having then been undertaken, on a scale adapted for wide Boats, and the *Oxford* and the *Coventry* Companies having engaged to widen their Canals between Braunston and Marston Bridge, and the Wirral part of the *Ellesmere* Canal being also undertaken on an enlarged scale. He proposed this Canal to commence in the *Chester* Canal (which is a wide one) at Nantwich, proceed Eastward to cross the *Grand Ridge* by a Tunnel (of greater width than that at Harecastle, a few miles NE), connect with the western part of the *Newcastle (Underline) Junction*, and I believe with *Gresley's* Canal also, and entering our Map (page 193) at i, proceed nearly as the blue dotted line shews, to cross the *Trent and Mersey* Canal near Burslem, and its Caldon branch near Bucknall also, I believe,

I believe, and tunnel through the *West Churnet Ridge* into the Cheadle Coal-field (I. 173), and lock thence down to Uttoxeter, occupying here for some distance, nearly the same track as the extension of the Caldon branch of the *Trent and Mersey* Canal has since done, and proceeding on the SW side of the Dove, to cross again and connect with the *Trent and Mersey* Canal, in its wide part below Horninglow, and proceed on to cross and join the *Trent* River (then navigable) below Burton; and hence this Canal was intended, I believe, to proceed SW and then SE, nearly along the track since proposed to be occupied by the *Swadlingcote and Newhall* Rail-way, as shewn by the line of blue dots, and to join the *Ashby-de-la-Zouch* Canal by its proposed Swadlingcote branch. Nantwich, the 205th Town, with 3,999 Inhabitants; Newcastle Underline the 118th, with 6,175 persons; Cheadle the 279th, with 3,191 persons; Uttoxeter the 283d, with 3,155 persons; Burton-on-Trent the 207th, with 3,979 persons, and Ashby-de-la-Zouch the 284th, with 3,141 persons, are situate on or near to this proposed line of *wide* Canal.

It was a very favourite idea with the late Joseph Wilkes, Esq. (as Mr. Pitt has observed in his Leicester Report, p. 314), and with many other spirited improvers of that day, that immense advantages would accrue to the commerce of the Country, if the same Barges which navigate the *Thames* River, were enabled to pass with their lading on to the *Trent*, the *Humber*, and all its connecting navigable Rivers, and on to the *Dee* and the *Mersey*, &c. and *vice versa:* and, in consequence, immense sums of Money were expended in making several of the *wide* Canals in the interior of the Country that have been mentioned above, where the

the extra expenses of deep and wide cutting, wide and higher arches for Bridges, extra expense in raising the approaches to such, Tunnels on larger scales, wider and more expensive Locks and Gates, &c. &c. have been as yet almost entirely thrown away, owing to the want of a thoroughfare being made for these large Boats: and it seems the opinion of several of the best informed Engineers and Commercial Men of the present day, that Canals with Locks, not much exceeding 7 feet wide, are best adapted, from the comparative cheapness of execution, and consequent lowness of Tolls, for all the interior parts of the Country. This line, between Horninglow and the Dilhorn Collieries, nearly coincides with that which the *Dilhorn* Canal had been intended to occupy in 1792.

Congleton Rail-way.—On the south-east of Congleton in Cheshire, about 2 miles, at the NW corner of Congleton Moss, a Coal-yard was established about the year 1807, for the supply of this Town (which is the 177th on the list, with 4,616 Inhabitants), and a Rail-way was laid therefrom S, about 2 miles, to Stone-trough Colliery in Woolstanton. It was laid with oval bars of iron, on the top of which, the pulley-formed wheels of the trams ran, see p. 288; but when I saw this Rail-way in July 1809, it seemed to be almost or quite disused, the reason of which I did not happen to learn.

COVENTRY *Canal:*—Acts 8th, 25th, and 26th of Geo. III.; the second of these being obtained by the *Trent and Mersey* Company, but relating principally to the northern part of this Canal. The general direction of this Canal, between the extreme ends of its main line,

line, is nearly SE for about 33 miles (but 5¼ intermediate miles have been sold, as will be further mentioned), in the Counties of Stafford and Warwick: it quits the Map (facing page 193) at *N*. Its southeastern part is considerably elevated, so as to cross the *Grand Ridge* in a deep-cutting, near Bedworth. Its general objects are, as part of the grand line of communication between London, Manchester, Liverpool, &c. the export of Coals from the Pits in its vicinity, and on its connecting Canals and Branches, and the supply of Coventry City, by the branch thereto; Coventry is the 31st on the list of population of the British Towns, with 17,923 Inhabitants; Nuneaton the 159th, with 4,947 persons; Atherstone the 318th, with 2,921 persons; and Tamworth the 304th, with 2,991 persons, are also near to it; and Litchfield, the 157th, with 5,022 persons, and Hinckley the 122nd, with 5,058 Inhabitants, are also at no great distance from this Canal.

It commences in the *Trent and Mersey* Canal at the Toll-house and Wharf on Fradley Heath; its detached part terminates in the *Birmingham and Fazeley* (purchased) Canal at Whittington Brook, very near to the commencement of the *Wyrley and Essington* Canal, in this: it commences again in the *Birmingham and Fazeley* Canal at Fazeley, and terminates in the *Oxford* Canal at Longford. At Griff it is joined by the late Sir Rodger *Newdigate*'s Canal, and near to this at Marston Bridge by the *Ashby-de-la-Zouch* Canal. From Longford, there is a branch SW to Coventry: at Shackleton, there is a short cut and Rail-way therefrom NW, to Bedworth Colliery: at Griff-hollow, there is a Cut and Rail-way therefrom W, to Griff Colliery: from near Atherstone a Railway

way branch SW, to Oldbury Colliery: near Grendon, there is a Rail-way branch SW, to Badesley Colliery, &c.;—S of Polesworth, Collieries are situate close to the line, and others very near it, at Two-gates, near Fazeley.

The detached part of the line, from the *Trent and Mersey* Canal to the *Wyrley and Essington* Canal, 5¼ miles, is level, and with the *Birmingham and Fazeley* Canal, through which the same level is continued thence, 5¼ miles, to the branching of the *Birmingham and Fazeley* line, at Fazeley; and this level is also continued thence, along the Vale of the Anker, to the S side of Atherstone Town, 10 miles, with a rise of 96 feet, by 13 Locks: whence to *Newdigate*'s Canal, about 8⅛ miles, is level: whence to the *Ashby-de-la-Zouch* Canal ¼ mile is level: and thence to the *Oxford* Canal at Longford, 3¼ miles, is level. The branch thence to Coventry, 4¾ miles, is level with the summit pound, and so is the cut of about ¼ of a mile towards Bedworth, and the cut of about ¼ of a mile long to Griff-hollow. The Rail-way branches all rise from the line; but I am not acquainted with their levels. The summit pound of this Canal and its branches, and on the *Ashby-de-la-Zouch* and *Oxford* Canal, forms together, the longest piece of level artificial Water in the Kingdom.

This is a narrow Canal, but the Company have bound themselves to the *Grand Junction* Company, to widen the same, between Longford and Marston-Bridge, to the width of their's and the *Ashby-de-la-Zouch* Canal, when thereunto required; but which now may, perhaps, never happen, as observed, page 331. It is provided by the *Oxford* Act, that a Stop-gate or Lock shall be maintained at Longford, to be kept shut when

when the supplies of water to this Canal begin to fail, in dry seasons.

The deep-cutting SE of Bedworth on the *Grand Ridge*, is in Red Marl (covering Coal-measures " unconformably"! as my friend Mr. Benjamin Bevan lately informed me) 600 yards long, and 36 feet deep in the middle. A considerable Aqueduct, of Bricks, of 2 or 3 arches, conveys this Canal over the Tame River on the NE side of Fazeley Town, and there is a smaller one over Whittington Brook. On the Coventry branch there are two considerable Embankments and Aqueduct Arches; near to one of these, I observed, 1 m. from Coventry, a weir and trunk through the bank, used for *irrigating* some Meadows below the Canal, in rainy seasons; see Vol. II. p. 493.

Mr. *James Brindley* was the original Engineer to this concern, and the level part of the line, from Longford to Atherstone, and the branch to Coventry, together 16¾ miles, and the shorter branches therefrom, were finished in 1776; when the further progress of the works were suspended for want of Money, for 10 years: at length the *Trent and Mersey* Company came forwards to assist, by completing 11 miles of the line, connecting with their Canal at Fradley Heath, and the *Birmingham and Fazeley* Canal at Fazeley; one half of which length, between Fazeley and Whittington Brook, was sold to the latter Company, by mutual consent; and for the remaining 5½ miles, between Whittington Brook and Fradley Heath, this Company repaid the *Trent and Mersey* Company, on the 4th February, 1787, and who thus came to have a detached part of their Canal. The whole was completed in June 1790, and all these important communications opened: this Canal effecting, at that period, the only water communication

nication between Birmingham and London; but a more direct one has since been opened, by means of the *Warwick and Birmingham*, and *Warwick and Napton* Canals, into the *Oxford* Canal, not far from the termination of the *Grand Junction* Canal.

This Company have been authorised to raise 120,000*l.* in 100*l.* Shares; for some years after the Canal was completed, and while the trade of Birmingham continued to pass through it, these Shares sold for 400*l.*; on the opening of the new lines above mentioned, they fell to 350*l.* and their annual Dividend to 8*l.* Since the completion of the *Grand Junction* Canal, this concern has again been more flourishing than ever: in 1805 the Dividends had risen to 16*l.*

The Tonnage allowed to be taken is, ½*d.* per Ton per mile for Lime and Limestone, and 1½*d.* per Ton per mile for all other articles (except Road and Paving Materials and Manures, on the level pounds, or when the water runs waste at the Locks). On the completion of the adjoining Canals, the Tonnage on several of the articles was, by general consent of these Companies, reduced to 1*d.* per Ton per mile: by the 9th of Geo. III. for the *Oxford* Canal, it was enacted, that this Company should be entitled to the Tonnage on Coals carried on the first 2 miles of that Canal, and they in return should be entitled to the Tonnage on all articles (except Coals), passing from their Canal, and carried the first 3¼ miles on the Coventry branch.

The Act of 34 Geo. III. for *Ashby-de-la-Zouch* Canal, granted to this Company 5*d.* per Ton on all Goods passing from this Canal to that, or from that to this, or that may pass on any part of the *Oxford* or *Grand Junction* Canals, after or before passing on the *Ashby* Canal, except Farming produce, Manures,

or

or Road-materials, or Iron or its Ores, produced or dug in the vicinity of the *Ashby* Canal; and a further sum per ton, equal to the Tonnage between Longford and Griff, on all Goods that may hereafter pass from the *Ashby* Canal to the *Oxford* or *Grand Junction* Canals, or *vice versa*, by any new communication, which these duties are intended to prevent: and for enforcing which, this Company may erect Toll-houses and Stop-bars, and place Collectors, on any part of the *Ashby* Canal.

In the northern part of Coventry Town, the branch from this Canal terminates on very high ground, at one of the completest Wharfs in the Kingdom perhaps: the Canal branches into two, with a spacious Yard between, just of the proper height for landing or loading Goods: at the lower side the Wharf is narrow, and of the usual height of Carts and Waggons above the street, so that they can be backed up within a sufficient distance, to throw Coals out of the Boats, or readily move Goods therefrom, into them, or the reverse of these. On the south-east side of the Company's Yard, there are large open Sheds constructed, for covering both the Boats and the Carts while they are loading, &c. as above: here is a very complete and simple *Crane*, with a Derrick and Chains at top, instead of a Gib; a Machine-house for weighing the Coals and other Goods, in the Carts, &c. Over the entrance, facing a wide main street of the Town, is a large Room for the Meetings of the Company, and others for their Agents and Clerk's use, which were built about the year 1784.

CROMFORD *Canal:*—Act, 29th Geo. III. The general direction of this Canal is about NW, by a bending

ing course of 14¼ miles, in the Counties of Nottingham and Derby: its northern parts are considerably elevated, penetrating the *East Derwent* Ridge by a Tunnel: its general objects are, the export of Coals, Limestone, Iron, Lead, Mill and Grind Stones, Freestone, Marble, Chert, Fluor, &c.; and the import of Corn, Malt, Deals, (or Raff), Coals, to the north-eastern end, &c.

Cromford, at its extremity, is the 658th Town on the British population list, with 1,259 Inhabitants;—Wirksworth the 245th, with 3,474 persons; Crich the 509th, with 1,828 persons; Belper the 129th, with 5,778 persons; and Alfreton the 253rd, with 3,396 persons, are also Towns at no great distance from the line of this Canal.

It commences in the *Erewash* Canal at Langley-Mill or Bridge, and there also connects with the *Nottingham* Canal, and terminates at the Town of Cromford: from Golden Valley or Codnor lower Park, there is a branch of Canal NE, 2¼ miles to Pinxton Wharf: and from which there is a short cut NW to Somercotes Furnace: from the E end of the Derwent Aqueduct, there is a cut of ⅜ of a mile N to Lea-wood Wharf: another from above the Locks to Codnor lower Park Lime Kilns, and another shorter Cut, and Lime Docks from the lower Canal (see Vol. II. p. 421): another short Cut SW, to Aldercar Colliery, attempted about 1794; and from out of the large Butterley Tunnel, a short Tunnel was made S, for small Boats, and formerly used for working the Butterley-car Coals (see Vol. I. p. 102).

The Rail-way branches to this Canal are, from near Langley-Mill E 1¼ miles, to Beggarlee Colliery, with a branch therefrom NE, ¾ of a mile to the Pumping Engine;

Engine; this is a very neat and perfect Rail-way, belonging to Thomas Walker, Esq.: it crosses a Railway branch from the *Nottingham* Canal to Old Brinsley Colliery. From Brinsley Wharf NE 1 mile, to New Brinsley Colliery: from Brinsley Aqueduct NW ¼ of a mile to Benty-field Colliery: from Codnor Wharf NW, ¼ of a mile to Codnor nether Park Collieries and Ironstone Pits: from Codnor lower Park Wharf, ½ m. S, to Codnor nether Park Colliery: from Golden Valley S, ¼ of a mile, to Codnor upper Park Collieries and Ironstone Pits: from ditto NNW, 1¼ mile, to Greenhill-lane Colliery: from Butterley Furnace (on the Tunnel) S, ¾ of a mile, to Butterley-car Colliery: from near Padley Hall SE, 1 mile, to Greenwich Colliery: from ditto NE, ¾ m. to Pentrich Colliery: from near Pentrich Mill SE, ¼ a mile, to Harts-hay Colliery; and from Bull-bridge Wharf N, 1¼ mile, to the Crich SE, or great Limestone Quarries (Vol. I. p. 409, and II. 429).

In the Village of Fritchley, this last, or Crich Railway, passes under a Stone Bridge, and ¼ mile north of this, over a private Road on a Wooden Bridge, and enters the Quarry by a Tunnel 100 yards long, that was driven in the L. Shale, until it penetrated the 1st Limestone, and where the Quarry was begun underground, about 1793, that in 1808 had been extended, to an open pit, that is, I suppose, 150 yards long, 70 or 80 yards wide, and 24 yards deep in many places! such has been the immense demand already for this valuable stone.

At Bull-bridge this Rail-way is continued to six Tipples or Machines on a high Bank, for overturning and shooting the contents of the Trams; 4 of them are adapted to the Wharf below, and 2 of them tipple
the

the Stone down an inclined plane, at once into the Boats, that are made entirely of wrought Iron plates, for sustaining this very violent mode of loading; stout planks of deal are laid along their bottoms for receiving the shocks of the first layers of stone, in thus loading these Iron Boats. The Trams used on this Railway have also plate-iron bottoms and sides, and hold about 34 to 35 cwt. of Stone, in blocks of ½ to 3 or 4 cwt. each: the wheels are cast with round holes in them instead of open spokes, and thro' these, short truncheons of wood are put, for locking the wheels, while the loaded Trams descend down steep hurries from the higher parts of the Quarry; into these holes they also put levers, to turn the wheels about, occasionally.

Five of these Trams were drawn by one horse; but it appears from the Monthly Magazine, Vol. 37, p. 62, that one of Mr. William Brunton's patent Propellers, worked by a Steam-engine, made at Butterley, was tried here in Nov. 1813, with success, and that they were intended to be established here, for entirely superseding the use of Horses on this Rail-way: this machine acts by legs or propellers stepping on the ground, behind the Engine, mounted on Tram wheels, and not by a cog wheel acting in cogs cast on the Tram-plates, as John Blinkinsop's patent "Iron Horses" do, which are made by Messrs. Murray and Wood of Leeds, that have been more than 2 years at work at Middleton Colliery, 2 m. S of Leeds; to which place one of them daily brings 400 Tons of Coals or more: at Willington Colliery, near Newcastle, one of them has been some time in work, and 2 others were now making (April 1814), for the same Coal-master: the general use of one or other of these machines, for avoiding the use of Horses, and more cheaply conveying Coals and other heavy articles on Rail-ways, would be a great thing for the Country.

Besides the 12 Rail-ways above enumerated, that branch from the main line of this Canal, there are from the Pinxton branch, others of ½ a mile W, from the Somercotes Furnace cut, to Riddings Collieries and Ironstone Pits: from near Pye-bridge W, ¼ of a mile, to Somercotes Colliery: from ditto NW, ½ a mile, to Nether Birchwood Colliery: from Pinxton Wharf NE, ¼ of a mile, to Pinxton lower, or South Colliery, on which Wooden Rails remained in use in 1808: and from ditto NE, ¼ of a mile, to Pinxton upper Colliery.

In the Act for this Canal, there is *no distance 'limited*, within which Rail-way branches may be made, to connect therewith (as observed, page 287), but such may be freely made by the owners of Mines, or their Lessees or Tenants; collateral branches of Canal may also be made in their own lands, or by consent, on erecting Stop-gates, and not injuring of the supplies of the Canal, or of any Mill or Furnace with water.

Butterley and Somercotes Iron Furnaces, Foundries, and Works (Rep. I. 397), and others more recently erected in Codnor Park, are on the banks of this Canal. In the Coke Yard on the E side of Butterley Furnace, two large Shafts descend to a recess for Boats, adjoining the Canal Tunnel, thro' which the large Tram boxes of Coals, Ironstone, Limestone, Fluor, &c. are drawn up, for the use of the Works; and Pig Iron and Cast Goods, &c. are lowered into the Boats below, to be sent off by the Canal. Formerly, a large water-bucket, supplied from a Reservoir, descended in another Shaft, as a counterpoise for drawing or lowering Goods in these Shafts, but a very complete *Whimsey Steam-*

Steam-engine has been substituted; guide chains descend the drawing and lowering Shafts, to steady the frames that suspend the Tram-boxes: which last are held suspended over the Shaft, while a Stage is slid over it, on which a pair of Wheels, and a Horse attached to them by shafts, have been backed; the Tram-box is then lowered and placed on the Wheels, the stage is slid again off the Shaft, and the Horse then proceeds with the Tram-box and its contents, to any part of the Works, and returns in like manner to the other Shaft, with Goods that are to be lowered in like manner and sent off.

Near to the line in Alderwasley, there are Lead-works, and others near to Lea-wood Wharf, and Cotton-Mills, a large Hat Factory, &c.: at Cromford also there are extensive Cotton-mills, &c. at the termination of the Canal: at the termination of the Pinxton branch, there has a considerable China Manufactory been established. The Shale Freestone of White Tor and Combs-wood Quarries, is carted to, and sent off in considerable quantities from Lea-wood Wharf, to the southward; Mr. Benjamin Bevan, the Engineer, has used this stone for the copings of Locks and Bridges on the *Grand Union* Canal, &c.: this stone in the vicinity of Leicester is sometimes called " Mansfield Stone ;" why this egregious misnomer I cannot tell. The Free Limestone of the 4th Rock, from Hoptonwood Quarry, is also carted to, and prepared for exporting, at a Saw-mill erected at Lea-wood Wharf. At Coddington the Canal approaches close to the famous Millstone Grit, or 1st Grit of this district, and considerable quantities of this stone are put on board, from Carr Quarry therein: some of the stone from the same

Rock, but of a Salmon-coloured tinge, are also put on board at Cromford, from Stone-house Rocks.

Shale Freestone is also dug at Bull-bridge Quarry, and has been tried at Butterley Furnace, as a Firestone: the same Quarry also produces some grey Slates or Tile-stones. Near to Pentrich-lane on the W, this Canal crosses the range of the celebrated *paving-stone* Rock of Ealand-Edge and Cromel-bottom (I. 164), on the *Calder* Navigation, whence, and by which means, London, and a great part of the South of England, has been supplied for years past with " Yorkshire paving;" the only other Navigations that cross this 4th Rock, are the *Huddersfield* (or *Ramsden's*) Canal, near that Town, and the *Leeds and Liverpool* near Kirkstal-bridge: Coburn Quarry, in South-Winfield Park, (I. 423) within $1\frac{1}{4}$ m. N of this Canal, produces paving-stone, no ways inferior to the Yorkshire, and some grey Slate or Tile-stones, and blocks of fine Freestone, some of which are now carted to the Canal at Pentrich-lane.

From the *Erewash* Canal (and *Nottingham*) to the Pinxton branch at Golden Valley, is $3\frac{1}{2}$ miles, with a rise of 80 feet; thence to Pentrich-lane, four miles is level: thence to the S end of Bull-bridge Aqueduct, one mile is level: thence to Lea-wood cut, at the E end of the Derwent Aqueduct, $4\frac{1}{2}$ miles is level, and thence to Cromford Wharf, $1\frac{1}{4}$ miles is also level: the Lea-wood branch has a Lock at its commencement, but at times, its water is not higher than that in the Canal. The Pinxton branch and the Somercotes cut are level with the summit pound. The width of this Canal at top is 26 feet, the Boats are 80 feet long, $7\frac{1}{2}$ feet wide, and $3\frac{1}{2}$ feet deep, and when empty, they

draw

draw eight or nine inches of water, and when loaded with 22 tons, they draw about 2¼ feet. The excellent regulations, as to the guaging and ascertaining the lading of the Boats, concerted and conducted on a uniform plan, on this and eight other adjacent Navigations, have been mentioned in Vol. I. p. 183 and 184.

The *Tunnel* at Butterley, nearly E, 2978 yards, is driven in Coal-measures, about 57 yards below the Ridge, and is lined with Bricks, except in such places as Rocks were perforated, which appeared capable of standing themselves; the Tunnel being nine feet wide inside at the water's edge, and the crown of the arch eight feet above this. Mr. William Jessop has a very curious Section of this Tunnel, from which I took the following measurements and particulars of the Strata penetrated, faults, &c. beginning at the E end, viz.

- 352 yards, have a dip W, of 2 in 3, in which length were five Tunnel-pits (No 1 to 5), and two Coal-seams appeared in some parts of these measures, in and above the Tunnel, of 18 and 15 inches thick, respectively, (the first mentioned uppermost: query, 12 and 30 inches thick?); a fault then crosses the line, and raises the measures 18 yards on its W side.
- 440 yards, dip E about 1 in 14, three Tunnel-pits (No. 6 to No. 8), and three Coals, of 12, 30, and 12 inches thick, respectively.
- 445 yards, level, three Tunnel-pits (No. 9 to No. 11), and three Coals, as above, the middlemost of which Coals, has a sough driven in it from the Tunnel, to Butterley-park Colliery (see

1237

Vol. I. p. 192): then a fault crosses, which lets down the measures on its W side one yard.

580 yards, level, four Tunnel-pits (No. 12 to 15), and three Coals, as above, through part of this length; the upper one bassets at 220 yards from the last fault, and the lowest of them lays nearly on a Gritstone Rock (the 16th?), which ranges along the lower part of the Tunnel, through all this and the two preceding lengths.

342 yards, dip W, 1 in 8 (on the average, there being a hollow in the middle), 4 Tunnel Pits (No 16 to No. 19), and two Coals, as above.

20 yards, in fault-stuff, that crosses obliquely, and which fault probably lets down the measures 22 yards on its S W side.

213 yards, dip E, 1 in 2, Tunnel Pits (No. 20 to No.   ) and two Coals, of 24 and 12 inches thick respectively, the first (and uppermost) of which bassets 115 yards from the last fault: another fault then crosses, and sinks the measures on the W side, 1 yard.

178 yards, dip E, 1 in 3, 2 Coals, of 12 and 30 inches, thick respectively, the latter being Cannel Coal: a fault then crosses, and lets down the measures on the W side, 15 yards.

200 yards, dip E, almost 1 in 2, 1 Coal of 24 inches (quere 30 inches?): a fault then crosses, and rises the measures on its W side 50 yards:—and lastly,

208 yards, dip E, about 1 in 1, Tunnel Pits (No.   to No. 33), and 2 Coals, 24 inches and 12 inches thick respectively, both of which basset over the Tunnel, the lower one, at about 19 yards from its W end.

2978 yards of Tunnel.

In

In six places near to the W end of this Tunnel, and one near the E end, the measures are stated to have run or crowned-in, or fallen, before the arch could be turned. According to the information that my friend Mr. William Smith collected in 1794, this Tunnel cost about 7*l.* per yard in length.

At the SE point of Crich Chace, ¼ *m.* W of Bullbridge, there is a short Tunnel, through Limestone Shale, for avoiding a loop in the line.

On the NE of Wigwell, this Canal is carried over the Derwent on a large Aqueduct bridge, 200 yards long, and 30 feet high, which was built in 1792; the River arch is 80 feet span, with a smaller one on the meadows on each side, for private Roads, one of which, from Belper to Cromford, has been mentioned, pages 226 and 271; two years after the Canal was opened, the dry or moss-laid rubble Wall, at the north-west corner of this Aqueduct, gave way, and the Canal burst: when I saw it in 1808, these dry walls had a very great batter or slope.

Over the Amber River at Bull-bridge, there is another very considerable Aqueduct, about 200 yards long, and 50 feet high, built of Shale Freestone, consisting of a large arch for the River, and a smaller one for a Mill-lead S of it, and also a *Gothick* arch for the Turnpike Road, which, owing to its improper shape, is bulged a good deal, and on the N of the River is another arch for a private Road under the Canal. These two Aqueducts are said to have cost together, 6,000*l.* On the W side of Brinsley, this Canal passes over a low Aqueduct on the Erewash River, between Nottinghamshire and Derbyshire.

There is a considerable Deep-cutting at the end of the great Tunnel near Padley Hall, the stuff from

which

which has been very judiciously disposed of, in forming the head for a large Reservoir of 50 acres, over the west end of the Tunnel; this head is 200 yards long, and 33 feet high in the middle, its base being there 52 yards wide, and its top is 4 yards wide; the cost is said to have been 1,600*l.*; the mean depth of water is 12 feet, and it contains about 2,800 lock-fulls of Water, which is let out, when wanted, by a large pipe and cock, in one of the Tunnel Pits.

The stuff from a smaller Deep-cutting in Golden Valley, has furnished the head for a smaller Reservoir on the E end of the Tunnel. Lower down in Golden Valley, there is another considerable Reservoir, and a smaller one at Swanwick Delves.

It was intended at first, to supply this Canal by a feeder from the Derwent in Matlock-Bath Dale, not exceeding $\frac{1}{70}$th of the stream at any time, weekly, from 8 o'clock on Saturday Evening, to the same hour on Sunday Evening, estimated to be 41,040 Tons of water, at least, weekly: but the plan was afterwards changed, for turning the very large stream of warm water from Cromford Sough (see the " Philosophical Magazine," Vol. 43, p.   ) into the head of this Canal, at such times as it is not used at Richard Arkwright's, Esq. Cotton Mill; and I believe some of this water goes into the Canal, during almost every night, the consequence of which frequent supply of *warm* water, is, that this Canal, to the W of Butterley Tunnel, very rarely if ever freezes, as I have lately been told, but often emits a steam from its surface, which some Writers have, without any reason, ascribed to volcanic fires under it: the place where the hot Spring originally vented itself, that has been long drained by Cromford Sough, is full two miles W of the head of the Canal, by the side of a great fault,

fault, Vol. I. pp. 64 and 505. The whole of the long level of this Canal is made one foot deeper than necessary, under the idea of acting as a Reservoir in dry seasons.

This Company is authorised, to require Mine-owners within 1000 yards of their line and branches, to lift their Water high enough to run by proper feeders into this Canal, and if extra expense of pumping is thereby occasioned, the Company are to contribute their share, for the extra height such water was lifted: but the Company are restricted from erecting or supplying any Mills with water: the Fisheries in this Canal are reserved to the respective Lords of the Manors, but they must not let off its waters, or restrain the Company from so doing.

Coal-masters are restrained from working under the Canal, until after giving notice to the Company, to purchase the Coals under the same, at a valuation; the Company's Agents may enter adjacent Coal-pits to survey, and if Coals have been worked under or too near the Canal, without notice, and refusal or neglect to purchase by the Company, they may effectually wall up and secure the same again, at the expense of the Coal-master.

It having been supposed, that this Company were liable to pay for pumping all the Canal Water that might leak down into a Coal-pit, in consequence of the Company refusing to purchase, and permitting the Coal-owner to work under their Canal, I was told, that this Company, several years ago (on the advice of Mr. Joseph Butler) paid to the Lessees of Codnor nether Park Colliery, a moderate price for their Coals under the Canal (not including the profits on getting), on condition of being exonerated from such

such pumping claim, and then leaving it to the option of the Coal-master, to work these Coals under the Canal or not, as he may think fit.

In consequence, however, of the little damage that the *Nottingham* and *Erewash* Canals have sustained, from Coals being wrought *(by the long-way*, Vol. I. p. 344, here almost universally practised) under those Canals, I was told, that this Company, a few years ago (on the advice of Mr. Thomas Walker) refused to purchase the Coals under their Canal, that were won by the new foundation put down S of the last, on the same Colliery, but left the parties to work the same, or not, as they may think fit: these appeared to me to be circumstances so interesting to Mine as well as Canal-owners, as to be worth recording.

The Hedges by the Towing-path of this Canal in Codnor-Park, and other places, are neatly clipt and kept. The Engineers employed upon this Canal, were Mr. William Jessop, sen., Mr. — Dadford, Mr. — Sheasby, Mr. Benjamin Outram, and Mr. Edward Fletcher, and it was completed about the year 1793.

The Tonnage allowed to be taken on this Canal, is not to exceed,

1*d.* per Ton per mile for Coals, Coke, Lime, or Limestone, intended for, or broken for burning.

1½*d.* per Ton per mile for all other Goods, which have not passsed *from* the *Erewash* Canal.

2*d.* per Ton per mile for all Goods which have so passed.

3*d.* per Ton extra, on all Goods (except Coals, Coke, Lime or Limestone for burning), passing from or to this Canal, and the *Erewash* Canal.

12*d.* per Ton extra, on Coals navigated between the Amber

Amber Aqueduct and Cromford, or within two miles east of that Aqueduct, passing towards it.

Fractions of a Mile, and of quarters of a Ton, are to be considered as whole Miles, and $\frac{1}{4}$th of Tons, in calculating Tonnage; 50 feet round measure of Oak, Ash, Elm, or Beech Timber, or 40 feet square-measure (see Vol. II. p. 319) of such Timber, or 50 cubic feet of Fir, Deal, or Poplar Timber, shall be deemed a Ton.

Gravel, Rubble or Paving stone for the Roads (not Turnpike), and Manures (except Lime) for use in the Parishes through which the Canal passes, are to go Toll-free on the pounds, between the Locks, and thro' them when the water flows over their weirs, on giving six hours notice to the nearest Toll-collector.

All Wharfs made by the side of this Canal, are to be public ones, and only 1d. per Ton for Coals, Lime, Limestone, Clay, Iron, Ironstone, Timber, Stone, Bricks, Tiles, Slate or Gravel, and 3d. for other Goods may be taken, as Wharfage by the owners, for six days laying, and longer for some articles that are specified in the Act: no Wharfage, is to be taken at the Company's Wharfs, until after six months.

At Cromford Wharf there are large Warehouses for Goods, others in Golden Valley, and at Pinxton.

This Company were authorised to raise 46,000*l.* in 100*l.* Shares, and 20,000*l.* more on Interest or Mortgage of Tolls; paying interest also on their Shares during the making of the Canal: the total cost of which is said to have exceeded 80,000*l.* The Dividend per share was 10*l.* per annum in 1810.

The price usually paid for cutting and wheeling of Clay and Earth in making this Canal, was $3\frac{1}{2}d$. per cubic yard, per stage of 20 yards: for Gravel or Rubble,

Rubble, 4¼d. per yard; for stony ground 4½d. per yard; besides 4d. per cubic yard, in each case, for all stones got out and stacked. At Brinsley, Bull-bridge (or Amber Wharf), Codnor lower Park, Cromford, Langley-mill, Pinxton, and Pye-bridge Wharfs, there are Lime Kilns, &c. see Vol. II. pp. 419 to 428.

About the year 1771, the *Chesterfield and Swarkestone* Canal was proposed, to cross the line now occupied by this Canal, near Pentrich-lane. Soon after the passing of this Act in 1789, and again in 1802, the *Cromford and Bakewell* Canal was proposed to connect with this Canal at Lea-wood Wharf. In 1801, the *Belper* Canal was proposed to join this at Bull-bridge. In 1810, the *High Peak Junction* was proposed to join this Canal at the Derwent Aqueduct. And in 1810 also, the *North-eastern* Canal was proposed, either to join this Canal at Pinxton, or near to Pentrich-lane.

When this Sheet was in the Press (in May 1814), Mr. Richard Wilson, Stone Merchant of Millbank-street Westminster, called on me, and mentioned, that he had formed the design of making a Rail-way branch from the head of this Canal, through Cromford Town, up Bonsal and Griffe Dales, and following nearly the line of the new Turnpike Road, to near New Haven, and thence skirting the Limestone heights, in the best practicable line (principally thro' the Duke of Devonshire's Estate) to Thirkelow Gate (4th) Limestone Quarries (Vol. II. p. 424); thence crossing on the inosculation between the Wye and Dove Rivers, and thro' a short Tunnel under the *Grand Ridge* at Thatch Marsh; crossing the Congleton Road on Goyte Moss (passing near its Collieries, I. 198), and across the Macclesfield Road W of Moss-houses, and thence

thence down to Goytes-clough (2d Grit) Paving-stone Quarry in Cheshire (Vol. I. p. 424 and 430), a distance altogether, of 27 to 30 miles.

And he mentioned also, that it was probable, that this line would be joined near to the Congleton Road on Goyte Moss, by a Rail-way branch from the Caldon branch of the *Trent and Mersey* Canal, near Leek Town. And also, that a Rail-way branch from the *Peak Forest* Canal at Whaley-bridge, would probably join these at Goytes-clough Quarry.

The same Gentleman mentioned other schemes that he had, in case the above should not succeed, of a Rail-way branch from the head of this Canal at Cromford, passing up Matlock-Bath Dale; thence following the Vales of the Derwent and Wye to Bakewell Town, and thence by Ashford (past the Chert Quarries, I. 273, and Black Marble Quarries, I. 231), and still up the Wye and Sherbrook Dales (passing near to Buxton), to a summit and Quarries in the 4th Limestone; and thence thro' the *Grand Ridge* NW of Edge-end House, by a Tunnel (that would cut the 1st Coals, or Thatch-marsh seam, I. 212), and thence across the Goyte to Goytes-clough Quarry, as above: And from this line, above Bakewell, he mentioned also a scheme for a Rail-way, to the Eastward, across the Derwent near Baslow, and following thence nearly the line described in Mr. Gratton's scheme for the *Baslow and Chesterfield* proposed Canal, p. 311, to join the *Chesterfield* Canal at that town.

Unlikely as these several schemes appear to be carried into effect, some of them possibly may be so, and I have thought, that it would not be right to let the present opportunity slip, of recording them, for future and more mature consideration, by the Parties most

most interested:—I confess myself by no means sanguine in thinking, that the compound route *by Canal and Rail-way*, that might by these means be opened between London and Manchester, across our Map from *F* to q, and from Manchester eastward, from q to *Q*, &c. would prove at all preferable to the more circuitous routes, *by Canals*, that cross our Map from m to *N*, and will do so from q to D, and I to *F:* because of the great objections to exposing packages of light and valuable Goods, to the very increased dangers of *damage and pilfering*, when exposed in separate Rail-way Trams, and in the frequent loading and unloading, consequent on this mode of conveyance, instead of being securely placed in a Boat, and covered and fastened down, at the Wharf at Manchester, and remaining undisturbed until they arrive in London, and *vice versa*. The Rail-way branch from the *Ashby-de-la-Zouch* Canal, that passes through that Town, was reported to me when there, as having almost wholly failed in the expected carriage of packages of Goods, such as the Waggons still are employed to convey, on account of the security they afford from damage and loss; and I doubt not but the result would be similar, of inquiries, as to the *general carriage of Goods* on the *Surrey Iron Rail-way*, and perhaps most others in the Kingdom.

Cromford and Bakewell proposed Canal:—In the year 1789, Mr. Benjamin Outram projected a Canal, and Mr. John Nuttall made a Survey, from the *Cromford* Canal (the works of which were then commencing), at the intended Aqueduct S of Lea-wood, passing through a branch from the *West Amber* Ridge, to near Tansley, and pursuing the Vallies of the Derwent and Wye Rivers to the Town of Bakewell:
about

about 11¼ miles, in a NNW direction; the northern end being considerably elevated: its objects were the import of Coals, Malt, Deals, &c. and the export of Lead, Marble, Chert, &c.

Cromford is the 658th British Town, with 1,259 Inhabitants, and Bakewell the 591st, with 1,485 persons.

Mr. Nuttall extended his levels northward by Buxton to Whaley-bridge, and to the Mersey River at Stockport Bridge, which last he found to be 52 feet lower than the Canal level at Cromford. In 1810, the *High Peak Junction* Canal was proposed to occupy very nearly the same line as this, see also page 351.

DEARNE AND DOVE *Canal:* Acts, 33d and 40th Geo. III.—The general direction of this Canal is about NW for 10 miles, in the West Riding of Yorkshire: its northern end is considerably elevated, crossing the *North Dove* Ridge by a deep-cutting near Ardsley, and the *South Dearne* Ridge NNE of Swinton Chapel: its general objects are the export of Coals, Iron, Grinding, Building, and Paving Stones; the import of Limestone, Deals (or Raff), and the forming of a communication for the great manufacturing districts of Sheffield and Rotherham on the *Don* (by means of the *Barnsley* Canal), with Wakefield, Huddersfield, Halifax, Manchester, Liverpool, Leeds, Bradford, &c. &c.—Barnsley, near to this line, is the 153th British Town, with 5,014 Inhabitants, and Rotheram, at no great distance from it, is the 315th, with 2,950 persons.

This Canal commences in a side-cut of the *Don* (or Dun) Navigation, near Swinton Chapel, and terminates in the *Barnsley* Canal at Eymings-wood, ½ m. E of Barnsley: from Knoll Brook in Womb-

'well, there is a branch SW, 2¼ miles to Elsicar Lower Furnace; and from near Swith-hall there is a branch W, 2 miles, to Worsborough Bridge and Furnace; and whence a branch is provided for, SW 1½ miles, to Rockcliff-bridge, but it has not yet been executed, I believe: and there are probably some Rail-way branches, to the Collieries near the line and its branches, such being provided for in the Act, to the extent of 1000 yards distance, and to 2000 yards near Wath.

From the *Don* Navigation to Knoll Brook Aqueduct, 4½ miles, has a rise of 41¼ feet: thence to the Elsicar (or Cobcar-Ing) branch, ¼ mile, has a rise of 2¼ feet: thence to Aldham-Mill Aqueduct, 2¼ miles, is level: thence to the Rockcliff (or Worsborough Bridge) branch, ¼ of a mile, has a rise of 59¾ feet: and thence to the *Barnsley* Canal, 2⅞ miles, is level. The Elsicar branch to Cobcar-Ing, 1¼ miles, is level; the remainder of this branch, ⅞ of a mile, to Elsicar Lower Furnace, has a rise by Locks, and was made about the year 1797 by Earl Fitzwilliam, on condition of being allowed to draw water for the Locks, from the Elsicar Reservoir, belonging to this Company. The Rockcliff branch to Worsborough Bridge, is level; and thence to Rockcliff Bridge, 1¾, is a rise of 56 feet.

The width of this Canal and its Locks are adapted for 50 or 60 Ton Boats, such as navigate the *Don* River, and for their accommodation, this Company has engaged to keep a depth of 4½ feet of water on their lock-sills, at all times. The Locks and Aqueducts, &c. on this Canal, are built with excellent hewn Stone. At Elsicar, this Company have a large Reservoir, and feeders from different Brooks; but Tumbling-bays or Guage-weirs are directed to be made by the Act, for supplying several Mills, without diminution of their
water

water in dry seasons. Stop-gates are provided at Ey-mings-wood, for preventing either this Canal or the *Barnsley* from drawing down the other's Water, when its own supplies may fail. Mr. John Thompson was an Engineer employed on this Canal, which was finished in the year 1804.

This Company were empowered to raise 100,000*l*. in 100*l*. Shares: their rates of Tonnage being various and complicated, I must refer for them to Phillips's 4to. " History of Inland Navigation," Appendix, p. 62 to 66, and to the Act of 40th Geo. III. which very considerably increased the Tolls there mentioned, I believe. Boats are to pay Tonnage for 6 miles of distance, however short their course may have been on this Canal.

DERBY *Canal:* Act, 33 Geo. III.—The general direction of this Canal is nearly NE, by a bending course to the NW of 14¼ miles, in the County of Derby, passing its County Town: it is not much elevated in any part; it crosses the *West Derwent* Ridge by a slight cutting, NE of Skcton-leys in Chellaston, and the *East Derwent* Ridge by a cutting equally slight, on the NW of Breaston: the general objects are, the supply of the Town of Derby with Coals, Building-stone, Gypsum, and other articles, and exporting Coals from the Pits on its northern branches, manufactured Goods, and Cheese and other agricultural products, and the forming a nearer conveyance for the Peak Limestone, into the south-eastern parts of the County, &c. Derby is the 45th Town on the British population list, with 13,043 persons, and is the only considerable Town near to this Canal.

It commences in the *Trent and Mersey* Canal, N of

the Town of Swarkestone, and terminates in the *Erewash* Canal, ¼ a mile S of Sandiacre: from near Derby a branch proceeds N to Little Eaton, 3 miles, and is continued thence up the Bootle Vale, 6 miles further, to Roby west-field Colliery in Denby: from this Rail-way extension, there is a branch W ¼ of a mile, to Denby-hall Colliery*: one was provided for in the Act, E to Smalley Mill, and to Horsley Collieries: and there are two short branches into Little Eaton Common Quarries (Vol. I. p. 419).

In the NE part of Derby Town, a short cut and a Lock, conducts Boats into the pound of the River above the Silk-mills Dam, near to St. Mary's Bridge, so that they can proceed N up the Derwent 1¼ miles, to Darley Mill. This Company has also a detached short length of Canal, between the *Trent and Mersey* Canal and *Trent* River near to Swarkestone, ¼ of a mile W from the commencement of their line, which detached part seems now of little use, since the discontinuance of the Navigation on this River: and they have also engaged to make another such detached junction between the *Trent and Mersey* Canal and the *Trent* River, at Weston Cliff, opposite to the proposed *Breedon* Canal and Rail-way, in case that the same should be carried into effect. At first, this Canal connected with the *Derwent* Navigation at the Town of Derby, I believe, but soon after, this Company purchased up the Shares of that Company, and shut up the River Navigation, as being very inconvenient, and now useless.

---

* After page 313 was in the press, I learned from Mr. Charles Sylvester of Derby, who called on me, that another Rail-way has been made from this Colliery, thro' the southern part of Morley-park to Belper Town, see *Belper and Morley-park* Rail-way.

From

From the *Trent and Mersey* Canal to the stop-lock at Cock-pit-hill Wharf in Derby, 5¼ miles, is a rise of 12 feet, by 2 Locks (at Skelton-leys): thence across the Morledge and the Derwent Rivers (thro' the lower Dam) to the Darley-Mill branch, and the Warehouses in St. Alkmund, ¼ of a mile, is level: thence to the Little Eaton branch, ⅜ of a mile, is level: and thence to the *Erewash* Canal, 8¼ miles, has a fall of 29 feet, by 4 Locks (2 of them near the Erewash, and 2 at Burrowash Mill). The detached part, ¾ of a mile long, has a fall of 3 Locks to the *Trent*: the Little Eaton branch, 3 miles, has a rise of 17 feet by 4 Locks. The lengths on the Rail-way extension are as follows, viz. from the Wharf at Little Eaton to the branches into Little Eaton Common Quarries, ⅜ of a mile: thence to the proposed Smalley Mill branch, 1¼ mile: thence to the Denby-hall branch, 3¼ miles; and thence to Roby west-field Colliery, ⅜ of a mile: all these Rail-way branches rise from the line, but I am unacquainted with their exact levels.

This Canal is 44 feet wide at top, 24 at bottom, and 5 feet deep, except the summit pound of the Little Eaton branch, about 1¼ miles long, which is cut 6 feet deep, in order to act as a Reservoir, after wet seasons. The Locks are 90 feet long, and 15 feet wide, inside.

A Market Boat, decked over, with seats, and a fire-place, for the accommodation of Passengers, starts from Swarkestone every Friday Morning, to carry Market-people to Derby, at 6*d.* each: and which again leaves Derby at 4 o'clock for Swarkestone: this plan is worthy of far more general adoption on the British Canals.

Over the small River Morledge, on the SE side of Derby Town, this Canal is conveyed in a low, Cast-

Iron Trough or Aqueduct, erected in 1795: near the E corner of Synfin Fen, there is a small Aqueduct Bridge and a high and long Embankment, over a small Brook, and other small Aqueducts over Chaddesden, Oakbrook, and Rislip Brooks: across the Derwent River on the E side of Derby, there is a high Weir, 100 yards long, for the joint use of the Mills, and this Canal, whose barges cross the Derwent in the pound above this Weir.

This Canal is fed by its Little-Eaton Branch, in part from the Bootle River, and in part by a feeder near a mile long (and is in places cut 13 or 14 feet deep) from the Dam of an old Mill on the Derwent below Duffield Bridge, where a guage-sluice, 2 feet wide, 2 feet 2 inches high, and which generally has a head of 6 feet 2 inches above its cil, was erected in 1806; but it is limited in the use of both these supplies, to 4 hours on Sunday and 4 hours on Thursday Evenings: and in order that no more water may be taken from the Mills at Derby (where there generally is such a profusion of it), stop-locks have been erected on both sides of the Derwent, the water in the Canal on both sides must be kept higher than the River, and a close trunk feeder ¼ mile long, has been laid from one side to the other, passing under the large Weir above mentioned, and by the side of the Iron Aqueduct, for supplying the summit pound on the southern side of Derby; which it does but imperfectly, at times, for on the 18th of May, 1809, I saw Boats laying at Boulton, in want of water to proceed: this seems too much allied to other instances, of the partiality shewn to Mill-owners, that I have alluded to in p. 291, who, with equal justice might, and ought equally to be *obliged to sell* a part, or even the whole of their *property in the use.*

of

of Water, as the *Land-owners to sell their Land for works* of public accommodation and benefit, like the present. Mr. Benjamin Outram was the Engineer to this Canal, which was finished in the year 1794.

Separate rates of Tonnage are limited by the Act, on different parts of this Canal and its branches, which renders them too long for insertion here: they may be seen in Phillips's 4to. " History of Inland Navigation," Appendix, p. 55 to 59. Manures are to pass toll-free, and Puncheons or Clogs of Wood for the adjacent Coal-pits (see Vol. I. p. 347, and II. p. 222, &c.), also all Road-materials, except for Turnpikes: and if the Derby and Mansfield Turnpike-road Tolls, are reduced below 4 per cent. on their debt, this Company is to make them up to that sum!; I am not acquainted, whether this applied beyond the *then existing term* of this Turnpike Act?. 5,000 Tons of Coals, annually, are to be allowed to pass to Derby, *toll-free*, for the use of the Poor thereof (or in aid of its Poor-rates!). The regulations for guaging of Boats, and ascertaining of the tonnage of Coals, adopted by this Company in concert with eight other Companies, have been mentioned in Vol. I. p. 182.

This Company were authorized to raise 90,000*l.* in 100*l.* Shares, on which the Dividends are never to exceed 8 per cent. annually; but when 4,000*l.* is accumulated as a Stock for contingencies, the Tolls are to be reduced. The Tolls are now annually let by Auction, from the 10th of October: Mr. John Curzon of Swarkestone, is Clerk to the Company.

In St. Alkmund in Derby, there are large Warehouses, under which the Boats pass, to load and unload: At Breaston and Draycot, Burrowash and Spondon, Chaddesden and Derby Wharfs, there are Limekilns

kilns (see Vol. II. p. 419 to 422), and Gypsum Kilns at Skelton Leys. At Derby there are numerous Manufactories, on its banks, Iron Mills at Burrowash, &c.

About the year 1771, the *Chesterfield and Swarkestone* Canal was in agitation, intended to occupy nearly the same ground southward of Derby, and north of it to Little Eaton, as this Canal does, and thence to proceed northward up the Vale of the Derwent: about 1793 the *Breedon* Canal and Rail-way was intended nearly to connect with this, and the *Derwent River* Navigation did so, I believe, when this Canal was first completed.

Derwent (Derby) Navigation, now discontinued.— The general course of this navigable part of the River was NW, by a crooked course of near 12 miles, in the County of Derby, commencing in the *Trent* River (where the *Trent and Mersey* Canal also commences) at Wilden Ferry, and terminated at the Town of Derby (the 45th, with 13,043 Inhabitants): its general objects were, the supply of that Town, and the export of its manufactures. It being expected, that the Trade and Tolls on this Navigation would decline greatly, on the completion of the *Derby* Canal, that Company engaged to purchase the Shares of this concern, for the sum of 3,996*l*.; and in 1794 the Navigation on this River was in consequence entirely discontinued. At the Mills at Little Wilne and Burrowash, there were Locks, I believe: see an account of the Strata crossed by this River, in Vol. I. p. 471.

Dilhorn proposed Canal.—In the year 1792, Mr. John Nuttall surveyed the line for a Canal, between the *Trent and Mersey* Canal, near Monks Bridge in Stafford-

Staffordshire, and Dilhorn Collieries, ¼ a mile NE of that Town; a distance of about 24 miles in a NW direction, in the County of Stafford. Its objects were the export of Coals, Gypsum, &c.; the supply of Uttoxeter and Cheadle, which are in its route (the former being the 283rd British Town, with 3,155 Inhabitants, and the latter the 279th, with 3,191 persons); which route closely followed the Dove River (passing near to the Gypsum Quarries at Fauld-Hill, Row-Bank, &c. Vol. 1. p. 151), on the E side of Uttoxeter, then for some distance occupying nearly the same ground as the Caldon branch of the *Trent and Mersey* Canal has since done, and thence up the SW side of the Cheadle River, past Eaves Colliery and the Brass-works, &c. to Dilhorn Colliery. Four years later, this same line, nearly, was surveyed again for part of the route of the intended *Commercial* Canal.

Don *Navigation*, or Dun; Act, 12th Geo. II., and 33rd Geo. III. (the latter, for *Stainforth and Keadby* Canal).—The general direction of the navigable part of this River, is nearly SW for about 40 miles, in the West Riding of Yorkshire (including what in some Maps is called the Dutch River, from Gool Bridge to New Bridge at its north-eastern end): it enters our square of Map (p. 193) at L: its north-eastern end is Embanked through a flat country, (where the Warping of Lands is practised, see "Annals of Agriculture," Vol. 33, p. 396), and the Tide flows above the mouth of the River Went, but the other end is considerably elevated. Its principal objects are, the supply of Sheffield with Coals, Limestone, Foreign Iron, Grindstones, Deals (or Raff), Corn, Malt, &c.; the export of its manufactured Goods, Coals, Limestone,

Pig

Pig Iron, &c. and forming part of the line of communication between the manufacturing Towns of Sheffield, Rotherham, Barnsley, Wakefield, Huddersfield, Halifax, Leeds, Bradford, &c.

Sheffield, 3¼ miles from its termination, is the 14th British Town, with a population of 35,840 persons, Doncaster is the 100th, with 6,935 persons, Rotherham the 315th, with 2,950 persons, and Thorne the 345th, with 2,713 persons, and Snaith is also a considerable Town near its course. It commences in the Tide-way in the York *Ouse* River at Gool-Bridge, E of Snaith, and terminates at Tinsley Wharf, 2¼ miles from Attercliffe Bridge, where, according to the Act, it was intended to terminate, but the smallness, great fall, and crookedness of the River, and the great value and importance of the Mills upon it, has hitherto prevented this part of the work from being carried into execution, and a paved Road, between Sheffield and Tinsley (which is free from hills) has been substituted.

At New Bridge, SW of Snaith, it connects with a cut or branch nearly N, of about ¾ of a mile, to the *Ayre and Calder* Navigation: with the *Stainforth and Keadby* Canal branch at Hangman-hill near Thorne, and with its line at Fishlake near Stainforth; and with the *Dearne and Dove* Canal, SE of Swinton Chapel. In several places there are side-cuts and locks for passing or avoiding the Mills and the shoals in the River. A mile above Rotherham Bridge, there is a cut about ¼ of a mile north east (with a Lock) to Massborough-holms Iron Furnace, and the extensive Works of Messrs. Walkers.

Near to the NW end of Rotherham Town, there is a Coal-yard and Tipples for loading Carts, at the commencement of a Rail-way NW, about ¾ of a mile, to Clough

Clough Colliery. At Darnal, not far from the intended extension of this Navigation, there is a Coal-yard and Tipples, and a Rail-way with an Inclined-plane, to High-hazles Colliery, and a branch to the Steam-Engine belonging thereto, yet neither of these Rail-ways connect with the Navigation, but are used for the supply of Rotherham and Sheffield Towns, principally, by means of Carts.

In 1803, notices were given of an intended application to Parliament, for a new Act for Weirs and side-cuts to this River, in Mexborough, Spotborough and other places, and for making a new course for the River near to the junction of the Dearne River: in February 1803, there was a plan on foot, and a Subscription of 30,000*l.* made, for extending this Navigation, by a Canal from Tinsley up to Sheffield Town: in 1810 the *North-eastern* Canal was proposed to join this navigation a little above Rotherham: and in 1811, the *Tinsley and Grindleford-bridge* Canal was proposed, to join this Navigation at the former place. In the present year (1814) Meetings have been held, and an application to Parliament has been determined on, for a separate Act for the northern part of the proposed *North-eastern* Canal, between the *Don* Navigation and the *Chesterfield* Canal near Killamarsh.

Erewash *Canal*: Acts, 17th and 29th *(Cromford)* Geo. III.—The general direction of this Canal is nearly N for 10¼ miles, in the Counties of Derby and Nottingham, following the course of the vale of the River Erewash, its northern end is rather elevated: its chief objects are, the export of Coals, Limestone, Iron, Lead, Mill-stones, Grindstones, Freestone, Marble,

ble, Chert, &c. and the import of Corn, Malt, Deals, &c. Heanor, Ilkeston, and Eastwood, are small Towns near to this Line: it commences in the *Trent* Navigation, at Trent Lock near Sawley (opposite to the *Loughborough* Navigation), and terminates in the *Cromford* Canal at Langley-mill or Bridge, where the *Nottingham* Canal also connects and terminates: NE of Stanton by Dale, it is joined by the *Nutbrook* Canal, and $\frac{1}{2}$ a m. S of Sandiacre, by the *Derby* Canal.

On Ilkeston-common, there is a short cut NNE to Benersley Colliery: from Cotmanhay, a Rail-way $\frac{1}{4}$ m. NW to Cotmanhay-wood Colliery: a little S of this is another short cut, whence a Rail-way led across the Erewash to Newthorpe-common Colliery, formerly, the working of whose coal-seams was continued under this Canal for a considerable distance, and occasioned no farther damage to the Canal Works, than a little labour to raise the Banks and Towing-path, see page 348.

From Shipley Old Wharf, near Newmanleys Mill, it had formerly a wooden Rail-way branch, W for $1\frac{1}{2}$ miles, to Shipley Colliery; which branch was discontinued about the year 1796.

E of Ilkeston there is a Pottery established on the Bank of this Canal; SW of Long-Eaton, a Boat-building Yard, &c.; Lime-kilns at Ilkeston-common, Langley-mill, Long-Eaton, Sandiacre, Sawley or Trent-Lock, and Shipley Old Wharf (see Vol. II. p. 425 to 429), &c.

From Trent-Lock to the *Derby* Canal is 3 miles: thence to the *Nutbrook* Canal is $2\frac{1}{4}$ miles; and thence to the *Cromford* and *Nottingham* Canals at Langley-mill, is $5\frac{1}{2}$ miles; the whole rise being $108\frac{1}{2}$ feet by 14 Locks:

Locks: and there are 25 Bridges over this Canal. Over Nutbrook, on the NE of Stanton, and over the Erewash above Newmanleys Mill, this Canal is conveyed on Aqueduct arches. At Langley-mill, and Ilkeston-common, there are feeders to this Canal from the Erewash River.

Mr. William Jessop was the Engineer to this Canal, who finished it in a few years. By the Act for *Cromford* Canal, 27th Geo. III. the Tolls allowed to be taken by this Company were reduced one-half, except on Coals and Cokes: by the *Derby* Canal Act, 33rd Geo. III. a further reduction of Tonnage was made, on the part between the *Trent* and the *Derby* Canal near Sandiacre, provided that no other junction between these Canals are made: and by the Act for the *Trent* River, 34th Geo. III. the annual Rent of 5*l*. payable by this Company to the Trent Proprietors, was commuted for a Toll of 6*d*. on every laden Boat, which shall cross the *Trent* between this Canal and the *Loughborough* Navigation. The regulations concerted between this and eight other Canal or Navigation Companies, for guaging of Boats, and ascertaining the Tonnage of Coals, &c. in one uniform manner, have been mentioned in Vol. I. p. 182: one of the Weighing-houses there mentioned, is situated at the S end of this Canal at Trent-Lock or Sawley-wharf, and is kept by Mr. Hopkins.

At one time, since the completion of this Canal, the Shares herein sold at three times their original price. In 1794, an Act passed for making a long side-cut (or *Trent* Canal) for the Trent Navigation, which was to cross this Canal N of Long-Eaton Town, but the same has not been executed.

GRANTHAM

GRANTHAM *Canal:* Acts, 33rd and 39th Geo. III.—
The general direction of this Canal is nearly E, by a
crooked course of 33¼ miles, in the Counties of Nottingham, Leicester, and Lincoln, passing off our
square of Map (p. 193) at X. Its eastern end is rather
elevated: it crosses the *West Dean* Ridge, near
Cropwell-Bishop, and the *East Trent* Ridge, NE of
Woolstrop, by deep-cuttings. Its general objects are,
the supply of Grantham and the Vale of Belvoir (thro'
which it passes), with Coals, Lime, Deals, &c. and
the export of farming products: the famous strata of
Blue *Lias* Limestone (Vol. I. p. 114) crosses this Canal, not far from Cropwell-Bishop, and might supply
a great deal of Tonnage thereon.

Nottingham is the 15th British Town in point of
population, with 34,253 persons; Grantham is the
231st, with 3,646 persons, and Bingham, the 645th,
with 1,326 persons, near the line of this Canal. It
commences in the *Trent* Navigation in West-Bridgeford, very nearly opposite to the commencement of
the *Nottingham* Canal, and terminates at the SW
corner of Grantham Town. From the NW of
Cropwell-Bishop, it has a branch NE 3 miles to
Bingham Town.

From the *Trent* River to Cropwell-Bishop, 6¼ miles,
is a rise of 82 feet: thence to Stainwith Closes near
Muston, 20 miles, is level: thence to Woolstrop-point,
1¼ miles, is a rise of 58½ feet: and thence to Grantham, 5 miles, is level. Over the Brooks that are
crossed by this Canal, there are many small Aqueduct
Arches, and one of considerable size N of Woolstrop,
I believe.

The greater part of the cutting of this Canal being
in clayey soils and strata, it has been made to depend
entirely

entirely on Reservoirs, for retaining the surplus waters of floods, for its supply: the largest of which is near Nipeton, in the head of the Dean or Devon Valley, which was originally made of 60 acres extent, and 9 feet deep, but the supplies proving inadequate, in 1804 the bank or head of this Reservoir was raised 4 feet higher; near Denton, in a valley leading N to the Witham, there is another Reservoir of 20 acres, and nine feet deep.

The Tonnage on all Goods passing to or from this Canal, and the *Trent* River, is to be 2½d. per Ton, and 1¼d. per Ton per mile for navigating on this Canal: Manures and Road-materials are to pass Toll-free, except Limestone, which is to pay ¼d. per Ton per mile: by the Act of 34th Geo. III. (for the intended *Trent* Canal or side-cut), the Trent Proprietors were to deepen the River between the entrance of this and the *Nottingham* Canal, and Boats passing between them, were in consequence, to be subject to additional Tolls: but Goods passing from this Canal on to the *Trent* River, were not to be subject to its new rates, unless they passed on to the intended long side-cut or *Trent* Canal, which has never been executed, as already mentioned. By the first Act for this Canal, Tolls are provided on Goods passing on to or from the then intended *Newark and Bottesford* Canal. In 1798, this Company concurred with eight other Canal and Navigation Companies, in establishing a uniform mode for the guaging of Boats, and ascertaining the Tonnage of Coals, &c. as has been mentioned in Vol. I. p. 182.

This Company were authorized to raise 124,000*l*. part in Shares of 100*l*. each, and part in new Shares of 120*l*. each: the works are said (by Mr. W. Pitt,

Leicester

Leicester. Report, p. 316) to have cost 100,000*l*. and in 1809, to pay five per cent. interest: the Proprietors are limited to a Dividend of eight per cent. per annum, but after 3000*l*. is accumulated as a Fund, the Tolls are to be lowered, as much as circumstances will admit. In 1793, the *Newark and Bottesford* Canal was proposed, to join this near Stainwith.

GRESLEY's *Canal:* Act, 15th Geo. III.—The direction of this Canal, or water-level, constructed at the sole expense of *Sir Nigel Bowyer Gresley*, Bart. is about NW, for 2¼ miles in the county of Stafford, passing off our square of Map (p. 193) at *h:* which letter is said at page 215 to mark the *Newcastle* (Underline) *Junction* Canal, by mistake. It is considerably elevated, and situated near to the *Grand Ridge* on its eastern side. Newcastle Underline is the 118th Town, with 6,175 Inhabitants, for the supply of which with Coals, it was at first principally intended: but since its extensions each way, its objects also have been, to export Coals, import Limestone, &c.

This Canal commences in the *Newcastle Underline Junction* Canal, at the NW end of Newcastle Town, and terminates in the detached part of the same Canal, near Apdale Colliery, and is all on one level. Sir Nigel was bound by his Act, to supply all the Inhabitants of Newcastle Underline with Coals, until the year 1796, at 5*s*. 6*d*. per long Ton (20×120 lb.), or 3¼*d*. by the single cwt., and for the ensuing 21 years to 1817, at 6*s*. per long Ton! In 1796, the *Commercial* Canal was proposed to connect with this Canal: in 1798, the *Newcastle Underline Junction* Canal was undertaken, to join this at both its extremities; and in 1805 it was said, that the *Nantwich and Newcastle Rail-*

Rail-way was intended to join this Canal at Dales-
pool.

High-Peak Junction proposed Canal.—At the
time that the *Cromford* Canal was in agitation, the
design seems to have originated, of conducting a Canal thence thro' the High Peak Hundred of Derbyshire, to connect with those about Manchester, the
Cromford and Bakewell proposed Canal, as already
mentioned, being a part of this design. In the spring
of 1810, this design was again renewed, and a line was
surveyed by Mr. Brown of Disley, and Mr. Johnson
and Mr. Meadows of Manchester; which was afterwards revised by Mr. Rennie, and whose Report
thereon was circulated, in a printed Letter dated 26th
October, 1810.

The general direction of this line is about N.W, by
a bending course to the north, of about 36 miles:
its northern part being very considerably elevated,
crossing the *Grand Ridge* by a long proposed Tunnel
under Bowden-edge, NE of Chapel-en-le-Frith, crossing the *East Wye* Ridge NE of Bakewell by another Tunnel, and a branch from the *West Amber*
Ridge by another Tunnel. Its general objects in view
are, the import of Coals, Grain, &c. the export of
Limestone, Lead, &c. and the opening of a more
direct communication for Derby, Nottingham, Leicester, &c. with Manchester and its populous neighbourhood. Cromford, the 658th Town on the list, with
1,259 Inhabitants, Bakewell the 591st, with 1,485 persons, and Chapel-en-le-Frith the 353d, with 2,667
persons, are on this Line.

It commences in the *Cromford* Canal, at the E end of
the Derwent Aqueduct, and terminates in the proposed

extension of the *Peak Forest* Canal (instead of its present Rail-way in this part) on the SW side of Chapel-Milltown. From the *Cromford* Aqueduct on the Derwent, past Lea-wood Cotton-mill and Cupola, &c. to near the W end of Lea village, 1¼ mile, has a rise of 108 feet; thence through a Tunnel of $1\frac{7}{16}$ mile north-ward, to Tansley Brook (¼ of a mile below the Village) is level; thence past the Bump-Mill, and above Matlock Town, thro' Matlock-Bank Village, past Toad-hole Cotton Mills, &c., across the Derwent above Great Rowsley Bridge, on an Aqueduct Bridge, to the Wye at the S corner of Haddon Park, $7\frac{1}{16}$ miles, is level; thence across the Wye, up its W bank, until past Haddon-Park, and then on its E bank, passing on the E side of Bakewell Town, and ascending the Vale NE of it (½ a mile beyond the Town), 2¼ miles, with a rise of 76 feet 8 inches: thence through a Tunnel NE, into the Pilsley Valley (¼ a mile from the Village), 1 mile, is level: thence near to Bubnell, and to Calver-Peak Lime-quarries. (Vol. II. p. 421), and across the Derwent again on an Aqueduct Bridge (51 feet high) below Stoke, 4¼ miles, is level: thence through Froggatt and Grindleford-bridge to Upper Padley, 1¼ m. is level: thence to Hazleford or Lead-mills Bridge (passing near the long celebrated Peak Millstone Quarries, in the 1st Grit Rock, at Old Booth Edge, &c. Vol. I. p. 221), 2¼ miles, is level: thence across the Derwent, on an Aqueduct, above Mytham-bridge, and on the E side of Hope, to Nether, or Lady Booth in Edale, 8 miles, with a rise of 345 ft. 9 in. (making a total rise of 530 ft. 5 in.); thence to near Barber-Booth, 2¼ miles, is level: thence 2¼ miles WSW of proposed Tunnel, is also level, to near Shire Oaks in

Malcalf:

Malcalf: and thence to the intended *Peak Forest* Canal extension, near Chapel-Milltown, 1¼ mile, has a fall of 148 ft. 9 in.

The width of the Canal to be adapted to 25 Ton Boats, such as are used on the *Peak Forest*, &c. Canals, but the Tunnels to be made 16 feet wide in the clear, so that the Boats may pass each other in them. Ample Reservoirs can be made, for supplying the Canal, and even benefiting the Mills on the Derwent and Goyte. Mr. Rennie's estimate of the total cost is 650,000*l*.

In 1789, the *Cromford and Bakewell* Canal was proposed to occupy nearly the same ground as the southern part of this line: in 1810, it was suggested, that the *Baslow and Brimington*, or the *Baslow and Chesterfield* Canals, might, one of them, join this Canal at the east end of the Stoke Aqueduct; and in 1811, the *Tinsley and Grindleford-bridge* Canal was proposed to join this, at Upper Padley, near the latter place; and the prospect of carrying these Canals into execution, seems yet not entirely despaired of by their promoters, I believe.

HUDDERSFIELD *Canal:* Acts, 33rd and 40th Geo. III. —The general direction of this Canal is SW, for 19¼ miles in the West Riding of Yorkshire, and in Lancashire, entering our square of Map (p. 193) at D: its middle part has a great elevation, crossing the *Grand Ridge* by one of the longest Tunnels in England, under Stand-Edge or Pule-hill, at near 250 feet below its surface, of 1st Grit: its objects are the supply of the middle parts of its line, with the Coals found near both its extremities, and with Lime from Marple Kilns (see Vol. II. p. 427), and others on the *Peak Forest* Canal, and from Fairburn and Brotherton by the *Ayre and Calder* Navigation: the forming of a more direct communication between *Hull* and *Liverpool*, by way of Wakefield, Huddersfield, Ashton Underline, Manchester, and Warrington.

Huddersfield is the 63rd Town on the British Population list, with 9,671 persons, Ashton Underline is the 26th, with 19,052 Inhabitants, including those of Staley-bridge, which is a considerable Town on this line. It commences in Sir John *Ramsden*'s Canal on the S side of Huddersfield Town, and terminates in the *Manchester, Ashton and Oldham* Canal, at Duckinfield Bridge, on the SE side of the Town of Ashton Underline, not far from the commencement of the *Peak Forest* Canal.

From *Ramsden*'s Canal to Rough-lee near Marsden, $7\frac{1}{4}$ miles, has a rise of 436 feet: thence through the Tunnel to Wrigley-mill near Saddleworth, 4 miles, is level: thence to the *Manchester, Ashton and Oldham* Canal, $8\frac{1}{4}$ miles, has a fall of $334\frac{2}{3}$ feet. The Locks on this Canal are 72 feet long and 9 feet wide, and their rise is generally about 10 feet. The Bridges over this Canal have very injudiciously, from motives of economy, been made without Towing-paths under them, even at crossing of some of the Turnpike Roads, and the inconvenience therefrom is found so great, that it is judged, that most of these Bridges must be taken down and widened.

The Tunnel under Stand-Edge, or Stanage, is $3\frac{1}{4}$ miles long; it commences in the 1st Coal-Shale (with an eastern dip), in which a thin Coal was worked near to Red Brook, for the use of a very large Steam-engine and Pumps (14 inch, with a 7 feet stroke) that were put down, in a shaft 164 yards deep, in the year 1794, for draining the middle parts of the Tunnel, and enabling the work of excavating the Tunnel to proceed

proceed from 4 points at the same time: the whole of the 1st or Millstone Grit was penetrated, with an E dip, and the remainder of the Tunnel westward, was drove in the Limestone Shale, dipping at first E, then becoming flat, and at length dipping slightly to W, from which Shale, the W end of the Tunnel emerges, in the curious denudated tract, in the Tame Valley (extending thence to Scout-mill, SE of Mossley), that I have mentioned, Vol. I. p. 236, as probably *having the Peak-Limestone under it*, at a practicable mining depth.

Nearly two-thirds of the length of this Tunnel, driven in 1st Grit or Shale Grit, or in the harder parts of the Shale, appeared so hard and sound, that no arch therein was judged necessary: but some parts which at first seemed so very secure, began on exposure to the air to moulder and fall, and the same has been since lined with an arch; and I wish that this may not in time prove necessary, to the interruption of the trade and endangering of the Tunnel, through all but the permanent Grit Rocks. This famous Tunnel, begun in 1794, was completed in May 1810, and is said to have cost 130,000*l*. Through the 1st Grit Rock, with a rapid south-western dip, there is another short Tunnel of 200 yards, for avoiding a sudden loop of the Canal, on the SE side of Scout-mill.

East of Duckinfield Bridge ¾ of a mile, this Canal is carried over the Tame on a cast-iron Aqueduct of 50 or 60 feet wide, built in 1801; a square trough put together with flanches, which are strengthened by brackets, cast at proper intervals for the rivets between them: the vertical flanches are widest at bottom, and have thin bracket ribs cast between them: the joints were filled with an iron cement, compound of iron turnings and borax, made into a paste with boiling water. A stage for the towing-path ought to have been attached to this trough, either of iron or wood; instead of which, a ridiculous stone Arch, rising much above the Aqueduct, has been built on its W side, 12 feet wide, for the use of the towing Horses and Boatmen. At Wright's Mill is a 2-arch Aqueduct over the Tame, in Yorkshire, and another SE of Dobs-Cross. There are also Aqueducts over the Coln below Skyer-bottom and Paddocks Mills.

This Company were authorized to make Reservoirs for the flood-waters, in the elevated Valleys near to their Tunnel, for containing 20,000 lock-fulls, each 180 cubic yards of Water, and even more if they should be found necessary: three large ones have accordingly been constructed on Wessenden Brook, SE of the Tunnel, and 2 on the North side of the Tunnel: one of these Reservoir heads, gave way in 1798, and another (called Driggles) of 28 acres, on the 29th of November, 1810, and the sudden Floods thus occasioned, did considerable damage to the Mills and Villages near and below Marsden: on Staley Brook, a large Reservoir N of Armfield (Vol. I. p. 497) broke down soon after it was made, and had not been repaired when I saw it in 1809: on the W of Slaugher-aite, and on the W of Saddleworth, there are other Reservoirs.

About the year 1798, that part of the line between Huddersfield and Marsden, was completed and opened: and sometime about 1806, the part between Ashton Underline and Woolrood near Dobs-Cross, was completed and brought into use: about the year 1800, the Company were unable to proceed with their Tunnel, many of the original Subscribers being found unable

able to complete the Calls on their Shares, and the same became in consequence, considerably depressed in value: they managed however to surmount these difficulties, and in 1810 completed their undertaking.

The rates of Tonnage allowed to be taken, are from 3d. to ¼d. per Ton per mile, on different descriptions of Goods (see Phillips's 4to. "History of Inland Navigation," Appendix, pp. 135 and 136), besides 1s. 6d. extra per Ton on all Goods that pass through the Tunnel: less lading than 15 Tons is not to pass any Lock, unless the Water runs waste thereat, without consent: Goods are to pass free of toll on the part of *Ramsden*'s Canal between the termination of this Canal, and Sir John R.'s Warehouses in Huddersfield, and in consequence, this part of his Canal is to be kept in repair by this Company: who are also to guarantee his Tolls, to the annual amount of the average of 3 years preceding the making of this Canal. This Company were authorized to raise 274,000l. in 100l. Shares: they are precluded from making any branch or extension eastward, for connecting with the *Barnsley* or any other Canal than *Ramsden*'s, under the penalty, of the Tolls on any such branch or extension being taken and divided, between *Ramsden*'s, the *Calder and Hebble*, and *Ayre and Calder* Proprietors. Mr. Edward Love, of Staley Bridge, is a resident Agent for the western part of the Canal, and the principal Agent is James Meadows, Esq. of Piccadilly-street in Manchester.

IDLE *Navigation*:—The course of the navigable part of this River is nearly west for about 10 miles in the County of Nottingham, entering our square of Map (p. 193) at N: it is but very little elevated above the tide: its objects are the supply of the Town of Bawtry

try with Coals, Deals, &c. and the export of Agricultural Products: it commences in the *Trent* River at West Stockwith (near to the commencement of the *Chesterfield* Canal), and terminates at the Town of Bawtry. Misterton Sas or Sluice, at ¼ a mile from the *Trent*, has two lock-doors 16 feet high (and forming a passage 17¼ feet wide) opening towards the *Trent*, for keeping its embanked floods, out of the low lands across which this Navigation passes. Mr. John Smeaton was consulted on this Navigation, and the drainage of Misterton Car (whereon a base line of the Government Trigonometrical Survey was measured in 1801), and other low lands, that are connected with it.

LEICESTER *Navigation:* Acts, 31st, 34th (for *Ashby* Canal) and 37th Geo. III.—The general direction of this Navigation is nearly S, by a bending course to the Eastward of 15 miles, in the County of Leicester, following the Vale of the Soar: its lower extremity, in the Canal Basin NW of Loughborough, is said to be elevated only 147 feet above low water of spring tides in the Ocean?, but its Cole-Orton and Barrow-hill Rail-way extensions of the Charnwood-Forest branch, terminate very nearly on the *West Soar* Ridge, at very considerable elevations.

Its general objects are, the supply of Leicester with Coals, Lime, Freestone, Paving-stones, Deals, &c. and the export of Coals, Limestone, Granite, and Slate from the Pits and Quarries near to its line and branches, and of manufactured Goods, and Cheese and other Agricultural Products; and the forming of part of the grand line of communication between London, and Nottingham and Derby Counties, which the *Grand Union* Canal, now on the point of completion (May 1814), will

will open, and which the proposed *North-eastern Canal* may be expected to extend, quite across our Map, from *F* to I.

Leicester, at its S termination, is the 21st British Town, with 23,146 Inhabitants; Loughborough, near its other end, is the 138th, with 5,400 persons; and Mount Sorrel near it, is the 587th, with 1,502 persons. This Canal commences in the Basin of the *Loughborough* Navigation, at the NW side of that Town, and terminates in the *Leicestershire and Northamptonshire Union* Canal Basin, near the West Bridge, on that side of Leicester Town. At Turnwater Meadow in Cossington, it is joined by the *Leicester and Melton-Mowbray* Navigation: from the opposite or western side of the Basin at Loughborough, the Charnwood-Forest branch of this Canal proceeds, nearly west, by a crooked course of about $11\frac{1}{2}$ miles (partly by Rail-ways, and partly by a long Water-level or detached Canal) to Cole-Orton and Swannington-common Collieries, and to Barrow-hill Lime Quarries.

The famous pozolanic, blue *Lias* Limestone Quarries of Barrow-on-Soar (I. 114 and 409), are near to the line of this Canal, so are the Sienite or Granite Quarries of Mount Sorrel (I. 152): the Charnwood-Forest branch passes through the Grace-Dieu Limestone Quarry (I. 158 and 410): the dark blue Slate, Paving, Grave-stones, shallow Cisterns, &c. of the Swithland Quarries (I. 153), are carted about $3\frac{1}{2}$ m. from the W, to Mount Sorrel Wharf, to be put on board the Boats on this Navigation: and at the bottom of p. 154 of Vol. I. I have hinted at the probability, that *Slate* of good quality might be had, on or near to the Water-level, on the Rail-way therefrom to this Canal, if sought for at a sufficient depth.

From

From the *Loughborough* Basin, to the entrance of the *Leicester and Melton Mowbray* Canal, is about 9 miles; and thence to the *Leicestershire and Northamptonshire Union*, is about 6 miles, the total rise being 43 feet. From the Loughborough Basin to Forest-lane Wharf, is $2\frac{1}{4}$ miles of Rail-way, with an ascent of 185 feet: thence to the foot of Barrow-Hill, NW of the village of Osgathorp is $8\frac{1}{4}$ miles, and level; and thence there is a Rail-way extension of 130 yards, rising into the Barrow-hill Limestone Quarries (I. 409); and a further extension of $\frac{1}{4}$ of a mile of Rail-way is provided for in the Act, to Clouds-hill Limestone Quarry; where, had this last extension been executed, junctions would very nearly have been effected, with the *Ashby-de-la-Zouch* Rail-way branch, and also with the proposed *Breedon* Rail-way. From the SE of Osgathorp Reservoir, a branch of $\frac{1}{4}$ of a mile proceeds almost S, and level, to Thringston-bridge Wharf and Warehouses; and thence a Rail-way proceeds SSW $\frac{1}{4}$ m. to Swannington-common Colliery, and another such branch SW, $\frac{1}{4}$ m. to Cole-Orton Colliery; and another branch from the last, of $\frac{1}{4}$ m. to another Colliery in Cole-Orton, was also provided for in the Act: and any other Rail-way branches are allowed to be made to Mines, &c. within 2000 yards: these Rail-way branches rise considerably from the Water-level, but I am unacquainted with the particulars: except that the descent is so considerable from Cole-Orton, that the Trams of Coals descended without Horses, regulated by means of a clamp or logger, acting on the wheels, by a man who rode on each, and the empty Trams were drawn up again by Horses.

The Rail-ways belonging to this Company are

single,

single, and have bars flat at top, and the wheels are cast with flanches, inside, for keeping the Trams upon them. The bodies of the Trams were made to lift off, or to be placed on their wheels, by means of Cranes erected on the Forest-lane and Thringston-bridge Wharfs, so that the bodies of the Trams only, stowed close together, could be carried in the Boats on the Water-level. The width of the side-cuts for avoiding the Mills, and the Locks, on the main line of this Canal, are adapted for the Barges that navigate the *Trent* river.

On the SW of Sheepshead, the S of Belton, and SE of Osgathorp, there are deep-cuttings across the points of Hills, for the Water-level. Across the Black-brook Vale, SW of Sheepshead, there is a very considerable Embankment, and an Aqueduct Arch, and smaller ones NE and N of Thringston.

At a branching of the Black-brook Vales, about ¼ of a mile above the Canal, there is a large Reservoir for supplying the Water-level, the head of which gave way, soon after the Works were completed, and occasioned such an inundation, that Mr. Jester's Farm-house and Premises were destroyed, and a Hay-cock was borne down by it, and wedged in the Aqueduct Arch, by which means, the Valley above the Embankment became filled, and by the pressure, a breach in the same was effected, which emptied the whole Water-level, whereby such an enormous flood was occasioned in this Valley (which passes through the late Mr. Robert Bakewell's Farm at Dishley), that a great many Sheep, &c. were drowned thereby, and other serious damages were done.

Mr. William Jessop was the Engineer to this Canal, and made the Survey for it in 1790, and Mr.

Christo-

Christopher Staveley, jun. was the resident Engineer during its execution. In December 1793, the part of the Line between Loughborough and Sielby, near Mount Sorrel, was completed, and in February 1794, the remainder of the same to Leicester was opened; and soon after, the Charnwood-Forest branch was completed, I believe, and the same continued in use until about the year 1800, when a dispute arising between Sir George Beaumont and Mr. Boultbee (his late Steward), concerning the Water drawn from their respective Collieries, which nearly adjoin, a tedious law-suit ensued, and, as has been mentioned in page 293, all these Coal-works were suspended, and the trade on these expensive Rail-ways, and the Water-level was entirely discontinued, and yet remain so, I believe: when I viewed the Forest in August 1807, the Canal was without any water in it, Cattle were rapidly treading in the banks, the Bridges were fast delapidating by mischievous boys, &c.

In the *Ashby-de-la-Zouch* Act, it is provided, as an indemnification to this Company, that in case of their Rail-way branches being extended to or near to Cole-Orton, Swannington or Thringston Collieries, any Coals from the same that might be carried through Blackfordby, should pay 2s. 6d. per Ton to this Company: the *Loughborough* Company were, by this Company's first Act, guaranteed in the receipt of 3000l. per annum, on condition of their taking 1s. 6d. per Ton or less (but not less than 10d.) Tonnage on Coals, passing between Loughborough and the *Trent* river.

The rates of Tonnage allowed to be taken by the Acts for this Canal, are various, on different kinds of Goods; this may be seen in Phillips's 4to. " History of Inland

Inland Navigation," App. p. 12. This Company are under an agreement with eight other Navigation Companies, as mentioned, Vol. I. p. 182, to follow an uniform plan of guaging of Boats, and ascertaining the Tonnage on Coals, &c. The Company were authorized to raise 84,000*l*.; and the interest paid on these Shares was 25 per cent. in 1807, notwithstanding the disuse of their Charnwood-Forest branches, as Mr. W. Pitt has stated in the Leicester. Report, p. 315.

LEICESTER AND MELTON MOWBRAY *Navigation*: Acts, 31st and 40th Geo. III.—The general direction of this Navigation is nearly E, for about 12 miles, in the County of Leicester, quitting our square of Map (p. 193) at *A*: it follows the course of the Wreak and the Eye Rivers, having its eastern end rather considerably elevated: its objects are the supply of Melton Mowbray with Coals, Lime, Deals, and Mount Sorrel Granite for Roads, &c.; the export of Farming Products; and forming part of the communication to Oakham. Melton Mowbray is the 444th British Town, with 2,145 Inhabitants, and Mount Sorrel, near the Western end of this Navigation, is the 587th, with 1,502 persons.

It commences in the *Leicester* Navigation, at Turnwater Meadow in Cossington, and terminates in the *Oakham* Canal, at the Town of Melton Mowbray. In Syston, near to this line, a coarse kind of Gypsum is dug, that is much used for Plaster Floors (I. 151); and a few miles to the NE of this, the famous blue *Lias* Limestone Strata (I. 114), cross this Canal, and this pozolanic Limestone is, or might be wrought, and furnish a considerable Tonnage thereon. This Canal
was

was completed in a few years after the passing of its first Act.

The original rates of Tonnage allowed to be taken by this Company, may be seen in Phillips's 4to. "History of Inland Navigation," Appendix, p. 13; but these have subsequently been varied, by their last Act, and several regulations of Tolls made with the *Oakham* Company. The guaging of Boats, and ascertaining the Tonnage of Coals, &c. are here conducted on the uniform plan with 8 other adjacent Navigations, that is mentioned, Vol. I. p. 182. This Company were authorized to raise 40,000*l*. At Melton Mowbray is a spacious Basin, Wharf and Warehouses, belonging to this Company.

LEICESTERSHIRE AND NORTHAMPTONSHIRE UNION *Canal*: Acts, 33rd and 45th Geo. III.—The general direction of this Canal (as it has been executed) is about SSE, by a bending course to the Westward, of 17 miles, in the County of Leicester, passing off our square of Map (p. 193) at *F*. It southern end is said to be elevated 350 feet above the sea ?, it crosses the *East Soar* Ridge by a Tunnel, N of Saddington. Its objects are, forming with the *Grand Union* Canal (that will be completed about Midsummer 1814), part of the grand line of communication between London and Hull, crossing our Map from *F* to V, and which may (by means of the proposed *North-eastern* Canal) be expected to extend the route, direct from the Metropolis northward, in a still more important line, across our Map from *F* to I; the supply of Market Harborough with Coals, Lime, Deals, &c. and the export of Agricultural Products.

Leicester is the 21st British Town, with a population of 23,146 persons, and Market Harborough is the 539th, with 1,704 Inhabitants. It commences in the *Leicester* Navigation, near the West Bridge of that Town, and terminates in the *Grand Union* Canal at Gumley Wharf: from which place there is a branch SSE, by a crooked route of about 3¼ miles, to Market Harborough. At Kilby Wharf, on the Leicester and Welford Road, there are considerable Quarries of the blue *Lias* Limestone, and Lime Works.

From the *Leicester* Navigation Basin to Fleckney, near the northern end of the Saddington Tunnel, is 12¼ miles, with a rise of 160 feet; and thence to the *Grand Union* is 4¼ miles, and level; and which same level is also continued along the Market Harborough branch. The Saddington Tunnel is 880 yards in length. Mr. John Varley, and Mr. C. Staveley, jun. were the Engineers to this Canal, which they finished to Gumley Wharf in March 1800; and about the year 1808 the Market Harborough branch was completed. The rates of Tonnage on various articles, and for different distances, may be seen in Phillips's 4to. " History of Inland Navigation," Appendix, p. 120. The Company were authorized to raise 300,000*l.* in 100*l.* Shares.

The design at first was, that the line of this Canal should terminate in the Upper *Nen* Navigation, at the Town of Northampton, where it was also to join a Rail-way branch from the *Grand Junction* Canal: the particulars of this unexecuted part of the line is as follows, viz. from Gumley Wharf to near Great Oxenden, 9¼ miles, and level, passing through a Tunnel of 1056 yards long on the W side of Foxton, in a short range of Hills E from the *Grand Ridge;* thence in ¼ a mile is a rise of 50 feet to the northern end of the summit level, which is 4¾ miles long, to near Maidwell, passing through a Tunnel of 286 yards, S of Great Oxenden, in the *South Welland* Ridge, and another Tunnel of 990 yards long, SW of Kelmarsh, in the *West Ise* Ridge, or Wellingborough branch from the *Grand Ridge:* from the southern end of the summit level, to the junction of the North River near Northampton, 11¼ miles, is a fall of 197½ feet; and thence to the head of the Upper *Nen* Navigation, ¼ of a mile, is level. Reservoirs were intended on the Oxenden and Kelmarsh Brooks, for supplying the summit pound, by reserved flood Waters.

About the year 1793, the *Uppingham* Canal was in contemplation, and in the first Act for this Canal a provision is made, for its junction herewith. The great difficulties which were found to attend the execution of the southern part of this Canal, occasioned, in 1803, a proposal by Mr. Thomas Telford, for executing a branch from Gumley Wharf, across to the *Grand Junction* Canal at Buckby Wharf; and about the year 1805, the same Engineer proposed, instead thereof, to shorten the main line, by an Embankment of 83 feet high, on the west of Market Harborough, and to leave the line near Maidwell, and proceed thence across for Buckby Wharf; but neither of these proposals being approved, in 1808 Mr. Benjamin Bevan proposed a new line between Gumley Wharf and Buckby Wharf, which he has since executed, as a separate concern, for the *Grand Union* Company.

LOUGHBOROUGH *Navigation:* Act, 16th Geo. III. —The general direction of this Navigation is nearly S, for about 8 miles, in or between the Counties of Leicester

cester and Nottingham, following the course of the Soar River, with occasional side-cuts and Locks, particularly at its southern or upper end, which is said not to be elevated more than 147 feet above Low-water in the Humber? Its objects are the import of Coals, Building and Paving-stones, Deals, Gypsum, &c. and the export of Barrow Limestone, Swithland Slate, Mount Sorrel Granite, manufactured Goods, Cheese and other Agricultural Products, &c. and the forming part of the new line of communication between the *Thames* and the *Trent*, &c. as already mentioned, p. 376. Loughborough is the 138th British Town, with 5,400 Inhabitants, and Kegworth, near the line, is also a considerable Town.

This Navigation commences in the *Trent* River at Red-Hill or Trent Lock, opposite to the commencement of the *Erewash* Canal, and terminates at the Rushes Wharf, at the north-west side of Loughborough Town: at the Basin, about 300 yards below Loughborough Wharf, it is joined, and indeed crossed, by the *Leicester* Navigation, whose Charnwood-Forest branch proceeds from the opposite or western side of the Basin, as mentioned page 377. Red-hill Gypsum Quarry, Vol. I. p. 150, is situate by the side of this Navigation, near to its northern end. The width of the Locks on this Navigation, is adapted to the *Trent* Barges, which pass through this Navigation in their way to *Leicester*. This Navigation was quickly completed, after the passing of its Act. On the making of the *Leicester* Navigation, that Company guaranteed the Tolls on this, to the amount of 3000*l*. annually, on condition of no more than 1*s*. 6*d*. or less than 10*d*. per Ton, being charged on Coals navigated hereon, between Loughborough and the *Trent* River.

Macclesfield proposed Canal.— In 1796, a Survey was made by Mr. John Nuttall, of the line for a Canal, from the summit pound of the Caldon branch of the *Trent and Mersey* Canal at Endon, to the summit pound of the *Peak Forest* Canal near Marple Chapel: and Mr. Benjamin Outram revised the same, and made a rough Estimate and a Report, which was printed March 10th, 1796. The general direction of this line is about north, nearly as shewn by blue dots in the Map facing p. 193, for about 27 miles, in the Counties of Stafford and Chester; at a considerable elevation, crossing the *Grand Ridge*, the *North Dane Ridge*, the *North Bollin*, and several other inferior Ridges, without Tunnels, or any very considerable deep-cuttings being necessary, owing to the deep Inosculations of the heads of the Valleys, that here fortunately occur.

Its general objects in view were, the supply of Macclesfield and the vicinity of the line, with Limestone, Coals, Deals, &c. the export of Coals, Building and Paving-stones, and grey Slate, from the Collieries and Quarries near the line, &c.: forming a communication between the towns of Macclesfield and Leek, and thence with all parts of the Kingdom, and opening a direct water communication between the important Pottery district of Staffordshire and the West Riding of Yorkshire, shorter than the present Turnpike Road. Macclesfield is the 49th British Town, with 12,299 Inhabitants; Leek the 223rd, with 3,703 persons; Stockport the 32nd, with 17,545 persons; and Congleton the 177th, with 4,616 persons, to or near to which two last places, branches were proposed.

From the *Trent and Mersey* Canal branch, to near Steel-house Green, about 2 miles, is a rise (of about 50 feet?)

feet?) for 5 Locks, and thence to the *Peak Forest* Canal, 25 miles, is level: from Harracles (near to the south-west corner of the large Rudyard Reservoir) a branch SE, of about 2 miles long, and level with the summit pound, was intended, to Leek Town; also from Rushton Marsh a branch Westward 5 miles, towards Congleton, might be made, on the same level; and another from near Marple NW 4 miles (with a fall), to Stockport, where it would approach within no great distance of the Heaton Norris branch of the *Manchester, Ashton and Oldham* Canal; making altogether, with the summit pound of the *Peak Forest* Canal, near 40 miles of level water.

This line would pass the Collieries of Bollington, Adlington, Worth, Poynton, and Norbury, and pass near to the celebrated Quarries of Freestone, Pavingstone and grey Slate at Kerredge, in Rainow Chapel, in the 4th Grit Rock (and not inferior to those of Ealand-Edge and Cromel-bottom, in Yorkshire), Vol. I pp. 165, 420, 425, and 430; and by the Styperstone-Park Freestone Quarries (Cheshire Rep. p. 18), and near to the Macclesfield Copper-Works, &c. It was proposed, to make the summit pound of this Canal half a yard deeper than necessary (as on the *Peak-Forest* summit), in order to have 1,100 lock-fulls of water, thus in reserve, that could in dry seasons be let down for supplying the lockage, but which is, indeed, but a very expensive, and mostly an inadequate substitute* for a separate Reservoir, above the Canal level.

---

* Because, as long as this extra height of water is maintained in the summit, more water than necessary is expended in the summit lock, at one or both ends of the summit pound; unless these locks are of less height than the others below them, in which case extra water must in most instances be let down, in order to pass these taller locks.

It was stated, that the line and branch to Leek would cost about 90,000l. and that connections northward from some of the Collieries to the *Peak Forest* Canal, and southward from others of them to Macclesfield Town, might be completed and become productive, in 1 year after obtaining an Act, and that the whole might be completed in 2 years; when the probable Tonnage was estimated, to produce 10,175l. annually; 5000l. of this on Coals and Lime.

The contending interests of the Land-owners and various Canal Companies, prevented this Canal being carried into effect, yet it has always appeared to me, particularly since the Caldon branch of the *Trent and Mersey* Canal was to be extended from Froghall to Uttoxeter, that this was a very eligible and important line of communication: and it seems much to be wished, that the Land-owners, and the *Trent and Mersey* Company, would concur in carrying it into effect, as also, in forming a new junction between the parts of the *Trent and Mersey* Canal, at Horninglow and Uttoxeter (nearly in the line of the proposed *Commercial* and *Dilhorn* Canals), by which means, a vastly shorter route would be formed, between Manchester and Derby, Nottingham, Leicester, &c. than at present exists, by way of Preston-brook, Northwich, Stone, and Fradley-heath.

In 1812, a proposed branch from the *Trent and Mersey* Canal branch at Norton, to *Bridgewater's* Canal at Sale Moor, was intended to approach very near to this line, E of Macclesfield Town.

MANCHESTER, ASHTON AND OLDHAM *Canal* (sometimes called the Ashton Underline Canal): Acts, 32nd, 33d, 38th, 40th, and 45th of Geo. III.—The
general

general direction of the main line of this Canal is nearly E, for 6¼ miles in Lancashire: its eastern end is rather considerably elevated, crossing the *South Irwell* Ridge by a moderate deep-cutting, E of Fairfield: its Stockport branch crossing the same Ridge, SW of Fairford, and its Hollinwood branch crossing the *West Medlock* Ridge, S of that Town: its objects are the supply of Manchester with Coals (and Stockport by means of its Branch), Building and Paving-stones, Lime, Corn, &c. the supply of Ashton Underline, and Stockport, &c. with Deals, Iron, &c. and forming part of the most direct water communication between Liverpool and Hull. Manchester is the 3d British Town on its population list, with 98,573 Inhabitants; Ashton Underline is the 26th, with 19,052 persons; Stockport the 32nd, with 17,545 persons; and Oldham the 36th, with 16,690 persons.

This Canal commences in the *Rochdale* Canal, in Piccadilly Street in Manchester, and terminates in the *Huddersfield* Canal at Duckinfield Bridge, on the SE side of the Town of Ashton Underline: at Walk-Mill, near Duckinfield, it is joined by the *Peak Forest* Canal; from Fairfield a branch proceeds northward by a crooked course to Hollinwood Wharf, at the S end of the Village (where it approaches within 3 furlongs of a branch from the *Rochdale* Canal), 2 miles from the Town of Oldham; and from Waterhouses Aqueduct (or Boodle Wood) on this branch, another proceeds NE up the Vale of the Medlock to New-Mill, passing Park Colliery in its course. From Clayton (or Droylsden) a branch proceeds S (passing near to Gorton Lime-works) to the Wharf at the N end of the Town of Stockport, in Heaton Norris parish (where it approaches within a short distance of the once proposed branch (of 7½ miles, with a rise of 60 feet) from Bridgewater's Canal, and also of the since proposed branch from the intended *Macclesfield* Canal); and from Taylor's Barn on this branch, another proceeds ENE to Beet-Bank in Denton, where it approaches within less than a mile of the line of the *Peak Forest* Canal.

From the *Rochdale* Canal to the Stockport branch, 2½ miles, has a rise of 114½ feet by 13 Locks: thence to the Hollinwood branch, 1¼ mile, is a rise of 48 feet, by 5 Locks: thence to the *Peak Forest* Canal is 2¼ miles, and level; and thence to the *Huddersfield* Canal is ½ a mile, and level. On the Hollinwood branch, it is 2¼ miles, between the line and the New-Mill branch, and level; and thence to Hollinwood is 1¼ miles, with a rise of 56¼ feet: the New-Mill branch is 1½ miles, with a rise of 96 feet. On the Stockport branch, it is 3¾ between the line and the Beet-Bank branch, and level, and thence to Heaton-Norris Wharf in Stockport, is 1¾ mile, and level; and the Beet-Bank branch is 2¾ miles, and level.

The Canal is 33 feet wide at top, 15 at bottom, and 5 feet deep in water, except the summit pounds, which are made 6 feet deep, under the idea of their acting as Reservoirs: the Locks are 80 feet long, and the Boats usually carry 25 Tons. There is a considerable Aqueduct Bridge over the Medlock River at Bradford (or Ancoats), near Manchester, and another over the same Rivulet at Waterhouses, and other smaller ones.

It is said, that under the Clause for varying their line, this Company appear authorized to buy or sell Land at any future time, but that the *Peak Forest* Company were refused this power by the Legislature,

after

after the completion of their line. About the Autumn of 1796, the main line of this Canal was completed, and in January 1797, the Stockport branch was opened. In August 1799, some of the works on this Canal suffered by a flood. The rates of Tonnage on this Canal, are given in Phillips's 4to. " History of Inland Navigation," Appendix, p. 21. The Company were authorized by their first 4 Acts, to raise 170,000*l*. and a further sum by the last Act, in 100*l*. Shares ;. which in 1802 were selling at 20*l*. below par, but have since much improved, I believe. Mr. James Meadows, of Piccadilly Street in Manchester, is the principal Agent for this Canal.

In 1796, the *Macclesfield* proposed Canal, was to have had a branch approaching near to the branch of this, at the S end of Stockport Town.

MERSEY AND IRWELL *Navigation :*—This has been mentioned, p. 215; but falling almost wholly beyond the limits of our Map, it must suffice here, to refer for further particulars to my article *Canal*, in Dr. Rees' Cyclopædia.

NEWCASTLE UNDERLINE *Canal:* Act, 35th Geo. III. —The general direction of this Canal is nearly W, by a very bending course to the S, of 3 miles, in the County of Stafford : it is considerably elevated, skirting round the point of a short Ridge, branching from the *Grand Ridge* on its eastern side: its objects are, the supply of Newcastle Underline, the import of Lime, &c. and the export of Coals, Farming products, &c. Newcastle Underline is the 118th British Town, with 6,175 Inhabitants.

This Canal commences in the *Trent and Mersey*

Canal, at Quinton's Wood in Stoke (near to the commencement of the Caldon branch), and terminates in the *Newcastle Underline Junction* Canal at the SW side of that Town. This Canal may take water from the Trent River (or rivulet rather here, which it crosses, on an Aqueduct Arch), between the 1st of December and the 1st of April. It was completed in a short time after the Act was obtained. The rates of Tonnage and Wharfage are, on Coals, Limestone, and Ironstone, $1\frac{1}{2}d.$ per Ton per mile, on all other Goods 2d.; but for less than a Ton of any article in a Boat, 6d. per mile.

This Company were authorized to raise 10,000l. in 50l. Shares.

NEWCASTLE UNDERLINE JUNCTION *Canal:* Act, 38th Geo. III.—The general direction of this Canal is NW, for a short distance, in two detached parts, in the County of Stafford, passing off our Map at h: its north-western part is much elevated, terminating almost upon the *Grand Ridge*, on its eastern side: its objects are the export of Coals and Agricultural Products, and the import of Lime, &c. Newcastle Underline is the 118th Town, with 6,175 Inhabitants.

This Canal commences in the *Newcastle Underline* Canal, at the SW end of that Town, and terminates at its NW side, in *Gresley*'s Canal: its north-western or main part, commences again at the other termination of *Gresley*'s Canal, and extends to Partridge-nest Collieries, with a branch to Bignal-end Colliery. Provisions are made in the Act, for Inclined-planes for Boats, and water-levels, with Engines to raise the Water to supply them, or Rail-ways and Inclined-planes, in case either of these should be found more eligible than

than a Canal with Locks. This Company were authorized to raise 12,000*l*. in 50*l*. Shares. In 1796, the *Commercial* Canal, for 40 Ton Boats, was proposed, which was to connect with, and occupy a part of the line of, this Canal, when sufficiently widened and deepened.

North-eastern (or Don and Cromford Junction) proposed Canal:—In the Spring of 1810, Mr. William Jessop, jun. surveyed the line for this much-wished-for Canal, in two detached parts; the northern part from the *Don* Navigation, a short distance above Rotherham (the 315th Town, with 2,950 Inhabitants), passing nearly S up the Vale of the Rother, about 8 miles, in the Counties of York and Derby, to join the *Chesterfield* Canal near Killamarsh: the other, or southern part, commencing in the Norbrigs branch of the *Chesterfield* Canal, $\frac{1}{4}$ m. E of Staveley, and proceeding nearly S about $13\frac{1}{2}$ miles in Derbyshire (near to Nottinghamshire) by Hardwick and Tibshelf, to the Pinxton branch of the *Cromford* Canal, on the S side of Pinxton Town; as I have been informed, but have to regret, that the correct information which I requested of Mr. J. by letter, has not been received.

I heard it mentioned, but whether correctly I cannot say, that this line proceeded about $8\frac{3}{4}$ miles up the Vale of the Dolce, with a considerable rise; then was to tunnel about $1\frac{1}{4}$ mile (through Coal-measures) under the *East Derwent* Ridge, about $\frac{1}{2}$ a mile E of Tibshelf Town, near to the Collieries, then proceed across the head streams of the eastern branch of the Amber River, on a level, to another Tunnel under the same Ridge, of about $\frac{1}{2}$ a mile long, about 1 m. E of South Normanton Town; and then descend in the Vale of the Erewash, about $1\frac{1}{2}$ mile (passing near to
Pinxton

Pinxton Collieries and China Factory) to the Pinxton Basin of the *Cromford* Canal branch.

It was likewise mentioned, that the same Engineer, about the same period, surveyed a different route for the southern part of this *North-eastern* Canal, which commenced in the basin of the *Chesterfield* Canal, on the N side of that Town (the 183d, with 4,476 persons), and proceeded nearly S, about 13¼ miles, in Derbyshire, by Wingerworth and South Winfield, to the *Cromford* Canal in Buckland Hollow, on the S of Pentrich-lane: the first 5¼ miles of this line ascends the Rother Vale, and that of its Woodthorp branch (passing Wingerworth Iron Furnace, I believe) with a rise of about 200 feet to a very moderate deep-cutting, a mile W of Tupton, that might convey it over the *East Derwent* Ridge, and whereon it might be amply supplied with Water, on proper purchases of Mill-properties, or arrangements with their owners, especially by the aid of sufficient Reservoirs in the Prass, Alton, and Holm-gate Vallies, in Ashover and North Winfield Parishes: and from this summit it descended the Vale of the Amber about 7¼ miles, to the *Cromford* Canal; passing near to (and it might even, I believe, go through) the important Paving-stone and Freestone Quarries at Coburn, in South Winfield Park*, and near several other Quarries.

It appears, that on the 7th of March, 1811, a meeting of the parties wishing to promote this important Canal, took place in London: who resolved to have Surveys prepared, by different lines, " the line

---

* See Vol. I. 426, 421, and 427, which Quarries are in the same, or 4th Grit Rock, and not inferior in quality, to the celebrated stone of Ealand-Edge, and Cromel-bottom in Yorkshire, I. 164, and of Kerredge in Cheshire, see page 387, herein.

originally

originally proposed (that by Hardwick, I suppose) having been objected to by several Land Proprietors," and to apply to Parliament for an Act, for that line which on the whole appeared the most eligible: but since then I have heard nothing material on the subject; except, that the northern part of the very important junctions that this proposed Canal embraces, was now pursuing separately, as mentioned in p. 363, whether by the same or different parties, I am unacquainted.

I hope, however, now that the *Grand Union* Canal may almost be considered as complete, and a direct water communication formed between the Metropolis and all the middle and southern parts of the kingdom, and with both Pinxton and Pentrich-lane, that the southern part of the projected Canal that I am describing (more than the northern one), may no longer be suffered to sleep, but that a further extension northward may speedily be made, from one of these points, so as to complete the line across our Map (p. 193) from *F* to I.

And I think it right to add, that to me, the advantages of the latter of these lines under consideration, appears in all respects to preponderate, excepting only, as to forming *the straightest general route;* for in point of *natural difficulties* to be encountered (and consequent *expense* of execution), *populousness of the districts* through which they pass, and their *trade and manufactures*, also as to valuable and various *Mineral products* that will be intersected, and as to the *collateral branches* they may admit of, all these are in favour of the *Chesterfield and Pentrich-lane* line.

Although it cannot be doubted that valuable seams of Coal might be discovered and wrought, along most of the other or Norbrigs and Pinxton line, yet

after

after serving the considerable adjoining district in Derbyshire and Nottinghamshire, that must from its elevation, be supplied by land carriage, the remainder of these Coals are perfectly adapted to descend on Rail-ways, to the Canals at the different ends of the line; and the other line is also not deficient in Coals, in any part of it.

The eastern line has no Iron Works upon it, nor has it any rich Ironstone Rakes, to supply such Works, the situations so far to the eastward, of the Codnor Park, Somercotes, and Duckmanton Ironstone Pits and Works, being owing to local denudations (Vol. I. pp. 163 and 164 and p. xlvi): but the western line has the two most productive Rakes of Ironstone of this whole Coal-field, the Black Shale and the Dog-tooth Rakes (in the 8th and 9th Coal Shales, about the middle of the Coal district, I. 393), ranging near to its whole length; on which Rakes there is now an Iron Furnace, and there might be others established: the 4th or Paving-stone Rock above mentioned, from different Quarries that might be opened in it, between Pentrich-lane and Ford, in which distance it closely accompanies the line; and Gritstone of valuable properties from other Rocks, as on Pentrich-common, might supply a very extensive Tonnage on this line, and on the *Cromford* and other Canals to the southward.

From the summit of this line, a level Rail-way branch might also be conducted, westward, for about 4 miles, to the valuable and extensive Limestone Quarries in Ashover (Vol. II. p. 416), along the south-western part of the line proposed in 1802, for the *Ashover and Chesterfield* Canal, the north-western part of whose line, between this summit and Chesterfield, this Canal would occupy, nearly: and this level Ashover

over branch, would admit of equal facility in carrying Coals, &c. to Ashover and its vicinity, as of bringing back Limestone, large coarse Grindstones, of the 1st Grit from Overton, &c. see Vol. I. p. 436, &c. The *Wingerworth and Woodthorp* Rail-way, belonging to Mr. Butler's Furnace, occupies at present, or nearly, a part of the route for this western line; which as long ago as 1771, was selected by Mr. Brindley for the northern part of his proposed *Chesterfield and Swarkestone* Canal.

NOTTINGHAM *Canal:* Act; Geo. III.—The general direction of this Canal is nearly NW, by a crooked and bending course to the SW, of about $14\frac{1}{2}$ miles in the County of Nottingham, as shewn in the Map facing page 193. Its north-western and middle parts are rather considerably elevated, crossing the *East Erewash* Ridge, by a considerable deep-cutting, 1 *m.* W of Wollaton: its general objects are the supply of Nottingham, and the export of Coals, Lime, Cast-iron, Lead, Millstones, Grindstones, Freestone, Marble, Cheese and some other Agricultural Products, &c.; and the import of Deals, Bar-iron, Corn, Malt, &c. Nottingham is the 15th British Town, with 34,253 Inhabitants, and Eastwood is also rather a considerable Town, near this Canal.

It commences in the *Trent* River, at the mouth of the Leen River (very nearly opposite to the commencement of the *Grantham* Canal) and terminates in the *Cromford* Canal, at Langley-mill, very near to the termination of the *Erewash* Canal, therein. From near Wollaton there is a cut to the NE, of $1\frac{1}{2}$ *m.* (level with the summit pound), and a Rail-way $\frac{1}{4}$ *m.* thence to Holy-wood Colliery in Billborough: from the deep-cutting

cutting W of Wollaton, there is a Rail-way branch NNW ¼ of a mile to Trowell-moor Colliery. In the deep-cutting at the point of the Hill SE of Trowell, the Canal passes through Trowell Heath Gritstone Quarry, and alluvial red Founder's Sand Pit, on the stone (I. 465). From the SE of Cossall, a cut proceeds E, ¼ of a mile, and thence is a Rail-way ¼ m. further, to Robinets Colliery: from the N of Awsworth there is a cut to the ENE, ¼ of a mile, and a Railway thence, 1 m. to Greasley Colliery; and from the same point of the Canal line, there is a Rail-way branch NE 1 m. to Limes Colliery; and from Langley-mill there is a Rail-way branch (crossing the Beggarlee Rail-way from the *Cromford* Canal) to the NE, ½ mile, to Old Brinsley Colliery, on which Railway, Mr. Joseph Wilkes's experiments were in part made, that are referred to in p. 289.

From the *Trent* to Wollaton, the distance is about 4¼ miles, and the rise   feet by   Locks; and from thence to the *Cromford* Canal, 9¼ miles, is level: the branches of Canal are all level with the summit pound, and the Rail-way branches rise therefrom. The width of this Canal, and its Locks and Boats, correspond nearly with those on the *Cromford* Canal; and this Company have concurred with that and 7 other Navigation Companies, in establishing Guaging-houses, and an uniform mode of measuring and ascertaining the Tonnage on Coals, &c. carried hereon, as mentioned in Vol. 1. p. 182. At the points of the Hills which terminate towards the Erewash Vale, which the Canal skirts at a considerable elevation, there are several deep-cuttings; and between these, are three considerable Embankments, with Aqueduct Arches for the Brooks; at Cossall Marsh, N of the Village, is a very large

large one, another SW of it, and another N of Awsworth, besides several smaller ones.

A large Reservoir has been made for the use of this Canal, ¼ m. NNW of Greasley, which is said to let down near 3000 cubic feet of Water per hour, regularly, for compensating the *Erewash* Canal and the Mills by its course, for the Lockage Water taken by this Company, out of the *Cromford* Canal Basin: this is effected, by means of self-regulating Sluices, at the Reservoir and at Langley-mill, the mathematical principles of which, may be seen in Quest. 767, in the "Gentleman's Diary" for 1798 and 1799. Newthorp Brook, and some smaller Streams, are also taken into this Canal, as feeders.

Mr. William Jessop was the Engineer to this Canal, and completed it before the year 1798.

Instead of purchasing the parts of the Wollaton Coal-seams, that lay under this Canal and its Locks, as is usually practised, see page 348, this Company permitted the Colliers, some years ago, to get them, in their usual or "long-way" of working (Vol. I. pp. 188 and 344), and so little damage to the Canal-works has ensued, from pursuing these works under great lengths of this Canal, (see also page 364) that Mr. Thomas Walker, an experienced Coal-viewer, and well acquainted with Canal-works, recommends it for more general adoption; and mentioned, that it would even answer, instead of buying Coals, to be lost to the Community, to take off the Crowns of the Bridges, and lay temporary Baulks and plank Bridges over them, while the Coals were working (with some additional care, as to expedition in getting the Coals, and not leaving Pillars or Props under the Canal, to occasion irregular settlement), and for 2 or 3 years after, until

until the earth had become solid again. At Nottingham and at Langley-mill, this Company has Wharfs and large Warehouses.

From Trowell-heath Quarry on this Canal, across the Erewash River to the *Erewash* Canal (near to the entrance of the *Nutbrook* Canal) is but ¼ a mile, and from which place northward, these two Canals pass up the two sides of the River for 5½ miles of direct distance, to Langley-mill: it appears therefore, that a junction here, by means of a low Aqueduct over the River, and Locks in this place to ascend to the *Nottingham* Canal, would be very useful, by saving no less than 13 miles of the present distance between Derby and Nottingham, unless by the Canal Boats passing down the *Trent* River, which it is not always safe to attempt. It would not, I think, be impracticable or difficult, from the excellent *data* that the Weighing-houses Books furnish, Vol. I. p. 184, so to apportion the Tolls, on Boats passing this proposed Junction, in different directions, as that considerable savings may be made to the Boat-masters and the Public, without loss to any of the parties: since the Mills might easily be compensated by an additional Reservoir, for the lockage Water that would be let down at Trowell-heath.

NUTBROOK *Canal:* Act, 33rd Geo. III.—The general direction of this Canal is nearly NW, for 4¼ miles, in the County of Derby: it follows the Nutbrook Vale, and its northern end is rather considerably elevated. Its sole objects are, the export of Coals from the Pits near it, and the import of Limestone, having no Town near it, but the very small one of Kirk Hallam. It commences in the *Erewash* Canal, ¼ m. S of Trowell, and terminates at Shipley Wharf, from whence

whence however there is a Rail-way extension ½ a mile NNW, to Shipley Colliery.

North of Lewcote-gate, there is a Rail-way branch W, ¼ of a mile, to West Hallam Colliery. East of West Hallam, there is a Cut near a ¼ of a mile to the W, whence a Rail-way branch formerly proceeded, to Ferneyford Colliery. At Hallam-bridge there is a short Cut, from whence a Rail-way formerly proceeded SW, ¼ of a mile, to Dale-Abbey Iron Furnace (1. 397). At Hallam-bridge in Stanton, and at Shipley Wharf, are Lime-kilns for burning the Crich stone, see Vol. II. page 429.

It has about 12 Locks, and 4 public Bridges on it. The Boats used on this Canal are similar to those on the Erewash, and the same regulations, as to guaging Boats, and ascertaining the Tonnage of Coals, &c., mentioned Vol. I. p. 182. A very large Reservoir in Shipley Park, and 4 smaller ones, in or near the Park, supply this Canal in part, which also has a feeder from Dale-Abbey Brook. Edward Miller Mundy, Esq. and the late Sir Henry Hunloke, Bart., were the chief Subscribers to this Canal, and were authorized to raise the sum of 19,500*l*. in 100*l*. Shares, on which the annual Dividend is not to exceed 8 per Cent., and the affairs are to be managed by a Committee, as usual.

The Proprietors of adjoining Estates are allowed to make Branches from it, and peculiar privileges were granted to Earl Stanhope, in the carriage of Iron-stone, Cokes, Limestone and Iron, free of Tonnage to or from his Dale-Abbey Furnace. The Rates of Tonnage, with the arrangements relating thereto with the *Erewash* Company, may be seen in Phillips's 4to. " History of Inland Navigation," Appendix, pp. 104 and 105.

PEAK FOREST *Canal:* Acts 34th, 40th, and 45th Geo. III.—The general direction of this Canal and its Rail-way extension is S E, about 20¼ miles, in the Counties of Lancaster, Chester and Derby: its southern end is very considerably elevated, terminating upon, or even beyond the *Grand Ridge:* it crosses the *East Tame* Ridge by a short Tunnel and deep-cutting N E of Hatherlow, in Cheshire. Its principal objects are, the export of the Peak-Forest Limestone from its southern end, of Coals from its northern end, Building and Paving Stones, Bar-Iron, &c., and the import of Coals to its southern end, Deals, Pig-Iron, &c.

Ashton Underline is the 26th Town, with 19,052 Inhabitants, and Chapel-en-le-Frith the 353rd, with 2,667 persons, very near to the line; and Stockport the 32nd, with 17,545 persons, and Disley, also a considerable Town, are not far from it.

This Canal commences in the *Manchester, Ashton and Oldham* Canal (within ½ a mile of the commencement of the *Huddersfield* Canal), near Ashton Underline, and its Rail-way extension terminates in Black-holes (Barmoor or Loads Knowl, or Daff-hole mouth, and Dove-hole) Quarries of 1st, 3rd, and 4th Limestone, in Peak Forest, see Vol. II. pp. 418 and 423.

The Canal terminates at Bugsworth Wharf, ¼ m. NE of Whaley-bridge; and has a branch from Bottom-Hall S, to Whaley-bridge Wharf. To the tops of the Marple Lime Kilns a short branch is conducted, and a short Rail-way branch to the bottoms or draw-holes of these Kilns, as mentioned Vol. II. p. 426. From near Jow-hole old Furnace, a Rail-way branch proceeds under the Turnpike-road, nearly SW 1 m. to Diglee (or Whaley Moor) Colliery, with Tipples on the Wharf, for loading Carts as well as Boats

with

with these Coals: and on the NE of Yeardley there is a short Rail-way branch, passing under the Turnpike-road, into a Quarry of 3rd Grit Building-stone, that is carried as far as Manchester, Stockport, &c. Hyde-lane, Woodley, Marple-wood, and several other small Collieries, are situated on or near to the line of this Canal. At Disley NE, Chapel-en-le-Frith, Dove-hole, Marple and Whaley-bridge, there are Lime-kilns on the line of this Canal, and its extension, as mentioned, Vol. II. pp. 423 to 432. Near Chapel-en-le-Frith are two Forges for preparing Bar Iron (and there were more formerly) and many other works thence to Marple.

From the *Manchester, Ashton and Oldham* Canal, to the S side of the Marple Aqueduct, 7 miles, is level; thence to Priest-Field, 2¼ miles, is a rise of 212 feet; thence to the Whaley-bridge branch, 5¼ miles, is level; thence to Bugsworth Wharf (Thomas Lees) ¾ of a mile, is also level; thence to White-hough, 1 mile of Rail-way (or rather, of the intended Canal extension, I believe) has a rise of 129 feet; thence to Chapel-Milltown, 1 mile (of proposed Canal?), is level, (to where the *High Peak Junction* Canal is proposed to join): thence to the bottom of the great Inclined-plane, 1¼ miles, has a rise of 56¼ feet*: thence to the top of the Plane, 512 yards, is a rise of 192 feet: thence to Bar-moor-clough (or Loads Knowl) about ¼ of a mile, has a rise of 24 feet (2 feet per 100 yards); and thence to Dove-hole Quarry, ¼ of a mile, has a small rise at first,

---

* Mr. James Meadows informed me, in July 1808, that the present Rail-way, thro' the three last distances, amounting to 3 miles and 150 yards, rises regularly at the rate of 4¼ inches in a yard, whence these rises are deduced.

then level through the Quarries, and then descends rapidly into Daff-hole-mouth Quarry, out of which the Trams are drawn by Horses, as will be mentioned further on. The Whaley-bridge branch is rather more than ½ a mile, and level; the Upper Marple Cut is level, and the Rail-way branches all rise from the line.

The width of the Canal is 30 feet at top and 15 at bottom, and the depth 5½ feet, except the summit pound, which is made ¼ a yard deeper, under the idea of its acting as a Reservoir: the Locks are 72 feet long and 8 wide, and their general rise about 13 feet each: over the tails of most of the 8 Locks in Lower Park, John Isherwood, Esq. near Marple Bridge, neat Horse Bridges are made, but which seem calculated, considerably to impede the towing of the Boats: 16 Locks occur in 1¼ mile below Marple Lime Works. The Boats carry 25 Tons, which are built at Marple, and when new (since Timber has so advanced) have cost 140*l.* each. The places of the Marple Locks, were at first supplied by a steep Rail-way, or Inclined-plane for Trams, from 1797 to 1806.

The large *Inclined-plane* for Trams, ¼ a mile E of Chapel-en-le-Frith, has a double Rail-way on it, the rises on different parts of which are unequal: the lowest $\frac{1}{7}$th part, or 16 yards, rises at the rate of 3 inches in a yard, and the succeeding portions, of 16 yards each, rise more and more by a regular increase, to the upper one, which rises 6 inches in a yard: a large inclined wheel or pulley, furnished with a brake for regulating the motion, has the large endless chain passed over it, to which the Trams are linked by short pieces of chain, so that 7 descend at once, and draw up as many empty ones, or partly loaded with Coals. The great chain is of the common kind, with 5 inch

links,

links, and cost at Birmingham 500*l.*; it had been 5 years in use when I saw it, and appeared none the worse for wear: a large Rope was at first tried, but soon broke; a patent twisted Chain was then tried, but being made too slender, it also broke. Blocks of Wood are laid across the plane every 10 yards, for the chain to run upon. This Plane is under the direction of Mr. Jerman Wheatcroft of Chapel-en-le-Frith, and James Hill resides in the House at top of the Plane, to attend and work it. The Inclined-plane into Daff-hole mouth Quarry is 33 yards long, and has a double Rail-way on it, and a Horse-gin wheel at its top, draws up the loaded Trams and lets down the empty ones. The tops of the Dove-hole Lime-kilns near this, are level with the bottom of the Quarry, and a very simple Wheel-barrow Plane is here used, for drawing up the burnt Lime from their draw-holes, as mentioned, Vol. II. p. 423.

At Disley Wharf, ¼ mile NE of the Town, there is a short Inclined-plane, on which a small Whimsey Steam-engine is used to draw up loaded Trams of Limestone and Coals, from the Boats on the Canal, to the tops of the Lime-kilns; whose bottoms or drawholes are above the Canal, for the convenience of loading Boats and Carts with Lime, see Vol. II. p. 423.

The Rail-way extension was at first made single, but in 1803, it was relaid, with a double Road, except about 100 yards, thro' the Stoddard Tunnel, the bars are 5 feet apart. The bodies of the Trams are made of rolled plate Iron, and carry 45 cwt. each: a door opens behind, for shooting their contents of Limestone into the Boats, which is effected at Bugsworth Wharf, by means of a moveable *Crane* or Tippling Machine, which runs on an outer Rail-way, laid on purpose, behind that on which the loaded Trams approach the sides of the Boats. A man, by means of a Winch-handle and Chain, which he fastens to the head of the Tram, first turned round, on a turning plate, with its tail towards the Boat, shoots out its contents.

Four Horses are employed, at length, to bring down 20 Trams linked together, containing 45 Tons of Limestone (which sells at Bugsworth at 2s. 1d. per Ton, Vol. II. p. 427), and they return with the empty Trams in the same way; a Horse can draw 2 Tons of Coals or Goods (in 2 Trams), up the Rail-way.

Near to Woodley, there is a Tunnel 200 yards long, with a Towing-path thro' it, but it is scarcely wide enough for the Boats and Horses to pass: at Hydebank in Chadkirk, there is another Tunnel, 300 yards or more long, thro' the point of a hill, 16 feet wide, without any Towing-path: near to the Marple Aqueduct, another Tunnel 100 yards long, with a Towing-path through it: and under Stoddard Houses in Chapel-Milltown, there is a Tunnel about 100 yards long, for the Rail-way.

North of Hatherlow in Cheshire, there is a long but not very deep-cutting, through alluvial Sand, principally, in an inosculation of the Valleys. Thro' numerous points of Hills there are shorter deep-cuttings, particularly between Duckinfield and Water-meetings, where, in less than 5 miles, 28 Embankments, several of them considerable ones, and Aqueduct Arches over small streams have been necessary, besides the Godley Brook Aqueduct of one Arch: these, on the E bank of the Tame River, across sudden alluvial Gravel, Sand, and Clay Knowls and Gullies, have been attended

tended with more difficulties than I have any where else seen in the same space, and the perfect execution of the Canal in this part, does great credit to the Engineers concerned.

The great Marple Aqueduct Bridge, over the Mersey, about a ¼ of a mile below the Water-meetings (of the Ethrow and the Goyte), where this River first takes its name, is among the most considerable of the works of this kind in the Kingdom: it consists of 3 equal semicircular Arches of 60 feet span, the middle one 78 feet high; the whole height of the structure being near 100 feet. The River, except in times of Floods, is confined to the middle arch; the lower half of the piers are constructed of rough red masonry of the 3rd Grit Rock, from Hyde-bank Quarry at its northern end: the upper part is of handsome white hewn masonry: 4 cylindrical holes are worked thro' the haunches of the arches, in order to lighten them. The abutments are properly splayed or widened, in handsome curves, and the walls batter or diminish upwards in the same manner, adding greatly to its strength and beauty. It was finished in 1797, and a view of this Bridge, and of the very fine sudden and wooded Vale that appears behind it to the West, has been engraven by Mr. Jaques.

Near to Ashton Underline, there is also a considerable Aqueduct for this Canal, over the Tame River, and a smaller one of two arches, one over the Goyte and the other for a bye Road, at Bottom Hall near Whaley-bridge. South-east of Turf-lee in Disley, a Road and small Brook pass under the Canal, and in several other places there are smaller Aqueducts and Embankments, as already mentioned.

This Canal is wholly supplied by flood waters, reserved in Combs-brook Reservoir, of 45 Acres, S of Tunstead-Milltown (the Fishery of which has been mentioned, page 203), whose head can, if found necessary, be raised so as to extend it to 90 acres. There are cuts along each side of this Reservoir, for collecting and carrying forward all the Brooks for the use of the Mills below: on the western and principal of these cuts, there is a Weir 10 feet wide (having a fall of 1 foot), with planks fixed across at $3\frac{1}{4}$ inches above the Weir, for turning all the Water that will not pass thro' this opening, into the Reservoir, over a side Weir $3\frac{1}{2}$ inches higher than the former one, and 16 yards long; whenever the Water is running over this last Weir into the Reservoir, the whole of the water in the eastern cut, is turned through a sluice into it, but at other times it runs past it, to supply the Mills below.

The water for the Canal is let out of this Reservoir by a large Cock, and carried in a feeder, separate from the Brook, near 2 miles, to Whaley-bridge, passing there, under the Goyte River, in cast-iron Pipes. By the side of the Bottom Hall Aqueduct there is a Weir, for discharging the surplus water of the Canal, consisting of arched openings in the Walls, with gates shutting together like those of Locks, but close together and contrary ways. Other Reservoirs have been provided for, but not found necessary, on Todds Brook, SW of Whaley-bridge, and above Wash, NE of Chapel-en-le-Frith.

Mr. Benjamin Outram was the Engineer to this Canal, and Mr. Thomas Brown the resident Engineer, who completed it, except the Marple Locks, in 1797. The Rail-way to Peak Forest was relaid in 1803, and the

the Marple Locks were opened, and the whole completed in 1806.

The Tonnage rates that this Company are allowed to take, may be seen in Phillips's 4to. " History of Inland Navigation," Appendix, p. 155. In 1808 the following were the rates that they took, viz. 1d. per Ton per Mile for Stone and Coal; 1½d. per Ton per Mile for burnt Lime, and 3d. Wharfage.

The Works of this Company are said to have cost 180,000l. By the first Act they were authorized to raise 150,000l. in 100l. Shares (which in 1802 bore a premium of 10 per cent.), and by their third Act, a further sum was permitted to be raised. At Whaley-bridge, Bugsworth, Disley, and Hyde-lane, there are Wharfs on this Canal, and a Rail-way yard at Townend, near Chapel-en-le-Frith.

In 1808 the quantity of Limestone brought down to Bugsworth Wharf by the Rail-way, was 50,000 Tons; and about 100 Tons of Merchandize was weekly carried up or down thereon, to or from Chapel-en-le-Frith. The Waggon or Tram hire, for carrying up Coals from Bugsworth Wharf to Black-hole Quarries, or for bringing down Lime therefrom, was then 7d. per Ton. Mr. James Meadows of Piccadilly Street Manchester, is the principal Agent to this Company, and Mr. William Bate the Clerk at Bugsworth Wharf.

In the year 1796, the *Macclesfield* Canal was proposed to join this Canal above the Marple Locks: In 1810 the *High Peak Junction* was proposed to join an extension of this Canal near Chapel-Milltown; for which proposed extension from Bugsworth Wharf (instead of the present Rail-way), notices were given of an application to Parliament by this Company, in the same year, and for amending their first Act as to Rates.

Rates. And at the present time (May 1814), there seems a scheme on float, for making a Rail-way branch from Whaley-bridge, S, 4¼ miles, passing Shallcross Colliery, to Goytes-clough Quarry; and perhaps meeting there, other Rail-way branches from the *Cromford*, and from the *Trent and Mersey* and the *Chesterfield* Canals, see pages 350 to 352.

ROCHDALE *Canal :* Acts, 34th, 40th, and 44th Geo. III.—The general direction of this Canal is nearly NE, by a very bending course to the NW, of 31½ miles, in the Counties of Lancaster and York, commencing at q, and crossing the corner of our Map to A. Its middle part is greatly elevated, crossing the *Grand Ridge* by a Tunnel N of Littleborough: and it crosses the *West Irk* Ridge by a deep-cutting near Manchet Hall, 2 m. S of Rochdale. Its general objects are, opening of a communication (the first direct one) between Liverpool and Manchester &c. and Halifax, Wakefield, Hull &c. and the export of Coals, Paving and Building stones, &c. found near the line.

Manchester is the 3rd British Town, with 98,573 Inhabitants; Rochdale the 108th, with 6,723 persons; and Todmerden is also a considerable Town, on the line of this Canal: and Halifax is the 69th Town, with 9,159 persons, not far from it.

This Canal commences in the Duke of *Bridgewater's* Canal, at Knot-mill (or Castle Key or Hill), at the SW side of Manchester Town, and terminates in the *Calder and Hebble* Navigation, at Sowerby-bridge Wharf, about 2¼ miles SW of Halifax: In Piccadilly Street, in the SE quarter of Manchester Town, it is joined by the *Manchester, Ashton and Oldham* Canal

nal; from near Failsworth it has a branch NE, to the Wharf on the W side of Hollinwood Village, 1½ mile from Oldham, which branch approaches here within ¼ a mile of a branch from the *Manchester, Ashton and Oldham* Canal; and another branch from Lower-place, nearly N, to School-lane, at the SE end of Rochdale Town.

From *Bridgewater's* Canal, to the *Manchester, Ashton and Oldham* Canal, $1\frac{1}{4}$ mile, is a rise of $75\frac{1}{4}$ feet: thence to the Hollinwood branch, $4\frac{1}{4}$ miles, is a rise of 181 feet: thence to the School-lane branch, $7\frac{1}{4}$ miles, is a rise of 120 feet: thence to near Clegg-Hall, $2\frac{1}{4}$ miles, is a rise of 62 feet: thence through the Dean Head Tunnel to Travis-mill ($1\frac{1}{4}$ m. S of Todmerden) $5\frac{1}{4}$ miles, is level: and thence to the *Calder and Hebble* Navigation, $11\frac{1}{4}$ miles, is a fall of 275 feet. The Hollinwood branch, near ¼ of a mile, is level; and the School-lane branch, near ½ a mile, is also level with the line: on which there are in all 49 Locks, of the same length and width as those on *Bridgewater's* Canal, adapted for 50 Ton Boats: over this Canal and its branches there are more than 60 Bridges.

The Dean Head Tunnel is $1\frac{7}{16}$ mile long, and a part of its cutting was in very hard Rock of the 1st Grit; and near Hallins-mill, NE of Sowerby, there is another Tunnel 70 yards long, through the point of a hill, which is arched 17 feet high, and 21 feet wide, with a towing-path through it. Over the Calder at Hebden Bridge, and on the S of Todmerden, this Canal crosses on Aqueduct Bridges: over Heelees Brook near Littleborough, Chadderton Brook, and Failsworth Brook, there are also considerable Aqueducts, and numerous smaller ones, and Culverts over other

other streams, accompanied by several large Embankments across the vallies in which they run.

The Mill-owners on the Roch and the Irk Rivers, which are branches of the Irwell, stipulated in the Act, that only the *surplus flood water* of the head streams of these Rivers, reserved in Reservoirs, should be taken for supplying the lockage on this Canal; and that Guage-weirs or Sluices should be constructed, for *constantly* discharging out of these Reservoirs or the Canal, into each of these Rivers or their branches, exactly as much Water as will yearly amount to the average quantity that is intercepted or taken therefrom respectively, into the Reservoirs or Canal during the year, with the exception only of *times of flood*; when more than certain quantities descend in those intercepted streams: by which it is plain, that these Mill properties must be *very highly improved*, in the constancy of their streams, and avoiding mischievous floods, *at the expense of the Canal Company*, as has been observed of another concern in p. 322, and the same remark might be made on cases in almost every quarter of the kingdom.

Mr. John Rennie the Engineer, was deputed to ascertain the quantity of Water necessary for the proper supply of the Mills, on all the streams that were proposed to be intercepted, as above mentioned; for which purpose, as I have been informed, he erected accurate Guages, which during a whole year, ascertained the exact quantity of Water that daily passed these Mills, noting very particularly the quantities on those days when the Mills were amply supplied, but without surplus discharged over their Weirs; and he found

found the general result to be, that during the year 1793, *sixteen times as much Water ran waste by the sides of the Mills, as they used!*

On Heelees Brook, falling into the Roch, two very arge Reservoirs were constructed (one of them 14 yards deep, I believe), but so much below the level of the summit pound, on the west side of the *Grand Ridge*, that a Steam-engine of 100 horse power was said to be necessary, for pumping up their water to the summit level; this, to me, seems to have been a very injudicious situation for Reservoirs, 'compared with Blackstone-Edge, on or near the Ridge; where also, one very large Reservoir was constructed at White-Holme, and three smaller ones on the west side, as well as one considerable one, and two smaller Reservoirs on the east side of the Grand Ridge, I believe:—owing, however, as I conclude, to *the undue proportion of Water furnished to the Mills* from these several Reservoirs[*], they have been found inadequate, as Mr.

Meadows

---

[*] It seems difficult to form any other conclusion than this, after reading pages 237 and 239 of Vol. 43, of the "Philosophical Magazine," where Mr. Thomas Hanson, Surgeon, of Manchester, details the experiments made by himself in Manchester, and by Mr. Matthew Leadbeater of Lane-side in Middleton, at White-Holme Reservoir, on quantities of Rain (and water from Snow and Hail) caught in Rain Guages, of exactly similar constructions, in these two places. Whence it appears, that in the last 9 months of the year 1813, the down-fall at Manchester was 29·928 inches, and on Blackstone-Edge 86·085 perpendicular inches, in the same period of time!. And whence the very extraordinary and important facts appear, viz. that the monthly falls of Rain, &c. in this period on Blackstone-Edge (when the proportion is least, in July), are more than $1\frac{1}{2}$ to (when largest, in December) more than $11\frac{1}{2}$ times the quantities, that fall at Manchester; the average of the 9 months being $2\frac{7}{8}$ times; and calculating for a whole year at this rate from Mr. Hanson's observations, we find that *more than* 100 *perpendicular*

Meadows informed me in 1808, and a new large Reservoir was then contemplated, on Blackstone-Edge; where doubtless, all the Water wanted by this Company ought to be collected during floods, above the level of their summit pound.

Mr. John Rennie was the Engineer to this Canal, and finished its eastern part, from Sowerby-bridge to Rochdale, on the 28th of December, 1798: on the 18th of September, 1802, it was continued in Lomeside Wharf, and on the 21st of December, 1804, the whole was completed. This Company were required to make a compensation in Tolls to the Duke of Manchester, for the diminution expected, in the use of his Warehouses at Castle Key in Manchester, and the same to the *Calder and Hebble* Company, for their Warehouses at Sowerby-bridge. The Tonnage rates that this Company are authorized to take on various Goods, Wharfage, &c. by their first Act, may be seen in Phillips's 4to. " History of Inland Navigation," Appendix, pp. 157 to 161; to which the second Act has made some additions. This Company were authorized to raise 391,000*l*. in 100*l*. Shares.

In 1791, a branch from this Canal was proposed, from near Todmerden (104 feet below the summit level) to the *Leeds and Liverpool* Canal, two miles beyond Colne, having a Tunnel of 1¼ mile long thro' the *Grand Ridge*, 3 miles N of Todmerden.

---

dicular inches of water falls annually!, on the extensive Bogs on this part of the *Grand Ridge*, and where consequently, a smaller space of Reservoirs than has already been made there, *ought* to supply, even a greater lockage than is likely ever to occur on this Canal: and which reasoning receives confirmation, from the supply of the *Peak Forest* Canal, mentioned page 408, and several others in England.

STAFFORD

STAFFORD *Railway:* Act, 49th Geo. III.—The general direction of this Rail-way is WNW, for 1¾ mile, ranging close on the N side of the Turnpike Road, from the *Staffordshire and Worcestershire* Canal at Radford-bridge, to the S end of the Town of Stafford, which is the 162nd, with 4,868 Inhabitants: its objects seem difficult to guess, as a Navigation, not very circuitous, already existed, as I was informed, between the same points. When I saw this Rail-way in November 1813, it was laying disused, and, on inquiring the reason, I was told, that Mr. Hall, at whose expense this work had been executed, had become a Bankrupt: the only issue to have been expected, I think, from such an undertaking.

STAFFORDSHIRE AND WORCESTERSHIRE *Canal:* Acts, 6th, 10th, 30th, and 45th Geo. III.—This is sometimes called the Wolverhampton Canal: its general direction is NNE for 46¼ miles in the Counties of Worcester and Stafford, entering our Map (p. 193), at *Z:* its middle part is very considerably elevated, crossing the *Grand Ridge,* about two miles NNW of Wolverhampton. Its objects are the export of Coals, and Agricultural Products; the import of Limestone, Deals, &c. and the forming of a communication (the first effected) between the *Severn* and the *Humber* and the *Mersey* Rivers, &c. Kidderminster is the 123rd Town on the British List, with 6,057 Inhabitants; and Penkridge is also a considerable Town, on the line of this Canal: and near to it is Bewdley, the 248th, with 3,145 persons; Stourbridge, the 201st, with 4,072 persons; Wolverhampton, the 39th, with 14,836 persons; and Stafford, the 162nd, with 4,868 persons.

It

It commences in the *Severn* River at Stourport (at the mouth of the *Stour* River), and terminates in the *Trent and Mersey* Canal at Great Haywood: near to Stewponey it is joined by the *Stourbridge* Canal: at Autherley (or Aldersley) by the old *Birmingham* Canal (about two miles from the termination of the *Wyrley and Essington* Canal): at Radford-bridge it is joined by the *Stafford* Rail-way. From near Baswich, the course of the Sow River is made navigable NW 1½ mile, to the Wharf at the S end of Stafford Town; from Latherford there is a Rail-way branch eastward, to Bushbury Colliery. The Company may make branches within 1000 yards of the line in any part, by consent of the Land-owners. At Radford-bridge Wharf, there are several Lime-kilns, for stone brought from Dudley or Caldon.

From the *Severn* River (at its usual or mean height, there being two basins at Stourport, one of which is always kept at the same level by flood Locks, and the other rises and falls with the floods and droughts in the river) to the *Stourbridge* Canal, 12¼ miles, is a rise of 127½ feet, by 13 Locks: thence to Tettenhall, 11 miles, is a rise of 166¼ feet, by 18 Locks; thence (along the summit pound, which is said to be 358 feet above the *Thames* at Brentford?) to the old *Birmingham* Canal, 1¼ miles is level; thence to the Streetway Turnpike Road, 8¼, is level: thence to the *Stafford* Rail-way, is 8¼ miles: and thence to the *Trent and Mersey* Canal, is 5 miles, the whole fall being 100¼ feet, by 13 Locks.

This Canal is 30 feet wide at top, and 5 feet deep, although the depth of Water on the Lock-sils is only 4 feet: the Locks are 74 feet long, and 7 feet wide; the Boats in general carry 20 Tons.

On the SW of Stewponey, there is a short Tunnel through the point of a Hill: S of Whittington there is another such Tunnel; and through part of the Town of Kidderminster, the Canal is arched over; N of Autherley there is a deep-cutting, another on Calf Heath, and numerous smaller ones. Over the Stour at Kidderminster, there is an Aqueduct Bridge: over Wordsley Brook another, at Prestwood: over the Sow at Milford, and the Trent at Harwood-mill, there are also other Aqueducts. On the SW of Chillington, this Company has a large Reservoir, and at Moseley, a smaller one, whose waters are conducted into the summit pound by feeders: in 1802, a morass of 500 acres was drained, by means of Cast Iron Pipes, 5 feet diameter and 135 feet long, laid under the Penk River near this Canal.

Mr. James Brindley was the Engineer to this Canal, which he began in September 1766, and finished it in 1772: the first Lock which this Engineer erected, was at Crompton on this Canal: several of the Locks were built with soft red grit stone, and were some of them were worn out in 1805, and the last Act was applied for, to enable the Tolls to be raised for rebuilding them.

The rates of Tonnage are stated in Mr. John Cary's excellent work on " Inland Navigation," page 32. By the *Dudley* Canal Act (16th Geo. III.), Coals brought from that Canal, and carried on this, may be charged 2*d*. per Ton per mile, but which rates, however, the Commissioners are authorized to lower. The usual charge made by Bargemen in 1796, for freight (including the Company's Tonnage) was, for perishable Goods 2½*d*. per Ton per mile, and for heavy unperishable Goods, 2*d*. This Company were

autho-

authorized to raise 100,000*l.* in 100*l.* Shares; in 1805, the Dividends thereon were 24*l.*

The *Stour* (Stourbridge) River, nearly along the southern part of this line, was made navigable a great many years ago, as high as Stewponey, but a violent flood having destroyed the works soon after, they were never renewed: although in 1758, Dr. Thômas Congreve proposed the same, with a Canal thence over the *Grand Ridge*, nearly in the line of this, to connect with the Penk, Sow, and Trent Rivers, which also were to be made navigable to the upper *Trent* Navigation at Burton: the *levels* stated on this occasion were, from the *Severn* to near Prestwood, a rise of 104½ feet; thence by the Smestal Rivulet to near Autherley, a rise of 181 feet: thence by the Penk Rivulet to Bull-bridge at Penkridge, a fall of 88¼ feet: thence by the Penk, Sow, and Trent Rivers to Burton, a fall of 100¼ feet, there to meet the Navigation on the upper part of the *Trent* River, that is now disused.

A Reservoir was intended to supply the Canal, of 456 acres, on the *Grand Ridge*, between Compton and Penford; its S head to be 10 feet high, and its N head 24 feet high: the Boats passing through it, by means of Locks. It was proposed by Dr. C. that only *Sailors* should be employed on this Canal, in time of peace. In 1765, a branch from the proposed *Ternbridge and Winsford* Canal was intended, to occupy nearly the same route as the northern end of this Canal, below Stafford.

Swadlingcote and Newhall proposed Rail-way.—In 1792, Mr. George Nuttall surveyed the line for a Rail-way, from the Trent River below Drakelow,
opposite

opposite to the navigable stream (or new Mill-dam) that connects with Burton Wharf and branch of the *Trent and Mersey* Canal; proceeding thence up the valley SE and then E, 4 miles, to Swadlingcote Colliery, with a branch therefrom NE, to Newhall Colliery. In 1809, this scheme was again revived, and the Survey for the Newhall branch, was extended up to Bretby Colliery. In 1796, the *Commercial* Canal was intended, I believe, to occupy nearly the line of this proposed Rail-way, in order to connect with the proposed Swadlingcote branch of the *Ashby-de-la-Zouch* Canal.

Ternbridge and Winsford proposed Canal.—In 1765, Mr. Robert Whitworth proposed a Canal from the *Severn* Navigation at Ternbridge below Shrewsbury, to the *Weaver* Navigation at Winsford-bridge, E of Middlewich, a distance of $63\frac{1}{2}$ miles, nearly NNE, with a bending course to the eastward, and crossing the *Grand Ridge* twice by deep-cuttings (of 25 feet each), in Offley Park near Eccleshall, and again in Madeley Park, SW of Newcastle Underline: from between which summits, the Trent branch, from near Great Bridgeford, was to extend eastward, 43 miles, to the *Trent* Lower Navigation at Wilden Ferry.

From the *Severn*, this line passed up the Vale, by the Tern River (crossing the line since occupied by the *Shrewsbury* Canal near Rodington) and up the Vale of the Strine River, passing Newport, and through Batchacre Park; and then turned eastward to the first summit, a distance of 24 miles, with a rise of $136\frac{1}{4}$ feet: thence into the Vale of the Sow to the Trent branch, $1\frac{1}{4}$ m. below Great Bridgeford, is $7\frac{1}{4}$ miles, with a fall of $54\frac{1}{2}$ feet: thence up the Vale of the Sow to the second summit, is 10½ miles, with a rise of 800¼ feet; and thence by Wyburnbury into the Weaver Vale near Nantwich (approaching there the *Chester* Canal, and crossing the proposed *Commercial* Canal line), and down the same to the *Weaver* Navigation, 22¼ miles, with a fall of 284 feet.

The Trent branch proceeded down the Vale of the Sow, entering our Map at b' (and proceeding nearly along a part of the course of the *Staffordshire and Worcestershire* Canal), and down the Vale of the Trent (nearly along the line of the *Trent and Mersey* Canal), to the lower *Trent* Navigation, 43 miles, with a fall of 209¼ feet. It was proposed to be a wide Canal for 50 Ton Barges, and the estimated expense was 99,800*l*. In 1760, the *Wilden and Kings Bromley* Canal was proposed to occupy nearly the same line as the southern end of the Trent branch, above described. In 1793, the *Sandbach* Canal was proposed to occupy nearly, the southern and middle parts of this line; and in 1797, the *Newport and Stone* Canal was proposed to pursue nearly the middle parts of its course.

Tinsley and Grindleford Bridge proposed Canal: (or Don and High-Peak Junction):—In the year 1810, Messrs. William and Josiah Fairbank, roughly surveyed this proposed line, from the *Don* Navigation at Tinsley Wharf, up the SE side of the River, and through the north-eastern part of Sheffield Town (the 14th, with 35,840 Inhabitants), to near the Ponds Colliery, which on my Map is a distance of about 3¼ miles, with a moderate rise: thence up the Vale of the Sheaf, past Heely-mill, Mill-houses, and the N end of Totley Village to the W side of Ronset Farm, 6¼ miles,

6¼ miles, with a very considerable rise: thence W under the *Grand Ridge* (covered here by 2nd Grit), and Fox-House into Burbadge Vale, by a proposed Tunnel of 2 miles: and thence down this Vale (near to the famous Peak Millstone Quarries, Vol. I. p. 221), 1¼ miles, with a considerable fall, to join the proposed line of the *High-Peak Junction* Canal at Upper Padley Mill, near Grindleford-bridge; making a length of 14 miles, nearly SW, in the Counties of York and Derby.

In July 1811, meetings were held in Sheffield, and a Subscription made, to which Mr. Peter Brownell was appointed Treasurer, for employing the above Surveyors, to make a more minute Survey and Section of the above described line, or one near it: and to extend the same down the side of the *Don* to near Rotherham (on account of the present inconvenient Navigation by the river and side-cuts, below Tinsley) the length of which would be 17¼ miles, it was said, but I have not been able to obtain the levels or other particulars of this Survey.

In 1803, there was a design of extending the *Don* Navigation up to Sheffield, and at various other periods a Rail-way extension instead of a Canal has been thought of (see page 352), through the northeastern part of the line above described.

TRENT *River*, lower (or present) Navigation: Act, 34th of Geo. III. (for *Trent* Canal).—The general direction of this Navigation, is nearly SSW, by a crooked and bending course to the SE, of about 89 miles; skirting Yorkshire for a short distance, thro' the counties of Lincoln, Nottingham, and Derby (for a short distance, at Sawley-Cliff Farm), through a part of Leicestershire (by the Sawley-bridge cut), and dividing that County from Derbyshire: entering our Map facing page 193 at V.

Its south-western end is perhaps not more elevated than 60 feet above low water at the *Humber* mouth? It is the only River Navigation that now enters Derbyshire, see Vol. I. p. 470. Its objects are the export of Coals, Iron, Lead, Gypsum, Salt, Granite, and other Paving, Freestone, Marble, Pottery-wares, and other manufactured Goods, Burton Ale, Cheese, Corn, and other Agricultural Products, &c.: and the import of Deals (or Raff), Bar Iron, Flints, Hemp, Flax, Cotton, Malt, &c. &c.

Gainsborough is the 147th British Town, with 5,172 Inhabitants; Newark the 95th, with 7,236 persons; Nottingham the 15th, with 34,253 persons, and Burton-on-Strather, a considerable Town on this River; and not far from it are also situated, Crowle, the 611th Town, with 1,424 Inhabitants; Southwell the 352nd, with 2,674 persons; and Bingham the 645th, with 1,326 persons. This Navigation commences in the Tide-way in the *Humber* River (at the junction therewith of the York *Ouse* River, and the *Market Weighton* Canal) at Trent-fall, and terminates in the *Trent and Mersey* Canal at Wilden Ferry, where formerly, the upper *Trent* and the *Derwent* River Navigations commenced, but both of which are now discontinued.

At Keadby it connects with the *Stainforth and Keadby* Canal, at West Stockwith with the *Idle* Navigation, and with the *Chesterfield* Canal: at Torksey, with the *Fos-dyke* Canal: at Crankleys, and also at Upper Weir, with the mouths of the *Dean* River, forming here a back stream of this River, pass-

ing

ing close to the Town of Newark, and which is often called "Newark-dyke:"—a little below Trent-bridge, S of Nottingham, it is joined by the *Grantham* and the *Nottingham* Canals: at Trent Lock or Sawley Wharf it is joined by the *Loughborough* Navigation, and the *Erewash* Canal.

From the *Humber* to the *Stainforth and Keadby* Canal, is about 9 miles; thence to the *Idle* Navigation, about 12 miles; thence to the *Chesterfield* Canal, about ¼ of a mile: and thence to Gainsborough-bridge, 4 miles: through which 25¼ miles, the tide flows, and it is navigated by Ships, to the port of Gainsborough. From this lowest bridge on the *Trent* to the *Fos-dyke* Canal, is about 12¼ miles: thence to the *Dean* River, or lower end of Newark-dyke, about 17 miles; thence to the upper end of ditto, three miles*: thence to the *Grantham* Canal, is about 18¼ miles: thence to the *Nottingham* Canal, about ¼ of a mile: thence to the *Loughborough* Navigation, and *Erewash* Canal, 11 miles; and thence to the *Trent and Mersey* Canal is 2 miles.

The last-mentioned 13 miles of this Navigation was formerly very imperfet, owing to the interruption of 21 shoals (over some of which there were Fords, p. 282), and 2 Bridges (since Harrington bridge has been built).

---

* The length of the back stream in this distance, called Newark-dyke, between Crankleys and the upper-end, is about 3¼ miles long, and has 2 Locks upon it, one at the SW end of Newark, and the other about a mile below the Town, which are adapted for 50 Ton Barges. The trade at Newark Wharf is considerable, in exporting Gypsum, and green and white Alabaster Chimney-pieces, &c. in imitation of Marble, from Beacon Hill (Vol. I. p. 151), blue *Lias* Flag-stones from Coddington (I. 115), Corn, Wool (for Yorkshire), &c.; and importing of Coals from the Derbyshire, Nottinghamshire, and Yorkshire Pits, Lime from Crich and Barrow-on-Soar *(Lias)* Limestone, Deals, Iron, &c. &c.

In 1761, when Mr. John Smeaton examined this part of the River, he found, that in its ordinary state in dry seasons, there was not more than 8 inches depth of water on the shoals in several places, and that at these times it was impossible for Boats to pass, but by the aid of flushes of water let down on purpose from Kings Mills (the lowest on the Trent, 4¼ miles above Wilden Ferry), and from Little Wilne Mill on the Derwent.

In order to remedy these defects, a side-cut 10 miles in length was proposed in 1794, from the mouth of the Leen River, and following, I believe, nearly the course since occupied by the Nottingham Canal, to near Lenton; and thence through the meadows below Beeston (where a short cut and Lock was to connect with the *Trent*), crossing the Erewash River on a low Aqueduct S of Toton, and crossing and connecting with the *Erewash* Canal, W of Long Eaton: crossing the *Derwent* near its mouth (on an Aqueduct?), and joining the *Trent and Mersey* Canal (above its entrance Lock?) at Wilden Ferry; in which distance the rise was stated to be 28 feet.

For this side-cut (sometimes called the *Trent* Canal) an Act was obtained, as already mentioned, and the Trent Proprietors were authorized, for carrying it into effect, to raise 23,000*l.* in 50*l.* Shares, but on which the profits were not to exceed 7 per cent. per annum, and were to be entitled to various Tolls or Tonnage Rates hereon, and on the River, that are specified by Phillips, in his 4to. " History of Inland Navigation," App. pp. 169 and 170; but these Tolls were not to commence until 13,000*l.* had been expended under this Act; which embraced also the improvement of the River below Nottingham, by deepening and narrowing the shallow places, so that there may at all times

times be 30 inches depth of water: the making of a *Horse* Towing-path (*Men* having been employed in large gangs, in dragging the Boats, and still are so in some places, I believe), purchasing the Nottingham Hauling-machine or Capstern, for pulling Boats up a rapid place of the stream, when increased by flushes of water let down, &c.

By the 33rd Geo. III. for *Grantham* Canal, this Company, on condition of effectually deepening the River between the commencement of the *Grantham* and the *Nottingham*, and *Trent* Canals, are entitled to take 1½*d*. per Ton for Lime, and 3*d*. per Ton for all other Goods (except Road-materials and Manure) which cross the *Trent* at this place: and by the Act of the same year for the *Derby* Canal, only one-half of the usual rates of Tonnage are to be charged by them on Goods passing only 3 miles on this River, in their way from or to their Canal.

This *Trent* Canal has not been carried into effect, nor did I hear the probability mentioned of its ever being so, as might well have been foreseen by those, who persuaded the Legislature to preclude the possibility of the Adventurers herein, ever reaping more than 7 per cent. from a laudable *speculation for the public benefit*, but whereon the probabilities greatly preponderate (without any such restriction) against the prospect of realizing half that rate of interest, on their expenditure, see page 292, line 1, where " limitations of profits," has by an error of the press, been omitted, before, " &c. &c."—and nearly all which has yet, I believe, been done, for remedying the defects of this important Navigation (see page 400) has consisted in the occasional removal of some Gravel from the shoals, by means of a sort of *dredging-machine* that they call a

Gravel

Gravel Plough, which I saw at work near Trent Lock, consisting of a large iron Shovel, suspended between 4 large Wheels; which machine is drawn by 4 horses, in dry seasons, a man riding on an elevated seat behind, to drive the horses over the shallow places in the River, until 18 or 20 cwt. of Gravel is collected in the Shovel, which is then drawn on to the sloping bank of the River, where its contents are shot out, by means of a Winch-handle, roll and ropes, that tilt up the hinder part of the shovel.

At the south-western termination of this Navigation, the Towing Horses, and Boy who guides them, have to cross a long and very dangerous Ford through this River at Wilden Ferry, where a towing Bridge ought by all means to be built, if the *Trent* Canal should longer remain unexecuted, and to which Bridge all other impediments, except the occasional floating of Ice (that might be warded off the wooden Piers, by slopeing Struts, as at Muskham Bridge), are now removed, by the shutting up of the upper *Trent* Navigation.

Hartington or Sawley Bridge, over this Navigation (II. p. 22) has six flat eliptical Arches for the River, besides one S of these for the Navigation side-cut, and 6 smaller ones for the Flood-waters on the Meadows; Nottingham Trent Bridge, has 17 arches. In 1809 there was an intention of building a Bridge at Gunthorp Ford: At Kelham and Muskham there are Wooden Bridges, and between the latter and Newark, there is a Water-Road, constructed by Mr. John Smeaton for the great North Road, across the flat Meadows and the *Dean* River, with 72 arches, which together, have 300 yards long, of clear Water-way for the Floods!

At

At Dunham there is a Bridge across this River, and another at Gainsborough, which was rebuilt not many years ago, under an Act of Parliament.

This River is subject to very great Floods, which, through all its lower part, are confined (as well as the Tide in some parts) by large Embankments on each side of it, made at a considerable distance apart, in some places; between which artificial banks, the stream winds, in a channel that shifts occasionally by the floods, as happened in the spring of 1800, when its course was considerably straightened thus, below Gainsborough: in some instances these operations, for keeping open and improving the navigable Channel, have been assisted and directed, by means of *Weir-hedges*, consisting of two strong stake-hedges placed parallel, and the space between them filled with stones, for acting as jetties to direct the current, as was practised at Dunham in 1787, see " Transactions of the Society of Arts," Vol. XIII. p. 143.

It happens on this, as on most other large Rivers, which meet the Tide in flat inland districts, that very copious deposits of slime or warp take place on their muddy banks in such situations: and which circumstance has been turned to account, in raising and recovering Land within the Banks of this River, by simply staking down rows of furze faggots thereon, to check the current and encourage the warping, as was practised by Mr. Samuel Crawley at Dunham in 1789, as mentioned in " Trans. Soc. of Arts," Vol. XIII. p. 141, with a Map of the Holts or new Islands thus obtained*.

A still

---

* Sir Thomas Hyde Page practised this mode, and Mr. Smeaton, as mentioned in his Reports; Mr. Anthony Tatlow did so also, on the

Sea

A still more advantageous use of the thick Waters of this River, has also been made in its vicinity, in Lincolnshire and Nottinghamshire, by the *warping* of Lands for their improvement (see Vol. II. p. 494), as particularly described in the " Annals of Agriculture," Vol. XXXIII. p. 383. So abundant is this deposit, that considerable care is necessary to prevent the Warping and Drain Sluices from choaking with Warp, as happened at Walkeringham Breaches, and Moreton Ferry, near Gainsborough.

In the year 1760, the *Wilden and Kings Bromley* proposed Canal, and in 1765, a branch of the proposed *Ternbridge and Winsford* Canals, were intended to join this Navigation at Wilden Ferry, and having nearly similar objects in view, as the *Trent and Mersey* Canal has since accomplished.

Trent Upper Navigation, formerly: Acts, 10th and 11th William III.—Has, in consequence of a purchase made of the Earl of Uxbridge, its sole proprietor, by the *Trent and Mersey* Company (whose Canal runs by its side), been shut up and discontinued since the year 1805, as mentioned, Vol. I, p. 470. The general direction of this Navigation was about WSW, by a crooked course of about 20 miles, skirting part of the County of Leicester, thro' the County of Derby, and skirting a part of the County of Stafford, being in no part very considerably elevated: Its general objects were, the supply of Burton (which is the 207th Town, with 3,979 persons) with Coals, Deals, &c. and the export of Salt, Gypsum, Earthen-

---

Sea Sands at Pembrey in Caermarthenshire, as recorded in the " Communications to the Board of Agriculture."

wares, Ale, Cheese, and other Agricultural Products, &c.

This Navigation commenced in the Lower *Trent* Navigation, at the commencement of the late *Derwent* Navigation and the *Trent and Mersey* Canal, and terminated at Burton Bridge, at the NE end of the Town; where, by means of the Mill-dams, it nearly approached a cut from the *Trent and Mersey* Canal, at the SW end of the Town. At Swarkestone, it connected with the *Derby* Canal. At Winhill there was an Iron Forge, and at Kings Mills Flints for the Potteries were ground, by the side of this Navigation. From the *Trent* Navigation, to Weston Cliff is about 6 miles: thence to the *Derby* Canal 2½ miles: and thence to Burton 11¼ miles.

On this Navigation there were two Locks at different Mill Weirs, prior to 1765; at which period, heavy complaints were made, of natural impediments to the Navigation on this River from more than 20 Shoals (many of them having Fords thereon, see p. 282), that were impassable by Boats in dry seasons, without flushes of Water from the Mills, and from other impediments, created, as has been publicly asserted, by the Lessees of this concern, in wilfully occasioning and suffering, a sunken Barge loaded with Stones, to lay for 9 years in the Lock at Kings Mill Weir, to compel the unloading and employment of other Barges above this place, from those belonging to general Traders below it. The Tonnage allowed to be taken, was 3d. per Ton, on Goods carried on any part of this Navigation.

Over this part of the *Trent* River there are 5 Ferrys, as mentioned page 282; and over it there are

are three stone Bridges; that at Burton on Trent is 1545 feet long, with 34 arches, having often been called the longest in England. The Bridge at Swarkestone, passes, however, over more than twice this width of water in times of flood; one of which, accompanied by floating ice in 1805, having swept away the northern end of this Bridge, the same has been rebuilt, with five large Arches over the River: and Cavendish or Shardlow Bridge, built about the year 1760, consists of three large and two small Arches. The Strata through which the Vale of the Trent and the other Derbyshire Rivers are excavated, have been particularly mentioned in Vol. I. p. 469, &c.

In the year 1750, Dr. Thomas Congreve proposed a *Staffordshire and Worcestershire* Canal, intended to connect with this Navigation at Burton; and so was the *Swadlingcote and Newhall* proposed Rail-way, nearly: and an extension of the *Ashby-de-la-Zouch* Canal, proposed in 1796. In 1797 the *Commercial* Canal was proposed to cross and connect with this Navigation, below Burton; in 1793 the *Breedon* Railway was proposed to join this Navigation near Kings Newton; and opposite thereto, at Weston Cliff, a short detached cut belonging to the *Derby* Canal, was intended.

TRENT AND MERSEY *Canal* (sometimes called the Grand Trunk Canal): Acts, 6th, 10th, 15th, 16th, 23rd, 25th, two of 37th, 42nd, and 48th Geo. III.— The general direction of the main line of this Canal is nearly SE, with a bending course to the S, of 93 miles, in the Counties of Chester, Stafford, and Derby, entering

entering our square of Map (p. 193) at m. Its middle part is very considerably elevated, crossing the *Grand Ridge* by a long Tunnel at Harecastle, E of Talk, and the *North Dean and Weaver* Ridge by a shorter Tunnel at Preston Hill; and its Caldon branch crosses the *West Churnet and Dove* Ridge by a deep-cutting SW of Endon.

Its principal objects are, the export of Coals, Limestone, Freestone, Gypsum, Salt, Copper, Bar-Iron, Pottery-wares, and other manufactured Goods, Burton Ale, Cheese, Corn, and other Agricultural Products, &c.; and the import of Deals, Pig-Iron, Flints, Chert, Malt, &c. and forming part of the grand inland communication (the first effected), between the Ports of Liverpool, Hull, Bristol, and London.

Northwich is the 626th Town in the British population list, with 1,383 Inhabitants; Middlewich the 667th, with 1,232 persons; Sandbach the 419th, with 2,311 persons; Newcastle Underline the 118th, with 6,175 persons; Stone the 418th, with 2,314 persons; Rudgley the 437th, with 2,213 persons; Litchfield the 157th, with 5,022 persons; and Burton on Trent the 207th, with 3,979 persons, are situated on or near to the line of this Canal; and Leek the 223rd, with 3,703 persons, and Uttoxeter the 283rd, with 3,155 persons are on its Branches.

This Canal commences in *Bridgewater's* Canal at Preston Brook, and terminates in the lower *Trent* Navigation at Wilden Ferry: At Quintons Wood in Stoke, it is joined by the *Newcastle Underline* Canal; at Great Haywood, it is joined by the *Staffordshire and Worcestershire* Canal; at Fradley Heath Inn and Wharfs, by the detached part of the *Coventry* Canal;

and

and near Swarkestone, it connects with and is crossed by the *Derby* Canal. For 16 miles at its south-eastern end, between Wilden Ferry and Burton Bridge, this Canal ran parallel with and near to the Upper *Trent* Navigation, from about the year 1770 until 1805, when the interest in that concern was purchased of the Earl of Uxbridge by this Company, and the River Navigation in this part was discontinued, see Vol. I. p. 470. Near to its north-eastern end, from Dutton to Anderton, this Canal runs for 5 miles by the side of the *Weaver* Navigation.

From Etruria, a principal branch (called the *Caldon* branch) proceeds SE by a very bending course to the NE, of 28 miles, to Uttoxeter; from Stanley-Moss on this Caldon branch, a branch proceeds nearly NE to Leek, and from Froghall, a cut $\frac{1}{4}$ of a mile, and Railway branch thence E about $3\frac{1}{4}$ miles, with 4 Inclined-planes, to Caldon-Low Lime Quarries, Vol. I. p. 409. From the line at Shapnall, a branch proceeds SW into Burton Mill-dam (and therein approached near to the termination of the Upper *Trent* Navigation): from Winchnor-Bridges a cut proceeds SW to Winchnor Iron Forge, I. p. 404: from Longport there is a short cut eastward to Dale-hall, and a Rail-way branch thence to Burslem Potteries: from Shelton a short cut to Cobridge: into Lane delph Colliery there is a cut and small Tunnel, E, $\frac{1}{3}$ of a mile; and from the Harecastle Tunnel there are also Tunnels for Boats, into the adjacent Coal-works, I believe.

From Red-Bull Wharf there is a Rail-way branch eastward to Trubshaw Colliery; another near, to Briary-Field Colliery, I believe; and others to Kidcrew Collieries. From Etruria there is a Rail-way branch NE, to Handley-Green: from Stoke a Railway branch S $1\frac{1}{2}$ m. to Lane-End: from Armitage

Wharf

Wharf a Rail-way branch SW, about ¼ a mile, to Bruerton Colliery, I. 192: and from Froghall a Railway branch of about ¼ of a mile to Cupola Flint-mill.

Near to the line of this Canal, there are considerable Salt-pits and Works (see Cheshire Report, p. 19), on the NE of Northwich, NW of Middlewich, near Wheelock SW, and Roughwood SE of Sandbach, NW of Church-Lawton (Vol. I. p. 147), and at Shirley Wich near Gayton Weston. Gypsum or Alabaster from the Chellaston Pits, is put on board Boats on this Canal at Cuttle-bridge Wharf in Swarkestone, I. 149, and some from Fauld-Hill and Horninglow Marl-pit, is brought to Horninglow Wharf (I. 151). The excellent coarse Gritstone of the 1st Rock is dug at Cunsal Wood, I. 418, and finer grained, but less durable Freestone out of the Red Marl, at Weston Cliff, and at Tixall. Near Great Haywood, there are extensive Red *Marl* Pits for supplying the Farmers, where this valuable Manure is less accessible.

At the following Wharfs, &c. there are Lime-kilns, as mentioned in Vol. II. pp. 418 to 433, viz. Aston, Cuttle-bridge in Swarkestone, Caldon Low, Froghall, Horninglow, Shardlow, Twyford, Uttoxeter, Weston, and Willington, and there are others at Chedleton, Cunsal, Leek, &c. Near Cunsal and Froghall, there are Mills for grinding Flints and Chert (from Kent and Wales) for the Potteries : Iron Forges at Clay-Mills (I. 404) and Cunsal : Copper Smelting Works at Whiston (Vol. I. p. 353, but which do not use the Rail-way by the side of them) : Copper and Wire Works at Oakmoor and at Alveton : Cotton-mills at Alrewas ; large Malt-houses at Shardlow ; Brewhouses at Branston near Burton, &c. &c.

From *Bridgewater's* Canal to Middlewich, 18 miles,

miles, is level; thence to near Talk, 11 miles, has a rise of 326 feet by 35 Locks: thence along the summit pound of the line (said to be 420 feet above the *Thames* at Brentford?), thro' Harecastle Tunnel to the Caldon branch at Etruria, 6 miles, is level: thence to the *Newcastle Underline* Canal, ¼ of a mile, is a fall of   feet, by 6 Locks: thence to the *Staffordshire and Worcestershire* Canal, 16¼ miles, is a fall of   feet, by 13 Locks (the total fall in the two last distances being 150 feet); thence to the *Coventry* Canal, 13 miles, is about 32 feet fall, by 4 Locks: thence to Horninglow Wharf, 12 miles, is about 86 feet fall, by 11 Locks: thence to the *Derby* Canal, 10 miles, is about 8 feet fall, by 1 Lock (at Stenson); and thence to the lower *Trent* Navigation, 6 miles, is about 40 feet fall, by 5 Locks.

From the summit level of the line at Etruria, to near Bagnall on the Caldon branch, 5¼ miles, is a rise of 75 feet by 7 Locks: thence to the Leek branch and feeder, at Stanley Moss, along the highest summit pound, 1 mile, is level: thence to Froghall, 7¼ miles, is a fall of 61 feet by 9 Locks: and thence to Uttoxeter, 14 miles, is a fall of   feet by   Locks. The Leek branch is 4 miles, and level with the summit pound: the short cut to the Lime-kilns and Inclined-plane at Froghall, is level, and therefrom the rise is very great to the Caldon Low Quarries, as will be further mentioned herein. The Burton branch is about 1 *m.* long, and falls by   Locks from the line: the Wichnor Forge cut, ¼ of a mile, falls by a Lock: the other short cuts are level, and the Rail-way branches all rise from the line, I believe.

The width of this Canal, from Preston Brook at its NW end to Middlewich Wharf, and from Wilden Ferry

Ferry at its S.E end to Horninglow Wharf, is 31 feet at top, 18 feet at bottom, and 5½ feet deep; the Locks here being 14 feet wide, and adapted for River Barges of 40 Tons burthen: but the middle parts of this Canal, and its branches, are only 29 feet wide at top, 16 feet at bottom, and 4½ feet deep, and the Locks only 75 feet long and 7 feet wide, adapted for Boats 70 to 74 feet long, 6 feet wide, and carrying 20 to 25 Tons of lading: at Shardlow and at Willington, Boats are built.

The tail-gates of the Lock at Stenson are taller than usual, and they shut against a strong angular frame of Wood at top, similar to the Lock-sill below, to support and give strength to the Gates: there is a Bridge over the tail of the Lock, against which the top frame or sill abuts, and is sufficiently high for tall loaded Boats to pass under it, in going in and out of the Lock. The Locks on the Uttoxeter branch are very well constructed, having screws to draw the Lock-paddles, which stand a yard above the Gates: the Lock-weirs are further from the Head-gates, and longer than usual; and the paddle-holes vent into the Lock, above the level of the lower Canal, by which means leakages in the paddles may be constantly seen, and remedied. At Cunsal Flint Mill there is a Lock with a rise of 15 feet, above which, for 1¼ miles the Boats proceed along the course of the Churnet River, in a very beautiful wooded Valley.

The Earl of Shrewsbury is Proprietor of the famous Caldon Low Lime Quarries, and on the making of this Canal, granted a long lease of them to the late John Gilbert of Worsley, the late Sampson Whieldon of Caldon, the late George Smith of Eaves in Whiston, and the late Richard Hill of Farley, who were called the "Caldon Lime Company;" and who had likewise leases of an extensive track of Coals (yet the seams thin and indifferent, in the 1st and 2nd Coal Shales) near to Froghall, but which Coal-works they have discontinued, and when I saw the Caldon and Froghall Lime-works in 1808, the Coals used thereat, were brought 22 miles along the Canal, from Mr. Gilbert's Kidcrew Collieries.

The first Rail-way laid from Froghall to Caldon Low, in 1777, was composed of Cast-iron Bars, spiked down upon sleepers of Wood laid across the Rail-way, the total cost being about 20*s.* per yard run, it is said: but this line appears to have been set out, before the true principles of this branch of Engineery was well understood, and was very crooked, steep, and uneven in its degrees of declivity, in different parts. In a few years after, a new line was chosen, thro' great part of the distance, and the old one abandoned; but this second Rail-way was also very defective in the above particulars:—it has been stated, that on this Rail-way, during 9 months of the year 1794, one horse made in each week 3 journies on 4 of the days, and 2 journies on the other 2 days, drawing down each time 66 cwt. of Limestone: which afterwards was forwarded to the line of this Canal at Etruria, 14 miles distant; the Lime Company found Boats, and paid the Boatmen 9*d.* per Ton, for finding Horses and Boys, Towing-lines, &c.

The inconveniences of these steep and imperfect Rail-ways, occasioned the application in 1802 for the Act for a new Line of Rail-way, to be laid double, on stone blocks, with a moderate and proper slope, and with intervening *Inclined-planes;* and in the following year the same were carried into effect under Mr. Rennie, and are among the most complete works of this kind in Britain. On the Wharf at Froghall, a store

of

of Limestone is kept, ready for loading the Boats, which is shot thereon from 10 Tippling-machines; and the draw-holes of 4 large Lime-kilns are also near at hand, for loading Boats with burnt Lime. Short Rail-ways from the Tipples above-mentioned, lead to the bottom of the lower Froghall Plane, which is 65 yards long, with a perpendicular rise of   feet: the Trams loaded with Limestone, are let singly down one of the Rail-ways hereon, and the empty Trams, or loaded with Coals, are drawn up the other, by Chains that wind contrary ways round a large horizontal Drum, furnished with a regulating Brake, at the top of this Plane.

From the top of the lower Plane, there are branch Rail-ways laid to Tipples, for emptying Limestone and Coals at the tops of the Lime-kilns*, mentioned above, that are situated on each side of this Plane: and from these and the top of the lower Plane, a Railway of about 50 yards long, with a rise of 8 inches per Chain (or $\frac{4}{11}$ths of an inch in a yard), conducts to the bottom of the upper or great Froghall Plane, which is 303 yards long, with a rise of   feet.

Under a large open Shed at the top of this Plane, there are fixed two strong wooden pulley Wheels turning on vertical axes, and round these, and round other pulley Wheels at the bottom of the Plane, a strong endless Chain (weighing 72 cwt.) passes, and over cast-iron guide pulleys, so as to conduct the Chains over the two Rail-ways on the Plane, on the

---

* I was rather surprised to see the large pieces of Limestone, some a cubic foot, I think, that were laid on the tops of these kilns, when heaped up with Coals and Stone, in the manner that I have described, without approving it, in Vol. II. p. 420 and 427.

middle parts of which, it drags, on smooth Cast-iron saddle-shaped blocks, fixed between the Rails, for the purpose; to the descending side of the chain 5 loaded Trams of Limestone are hooked, by short lengths of chain, and to the ascending side as many Trams, in part loaded with Coals, or empty, are hooked; and their too rapid motions are prevented, by a brake, acting on one of the pulley Wheels.

At the top of this Plane, a person resides to take care of it, and has at his door a Weighing-machine, on which each Tram of Limestone is weighed, before it is attached to the great chain, and launched on to the Plane, by removing a stop that prevents its descending by accident: from the accounts that I saw here, it appeared, that the weight of Limestone in each Tram vary from 22 to 30 cwt. average 25 cwt. (of 120 lb.)

From the top of this Plane a Rail-way proceeds $\frac{1}{4}$ of a mile (passing Lees Colliery) with a rise of $\frac{4}{11}$ths inch per yard, to the foot of the Whiston Plane, which is just NE of the Copper Works: this Railway, at 60 or 70 yards from the last Plane, began to be cut some yards deep in the 1st Grit Rock, with a SE dip into the Whiston Coal-trough.

The Whiston Plane is   yards long, and rises   feet, having similar pulley Wheels, a Chain and brake, to those above described: from its top a Rail-way again proceeds about 1 mile, with a rise as before, to the bottom of the Upper Cotton Plane, which is 294 yards long, and rises   feet, and whose endless chain passes round a single large pulley Wheel, whose axis is so inclined, as to suit the inclination of the Chain, and has a brake, &c. as before. In order to obtain an uniform slope for this Plane (a thing no ways essential,

see

see page 404), a deep-cutting, 30 feet deep, has been made at its lower end, and its upper end is banked up, to the height of 15 feet.

From the top of this upper Plane, the Rail-way is again continued, with the same degree of rise as before, 1¼ mile, into Caldon Low Lime Quarries. Near the top of this Plane, the Rail-way again crosses the 1st Grit Rock with a WSW dip, and under the Turnpike Road it is deep cut in Limestone Shale: at Haughton Cross there is a Sale Coal-yard, with 4 turning-plates and Tipples, for shooting Coals into Carts, for the use of the Inhabitants and the private Lime-kilns of this elevated district.

The Trams used on this Rail-way and Plane, have stout lower side-pieces of wood, which project at each end, and are hooped with iron, which just meet together, and receive the shock when the Trams overtake each other, at the bottoms of the Planes, and on other occasions of their striking each other: one Horse draws 12 of these Trams, loaded, down these Railways, and as many empty ones up, but extra Horses are necessary, I believe, in bringing up Coals. The hours of working at these Planes daily, are from ½ past 5 o'clock in the morning to the same hour in the afternoon, in which time 18 dozen of Trams of Limestone are usually let down, perhaps containing 270 Tons of Limestone.

The number of Road and Foot Bridges over the Canal (exclusive of the Uttoxeter branch, I believe) is 258: near Red Bull Wharf, I saw a Bridge at the tail of a Lock, without a Towing-path, but having a separate small arch for the Towing-horses, under the landing-up of the Bridge.

The Harecastle Tunnel is 2,888 yards long, through

Coal-measures, at 70 yards beneath the Ridge: it is arched 12 feet high, and 9 feet wide: in its course it intersects several valuable seams of Coal, some of which are worked, by means of small branch Tunnels from this. The cost of driving this Tunnel was 70s. 8d. per yard run, in the year 1776, and being the first public Canal Tunnel that was constructed in Britain, it attracted for several years, a great deal more of attention. than it now deserves. At Preston Hill, near to the commencement of this Canal, is another Tunnel 1,241 yards long, 17¼ feet high, and 13¼ feet wide. At Barnton near Northwich, the Canal is tunnelled, of the dimensions above mentioned, thro' the point of a hill, 560 yards in length; and thro' another point of a hill at Salters Ford, 350 yards long, not far from the last. At Armitage there is a fifth Tunnel, on the line, 130 yards long. At Soils Bank on the Leek branch, there is a Tunnel, through the point of a gravelly hill, 200 yards long; and at Froghall Old Wharf, another 80 yards long, on the Caldon branch.

At Monks Bridge, between Derbyshire and Staffordshire, this Canal is carried across the flat Meadows of the Dove Valley, on an Embankment 13 feet high, for 1¼ miles, with Aqueduct Bridges containing 23 arches, from 15 to 12 feet span, 12 of which arches are over the main stream of the Dove: over the Dane River NW of Middlewich, is an Aqueduct Bridge of arches of 20 feet span each: others over the Trent ¼ m. N of Rudgley: over the Churnet 1 m. E of Alveton for the Uttoxeter branch, and over the Teen, 1¼ m. NW of Uttoxeter: over the Churnet at Well-Grange for the Leek branch: with numerous smaller Aqueducts and Culverts, under the line and its branches.

An Aqueduct of 6 arches was at first intended at
Alrewas,

Alrewas, but instead of which, the Canal enters the branch of the Trent River leading to Wichnor Forge, and near to Alrewas, connects with and crosses the Trent, having long Weirs on its lower side, defended by a row of strong piling, to prevent the Boats being forced over them in great floods, and the Towing-path, on the upper side of the Canal line, is carried over a long narrow wooden Bridge, supported on Piles in the River.

The waste Weirs on the Uttoxeter branch, appeared to me well constructed; where the Towing-path is on the lower side, large Culverts are laid under it, and the water rises up, to flow over a Weir, of proper height and length, placed some yards beyond the Towing-path, instead of being under it, as is more usual.

In the Rudyard Valley, 2m. NW of Leek, this Company has a *Reservoir* of 70 acres, and which might be extended to 160 acres if wanted, which is situated in a deep and remarkable inosculation of the Churnet and Dove Valleys on the *Grand Ridge;* where, although it might, from the Valleys in higher parts of the Ridge on each side, be supplied with almost any quantities of Water, the same might with ease be let out, to descend to either Sea, with only 13 feet cutting for a short distance at its northern end. A pretty considerable stream called Radbrook, which now enters the Reservoir from the NE, at its upper end, formerly ran to the Dane, but a trench or feeder from Ryecroft-gate, only 2 feet deep in any part, sufficed for turning this singular stream into the Reservoir, and thence into the Churnet River. In order to provide for the lockage down to Uttoxeter, and the increasing trade on the Line of this Canal, on each side of the Harecastle summit, the Act of 1808 provided for a new Feeder to this Reservoir, for diverting flood-waters of the Dane River, at Dane-bridge Paper-mill in Wincle Chapel, after its dam has been raised 6 inches; this Feeder is to descend the S bank of the Dane, to turn Wall-Hill near to Hug Bridge, and thence on the E side of Rushton Marsh to the Ryecroft-gate feeder, already mentioned.

This grand Reservoir, which is near 2 miles long, and near ¼ of a mile wide in the middle, was constructed under the direction of Mr. John Rennie, who began it in 1797, and it was completed in 1799 by Thomas Peak and John Mansfield, the Contractors, Mr. Potter being the resident Engineer, and Samuel Whiston, who now occupies Phillips Hay-house at its SW corner (to take care of it, and regulate the discharge of Water therefrom), an Overlooker throughout its construction. The embanked head, is at top 280 yards long, and at bottom 220 yards, and is 36 feet high: its width at bottom on the flat Meadow is 100 feet, and at top 60 feet; it is not straight across the Valley, but bowing upwards towards the Reservoir 40 or 50 in the middle. It was begun by sinking, a ditch for puddle, 5 yards wide and 12 feet deep, quite across the Valley, the first 6 feet of which sinking, was in silt and peat, then 2 or three inches of gravel, and then 5½ feet in blackish clay, becoming more regular as they descended, and which is the top of the Limestone Shale, I believe, for Cliffs of 1st Grit are seen on both the banks of this Reservoir, becoming exceedingly coarse and irregular, like Gravel Rock, near its SW corner.

A stupendous Weir of hewn masonry, 60 feet wide, has been constructed at the E end of the head, for discharging occasional large Floods, or smaller ones when

when the Reservoir may happen to be nearly full; such as had only happened once in the 10 years before I saw it, viz. on Shrove Tuesday 1807, when, and for 44 days afterwards, the Reservoir continued to discharge over this Weir, at one time 3 inches deep. In constructing this Weir, which is on a circular plan, bowing towards the Reservoir, an inverted Arch 60 feet wide, (between the wing-walls, which are parallel) and 70 or 80 feet long, has been made, of jointed masonry in large blocks (of 1st Grit, got 100 yards off NW) for the water to fall upon. The walls are plain, except the upper 16 or 18 feet, which are of rusticated masonry, the Weir being coped with stones, a yard wide and 14 inches thick, and paved for many feet back, with nearly similar blocks, into which four stout and long iron bars are sunk, and filled up with lead, quite across the Weir, for tyeing the whole together.

The head of the Reservoir is 4 feet higher than this Weir, defended by splaying Walls at each end of it: the inner side of the head is faced with rough stone-work, as low down as the water's surface usually varies: an inclined shuttle in the face of the head, moved by a wheel and pinion and rack, occasionally shuts the mouth of a 5-feet Culvert, at bottom of the Reservoir, that leads to two Iron Pipes, 12 inches diameter and 12 yards long, well secured in the puddle ditch of the head, which discharge the Water, for the use of the Mills on the Churnet, and the supply of the Canal and its branches. Through the whole height of the centre of the head, an elliptical well Stair-case of hewn masonry, 10 feet long and 7 wide, has been carried up, with a wall round it 10 feet high, over which a grating of iron bars is fixed, to prevent

persons

persons descending this stair-case, of 69 steps, except through a door in this wall, which is kept lockt. At the bottom of these stairs, two large Cocks, or rather conic plugs, for the ends of the Pipes above mentioned, are contained in square iron boxes; the plugs are fixed on the arms of vertical axes, that pass up thro' the lids of these boxes, and have cog-wheels on them, in which endless screws work, on whose spindles hand-wheels are fixed: the friction of the screws being sufficient, to retain the plugs in any exact place that they are set, either wholly, partially, or not at all impeding the efflux of the water from the Pipes, through 9 inch passages. Short 12-inch pipes discharge the water from the iron boxes into a sump, at bottom of the oval Well, from which a large Culvert thro' the outer part of the head, conveys it in a quiet and almost level stream, to a rectangular walled cistern or dam; at the SW corner of which cistern the Weir is placed, for supplying the Mills, by means of the former channel of the River; it is 60 inches long, of wrought iron, and painted, set in and covered, by large hewn stones, forming a horizontal slit 1 inch high, above which there is commonly about an inch pen of water; the remainder of the water flowing over the W side and S end of this Dam, which acts as a Weir, 60 yards long, and level with the top of the slit, for supplying the Feeder, which is 3 miles in length, down the W bank of the Churnet, to the Leek branch, near to the Well-Grange Aqueduct. The Cocks are so regulated by the Attendant, as always to keep the Miller's Weir fully supplied, and to discharge such a surplus over the feeder Weir, as is wanted from time to time in the Canal.

A spring of Water, which broke out under the great Reservoir

Reservoir Weir, seemed to threaten the bulging of its walls, until a sough was driven into the bank, on its E side, to collect these springs, in part leaking from the Reservoir, I suspect: and in order that the Water so collected, may not descend to the Mills, in addition to the guaged quantity that they are entitled to, as above mentioned, it is carried along a cut at foot of the Reservoir head, into the feeder; a guage-weir 40 inches wide, on this cut, usually in the winter season runs 1 inch deep, and is therefore then, near a 3rd of the stream to which the Mills are entitled. Some distance below the inverted arch or seat of the great Weir, a low flat grassy bank of earth, is made to act as a Weir, for diverting the spring and leakage water to the feeder, but over which, flood waters can freely range, in their way to the former channel of the River. In a branch of the valley above Norton, at Grena wood, is another Reservoir of about 20 acres, called Knipersley-pool, which is assisted by a feeder of some length from the SE, from Lionspaw Brook: and three other Reservoirs in this district, were provided for, in the Acts for this Canal.

Mr. James Brindley, Mr. Hugh Henshall, Mr. John Smeaton, Mr. John Rennie, Mr. Potter, &c. were the Engineers employed or consulted, on the works of this Canal or its branches; which works were begun in July 1766: in April 1773, the line eastward of the Harecastle Tunnel, was completed; and in May 1777, the whole line, and the branch to Caldon Low was completed and opened. The Leek and the Cobridge branches were undertaken since 1797: the Lane-End, Handley-Green and Burlsem Rail-way branches, were projected in 1802. In 1807 the Uttoxeter

eter branch was undertaken, and extended from Froghall to Oakmoor in August 1808, to Alveton in May 1809, and to Uttoxeter September 3d, 1811.

The Tonnage allowed to be taken, is $1\frac{1}{2}d.$ per Ton per mile, with reasonable Wharfage after 24 hours, on all kinds of Goods; but Paving-stones and Road-materials (Limestones excepted), and Marl and other Manures may pass toll-free on the pounds, and through the Locks, when water runs waste thereat. The usual price of Freight has been mentioned at $1d.$ per Ton per mile. Until about the year 1785, Men were employed in large gangs, to drag the Boats on this Canal, and on the *Trent* River near it, instead of Horses, which are now so universally used for towing.

The Act 33rd Geo. III. for the *Derby* Canal, granted some rates to this Company, on Goods crossing this Canal, or passing out of it into the *Trent*, by the detached parts of the *Derby* Canal, see Phillips's 4to. " History of Inland Navigation," Appendix, pp. 58 and 59.

This Company have been authorized by their different Acts, to raise $334,250l.$; the amount of their Shares was $200l.$ each, until 42nd Geo. III. which divided them into $100l.$ Shares.

The late Duke of Bridgewater availed himself of a Clause in the first of this Company's Acts, for powers to complete the junction of his Canal with the *Mersey* at Runcorn: and the 6th of their Acts, authorized their executing 11 miles of the northern end of the *Coventry* Canal line, and the transfers that they afterwards made of this, in equal moieties, to the *Coventry* and the *Birmingham and Fazeley* Companies, as already mentioned.

At Shardlow, Horninglow, Fradley Heath, Leek, &c. &c. there are large Warehouses for the accommodation of the Trade, and near 20 public Wharfs.

In the year 1760, the *Wilden and Kings Bromley*, and in 1765 the *Ternbridge and Winsford* Canals were proposed, to pass through parts of the track now occupied by this Canal: in 1792, and again in 1809, the *Swadlingcote and Newhall* Rail-way was proposed to connect with the Burton branch of this Canal, by means of the new Mill-dam. In 1793, the *Sandbach* Canal was proposed to join this, near that Town; and the *Breedon* Rail-way was intended to be connected herewith at Weston Cliff, by means of a short cut to the *Trent:* in 1794, the Trent Canal or side-cut to the lower *Trent* Navigation, was intended to join this at Wilden Ferry, and thus extend a Canal Navigation, down to Nottingham, instead of the present imperfect one by the River.

In 1796, the *Macclesfield* Canal was proposed to join the Caldon branch of this, at Endon, for effecting a junction with the *Peak Forest* Canal; and thereby opening a more direct route to Manchester; in p. 388, I have endeavoured to recommend the revival of this scheme, connected with a new junction to be made between the branch of this Canal at Uttoxeter, by Tutbury, to the line near Monks Bridge: and I would further remark here, connected with this wished-for junction, that it appears to me, a Rail-way branch might be made from the N end of the Alveton Aqueduct, on the Uttoxeter branch, first E, and then N, up the Valley, to reach the 4th Limestone Rock on the NE of Ramsor, without so many Inclined-planes, and such great expenses, as attend the Caldon-low Railway, and perhaps, without any Inclined-plane, by
tunnelling

tunnelling some distance at the head of the valley, to cross the Great Fault (Vol. I. p. 284), and reach the Limestone, to begin a Quarry, as was done at Crich, see page 338.

In 1796, the *Commercial* wide Canal for River Barges, was proposed to cross this Canal at Horninglow, and again at Bucknall and Burslem; and in 1797, an extension of the *Ashby-de-la-Zouch* Canal to join the Burton branch of this, was proposed; and connected with which last, it was proposed, to widen the middle parts of this Canal and its Bridges, Locks, and Tunnels, between Horninglow and Middlewich, so that the River Barges might pass through its whole length.

In 1812, a survey was made for a branch, from this Canal branch at Norton, to pass by a Tunnel under the *Grand Ridge* (through Coal-measures), into Dane Henshaw Vale (I believe); across the Dane, and thro' another Tunnel under the *North Dane* Ridge S of Macclesfield; passing E of that Town (and there approaching near to the proposed line of the *Macclesfield* Canal), and near to Prestbury, to join *Bridgewater's* Canal on Sale Moor, a distance of near 30 miles, nearly N, in the Counties of Stafford and Chester.

In 1814, this Company (as I have been informed), consented to make a Rail-way branch from their Canal branch near Leek (passing through soft reddish fine Gritstone of the Red Marl? near Bridge End), and up the valley thence to Bramcoate, where an Inclined-plane is intended, to ascend and cross Gold-sitch Moss, by the Roches Rocks of 1st Grit, and its Collieries, and those near Notbury, and ascend again, to cross the *Grand Ridge* on Goyte Moss (near its Collieries), and to descend thence to Goytes-clough Quarry,

Quarry, of Paving and Freestone (in the 2d Grit Rock, I. 425 and 429), which is leased by Thomas Pickford, Esq. to Mr. Richard Wilson, who, on condition of this Rail-way branch being made, by this Company, and they (in conjunction with other Canal Companies southward to London), agreeing to charge only 0½d. per Ton per Mile Tonnage on this Stone, in its way to the Metropolis, he is to guarantee an annual income in Tonnage, of 7 per cent. on the money they should expend, in making this Rail-way branch: and it seems to be part of the further scheme of Mr. W. to endeavour to connect the Rail-way branch at Goytes-Clough, with a Rail-way branch from the *Peak Forest* Canal, and with others from the *Cromford* and *Chesterfield* Canals, as mentioned pages 350 and 351. Mr. Joseph Cubley is the Agent for this *Trent and Mersey* Canal, at Swarkestone.

Wilden and King's Bromley proposed Canal.—In 1760, Mr. James Brindley and Mr. John Smeaton, surveyed the line for a proposed Canal, from the lower *Trent* Navigation at Wilden-Ferry (at the junction therewith, of the *Derwent* and the upper *Trent* Navigations), proceeding up the Vale of the Trent, passing King's Bromley (5 m. N of Litchfield), to Longridge near Burslem, situate near to the *Grand Ridge*, on its SE side. The direction of this line was about WNW, by a very bending course to the southward, of 55¼ miles, in the Counties of Derby and Stafford. This line passed near to Burton on Trent, the 576th Town, with 1,536 persons; Rudgley, the 437th, with 2,213 persons; and Stone, the 418th, with 2,314 persons: and from this line, branches were proposed S, 10¼ miles, passing Tamworth (the 304th Town, with

2,991 persons) to the Tame River, half a mile SE of Fazeley: another SW, three miles, to Litchfield (the 157th, with 5,022 persons); and another branch, W .3¼ miles, to Newcastle Underline, the 118th Town, with 6,175 persons.

From the lower *Trent* Navigation to King's Bromley, 25 miles, is a rise of 110 feet, for 19 proposed Locks; and thence to Longridge, 30¼ miles, a rise of 166¼ feet, for 28 Locks. The branch from the line to Fazeley, 10 miles, of level, and thence half a mile to the Tame, a rise of 17 feet, for three Locks; the proposed branch to the Mill-pool near Litchfield, 2½ miles, rises 18 feet, for three Locks; and thence half a mile, rises 30 feet, for five Locks: and the Newcastle branch, level with the line in that part. This Canal was intended to be 24 feet wide at top, and 2½ feet deep, with *Fords* instead of Bridges: the estimate was 100,200*l*.

Mr. Smeaton suggested the extending of this line over Harecastle-hill, by deep-cutting, with Reservoirs and steam-pumping Engines, for supplying the summit, in order to extend it to the *Weaver* Navigation.

The *Trent and Mersey* Canal, has since been effected, nearly along this proposed line; previous to which, the *Ternbridge and Winsford* Canal was proposed, through its south-eastern part: the *Newcastle Underline* Canal has supplied the place of the branch to that place; the detached part of the *Coventry* Canal, and northern end of the *Birmingham and Fazeley* Canal, have occupied the route of its Tame branch; and from the last line, Litchfield is now nearly approached, by a part of the *Wyrley and Essington* Canal, instead of the branch here proposed.

WINGERWORTH AND WOODTHORP *Rail-way.*—
A private one, constructed about the year 1788, by
Mr. Joseph Butler, for the use of his Iron Furnace in
Wingerworth, as has been mentioned, page 288: its
direction is nearly S, for about a mile, from the Bridge-
loft of the Furnace, to Woodthorp-end Ironstone
Pits (I. 218): the bars are four feet long, weighing
about 32 lb. and are spiked down to wooden bearers,
across the Rail-way, at 20 inches apart: this is said to
have been the earliest use of flanched Bars, except
under-ground in Mines, and although the ground was
very unfavourable, chiefly along an old bell-work, not
yet thoroughly settled, yet it answered perfectly, in
reducing the cost of Team-labour, as mentioned at
bottom of page 289.

In 1771, a part of the *Chesterfield and Swarkestone*
Canal, in 1802, a part of the *Ashover and Chesterfield* Canal or Rail-way, and in 1810, a part of the
western route of the *North-eastern* Canal, were severally proposed to pass, through very nearly the same
ground as this Rail-way occupies.

WOOD-EAVES *Canal*—Is a small private one, constructed about the year 1802, as a long Dam for Mr.
Cooper's new Cotton Mills, three-quarters of a mile E
of Fenny-Bentley. Its direction is nearly NE for 1¼ m.
On the NW side of the Brook, the small Boats hereon
have principally been used for bringing Limestone
to the Mills, from the curious rib of contorted black
shale Limestone, which crosses this Dam, about a mile
from the Works, as mentioned, Vol. I. p. 231.

WYRLEY AND ESSINGTON *Canal:* Acts 32d and
34th Geo. III. The general direction of this Canal, is

nearly

nearly SW by a very crooked course of 23 miles, in the County of Stafford, quitting our square of Map (p. 193) at *W*. It is considerably elevated, crossing the *North Tame* Ridge twice, by deep-cuttings, S, and again SW of Litchfield (and its Lord's-hay and Wyrley-bank branches also cross this Ridge), crossing the *South Bourne* Ridge, by a deep-cutting NE of Cat's-hall, and terminating very near to the *Grand Ridge**, near the E end of Wolverhampton Town, where it is said to be elevated 566 feet above the Thames at Brentford.

Its objects are, the export of Coals, Limestone, Iron, &c.; the import of Deals, Corn, Malt, &c.; and forming part of the shortest communication yet opened†, between the *Trent* and the *Severn* navigable Rivers. Wolverhampton is the 39th British Town, with 14,836 inhabitants; Litchfield is the 157th, with 5,022 inhabitants; and Walsall, the 136th Town, with 5,541 persons, near to this Canal. It commences in the detached part of the *Coventry* Canal (very near to the commencement of the *Birmingham and Fazeley* Canal), at Whittington Brook, and terminates in the old *Birmingham* Canal, near the E end of the Town of Wolverhampton.

From near Cat's-hall, a principal branch proceeds nearly S, 5¼ miles to Hay-head Lime-Quarries, SSW of

---

* The south-western part of this Canal and its branches, and the north-western part of the old *Birmingham* Canal, and its branches, form together, a very curious and extensive system of level Canal, much elevated, and skirting the several hills near their tops; while at the same time, the Walsall main branch of the latter Canal, and its collateral cuts, (all on one lower level), follow the valleys between these hills, in several instances; and near Bromwich, and near Tipton-green, there are cuts, with Locks, for connecting these upper and lower levels of the old *Birmingham* Canal.

† The *Worcester and Birmingham* Canal was since opened, Nov. 1814.

Aldridge;

Aldridge; from the NW side of Pelsall, a branch proceeds nearly NW, about 2¼ miles to Lord's-hay Colliery, E of Great Wyrley; from the W of Bloxwich, another principal branch proceeds nearly N, 3¼ miles, to Wyrley-bank Colliery: and from Bloxwich-lane, there is a cut southward, ¼ a mile, to near the north end of Walsall Town; where it approaches near to the termination of the Walsall branch of the Old *Birmingham* Canal; and so does a Rail-way branch, from the line at Bloxwich-lane Wharf, ¼ m. SE, to Mr. James Adam's Butts Lime-Quarries, and Kilns, near to the NE end of Walsall: from near Bloxwich-lane, there is likewise a Rail-way branch, (made in 1806), SW ¼ a mile, to Birch-hill new Iron Furnace, Colliery and Fire-Brick Kilns: and from Brown-hill, another Rail-way branch ¼ m. NW, to its Colliery.

From the branches above described, others proceed, viz. at Linley*, a short cut nearly S, and Rail-way therefrom by 3 branches into the Limestone Mines, (deep beneath the surface), in the Rock which dips northward and underlies the Pelsall, &c. Coal-Field;

---

* At the Wharf, and large Pye-Kilns here, Lime was delivered in 1809, into Carts at 10s. 6d., and into Boats at 10s., and Limestone at 3s. 9d. per Ton, (20 × 120lb.); the labour of getting and bringing the Stone to the Wharf, being contracted for at 2s. 4d., and of burning the Lime, (including loading), at 5s. 3½d. per Ton, in addition thereto. At Daw-end Wharf, the expenses on the Stone were 2s. 7½d. and of burning 5s. 3d.; and at Moss-close Works, in Rushall, near to Walsall, Stone 2s. 8d., and burning 5s. 7d. per Ton.

The Pye-Kilns used here, (see Vol. II. p. 440, and 435), are some of them so large, as to hold 450 Tons of Stone, and make about 300 Tons of Lime at once, with about 150 Tons of Coals: the only opening is at the end, where the Kiln is lighted, and if the wind sits that way, in four or five days, they begin to draw, while the middle and further parts are burning, and which sometimes they continue to do for four or five weeks.

to Upper-Park Lime-Quarries there is a short cut from Daw-end Wharf; there is a short Rail-way branch, nearly E, to the tops of the Limestone-Mine Shafts:—on the Wharf here, are Tippling machines, for loading Boats with Limestone; the Trams are suspended by Chains and Balance-weights, and tilted sideways, by means of a winch-handle and roll; and in 1808, a Rail-way branch was begun, by the late Mr. Wilkinson, S ¼ a mile, to his new Hay-Head Lime-Quarries. Also, from the Wyrley-bank branch, a cut proceeds W, to Essington New Colliery. Other branches may be made to any other Collieries, Mines, &c. within 5 miles of the line or branches mentioned in the Act, provided they waste none of the Company's water. Radley and Winterly Limestone Quarries and Works, in Rushall, are situate by the Hay-head branch; and Goscot, Birch-hill, and some other Collieries, are near to the Lime. At Gallows Wharf, SSE, and at another Wharf S of Litchfield, on the Colshill and Sutton Roads, there are large Warehouses, &c., and Pye Lime-Kilns, for the Rushall Stone. From the *Coventry* Canal to Cannock-Heath 7¼ miles, is a rise of 264 feet, by 30 Locks: thence through a deep-cutting to the Hay-head branch ½ a mile, is level: thence to the Lord's-hay branch 2¼ miles, is level; thence to the Walsall branch 4 miles, is level; thence to the Wyrley-bank branch 2¼ miles, is level; and thence to the Old *Birmingham* Canal, nearly 6 miles, is also level.

The Hay-head, Lord's-hay, and Walsall branches, and the Linley, and the Upper-Park cuts, are all level with the summit pound; the Wyrley branch, in the first mile rises 36 feet, by 6 Locks, and thence 2¼ miles, is level; the Essington branch rises from the last pound

24 feet, by 4 Locks. The Butts Rail-way branch falls, and then rises again; the Birch-hill R. w. branch falls from the Canal; the Daw-end branch rises; and the New Hay-head Rail-way branch, rises from the Canal.

This Canal is 28 feet wide at top, 16 feet at bottom, and 4¼ feet deep. There are aqueduct arches over rivulets, ¼ m. NE of Pelsall, and ¼ m. W of Aldridge.

A Reservoir on Cannock-Heath, with the water lifted from the several Mines near this Canal, serve to supply it with water: but in order to guard against its drawing any water from the Old *Birmingham* Canal, (which is wholly supplied by pumping), a Stop-Lock is placed near to Wolverhampton, for keeping this Canal 6 inches higher than that, and a man stationed there to attend it: this Company being also required, to vent all their surplus-water into that Canal. The Water-pipes for the supply of Litchfield, were particularly guarded in the Act, during the cutting of this Canal across them.

Mr. William Pitt was the Engineer to this Canal, and the works were in a few years completed. The rates of Tonnage, will be found in Mr. John Cary's " Inland Navigation," pp. 47 and 48, with a Map of this and the other adjacent Canals. Less than 20 tons in a Boat are not to pass the Locks, without paying for that lading, except empty Boats on their return. The Company were authorised to raise 160,000*l.*, the first 35,000*l.*, of this in 125*l.* shares: on the extension of this Canal in 1794, the Company were required to purchase the shares of certain discontented Proprietors; the new shares are 100*l.* each.

When I was examining this District, in the spring of 1809, the Walsall branch had been dammed off from the line, and been dry some years; the water therefrom, having broke down into Birch-hill Colliery Works, under it. The Hay-head Old Lime-Works had been some time disused, and owing to Mr. Wilkinson's death, a short time before, the New Hay-head Rail-way remained unfinished, and no use was making of the Canal to the southward of Daw-end Wharf.

In order to dredge up pieces of Limestone, and Coals, dropped into the Wharf-Basins, Locks and Bridges, or Mud settled therein, I saw a large Plate-iron Shovel, having a long crooked handle, like that of a Breast-Plough, to guide it by, used at Daw-end Wharf: it had a bail handle across it, and a chain from a swivel therein was used, to drag it, and haul it up, by men in a Boat, moored for the purpose. In 1792, two Rail-way branches were proposed from the line of this Canal, to Ashmore-Park Collieries in Wednesfield; and another was proposed to Stow-heath.

Having now gone separately thro' all the important Establishments for *Improved Communication*, (as proposed in page 206), and with considerable pains arranged the many important, useful, and curious particulars, that I have been able to observe or collect, respecting each Concern, which falls wholly or in part within the space of our Map, in any part of its course; I beg leave again to call the attention of my Readers, to my design, mentioned in page 284, of publishing at some future time, (by the assistance of my Sons), a far more complete and useful work on *Canals and Rail-ways*, than has yet appeared, with an introductory Treatise, on all the branches of *Engineering* connected therewith, proper Maps, Tables, &c.; and to request, that they will communicate to me, such authentic facts and documents, and use their influence with the

Engineers

Engineers and Agents of the several Canals, &c., to furnish such as may seem wanting, or as may appear to be incorrectly stated in the foregoing accounts, and in the accounts of the other British Establishments of the same kind, which are described in Vol. VI. Part I. of Dr. Rees' "New Cyclopædia."

SECT. IV.—FAIRS.

The *Fairs* of this County, in alphabetical order, have been stated to be held as follows, viz.

Alfreton. July 31, and Nov. 22 (for Horses and horned Cattle).

Ashburne. The first Tuesday in Jan. and Feb. 13 (Hor. and Cat.); April 3, May 21, and July 5 (ditto, and Wool); August 16 (Hor. and Cat.); Oct. 20 (ditto); St. Andrew's Eve, Nov. 29, or Saturday before (ditto, the Horses assembling 3 or 4 days previously).

Ashover. April 25, and October 15 (Cat. and Sheep).

Bakewell. Easter Monday, Whit. Mon. and Aug. 26 (Hor. and Cat.); Monday after Oct. 10, and Monday after Nov. 22 (ditto).

Belper. May 12, and Oct. 31 (Cat. and Sheep).

Bolsover. Easter Monday.

Chapel-en-le-Frith. Thurs. before Feb. 13, Mar. 24 and 29, Thurs. before Easter, April 30, Holy Thurs., and 3 weeks after (Cat.); July 7 (Wool); Thurs. before Aug. 24 (Sheep and Cheese); Thurs. after Sept. 29, Thurs. before Nov. 11 (Cat.)

Chesterfield. St. Paul, Jan. 25, or Satur. before (Cat.); Feb.

Feb. 28 or Sat. before, first Sat. in April, May 4, July 5 (Hor. Cat.); Sep. 25 (Cheese, Onions, &c.); last Sat. in Nov. (Cat. Sheep, &c.).

Crich. Old Lady-day, and Old Michaelmas-day.

Cubley. Nov. 30 (fat Hogs).

Darley-flash. May 13, and Oct. 27 (Cat. and Sheep).

Derby. Jan. 25, Mar. 21 and 22 (Cheese); Friday in Easter Week (Cat.); Frid. after May-day, Friday in Whitsun Week, and St. James, July 25 (Cat.); Sept. 27, 28, and 29 (Cheese); Friday before Old Michaelmas (Cat.); St. Luke, Oct. 18 (Cheese, see p. 63).

Dronfield. April 25 (Cat. and Cheese); Aug. 11.

Duffield. March 1 (Cat.).

Higham. First Wednesday in the Year.

Hope. May 12, and Sept. 29 (Cat.).

Matlock. Feb. 25, May 9, July 16, and Oct. 24 (Cat. and Sheep).

Newhaven. Sept. 11 and Oct. 30 (Horses, Cat. and Sheep, and the most celebrated holiday or *Gig* Fair of the County).

Plesley. May 6, and Oct. 29 (Hor. Cat. and Sheep).

Ripley. Wednesday in Easter Week, and Oct. 23 (Hor. and Cat.).

Sawley. Nov. 12, or Saturday before (Foals).

Tideswell. May 3 (Cat.); Second Wednesday in Sept. and Oct. 29 (Cat. and Sheep).

Winster. Easter Monday.

Wirksworth. Shrove Tuesday, May 12, Sept. 8, Oct. 4 and 5 (Cat.).

SECT. V.—MARKETS.

THERE are in Derbyshire eight ancient *Market Towns*, which are still well attended, and three other Markets of modern establishment, which seem increasing; while six others of its ancient Markets, have either greatly declined, or have been wholly discontinued, viz.

Alfreton, on Monday (originally was on Friday).
Ashburne, Saturday.
Bakewell, Friday.
Belper, Saturday (new).
Bolsover, (on Friday discontinued).
Buxton, (new), Preface to Vol. II. p. 20.
Chapel-en-le-Frith. Thursday.
Chesterfield, Saturday.
Crich, (quite declined, but attempted to be restored in 1810).
Cromford, Monday, for Corn, Saturday (new).
Derby, (All Saints), Wednesday, Friday (chief), and Saturday.
Dronfield, Thursday (much declined).
Higham, Friday, discontinued).
Matlock, (, discontinued).
Tideswell, Wednesday.
Winster, Saturday (very much declined).
Wirksworth, Tuesday.

In conversing on the subject of *Markets for Corn*, with the late Earl of Chesterfield, he mentioned it as his opinion, that it would tend to the benefit of the Country, if all such were held on *Saturday* only. About Barton on Trent, he said it was not the practice
to

to openly shew Samples, but those who had corn to sell, stood in the market, nibbling corns of oats, wheat, &c. between their teeth, as the signal to their customers!

Connected with this subject, it may be proper to notice what was mentioned to me, as the usual practice of the County, as to *Auction Sales*, viz. That the Seller being known to employ Puffers, or Sweetners, rendered the sale void; altho' he may appoint a person, either to bid *once*, or instead thereof, to deposit a folded paper on the Table, mentioning the lowest sum the Lot should go at; but in order to claim exemption from the Auction-duty, it is necessary publickly to read this paper before the Company, as to any Lot which is put up and bid for. Mr. *Boot* and Mr. *Shaw* were the *Auctioneers* whose Sales I most commonly saw advertised, while in the County.

The mode of *Sale by Ticket*, has already been described in Vol. II. p. 229, and it seems only necessary here further to add, that in case of the Vender's Ticket stating the highest sum, of any that are put into the Glass, and the Sale being void in consequence, it is considered, that he whose Ticket comes nearest to it in amount, is entitled to the preference, in treating for a purchase *by private contract*, if he declares himself so disposed.

SECT. VI.—WEIGHTS AND MEASURES.

ALTHO' this County seems less perplexed and disgraced, by the use of incongruously diversified, and uncertain Weights and Measures, of Commodities and Labour, than some others of the Northern and Western Counties,

Counties, yet very considerable reforms therein are wanting, and will not be neglected, I hope. The increasing use of Weighing-engines has been mentioned, p. 341 of Vol. I., p. 64 of Vol. II. and p. 245 herein, and their far more extended use cannot be too much recommended.

1. *Land.*—In the very numerous Maps or Plans of Derbyshire Estates, which I have had the opportunity of inspecting, as mentioned, Vol. II. p. 3, I do not remember the use, in any instance, of any *customary Acres**, &c.; but the statute *Acre, Rood,* and *Perch,* or Pole (of 4810, 1210 and 30¼ square yards, respectively), were exclusively used therein: and it seems only in a few instances, of paying for labour, that different meanings are given to the terms Acre and Rood, in stating quantities, as I will mention presently.

The statute *Acre of Land,* is in so much more *general and permanent use,* as applied to the same identical thing, than any others of the numerous Measures used in this Country; and is, by means of the Maps or Plans of Estates (which are now so common every where), so much better defined and preserved from time to time, than the measurements by any other denomination, or of any other thing, through the whole range of our common Measures and Commodities; and besides which, its Gunterian *decimal* measures of length, the *Chain* (of 66 feet), and its 100 *Links* (which Edmund Gunter contrived in 1624), the

---

* In cases where agricultural details may continue to be given, in these antiquated and barbarous denominations, the Tables for converting various of such measures into *statute* ones, which Mr. Layton Cooke calculated and published in the Farmer's Journal Newspaper of the 14th June, 1913, p. 299, will be found very useful.

Furlong

Furlong (of 10 Chains), and *Staff* (of 10 Links); as also, in its denominations of superfices, the square *Chain*, the *Acre* (of 10 Chains), *Acreme* (of 10 Acres*), and the *Hide* (of 100 Acres) of Land, are most of them so well known and established, that I have for more than 20 years past been anxious, to recommend *such a plan of generally reforming our National Measures, as shall leave to us all these denominations of Measures, in use;* instead of *changing every one of those in use*, as has been proposed by a majority of the Writers on this subject, who have proposed their *unit* of length to be, a second's Pendulum, a degree of Latitude, or some other *natural standard*, as they were called; or, instead of *changing all but one of our measures in use*, as all the Writers have in fact proposed, who adopted *any one of our present Measures (except the Foot)*, and proposed at the same time *a decimal scale* therein, or divisions, and multiples *by tens* only (which last *is an indispensible condition of any scheme of reform of Measures and Weights);* because of the curious fact, that amongst the many scores of relations, between the different general and local denominations of measures of length, superfice, or capacity (lineal or running, plane or square, or solid or cubic), which are more or less generally used in this Kingdom, as I have collected and calculated them, the following appear to be, *the only decimal relations*, except the *Gunterian measures* above mentioned, viz.

100 square *Feet* = 1 *Square*, of the Slater, Pavier, &c.
10 Wine *Gallons* = 1 *Anker* of Brandy, &c.
10 *Sacks* = 1 *Load* of Corn, in some places.
10 *Quarters* = 1 *Last* or *Wey*, ditto.

---

* See Vale's Agriculture, folio, London, 1675.

The advantages are many and considerable, which follow (as now in France) from having all *the Measures of capacity*, raised *decimally* from the *lineal unit*, as well as those of superfices; but notwithstanding the considerable use which some Artificers, Tradesmen, and Calculators make, of the superficial and the cubic Inch, Foot, and Yard, I have not, I believe, met with any instance, of *a Vessel used for measuring liquids*, or any loose granulous articles, being *adjusted to any decimal number of Inches, Feet, or Yards!*, altho' this would not be more difficult, or so much so, to execute, as to adjust these measuring Vessels to true Quarts, Gallons, Bushels, &c. containing, often, a fractional number of the cubic Inches, by which alone most of them can be defined, or their standards renewed.

I have therefore been, for abandoning all our existing denominations of measuring vessels, instead of retaining *some one of them*, as most of the proposers of a decimal scale for such measures, have wished to do: and my proposal with regard to Measures, is as follows, viz.:

Measures of Length, in Links.

$\frac{1}{1000}$, $\frac{1}{100}$, $\frac{1}{10}$ (Ring), 1 (*Link*), 10 (Staff), 100 (Chain), 1000 (Furlong), 10,000, 100,000, &c.

Measures of Superfice, in Links.

$\frac{1}{1000}$, $\frac{1}{100}$ (□ Ring), $\frac{1}{10}$, 1 (□ Link), 10, 100 (□ Staff), 1000, 10,000 (□ Chain), 100,000, 1,000,000 (Acre), &c.

Measures of Capacity, in Links.

$\frac{1}{1000}$ (cu. Ring), $\frac{1}{100}$, $\frac{1}{10}$, 1 (cu. Link), 10, 100, 1000 (cu. Staff), 10,000, 100,000, 1,000,000 (cu. Chain), &c.

With

With respect to *Weights*, after having collected all their denominations which I could meet with, or read of, as in use in the Kingdom, and calculated their relations, it still more curiously appears, that the following are the only *decimal* relations among them all, viz.

100 *Pounds* avoirdupois = 1 *Kintal*, Quintal, or Centner.

10 *Pounds* ditto = 1 *Ration*, of Oats.

The *Pound avoirdupois*, here mentioned (which, to distinguish it from the Troy and the Trone pounds, I would call *Libra*, *Lib.* or *lb.*), appears incomparably in more frequent and general use, than any other of our very numerous denominations of Weights. In the Custom-house, an immense number of articles of import are charged to the duties by this Pound, and more than 70 of them, I believe, by 100 such Pounds, or the *Kintal* above mentioned. My proposal, therefore, has been, to adopt the Pound avoirdupois* (*Libra*), as the

---

* On this part of my scheme, I can refer for support, to the proposal of the Committee of the House of Commons in 1814; altho' their scheme falls short of *a decimal division* and multiplication of this unit; which last point, and another (to which they have been inattentive) *the giving of an entire new Name to every new quantity* or denomination of Weight or Measure to be introduced, are both of them *of such essential consequence, towards the perfection or the adoption of any reformation*, that I most sincerely hope the good sense and intelligence of the Legislature, will cause them *to reject every proposition for materially disturbing our present Weights and Measures*, barbarous as their arrangements are, which does not proceed on these bases, viz. *a decimal division throughout, and new names for all new things*; and which does not at the same time, adopt *some one Measure* (whether lineal, plane, or solid), and *some one Weight*, which, respectively, *are most generally known and in permanent use* in the Kingdom, as *the two Units, from which all other denominations are to be raised*, respectively.

To the nation at large, it is almost wholly unimportant, and concerns only

the unit of a completely *decimal scale of Weights*, as follows, viz.:

Weights, in Pounds Avoirdupois.

$\frac{1}{10,000}$, $\frac{1}{1000}$, $\frac{1}{100}$, $\frac{1}{10}$, 1 *(Libra)*, 10 (Ration), 100 (Kintal), 1000, 10,000, 100,000, &c.

Lastly, as to *Money*, we fortunately now use in Accounts, only the four denominations, *Pounds*, *Shillings*, Pence, and Farthings. Amongst these, *the Pound Sterling* is so much more generally and importantly used than either of the other three denominations, that no hesitation can take place, in proposing this as *the unit of Money*, or circulating medium of value; and fortunately, *Two Shillings* is the exact $\frac{1}{10}$th of this unit, as follows, viz.

$\frac{1}{1000}$, $\frac{1}{100}$, $\frac{1}{10}$, (Two-shilling), 1 *(Pound* Sterling), 10, 100, 1000, 10,000, 100,000 (Plum), 1,000,000 (Million Sterling), &c.

One-hundredth of a Pound ($\frac{1}{10}$th of Two Shillings, or

---

only, in a slight degree, the few Artists, Men of Science, and Calculators, who would be called in, to prepare the new Standard Measures and Weights, of all the different kinds and denominations, and to prepare Tables to facilitate their use, and mark their exact relations to old denominations, whether any integer or *simple proportion* exists or not, between the two Units of Measure and Weight, above-mentioned; whether, for instance, *any even number of pounds of distilled Water* at a certain heat, fill exactly *any one of the solid measures* of a certain metal or substance, and proportions in its dimensions?, that may be proposed to be introduced. To the many, it would, on a reformation taking place, be thereafter of infinitely greater consequence, that *every Measure* of which they had the measuring Rules or the Vessels furnished, were clearly and perfectly related, *according to the decimal notation and mode of calculating*, common through civilized nations, to *a previously well known Measure*, and of which the standard would still exist; and that in like manner, *every new* (weighing) *Weight* they were so furnished with by authority, was but a decimation of *a previously well known Weight*, having its standard preserved.

of 48 half-pence), is so near to 5 half-pence, and $\frac{1}{5}$th of these last being a Farthing, no serious inconvenience or injustice could follow, on the enacting of the above decimal divisions of the Pound, by appropriate names, and issuing coin corresponding thereto, to permit, for one or two years, a Farthing to pass *(in change only)* for $\frac{1}{1000}$th of a Pound, and 5 half-pence (or $2\frac{1}{2}d$. of the present coin), instead of the $\frac{1}{100}$th of a Pound, before the present Copper Money need wholly be called in; and so, without any harm, Sixpences, Shillings, Half-crowns, and Three Shilling-pieces, might, for a time at least, continue to circulate with the new Decimal Coin, as the $2\frac{1}{2}$-100ths, 5-100ths, $7\frac{1}{2}$-100ths, $1\frac{1}{2}$-10ths, and $1\frac{1}{2}$-10ths, respectively, of the Pound Sterling.

This reformation of our Money, and enabling all Accounts to be kept *in one column,* instead of three, just as the Pounds are at present, and rendering unnecessary " Reduction of Money," now so formidable a rule to youths, in our Elementary Books of Arithmetic, and so troublesome in business, is indeed so easy to be accomplished, that it were extremely desirable Government would take it up, separately from, and previously to reforming the Measures and Weights, on similar principles.

It would be improper to go much into details in this place; I would however beg further to suggest, that Standards of each denomination of Measure and Weight, being carefully settled, as above-mentioned, and deposited in sufficient numbers in different public buildings, to prevent their ever being all lost. A very considerable number of Rules, Rods, measuring Vessels, and metal Weights should be prepared (partly at the public expense), and issued, as much as possible by, and thro' the present makers and venders of Measures and

Weights,

Weights, in order not to oppose their interests to the circulation of them.

And, that with these Measures and Weights, correctly printed Tables should be sold, at very easy prices, not so much for reducing old Measures or Weights, into new ones, as *for reducing Prices in the old denominations, and money, into equivalent Prices in the new denominations, and money*: these Tables to be so contrived, on separate pages, that sets of them might be done up together in a Book, for each different Trade.

The Draper, for instance, or other Tradesman, who sold hitherto only *by the yard run*, or in length, would have a Table, beginning with the lowest price *per yard* of any of his articles, and proceeding by pence, or farthings, to the highest price per yard, of any other of his articles for sale, and against each would be set down in the Table, the equivalent price of the new decimal Measure, less than the yard, and the new Measure next greater than a yard, each calculated to *the nearest* $\frac{1}{1000}$th part of a Pound, which is so much more exact than Farthings, as 1000 is larger than 960; and against that new Measure of the two, whose new price came nearest to the exact equivalent with the old price, some mark should be put, in order to enable bargains, measurements, and calculations to be made, with the least possible deviation as to resulting value, from the present ones. Other Trades, who now charge by *the foot run*, and by *the yard super*, for different kinds of their work or commodities, might have price Tables of these two kinds, and so on, to the greatest number any one Trade or person might require, without burthening himself with Tables he never would want.

By these arrangements, it would be easy for every Tradesman, having by him the new *Measures*, and the old also (as for a time they should be required to do), to satisfy any doubting or objecting customer, that he made no unfair advantages by the use of the new measures and money, by measuring and calculating the amount of the goods in question; first by the old and usual measures and prices, and then by the new measures and prices (as per Table), and such would in every instance, where no mistake was committed, *come to the same, or an equivalent amount*, in old and new *money;* of which monies also, short Tables should be circulated, in very considerable numbers.

Having thus proposed a sufficient alteration of our national Measures, Weights, and Money, to as many *decimal* scales, one constant *multiplier* or *divisor* for each of which scales, would suffice, for reducing our Measures, &c. to those of France, China, or any other nation having decimal measures, &c. or *vice versa*, almost as readily as if the two nations used the same identical measures; here I stop, and beg to deprecate in the strongest terms, any attempts to go further, and imitate those in France, who, in the wildest spirit of thoughtless innovation, attempted to alter the divisions of *time*, and of the *circle*, where no inconvenience previously experienced (as in the case of our multifarious and discordant measures, &c.), called for reformation, and where no earthly advantage followed the innovations, which were wantonly made, as long as the phrenzy lasted, on the venerable institutions of months, weeks, hours, minutes, and seconds, and on those of signs, degrees, minutes, and seconds, of motion in a circle; which already had their artificial lines, the

the sines, tangents, &c. *decimally* expressed, and beyond which nothing could be desired, on the score of utility.

From this digression, I will proceed to mention, as nearly in the order prescribed in the printed " Plan," as may be, such *local* Measures and Weights of different articles, as I met with in, or near the County, viz. in

1. *Building.* Slaters sometimes charge by the *Bay*, of 500 square feet; which is five *Squares* of some other slaters, thatchers, paviors, &c. In Glossop, and some other places, slating is measured by the *Rood* of 44 square yards (See Vol. I. p. 430).

Road-making. At Markeaton, &c. the *Load* of sifted Gravel is 20 heapt bushels, see p. 351. About Bakewell, the *Ton* of small broken stones is 20 stricken bushels, p. 261. In Stratford, Middlesex, the *Ton* of sifted Gravel (or half *Load*) is 23 cubic feet, see p. 253, note.

Fencing. In Ashover and other places, fence walling is measured by the *Acre* of 28 yards run, or the *Rood* of seven yards in length, see Vol. II. p. 85. In Foremarke, &c. the *Acre* of 32 yards run, and *Rood* of eight yards run, is used for hedging, ditching, &c. See Vol. II. p. 88.

Draining. Making of covered drains, is not uncommonly contracted, or paid for, by the *Rood* of seven yards run. See Vol. II. p. 385.

Trenching. About Matlock, digging of land is usually paid for by the *Rood* of seven square yards. See Vol. II. p. 347.

2. *Corn.* At the time I was surveying the County (between

tween September 1807, and December 1809) the Farmers of the southern part of the County were shamefully perplexed and inconvenienced, by the diversities of *Bushel* or strike measures, then most commonly in use at the surrounding Market Towns: according to the account given me by the late Earl of Chesterfield and others, these were as follows, beginning with the least, viz. the *Leicester* bushel was considered to be 32 quarts*, as by law all should be; that of *Derby*, 34 quarts (II. 119): in the estimations of Crops, &c. at Foremarke, it was stated at 35 quarts; at *Ashby-de-la-Zouch* Market, 35¼ quarts; at *Burton* on Trent, 36 quarts (II. 121, 128; III. 187); at *Uttoxeter*, 37 quarts; and at *Stafford*, 38 quarts.

I have been happy, however, since to learn, from the "Farmers' Journal," of the 18th of November, 1810, and 9th of November, 1811, that Meetings had been held at Derby and Chesterfield, for putting a stop to these shameful breaches of a positive and useful Law, and (10th Feb. 1812) that these laudable exertions had been crowned by success, in Derbyshire. A *Quarter* is eight Bushels: sometimes I have heard three bushels of wheat spoken of as a *Load*. The selling of Corn *by weight*, rather than measure alone, has been very properly recommended, in the West Riding of York, Report, p. 240. A practice worthy of imitation in this County, is stated to prevail in the Market of Stockton on Tees, that of considering a proper Winchester *Bushel* of Wheat, to weigh 60lb. and of

---

* Yet the Leicestershire Report, printed in 1809, p. 322, says, 34 quarts at Leicester, and 36 quarts at Ashby-de-la-Zouch.

Oats, 33 lb. and to allow proportionally for all weights per bushel over or under these. See Durham Report, 283.

Reaping of Corn, is usually let, and Crops estimated, by the *Thrave* of two Kivers or *Shocks*, or 24 *Sheaves*. See Vol. II. p. 122; III. 128.

Flour, and Meal of Oats, is principally, I believe, retailed by the *Stone* of 14 lb. avoirdupois (of 16 oz. each). In Vol. II. p. 65, an imperfect steel-yard, sometimes used for weighing Flour, is mentioned.

Hay, At Ashburne and other places, is sold by the *Hundred-weight* of 120 lb. avoirdupois, and *Ton* of 2400 lb. Straw and Dung are also sold by the same *Cwt.* and *Ton*. II. 183, 453, 470.

Potatoes, Are often sold by the *Bushel* of 90 lb. avoirdupois. See Vol. II. p. 155.

Butter, Was formerly sold about Derby by the *Pound* of 17 Ounces avoirdupois. See Vol. II. p. 42; but before now, I hope that the Pound of 16 Ounces, prevails universally.

Cheese, Was by the Dairymen universally sold by the *Hundred-weight* of 120 lb. avoirdupois. See Vol. II. pp. 41, 62; but as the Cheese Dealers, who bought of them, as invariably, I believe, sold again by the *Statute Hundred*, and *Ton* of 112, and 2240 lb.; I hope this latter practice has, or will, speedily be adopted by the Dairymen.

3. *Liquids*. The Ale, or Winchester *Gallon*, of 282 cubic inches, and Wine *Gallon* of 231 cubic inches, and their subdivisions of *Quarts*, *Pints*, &c. are here I believe in use generally; as well as the Dry or Bushel *Gallon* of 268.8 cubic inches, and its subdivisions, for loose articles, not liquid.

4. *Wood.*

4. *Wood.* Timber is here sold by the *round,* and *cubic* (and but rarely by the square or *caliper*) measures or denominations, of *Load* (40 or 50 feet), and *Foot,* see Vol. II. 319, and 320 note. *Billets* and *Roots,* stackt, are sold by the *Cord,* of 128, 155, and 162½ cubic feet, in different places and circumstances. See Vol. II. 236 and 237.

Bark of Oak Timber, seems to have been formerly sold, in most parts of the County, by the *Load* and *Rood,* of 70, and 7 yards run, as *set* up in the Woods to dry. See Vol. II. p. 334. The London *Load* of shaved and hatcht Bark is 45 cwt. or 5040 lb. The *Load* and *Quarter,* in Hope, &c. is 70 yards, and 7 yards run, of *set* Bark, II. p. 333. The *Quarter* of Bark, near Wakefield, Yorksh. is 9 heapt Bushels of *chopt* Bark, Vol. II. p. 338. The *Ruck,* in Glossop, is 5¼ cubic yards *stackt,* II. 335. The *Ton* of Bark, in the southern parts of the County, is 2400 lb.; but in Mellor, and some other places in the north, 2240 lb. See II. p. 334.

Charcoal, about Chesterfield, is sometimes sold by the *Load* and *Dozen,* of 144 and 72 level Bushels. See Vol. II. p. 236.

Coals. The formal *measuring* of these in Bushel, or other Measures, is wholly unknown in the County; although at some few of the Coal-pits, and formerly more than at present, the *Corves* or Boxes, Baskets or *Tubs,* in which they were drawn up the shaft, were reputed to hold 2¼ level Bushels, or nine Pecks; but this, or any other dimensions of the Corve, was seldom kept to, and this proved a very uncertain mode of selling Coals, as hinted, Vol. I. p. 340. At Tibshelf,

Tibshelf, and, I believe, some other Pits, they are sold by the *Three-quarter Stack*, of 105 cubic feet, I. 341. On some Pit-hills, the *Corves* are said to hold 2 cwt. or 240 lb.; but more commonly, so many Corves are reckoned to the *Ton*, without any accuracy or certainty. This practice proved very troublesome and vexatious to the dealers and consumers of Coals, carried on the Canals, prior to 1798, since which, a very regular system has prevailed, of selling and carrying Coals thereon, only by the *Ton* and *Cwt.* of 2400 and 120 lb. avoirdupois (Vol. I. 183); in some other situations, other *Cwts.* seem yet in use, in weighing Coals, as at Ashburne, 128 lb. and at Marple, 112 lb. Vol. II. p. 427.

Lime.—The *Load* or horse-load, of three Bushels or Strikes, heapt, seems the most common denomination, under which burnt Lime is sold or estimated, in or near the County, as at Birchwood-Park, Breedon, Newhaven, Stanton, &c. Vol. II. pp. 419, 441, 444, &c.: but sometimes, the *Bushel* of Lime is made level full, and not heapt, as at Ashover, Caldon, Matlock, Peck's-mill, &c. pp. 417, 420, 428, 438, &c.; and more rarely, as at Pilsbury, one upheaped and two level Bushels of Lime go to the *Load*, p. 405. The *Load* at Marple-bridge is $2\frac{1}{4}$ level Bushels, or 10 level pecks, Vol. II. p. 427. At Knitaker, Lime is sold by the *Chaldron* of 32 heapt Bushels, Vol. II. p. 425; at Aston, and the various other Canal Wharfs, a *Quarter* of Lime is eight level Bushels, Vol. II. p. 109, 418, &c.; but at Ticknall, and some other Kilns, it is eight heaped Bushels, p. 431. The *Score*, at Birchwood-Park, Hognaston, Wild-Park, &c. is 20 heapt bushels, Vol.

SECT. VIII.—MANUFACTURES.

As a Manufacturing County, Derbyshire will, I think, be found to rank higher than has generally been supposed, it being probable, that Lancashire, Staffordshire and Warwickshire, only, of the English Counties, excell it, in the extent or variety of their manufactured goods, which become the objects of the supply and general trade of the County, and of England, or of Foreign Export; whether we compare the value of manufactured Goods in each County, with the Acres, with the Rental, or with the Population of the same: in which points of view, the great manufacturing consequence of the West Riding of Yorkshire, (and of some other parts of other Counties), will be found much lowered, by the other Ridings or parts of such Counties, bearing a different character.

In my progress through the County, in the years 1807, 1808 and 1809, I bestowed no ordinary pains, when visiting the very numerous Towns, Hamlets, and Villages, contained in the Lists given in Vol. I. p. 78, and Vol. II. Preface p. 1, to obtain a List of the various species of *Manufactures* carried on in each place:—to have inquired, in more than a few cases, into the number of the different Establishments, much less into the Names of parties, and extent or Value of their products, in each place, appeared more than I ought to attempt, in justice to the very numerous other heads of inquiry, which I had in view, or with which I was charged by the Board.

The information which I have collected and arranged below, as to *the local situations of each species*

MANUFACTURES. 477

cies of Manufacturing Industry in the County, will I trust have its uses, particularly, if in consequence of more liberal means afforded to the Board, they should see it right, hereafter, to adopt the suggestion of Mr. John Holt, in his Lancashire Report, p. 211, for a particular Re-survey of the Manufacturing Counties, by persons competently skilled in that line, and competent also, to investigate and shew, " the actual effects of Manufactures upon Agriculture," in all their varieties of combinations, in each County.

In such case, I say, these arranged Lists of the Localities of the Derbyshire Manufactures, would have their important use; and more especially so would they, towards any future Re-surveys and revisals of the County Reports, (by one or more Surveyors to each), together with the Manuscript Lists of all the *Places* in the County, alphabetically arranged, with the *natural and artificial circumstances* of each place, briefly noted, the *Persons* of Note, for their knowledge and useful activity in each place, the kinds of *Cultivation, Manufactures, Mines, Quarries*, &c. &c. carried on or existing therein, which myself, and doubtless others, of the Board's Reporters, have formed, as our *Travelling Memorandums**, and texts for inquiries and inspections, while visiting each place.

1. *Species*

---

* Several persons who have seen my Derbyshire Common-place Book, here alluded to, have been sanguine in thinking, that if its multifarious facts, and some others of general use and interest, were arranged in the form of "*Notitia Villaris*," or a Derbyshire Gazetteer, of useful and curious local information to Travellers, in that County, and the same was printed in a portable form, (which Travellers might have interleaved, for making their own remarks), the sale of the same would defray its expense, and leave some remuneration for the trouble of preparing and putting

1. *Species of Manufactures.*—The various Trades and Products which I noted in the County, I have for the present purpose, divided into three principal heads, according as the raw materials of each, belong either to the *Animal*, the *Vegetable*, or the *Mineral* Kingdoms of Nature; and each of which I have again subdivided, according as these raw-materials are wholly, or principally, *the produce of the County;* or are the produce of some other County, or are *foreign;* the *Trades*, Products, &c. under each of these six heads, being alphabetically arranged, and the *Places* where each are carried on, are also alphabetically arranged, under each of these several heads.

---

ting it to press; I cannot however, find any Book-publisher of this opinion; unless, as one of them suggested, the Board would order a revised and fair Manuscript Copy of such a work, to be prepared, and lend its sanction to the publication.

I have taken the liberty of here recording this suggestion: and beg to add, that if the Board were to engage such of its Reporters, who appear to have been most circumstantial and regular in their *Notes*, to revise complete, and arrange them, as above-mentioned, and fairly transcribe them into Books, these *Manuscripts*, if no present opportunity offered for their *Publication*, when preserved in the Library of the Board, might prove of the most essential use hereafter, in that General Re-survey of the several Counties, which I think will ere long be seen desirable: particularly, towards inducing persons to undertake County Surveys, who from the value of their time in other pursuits, would shrink from the task of preparing these preliminary formula, so essential, for suggesting and arranging their subsequent and laborious inquiries and observations. For myself, I may say, that the labour of research into Books, and Maps, and by tedious inquiries of several Persons, thought to be best informed as to the local facts of the County, and in preparing such a sketch of Village Notes, as enabled me to commence my Survey, with any effect, was so great; that the meeting with such a document, ready prepared to my hands, as to any County, would go further towards inducing me to undertake its Survey, than almost any pecuniary terms that could be offered me: and this, I doubt not, would prove the case, with many other professional or practical men.

MANUFACTURES. 479

Other persons there doubtless are, who would have preferred some others of the various classifications of these species of Manufactures, &c. which might have been made; such will, however, I trust be satisfied, on finding the whole of them referred to in one series, in the alphabetical Index at the end of this Volume.

1. *Trades, &c. depending on Animal Products, of the County.*

Blanket-weaving, and scouring, is carried on in,

      Bramington.

Bone-crushing Mills; nine situations of these, are enumerated in Vol. II. p. 449.

Butter: Pretty ample local details of this domestic manufacture, will be found in pages 1 to 68, of the present Volume.

Button-moulds, of Horn and Bone, are made, (see Vol. II. 452), at,

      Whittington.

Candle-making, of Tallow.

| Allsaints, | St. Peter, | Wirksworth, |
| Matlock, | Ticknall, | &c. |

I am aware that this last is a very imperfect List: and so will those of some others, of the more common Trades, be found: the persons who gave me information, often omitted to mention these kinds of home Manufactures: I had intended to have applied to the Collectors of Excise, in order, if possible, to obtain complete Local Lists of the Trades subject thereto: but my time would not permit.

Carpet-weaving.

    Chesterfield.

Cheese: The local details regarding this staple Manufacture

facture of the Farmers, will be found in pages 1 to 63, of the present Volume.

Curriers, or Leather-dressers:

| Alfreton, | Chesterfield, | Wirksworth, |
| Belper, | | |

Felmonger.

Sudbury.

Fulling-Mills.

| Brassington, | Hartshorn, | Simondley, |
| Glossop, | Hayfield, | Whitfield. |

Glue-makers.

| St. Alkmund, | Chesterfield, | St. Peter. |
| Brimington, | | |

Leather-mill, for Oiled and Shammy Leather.

Hartshorn.

Meat, Beef, Lamb, Mutton, Pork, Veal, see pages 270 to 357, Poultry, 358, Game and Fish, 370 to 374.

Shoe-factory.

Chesterfield, &c.

Skinners, or Leather-dressers, Shammy, &c.

Alfreton,	Baslow,	Repton,
St. Alkmund,	Chesterfield,	St. Werburgh,
Ashburne,	Killamarsh,	Wirksworth.

Soap-Makers.

| Allsaints, | Staveley, | St. Werburgh. |

Stockings, of Worsted;—but very few if any of these are hand or kneedle *knit*, in the County, I believe; considerable quantities are *wove*, or made on Frames, but I have not been able to distinguish these, from the Weavers of Cotton *Stockings*, which are by far the most numerous, I believe.

Tan-

MANUFACTURES, &c.

Tan-Yards.

Ashburne,	Coxbench,	Olerenshaw,
Bakewell,	Crich,	Repton,
Baslow,	Dronfield,	Spondon,
Belper,	Goldcliff,	Toad-hole Furnace,
Chapel-en-le-Frith,	Hayfield,	St. Werburgh,
Charlesworth,	Langley,	Whittington,
Chesterfield,	Little Eaton,	Whittle,
Clifton,	Measham,	Wirksworth,
Codnor-Park,	Millthorpe,	&c.
Compton,	Milton,	

Tawers, Codders, or White-leather Makers.

Alfreton,	Crich,	Wirksworth.
Belper,	Matlock,	

Woollen Cloth Factories, Yarn-Spinning, Weaving, and Cloth Dressing.

Brassington,	Hayfield,	Simondley,
Chunall,	Ludworth,	Whaley-bridge,
Glossop,	Phoside,	Whitfield.

Worsted Spinning, for the Hosiers, by hand Jennies.

Litton,	Melborne,	Tideswell;

And in St. Werburgh, there are 2 Mills for this spinning.

2. Trades, &c. depending on Animal Substances, Imported.

Hat-making*.

Alfreton,	Hadfield,	Ludworth,
Belper, 2,	Holy-moor-side (Dog-	Matlock,
Bradwell,	hole),	Newton in Bla. 2,
Chesterfield,	Hope,	Wirksworth, 2.
Glossop,	Lea-wood,	

---

* At Mosborough and Ridgeway in Eck., rough, unsplit *Straw Hats* are made for Sale.

Silk-Spinning Mills.

St. Alkmund, 3,   Chesterfield,   St. Peter,
Allsaints*,   St. Michael,   St. Werburgh.

Silk-stocking Weaving.

St. Alkmund,   St. Michael,   St. Peter.
Allsaints,

3. *Trades, &c. depending on Vegetable Productions, of the County.*

Basket and *Whisket*-making.

Allsaints,   Holy-moor-side,   Walton,
Chapel-en-le-Frith,   Kelstedge,   Wingerworth,
Chesterfield,   Repton,   &c.

For these places, I might have referred to Vol. II. p. 262, but for the omissions there made.

Beesam, or Broom, making, see Vol. II. p. 234.

Barlow,   Whaley, &c.

Boat or *Barge-building*, for the Canals.

St. Alkmund,   Langley-mill,   Shardlow,
Bull-bridge,   Measham (of Iron),   Willington.
Butterley, in Pen.   Pye-bridge,
   (of Iron),   Sawley,

Breweries, see Vol. II. p. 127.
*Chamomile-*flowers, cultivated, see Vol. II. p. 169.
Charcoal-burning, and *Grinding,* see Vol. II. 235.

Alderwasley,   Hassop,   Wingerworth,
Chatsworth,   Lea-wood,   &c.

Charcoal Mills, for grinding it.

Adelphi,   Butterley, in Pen.   Lea-wood, &c.

Corn, Barley, Beans, Oats, Pease, Wheat, see Vol. II. p. 113 to 133.

---

* The large original Mills of Sir Thomas Lombe.

MANUFACTURES, &c.

Hoops for Casks, of Wood, see Vol. II. p. 233.
Malt-makers, see Vol. II. p. 127.
Mattrasses, Chair-bottoms, &c. of Straw.

    Upper-Town in Ash,   Tibshelf.

Millers, Flour or Meal-makers, see Vol. II. p. 492.
Scives, or Riddles for Corn.

    Repton, &c.

Shelling, or Oat-meal Mills, see Vol. II. pp. 129 and 457.

Timber, Abele, Alder, Ash, Aspen, Beech, Birch, Cedar, Cherry, Crab, Elder, Elm, Fir, (Balsam, Pine, Silver, Scotch Spruce), Hawthorn, Holly, Hornbeam, Horse-chesnut, Larch, Lime, Maple, Mountain-ash, Oak, Plane, Pear, Poplar, Sallow, Sweet-Chesnut, Sycamore, Willow, Yew, see Vol. II. pp. 244 to 268.

Turning-Mills, for Wood, Bobbins, Bowls, Cheese-vats, Dishes, Tool-handles, &c. see Vol. II. p. 234.

4. *Trades, &c. depending on Vegetable Substances, Imported.*

Bleaching-houses and Grounds*.

Bakewell,	Kelstedge,	New Brampton,
Clown,	Makeney,	Tansley,
Duffield,	Marple-bridge,	Thornsett, 2,
Hayfield,	Measham,	St. Werburgh.
Higham,	Milford,	

Calico

---

* As to Bleaching, Dyeing, &c. it may perhaps be said, that *vegetable* matters are sometimes not used therein; that Hatters may sometimes use only *Derbyshire* Wool (Vol. III. p. 93), and Flax-spinners, Flax grown in the County (II. 168); and that Edge-tool makers, Machinists, &c. may principally use the Iron of Derbyshire, or steel made therefrom,

MANUFACTURES, &c.

Calico Printing.

Glossop,   New Mills,   Thornsett.
Mellor,

Calico Weaving.

Alfreton,	Cow-way,	Ollerset,
Ambaston,	Draycot,	Osmaston,
Appleby,	Fritchley,	Padfield,
Ashford,	Glossop,	St. Peter,
Ashover,	Great Longsdon,	Repton,
Bakewell,	Hadfield,	Sawley,
Beard,	Hathersage,	Simondley,
Belper,	Hayfield,	Tideswell,
Brailsford,	Heage,	Toad-hole Furnace,
Breaston,	Little Longsdon,	Wardlow,
Car-meadow,	Ludworth,	Whaley,
Chapel-en-le-Frith,	Matlock,	Whittle,
Charlesworth,	Measham,	Winksworth,
Chevin-side,	Mellor,	&c.
Chisworth,	New Mills,	

Cambric Weaving.

Glossop,   Hadfield,   Ludworth.

Candle-wick, Bump or Bomp-spinning Mills.

Clown,   New Brampton,   Tansley.
Kelstedge,

---

from, in their fabricks; and that in all such cases, my arrangement of these Trades is improper; in strictness, this must be admitted, yet it might perhaps be difficult, to alter the arrangement with real advantage: the general *Index* to the Volume, will enable the finding of any of the Trades, &c. which I have noted.

Cotton-

Cotton-spinning Mills*.

Bakewell,
Bamford,
Belper, 4,
Bradwell,
Brough,
Brown-side,
Bugsworth,
Calver,
Castleton, 2,
Chapel-en-le-Frith, 2,
Charlesworth, 2,
Chisworth, 6,
Chunall,
Cressbrook,
Cromford, 2,
Darley-Abbey,
Dinting,
Dove-hole,
Draycot,
Edale-Chapel (Oler-Booth),
Fenny-Bentley (Wood-Eaves),
Fritchley,
Gamesley, 2,
Glossop, 7,
Great Longsdon (Cross-well-dale),
Hadfield, 2,
Harts-hay,
Hayfield, 2,
Kelstedge,
Lea-Wood,
Little Wilne,
Litton, 2,
Ludworth, 3,
Lumsdale,
Matlock, 2,
Matlock Baths,
Measham, 2,
Mellor, 5,
Milford,
Nether Booth,
New Mills, 7,
Ollerset,
Padfield, 5,
Peak Forest,
St. Peter,
Pichard-green, 2,
Plesley, 5,
Raworth, 3,
Simondley, 3,
South Winfield (Park),
Tansley, 4,
Tideswell,
Toad-hole, 3,
Toad-hole Furnace,
Tor-side,
Whaley-bridge,
Whitfield, 3,
Winshill,
Wirksworth.

Dye-Houses.

Alfreton,
Allsaints,
Alport in Yol.
Chunall,
Cromford,
Great Rowsley,
Hayfield,
St. Peter,
Tansley,
Upper-Town, in Ash.
Whitfield,
Wirksworth.

Flax Spinning Mills, Linen-Yarn Mills.

Holy-moor-side,
Kelstedge,
Matlock,
New Brampton,
Tansley,
Toad-hole.

Fustian-Weaving, Thicksets.

Ollerset,   Tideswell.

Hop-bag Spinning and Weaving, Wool-bags, &c.

Clown.

---

* In 1794, Mr. Robert Lowe states 34 *Cotton-mills* to have been erected in Nottinghamshire. See p. 170 of his Report.

Lace-weaving, or Warp Frame-lace making.

St. Alkmund,	Melborne,	St. Peter,
Allsaints,	St. Michael,	St. Werburgh.

Lace-working, or Needle-working of Frame-lace.

St. Alkmund,	Marston-Montgomery,	Roston,
Allsaints,	Matlock,	Sandiacre,
Ashburne,	Measham,	Sawley,
Bretby,	Melborne,	Shardlow,
Chesterfield,	St. Michael,	Snelston,
Cromford,	Milton,	Somersall-Herbert,
Cubley,	Norbury,	Stapenhill,
Doveridge,	Osmaston,	Swarkestone,
Findern,	Penter's Lane,	Ticknall,
Long-Eaton,	St. Peter,	Waldley,
Lullington,	Repton,	St. Werburgh,
Mackworth,	Rosleston,	Wirksworth, &c.

Linen-weaving, Sheeting, Checks, &c.

Belper,	Kelstedge,	Turnditch,
Butterley, in Ash.	New Brampton,	&c.

Muslin Weaving.

Glossop,	Lea-wood,	New Mills,
Hadfield,	Marple-bridge,	Raworth.
Hayfield,	Mellor,	

Night-caps, of Cotton Frame-knitting, are made occasionally, in a great portion of the places mentioned, p. 487, where Stocking-weaving is practised: in Tibshelf, and I believe several other of those places, Woollen Night-caps are made, in Frames or Knitting Looms.

Pack-thread Spinning, String, Twine.

Ashover,	Castleton,	Heage.

Paper-making.

Alport, in Yol.	Duffield,	Matlock Baths,
Brookfield,	Hathersage,	Toad-hole,
Darley-Abbey,	Hayfield,	Totley.

Rope-

Rope-making, Cords, Halters.

Ashover,	Heage,	Smalley,
Bakewell,	Holbrook,	Stanley,
Brampton,	Hope,	Upper-Town, in Ash.
Castleton,	Ilkeston,	
Draycot,	St. Peter,	

At Draycot and Upper-Town (Bower's Mill), Cotton Ropes are made, of *waste* or knotted ends of threads, from the Cotton Mills.

Sacking-weaving, Corn Bags.

Clown,	Hope,	Kelstedge.

Sail-cloth Weaving.

Clown.

Stocking-weaving; principally of Cotton, some of Worsted, Frame-knit.

Alfreton,	Denby,	Newton in Black.
St. Alkmund,	Draycot,	Ockbrook,
Allestry,	Duffield,	Packington,
Allsaints,	Duffield Bank,	Pentrich,
Ambaston,	Farlow-green,	St. Peter,
Appleby,	Heage,	Pilsley, in N.W.
Ashford,	Heanor,	Pinxton,
Ashover,	Holbrook,	Ravenstone,
Bargate, in Duf.	Holy moor-side,	Riddings,
Ditto, in Hors.	Hopping-hill,	Sawley,
Belper,	Horsley,	Shackle-cross,
Belper-gutter,	Horsley-Woodhouse,	Shipley,
Belper Lane-end,	Ilkeston,	Shirland,
Blackwell, in K.I.	Kedleston,	Smithsby,
Bonsal,	Kilburne,	South Normanton,
Breaston,	Langley (Kirk),	South Winfield,
Bredsall,	Little Eaton,	Spondon,
Bubnell,	Litton,	Tag-hill,
Chaddesden,	Long-Eaton,	Tibshelf,
Chesterfield,	Makeney,	Tideswell,
Chavin-side,	Mapperley,	Upper-Killis,
Clown,	Matlock,	St. Werburgh,
Coxbench,	Measham,	Whitwell,
Crich,	Melborne,	Wirksworth,
Cromford,	St. Michael,	&c.
Dale Abbey,	Nether-end,	

Tape-weaving Mills.

 Horridge-end,   Whaley-bridge.

Thread Spinning, Sewing Linen Thread.

 Kelstedge,   Matlock, 2,   Raworth.
 Marple-bridge,

At Draycot, Belper, and many others of the Cotton Mills, in the List, page 485, Sewing-cotton Thread is spun, and wound into neat Balls.

Whipcord-spinning.

 Kelstedge,   Ravenstone,   Upper-Town, in Ash.

5. *Trades, &c. depending on Mineral Products, of the County.*

Bakestone making, see Vol. I. pp. 431 and 444.

Boiler-making, of Wrought Iron, for Steam-Engines, &c. see Vol. II. p. 493.

 Chesterfield,   Measham,   Ripley.

Brick-making, Building, Draining, Fire, Paving, &c. see Vol. I. pp. 445 and 451, to 453.

Building-stones, or Freestone, Ashler, Coping, Eaves-slates, Gable-stones, Paving, Ridging, Grey Slates, or Tilestones and Stack-posts, see Vol. I. pp. 415 to 432.

Cannon-balls, or Shot and Shells.

 Butterley,   Somercoates.

And, I believe, at some others of the Iron Furnaces, Vol. I. p. 397.

Cannon-casting and Boreing.

 Butterley, in Pen.   Chesterfield.

Chain-making, Iron.

 Dronfield,   Killamarsh,   Openwood-gate,
 Duffield,   Measham,   Unston.

The Chains made at Dronfield and Unston are of Cast-Iron; a cheap kind, for certain purposes, where great strength is not required.

China-stone, or White Potter's Chert Pits, see Vol. I. pp. 272 and 415.

Cisterns and Troughs of Stone, to hold water. See Vol. I. p. 432.

Clay-pits, Brick, China, Fire, Pipe, Pottery and Tile. See Vol I. pp. 445, and 447, to 453.

Coal-pits. See Vol. I. p. 188.

Coke-burning. See Vol. I. p. 399.

Adelphi,	Hart's-hay,	Staveley (Forge),
Butterley, in Pen.	Morley-park,	Temple Normanton
Chesterfield,	Renishaw,	(Lings),
Codnor-park,	Riddings,	Troway,
Dunston,	Staveley (common spot),	Wingerworth.
Grass-moor,		

Copperas-stone, Brasses or Pyrites, Pits, see Vol. I. p. 219.

Copperas Works, Green Vitriol, see Vol. I. p. 218.

Cylinders, cast and bored, for Engines.

Chesterfield,	Butterley, in Pen.

Filtering-cisterns, see Vol. I. p. 434.

Fire-stone, Hearth or Furnace-stone Pits, see Vol. I. p. 221 and 431.

Frying-pans, of Iron.

St. Peter.

Grind-mills, Blade-mills, Grindstone-mills.

Abbey-dale,	Eckington,	Woodseats, in Eck.
Beighton,	Totley,	Woodseats, in Nor.
Dronfield, 3,	Upper Padley,	

Grindstones, see Vol. I. p. 435.

Gypsum

Gypsum Pits, Alabaster, Plaster, see Vol. I. pp. 149 to 151.

Hammer-mills; Forge, Tilt, Skelper or Planishing Mills. See Vol. I. p. 403.

Bugsworth,	Ford, in Eck.	New Brampton,
Chapel-en-le-Frith,	Heely-mill,	Norton-leys,
Chapel-milltown,	Killamarsh,	St. Peter,
Codnor-park,	Marple-bridge,	Staveley.

Hoops for Casks, of Iron.

Allsaints,	Burrowash,	St. Peter, &c.

Wooden hoops are made at Moor-hole, see Vol. II. p. 233.

Iron Forges and Furnaces, see Vol. I. pp. 397, and 403.

Ironstone Pits, argillaceous Ore, see Vol. 1. p. 217.

Lead Mines, or Veins of Lead Ore, see Vol. 1. p. 252.

Lead Smelting Cupolas, and Slag-mills, see Vol. I. p. 385.

Lime Kilns, see Vol. II. pp. 415 and 433.

Limestone Quarries, see Vol. I. p. 408.

Malt-kiln Plates, of perforated Cast Iron, see Foundries, Vol. I. p. 404—of Pottery, see Vol. I. p. 449.

Marble Quarries, see Vol. I. p. 413.

Marble sawing and polishing Mills, see Vol. I. p. 412.

Marl Pits, for manuring, see Vol. 1. p. 456.

Mill-stone Quarries, see Vol. I. pp. 221 and 272.

Nail-making, of Cast Iron.

Dronfield,	New Brampton.

————————, Clasp (or Carpenters'), and Spikes, &c.

Belper,	Coxbench,	Mill-close Lane,
Brentwood-gate,	Eckington,	Smalley.
Cartlidge,	Heage,	

Nail-

Nail-making, Horse-shoe.

    Belper,   Eckington,   Staveley,
    Belper-gutter,   Heage,   White-moor.
    Belper Lane-end,   Holbrook,

————, Shoemakers'

    Horsley-woodhouse,   Smalley.

Ore-dressing, washing, buddling, see Vol. I. p. 372.

Patten-rings or Clog-irons.

    St. Alkmund,   Duffield.

Pipe-making, Tobacco-pipes, see Vol. I. p. 448.

Pipes, of Earthenware, hollow Bricks, for conveying Water, see Vol. I. p. 449, 456.

    Chesterfield,   Smalley.

———, of Lead, drawn.

    Allsaints.

———, of Zink.

    St. Werburgh.

Plaster of Paris, Kilns and Works, Gypsum, see Vol. I. p. 150.

Potteries, earthen-ware, stone-ware, see Vol. I. p. 449.

Pot Stones, pye or lump stones for the Iron Forges, see Vol. I. p. 431.

Puncheons, stauncheons or props for the Coal-pits, see Vol. I. pp. 347, 348; and II. pp. 222, 224, 226.

Red-lead Works, Minium, see Vol. I. p. 591*.

    Alderwasley,   Holy-moor-side (Cat-   St. Peter,
    Allsaints,   hole),   Walton, in Che.
    Darley-abbey,   Lea-wood,

---

* Formerly, there were other Red-lead Furnaces, at Loads, Oakerthorpe, Totley, and Wingerworth.

Rivets, of Iron, softened, for Coopers, Boiler-makers, &c.

<small>New Brampton (cast).</small>

Rolling and Slitting Mills, for Iron Bars, Plate-iron, Nail-rods, &c. see Vol. I. pp. 403, and 404.

Rottenstone, or polishing earth, see Vol. I. p. 231.

Sand-pits, Casting or Founders, House-floor, Masons'-mortar, Scouring, and Scythe-stick Sand, see Vol. I. p. 463.

Saw-mills, for Stone and Wood, see Vol. I. p. 423; and Vol. II. p. 235*.

Screws, Carpenters', for Wood.

<small>Hartshorn.</small>

Scythe-sticks, and stones, for sharpening Scythes, Hay-knives, &c. see Vol. I. p. 439, 437.

Sheet-lead, Milled-lead, Rolled-lead.

<small>Alderwasley,   Lea-wood.</small>

Common Sheet-lead, is cast by most of the Plumbers and Glaziers of the County.

Shot, Leaden, see Vol. I. p. 391.

<small>Allsaints,   St. Peter (patent).</small>

Slitting Mills, see *Rolling*.

Spar-workers, Petrification-workers, Gypsum, Calc-spar, Fluor, see Vol. I. pp. 150, 459, and 461.

Sulphur-works, annexed to Lead-smelting houses, see Vol. I. p. 468.

Tenter-hooks, of Cast-iron, softened.

<small>New Brampton.</small>

Tile-kilns, Draining, Gutter, Hip, Pan, Plane and Ridge, see Vol. I. p. 451 to 455.

---

* Besides which, most of the Wood *Turnin-gmills* mentioned Vol. II. p. 234, have now, I believe, erected circular *Saws*.

Tire,

Tire, for Carriage wheels.
>Duffield.

Whetstones, Rubbers, Hones, see Vol. 1. p. 437 to 440.

White-lead Works.
>Allsaints, St. Peter, St. Werburgh.

Wire-drawing, Steel.
>Hathersage.

Wire-working, Safes, Scives, Screens, &c.
>Bakewell.

Zink Mines, Blend and Calamine, see Vol. I. p. 406.
Zink-work, maleable, Plates, Wire Pipes, &c.
>St. Werburgh.

6. *Trades, &c. depending principally on Mineral substances, Imported.*

Axes, Hatchets, Bills, Adzes.
>Dronfield, Marple-bridge, Staveley,
>Eckington, New Brampton (Cast-I.)*

Brass Foundry.
>Ashburne.

Bridle-Bits and Buckles.
>Bolsover (Steel) New Brampton (Cast I.)

China-Factories, see Vol. I. p. 447.

Chisels, Gouges, Plane-Irons, and other Edge Tools.
>Dronfield, Marple-bridge, Staveley,
>Eckington,

---

* By a modernly discovered process, various articles made here, (and at Dronfield, and Unston), of *Cast Iron,* are afterwards, either rendered as soft as the best Wrought Iron, or are converted to Steel.

Clock

Clock and Watch making.

 Allsaints,   St. Peter,   Walton on Tr. &c.

Colour-grinding Mills, Paint, see Vol. I. pp. 402 and 407.

 Allsaints,   St. Peter,   St. Werburgh.
 Bonsal (Dale),

Cotton-machinery Makers, for the Cotton-spinning Mills.

 Allsaints,   Glossop,   Kelstedge.
 Belper,   Hadfield,

Besides which, several of the larger Cotton-Mills, in the List page 485, have shops attached thereto, and keep workmen who repair or renew their Machinery.

Cutlery, Knives, Forks, &c.

 Alfreton,   Cole-Aston,   Little Norton,
 Allsaints,   Dronfield, (Cast I.), New Brampton, (Cast L),
 Chesterfield, Green-hill,   Unston, (Cast. I.).

File-Making, Rasps.

 Belper,   Norton-leys,   Woodseats, in No.
 Mosborough,

Flint-grinding Mills, for Pottery Glazing, see Vol. I. p. 447.

Frame-smiths, Stocking Loom Makers.

 Alfreton,   Kilburne,   St. Peter, 2,
 St. Alkmund, Loscoe,   Smalley.
 Heanor,

Glass-making.

    Glass-house Common.

Gun-powder Making.

    Ferneylee.

Hoes, (Garden, Turnip), Paring-shovels, Trowels, &c.

 Dronfield,   Ford, in Eck. New Brampton, (Cast L),
 Eckington,   Marsh-lane, Rigdeway, in Eck.

Imple-

MANUFACTURES, &c. 495

Implement-makers, Agricultural Tools, &c., see Vol. II. pp. 43, 50, 56, 62, 65, and 493.

St. Alkmund,	Glapwell,	Turnditch,
Allsaints,	Hartshorn,	Walton on Trent,
Baslow,	Linton,	St. Werburgh,
Brockhurst,	Longford,	Weston on Trent,
Calver,	Matlock,	Whittington,
Catton,	Repton,	Wirksworth,
Cross o' th' Hands,	Ridgeway,	Yolgrave,
Foston,	Swarkestone,	&c.

Malt-Mills, Steel-mills, see Vol. II. p. 58.

Chesterfield.

Mangles, for Linen Cloaths.

Linton.

Mechanists, Machine, Tool and Engine Makers, see Vol. II. p. 493.

| St. Alkmund, 2, | Butterley, in Pen. | Walton on Tr. |
| Allsaints, | Matlock, | St. Werburgh. |

Millwrights, see Vol. II. p. 493.

Needle-making.

— Hathersage.

Reaping-hooks, smooth edged, see Vol. II. p. 122.

| Bole-hill, in Eck. | Ford, in Eck. | Ridgeway, |
| Dronfield, | Marsh-lane, | Woodseats, in Nor. |

Scissars, of Cast Iron, cemented to Steel.

| Dronfield, | New Brampton, | Unston. |

Scythe-smiths.

Abbey-dale,	Green-hill,	Norton-leys,
Beauchief,	Hackenthorp,	Oaks,
Bole-hill, in Nor.	Hill-top,	Ridgeway,
Bradway,	Lightwood,	Stubley,
Calow, in Ches.	Little Norton,	Totley,
Cowley, in Dro.	Lower Birchett,	Woodhouse, in Dron.
Dronfield,	Mosborough,	Woodseats, in Nor.
Ford, in Eck.	Norton,	

Sickles,

Sickles, toothed Reaping Tools, see Vol. II. p. 122.

Bole-hill, in Eck.	Marsh-lane,	Slode-lane,
Bramley,	Moor-hole,	Staveley,
Dronfield,	Mosborough,	Troway,
Ford, in Eck.	Mosborough-moor,	Woodseats, in Nor.
Geer-lane,	Norton,	
Hackenthorp,	Ridgeway,	

Snuffers.

    New Brampton.

Soda-water Maker.

    St. Alkmund.

Spades, Shovels.

Bramley,	Killamarsh,	St. Peter,
Eckington,	Marple-bridge,	Staveley.

Spurs, of Steel.

  Bolsover, New Brampton.

Stirrup-irons, of Cast Iron, cemented.

    New Brampton.

Tin-plate Workers, Tin-men.

  St. Alkmund, Allsaints, 2, St. Peter.

Washing-machines, for Cloaths.

    St. Alkmund.

Worsted-machinery Maker, for the Worsted Spinning Mills.

    Allsaints.

Notwithstanding that many of the Manufactures and Productions above enumerated, are separately of small importance, and may contribute little or nothing towards an *Export* Trade from the County; yet taken in the aggregate, they must be admitted to present a most
flattering

flattering picture of the varied and great manufacturing industry of the County; shewing it to contribute, far beyond most other Counties, towards *the supply of all its own wants*, and contributing at the same time in no small degree towards, *the supply and general Trade of the Kingdom* at large, as I have already intimated at the beginning of this Section.

The vast variety and number of objects, claiming almost alike my attention on this Survey, prevented me, as already intimated in p. 476, from more than in a few instances, comparatively, taking those Notes of the numbers and the names of Manufacturers or Work People of different kinds in the several places, I have mentioned, and regarding the peculiarities, processes, and prices of their several Productions, the prices thereof, and of Labour, &c. &c. on all which, I generally found the Parties ready, and many of them even anxious to give information: and which at no distant period, they may, I hope, have the opportunity of communicating, thro' some more able hand (as is suggested in p. 477), towards a far more *minute and accurate View of the Manufacturing state of this and other Counties*, than can at present be presented. To have given herein, the few further particulars which I collected, as to Names, Products, and Prices, might have appeared partial, and on that score given offence to some; on which account, as well as to avoid further swelling this Report, I shall leave them still on my Manuscript Notes.

The chief *Articles of Export* or sale beyond the limits of the County, among its many natural and artificial Productions, enumerated in the previous pages, appear to be, Cotton-twist and Stockings, Silk-thread and Stockings, Calicos and Muslins, Frame-lace, Hats: Coals,

Coals, Iron, Edge-tools and Implements, Nails, Lead, Red and White Lead, Building-stone and Marble, Lime, Gypsum, Calamine, Chert, Fluor Spar, Copperas, Grind and Mill-stones, Fire Clay, Bricks, and Stones; and Cheese, among its agricultural products.

It is surprizing with what rapidity the different manufacturing Establishments were accumulated, in some places, in the parish of Glossop, at the NW extremity of the County, in particular; in which Parish, I was told, that about 25 years before I visited it, in 1809, only one old Mill existed, appropriated to the making of Oatmeal for its few Agricultural Inhabitants; yet at the time of collecting my Notes, out of the 112 *Cotton Mills* of the County, which are enumerated in p. 485, 56, or just half of these, were found in this Parish*! more than half of these erected in the 4 to 10 preceding years: besides which, I found scattered thro' 24 of the subdivisions of this Parish, which I have made in p. 81 of Vol. I, an immense number and variety of other Factories, and Manufactures, and new Trades, viz. Calico-weaving in 11 of these places, and Calico-printing in 4 places, Muslin-weaving in 7, Cambric-weaving in 2, Linen-thread Spinning in 2, Fustian-weaving in 1, Woollen-cloth Spinning, Weaving and Dressing in 7, Bleaching in 4, Dyeing in 4, Hat-making in 3, Paper-making in 1, Tanning in 3, Iron Forge and Tilt Hammer-mills in 2, Edge-tools, Spade, &c. making in 1, &c. The greater part of

---

* Without reckoning five other such Mills, close on its borders, beyond its boundary Brooks! One of these Mills in Glossop, Mr. Oldknow's, at Mellor, is so considerable, as within a single Building, 113 yards long, to employ from 450 to 500 hands, and spin weekly 150,000 hanks of Twist or Thread!, which is almost $3\frac{1}{4}$ millions of Miles of Thread per annum!

the

the erections and establishments for these Manufactures, have, I believe, been made, under Leases granted by Bernard E. Howard, Esq. now Duke of Norfolk.

2. *Earnings* of Manufacturing Labourers;—for the reasons already mentioned in page 497, I find myself unable to redeem the pledge I may be supposed to have given, in concluding my account of Agricultural Day-labour, in page 188 of this Volume: in general, I may however remark, that the earnings of the operative Manufacturers, except perhaps during very depressed states of their respective Trades, considerably exceed those of Agricultural Labourers, often very much so, and so do their style of living and means of comfort, except, as is much too common, an addiction to drinking, in a lesser or greater degree, diminishes their application to business, and considerably absorbs their earnings.

I have always been at a loss to discover the policy or justice, of affording to Manufacturing Labourers, greater pay or better food, cloathing and houses, than to those engaged in the most permanent and important of all Manufactures, for supplying Meat and Drink to the other classes, as well as themselves. At the time I am writing (Feb. 1816), I much fear that this comparative depression of the lower agricultural class, is greater than was ever before known: and that such an enormous abstraction has been made, by low prices and high Taxes, in the two past years, from the profits and capitals of the middle part of this class, as must long and distressingly keep the persons they employ to labour their Farms, in their present very depressed state.

3. *Rise*

3. *Rise of Rents* occasioned by Manufactures?—Undoubtedly on the first introduction of any considerable Manufactory, and for a considerable period afterwards, while the same continues flourishing and on the increase, considerable advances of the Rents of the surrounding Lands and Houses, take place; but as the Poor Laws (to be spoken of under the next head of this Section) make the greater part of the *extra* population collected together from other parts, or bred up for the sole purpose of the Manufactory, permanently chargeable on the Lands, the increase of Poor's Rates, in such places, at no distant period balancing the agricultural improvements made, occasion the Rents to become and remain stationary, while those of the Country in general, are gradually advancing, owing to the decrease of the value of Money, or even occasioning the Rents, in process of time, to decrease, as the demand for the goods manufactured may lessen, or be better or cheaper supplied elsewhere; or otherwise, when the place becomes over populated; towards which last, continued success in any manufacturing adventure, has an almost irresistible tendency to contribute.

The less that Goods manufactured in any place, depend on foreign countries for their sale, or on mere fashion, or on local distant consumption in our own, for their demand, the less fluctuating and more steady will be the employ afforded to the people engaged on such Manufactures, and the more permanent and real will be the advance of Rents, and other signs of stable prosperity attending them.

In proportion also, as the Manufactories afford steady employ, for young, robust, middle aged, and old Persons, of both sexes, in the ratios which Society locally

locally supplies these different classes of the Population, will their beneficial effects be permanently experienced, or the contrary.

A district, for instance, wherein only the most robust and able Men find profitable employ, even if they are not quickly worn out by the excess of their toil, can scarcely advance in real prosperity, whatever may be the wages earned by those actually employed, at least not under the baneful and demoralizing system of our present Poor Laws. Still less can a district, where mere children and the young only are principally employed, really advance in prosperity; especially when, as with most newly erected Cotton-spinning Mills of this and adjacent Counties, the demand *for Children's labour* therein, exceeds even the inordinately excited increase of Population in the place, and Children are not only sought for thro' the adjoining districts, but in many instances have been imported by scores at a time, and by hundreds in a year, from London, Bristol, and other distant great Towns; yet only making way there, for the new helpless victims, to the idleness, dissipation, and vices, which people the Work-house Nurseries in these Towns, who are to form the subjects of future supplies of Cotton-Mill Apprentices.

I am far from intending here to insinuate, that great care, and even kind attention, is not bestowed on the Cotton-Mill Apprentices in general, throughout this County; in several instances I have seen this to be the case, and a rather sedulous inquiry on this head from others, has not disclosed even suspicious hints to the contrary, in any instance, as far as I recollect; nor am I disposed to think or represent, that any very considerable or remediable degrees of vice or immorality, exist in these Apprentice Houses or Mills; nor that their employ is so unhealthy, as some have represented; yet I cannot bring myself to approve a system, too nearly allied to that of Chimney-sweepers' Boys (when under humane Masters), in one of the most deplorable features of their trade, that of *its failing them almost entirely when they grow up*, and leaving them a calling then to seek.

A considerable majority of the Cotton-Mill Apprentices appear, I think, to be Girls, particularly of those born in distant places; and except in a few instances, I could not learn, that more than a few of these were retained on Wages, at particular Mills, after the expiration of their Apprenticeship: but too often, such truly unfortunate young Women, disperse themselves over the Country, and for want of friends or employ, prematurely and inconsiderately get married to, or more improperly associate themselves with Soldiers, or other loose and unstationary Men, and at no distant periods, are (as observed in p. 33 of Vol. II.) passed home to the Parishes they were apprenticed in, for want of any other settlement for them being known, with several Children, to remain a burthen thereon.

An Agricultural Lordship (maintaining its own Poor) was pointed out to me, in an extreme corner of which, near adjacent to a large Cotton Mill, of several years standing, in the adjoining Parish, *its Apprentice-house had been built* (on a very small piece of Land purchased for such purpose, as it was represented), by the owner of the Cotton Mill, who was himself a considerable Land-owner and Occupier in the Parish in which his Mill stood, but of little more than the site of his Apprentice-house in the other; and it was mentioned, that in one year, seven women, who formerly had been Apprentices, or were Wives of such, several of them having children, were passed home to

and

and remained burthensome to their Lordship, which, comparatively speaking, had received scarcely any kind of benefit from the trade carried on at the Mill; and that the number of Paupers already thus accumulated, and with which they seemed inevitably fixed, was so great, as to threaten an almost entire absorption of the Rental of the Lordship!

Had I not previously seen and learnt from inquiry, that the system pursued at the Mill and Apprentice-house alluded to, was very regular and commendable, as to the care and attention to the morals and behaviour of the Apprentices therein, I might have been led to suppose, that a culpable laxity, and consequent depravity of the Girls and Boys, before the time of their leaving the Mill, had principally contributed to the evil so justly complained of; but my informants, on being questioned on this head, declared, that the system and state of morals among the Apprentices alluded to, had, from the erection of the Mill, always been good, and the conduct of the Girls rather exemplary than otherwise, as long as they remained in the House, either as Apprentices, or as Boarded Servants, some of them, for a time, after the expiration of their Apprenticeship. Which circumstances, too strongly, I think, point out *the inherent evils of a system of Child employ, accompanied by the want of it in maturer age*, for me to have omitted mentioning them, altho' not feeling myself at liberty to mention names, or more precise particulars.

An almost incessant state of extended war in which we have been kept, since most of the Spinning Mills were erected, and the system of *Child-labour* began, which I am now deprecating, has hitherto made the evil, as far as regards Boy Apprentices to such establishments

blishments, less apparent than as to Girls, owing to the insiduous arts and industry of Recruiting Parties stationed around them, who have succeeded in taking away these Youths, almost as soon as their engagements ended: yet, besides the evils already experienced, in the burthens which the deserted Wives, or Widows and Children of such, have brought on the parishes wherein large Apprentice-houses have existed, if now we shall be permitted for any period to enjoy a state of Peace, the return of those Soldiers, either as disabled, or as unemployed and supernumerary hands, and more so, the marriages and consequent increase of Women and Children, who almost inevitably will return to burthen the Lands which have been thus parochially associated with manufacturing Apprentice-houses, the evils to the Agricultural Interests arising from the male part of these Apprentices, will perhaps be found as great, as from the female part of them.

In some instances in Derbyshire, I found the Land-owners, as in other Counties I also have done, had for years past been aware[*] of the evils in the long-run, to be apprehended to their Agricultural Tenants and Estates, from admitting "Stockingers" and other manufacturing people, Apprentices to such in particular, who were employed in the large way for trade beyond the County, from settling on their Lands, on account of the unfair operation of the law and system of Parish Settlements, in such cases: and where, owing to the Lands in particular Parishes or Townships, belonging wholly to Proprietors of this way of thinking, or who, from other motives, have refused to grant

---

[*] Like the Duke of Rutland and others in Leicestershire, as is mentioned in p. 324 of Mr. William Pitt's Report.

Leases for erecting Manufacturing Establishments, I cannot doubt, but a minute and extended inquiry and comparison (such as I have hinted at in page 477) into the present state and future prospects, of places from which such exclusions have been made, and others where they have operated only in a partial way, or not at all, will be found to shew, the very injurious operation on the Rents and permanent Interests of the Land-owner, of the system under which so many fortunes have been rapidly made, by the considerable Manufacturers, as hinted in p 40 of Volume II.

It has appeared to me, that many sensible Persons, who hold contrary opinions to those I have ventured to express above, have been too much influenced in drawing their conclusions, by the avidity with which Manufacturers are mostly found, to invest their acquired wealth in the purchase of Lands around them, at higher prices than other purchasers could be found to give, although previously saddled with very numerous settlements of Cotton-mill Girls and Boys growing up, or with families from such or other Manufacturing Labourers, who had been previously attached to these Lands, by the Parochial Laws.

Whereas it would, I think, appear on a sufficiently enlarged view of the case, that the fact of successful Manufacturers, having in a few years past, become possessed of a considerable part of the Lands surrounding their Works, of nearly all of those which have been obtainable, has arisen, *first*, from the gradually declining state and property of the Landed Interest, obliging the Yeomen and lesser Land-owners to sell their Lands; *second*, to the aversion, joined to the inability of most of the great Land-owners, and others whose views were limited to incomes from Rents, to possess them-

themselves of Lands so incumbered and circumstanced: and *lastly*, that the wealthy Manufacturers, in making such purchases, contemplated, and had it in their power, to pursue schemes of varied improvement, and to adopt measures to counteract, in a degree at least, the ruinous tendency of the Poor-laws on Landed Property, thus over-peopled by manufacturing Families.

Accordingly it will be seen, that many of the most striking agricultural and other rural Improvements mentioned in these Volumes, have been achieved, (much to their credit) by Manufacturing Proprietors, bending their attention and applying their Capitals to the improvement of their newly acquired Lands, on the same spirited and systematic plans, as had previously succeeded in their trading concerns: and to me it seemed pretty evident, that the Manufacturing Establishments of the owners of the surrounding Lands, were managed considerably different from those of others, and so as *to limit child labour to the young of the resident population*, or nearly so, and as much as possible, in the various departments of their Spinning Mills, and in other species of Manufacture, which have been introduced, *to find steady employ for the individuals of all ages*, who inhabit the comfortable Houses with Gardens, which have been provided on the Estates of these Manufacturers, see p. 21 of Vol. II.

4. *Poor-rates* increased by Manufactures?—Without doubt, every considerable increase of the English population, under the demoralizing effects of its Poor-laws, must be soon followed by an increase in the actual amount, raised for relief of the Poor among them; but in the first and increasing stages of Manufacturing Establishments, it has often, for a time happened, that

that the increased new Rents of Lands, and the Assessments upon the Manufactories themselves, have prevented any considerable increase in the pound-rate of the poor-levies from appearing, and the improved price, or at least the increase of demand on the spot, for almost every article of Farming produce, have prevented the Farmer, and of course his Landlord, from feeling the burthen of the increase of Poor-rates.

Yet, ere long, even if the manufactory only becomes stationary, as to success, much more so if it begins at all to decline, in steadiness or quantum of employ, the Poor-rates begin rapidly to increase: such at least has been the representations made to me, in every place where Manufactories have some years existed, and I have inquired on this head.

In the singular case of Glossop, mentioned p. 498, it was mentioned, that on the first erection of a Cotton-mill in that parish, in 1784, the Poor-rates were about 1s. 6d. in the pound, on the actual Rents of the Lands; from which they gradually increased, as the Factories increased, until about 1807 or 1808, when they were 6s.; and even after a new letting of all the Lands had taken place, with a considerable increase, the Farmers in this parish were, in 1809 paying 5s. in the pound on their Rents, to the Overseers! In Holbrook Township it was mentioned to me, that the introduction of several " Stockingers" a few years before, had greatly increased their chargeable Poor, and the rates for their maintenance.

In Belper Township, the Poor's Rates had risen to 20s. in the pound on the old Rents, as is mentioned in Vol. II. p. 34, and to near 16s. on the improved Rents, of very highly improved Lands, and greatly increased and improved Buildings, the latter of which
were

were fully rated, as but seldom I believe happens, in proportion to Lands surrounding them.

To this extreme case of Poor-rate burthens, in Belper, a considerable manufacture of *Nails*, of many years standing, I believe, appears principally to have contributed, through the means of one of the most mischievous of the operations of the late Apprentice Laws, which were thought here to authorise, the lowest rank of Master Nailers, and even their working Journeymen of the lowest description, to take each as many Apprentices as they pleased, and to restrain any others from working at this very simple Trade*. Among this fraternity of working Nailers, such a system of combination was established and kept up†, as enabled expert

---

* See Resolutions to this effect, entered into by the Nailers of Horsley Woodhouse, in March, 1814, in the " Monthly Magazine," Vol. 37, p. 274.

† Under this system, the more considerable Master Nailers, have not been able to execute their orders, whenever trade was brisk, without permitting all their best Hands *to draw Money on account, on the Wednesday Evening*, although frequently, " Saint Monday," " Saint Tuesday," and, perhaps, " Saint Wednesday" also, had been religiously worshipped in the Ale-house, and few, if any, Nails had then been made; and when Saturday night came, a part only of the Wednesday advance could be set off, without instantly losing the Man; and the same again next week, and so on, until many of the best Nailers were 20*l.* in debt to their Masters: which debt, although not the least intention appeared of ever discharging it, was made the pretence, to any other Master who happened to have pressing orders for goods, to advance the first Master's debt, and a Guinea or sometimes more to the Nailer in debt, to be spent " in a good drunken bout," before he would lift a hammer towards executing such orders!

The Cutlers of Sheffield, in Yorkshire, appear to have acted still further on this monstrous system, and nearly all the expert Journeymen were represented, to be thus held to their Masters by debts, varying from 10*l.* to 50*l.*; and that when different Masters received pressing or large

expert workmen among them, to earn a shilling or more an hour; notwithstanding which, the Families of the Nailers appeared to me, very poor and distressed, when compared with the Cotton and other Manufacturing Families of the place.

SECT. IX.—COMMERCE.

ALTHO' Derbyshire, as an Inland County, is inferior to many of the Maritime Counties, in its extent of what may properly be denominated Commerce, yet the *Trent*, and *Trent and Mersey* Navigations crossing it, and the several other Canal Navigations and Rail-ways which enter it, and good Turnpike Roads, which cross it in various directions, (as shown in Plate III. and described p. 208 to 456), confer on this County a respectable share of those advantages, which result locally, from the general Trade and transit of Commodities, of the Kingdom at large: besides the Trade connected with the export from the County, of the very varied articles, the produce of its own Manufacturing, Mining, and Agricultural industry, which are enumerated pages 479 to 496, the import of raw mate-

---

large orders for Cutlery at the same time, a sort of general bidding often took place among them, as to *who would advance or buy up the largest debt*, to obtain a Man, and at the same time increase it, by a fresh advance, and allow ample time for *the spending of the same*, without grumbling! Ere this, the patent Nail-making Machines of Birmingham, introduced from America, and the alterations in the Apprentice Laws, have, I trust, in a considerable degree, brought these Knights of the Hammer to reason: but not, I fear, without imposing fresh and grievous parochial burthens on the Lands, on which by Apprenticeship they gained the privileges, which have been thus so much abused, as above.

rials

rials for these Manufactories, (which comparatively speaking are but few), and the import of Groceries, Wines, and other articles of foreign growth, and of a few manufactured goods, and articles from the adjoining and other Counties.

Respecting the Commerce and Trade of the County, nothing material occurred, for me to note, except as to the *circulating medium* thereof being, almost entirely *promissory Paper**, of private Bankers resident within the County, and near its borders, and of the Bank of England. Among the persons with whom I had money transactions while on this Survey, and numerous others with whom I conversed on the subject, as one of considerable importance, I found a very decided preference given, by receivers or changers of Notes, to those of the local Country Bankers, above those of the Bank of England.

On inquiring the reason of this preference, most of the persons alluded to, allodged, the vastly greater proportion of *forged* Bank-Notes in general circulation, than of Country ones in local circulation, their own inability to distinguish these from genuine ones; and to which several added, that they understood it to be

---

* On this head Mr. James Dowland informed me, that the receipt of the Rents of a large Estate in 1794 and 1795, was, in Paper and Specie, in the proportion of 395 of the former to 599 of the latter: that in the 10 years that followed, Specie had so much decreased, that the Receipts from the same Estate in the years 1804 and 1805, were in the proportion of 952 in Paper and 42 in Money: shewing an increase of more than 56 per cent. of *Paper*, in the whole circulating medium of the County, in this short period: and I think it probable, that the increase in amount of *Paper acknowledging Debts instead of paying them*, was still more rapid, in several of the years which followed this period. See the *increased price* of Wheat, Timber and Bark, mentioned, p. 226 of Vol. II.

the practice of the Bank, *to detain all Notes which they alledged to be forged*, thereby depriving the persons to whom they belonged at the time of presenting them, of the just right they had, *of returning these Notes*, and demanding repayment, of the persons they had taken them from, and could prove the same, by having noted the transaction on the Note, or by a proper description of it elsewhere, as Bankers do in their Books: and further, that they understood, in the numerous stoppages of Notes which took place at the Bank, the particular grounds or reasons on which Notes were condemned and *taken from the parties*, were rarely if ever stated; such, for instance, as that already the genuine Note had come in, of which the imitation was now stopped, nor were *the differences of such two Notes* attempted to be explained to the parties; whereas, the Country Bankers readily granted every required information, and facility, towards the returning and retracing of the few Notes, on which they saw reasons to demur payment, and to write, "suspected to be a Forgery:" advertising, in many instances, *the particular marks* of Forged Notes, which they had discovered, or believed to have been issued.

To the above reasons, not a few Persons added, that they greatly preferred the nature of the security afforded them, by the Estates and other visible Property of the neighbouring Bankers, whose Notes they held, to the "Assets" of the Bank, consisting very principally, as they believed, of engagements by the Government to repay advances to them, out of Taxes, *yet to be raised*, if not granted, also, to which Taxes they must themselves contribute.

The names of the several *Country Banks* in the County,

County, and of the Bankers in London, on whom they usually draw, have been stated, as follows: viz.

Buxton,	Goodwin and Co.	Austen and Co., Henrietta Street.
Chesterfield,	Jebb and Co.	Down and Co., Bartholomew Lane.
Derby,	Bellairs, Son and Co.	Down and Co., Bartholomew Lane.
	Crompton and Co.	Lees and Co., Lombard Street.
	Evans and Son, (formerly),	
	Smith and Co.	Smith, Payne and Co., George Street.
Measham,	Mammatt and Co.	
Wirksworth,		

Silver *Tokens*, purporting to be of One Shilling value, issued by different Manufacturers in Sheffield and other places, had a considerable and increasing circulation, while I was examining the County; and such are not yet entirely withdrawn I believe, notwithstanding the Acts that have been passed to suppress them.

The Soho Coinage of *Copper*, was almost exclusively in circulation, and the Tower Half-pence, however thick and good, were mostly refused: without any legal authority to do so, I saw the Toll Collectors on several Roads, peremptorily refuse these still legal Coin of the Realm, and shut their Gates against the Traveller, until he produced a Sixpence or Shilling, or more probably a *Token*, for change: this last circumstance, I intended to mention in p. 243 of Vol. II., but overlooked it in my Notes.

Effects of Commerce on Agriculture:—it will be gathered from articles 3 and 4 of the last Section, and the beginning of the present are, that the Farmers and Land-owners of this County, are much more affected

in

in their interests by Manufactures than by Commerce: that both of these operate, to stimulate agricultural industry, and increase its profits, for a time at least, cannot be doubted'; yet I maintain, that the profits or gains of Manufacturers have, even in these times of prosperity to both, greatly exceeded those of Farmers, as observed p. 40 of Vol. II.; and it is the same, though in a less degree I believe, with the Commercial and Trading Class; and through the subsequent operations of the Apprentice and Poor Laws, there cannot be a doubt, I think, if the present system of Parish Settlements and maintenance should be continued, but the Land-owners will, in too many instances, have reason sorely to rue the introduction or increase of Manufactures for Foreign Trade, on their Estates.

I saw nothing during my Survey, to induce me to concur with Mr. Thomas Brown, in p. 46 of the Original 4to. Report, in thinking, that any harm to Agriculture has, or is likely to result, from Manufacturers and Traders engaging also in Farming: very obvious good effects have on the contrary appeared, in numerous instances in the County, to result, from the spirit of enterprise and habits of vigilance and observation, as well as from the Capital, which trading people in general carry with them, into any new concerns in which they embark: and I have uniformly seen, here and elsewhere, as an able Land Valuer of my acquaintance used often to remark, that however superior at first, the knowledge of the *regularly bred Farmer,* educated at home*, may have been, to the Manufacturer

---

* In my former employ as an Agent, and since, I have seen such numerous and serious evils to result, from the narrow and contracted views, the prejudices and the obstinacy of Farming Youths, who have been

turer or Trader, entering on a Farm, yet, that in most instances, the latter, from superior vigilance and industry, in studying and profiting by his own practice, and that of his most successful Farming neighbours around him, have soon overtaken, and even excelled the hereditary Farmer, in every useful point of Farming knowledge and practice.

SECT. X.—THE POOR.

RICHES or Property, in whatever they may be defined to consist, whether in Lands, Houses, Goods, accumulated Clothing, Food, Money, &c., can never in any Community, remain for a length of time so *equally divided* among all its Members, as to exempt any considerable majority of its Individuals, from almost incessant care and toil, to procure a mere sufficiency for their wants, of food, raiment, and shelter; and, such a degree of *Poverty*, as shall compel a considerable portion of the Community to labour

---

been educated upon, and never left their Father's Farm for instruction; that I have often been desirous of recommending to the Board of Agriculture, and to the more considerable Provincial Societies, to endeavour to promote the placing out of Youths designed for Farmers, for a sufficient term, under skilful and enterprising Agriculturists, like Mr. Runciman of Woburn, (whom the late Duke of Bedford brought thither almost on purpose), and several in Norfolk (where the practice greatly prevails, Farmers mutually instructing each other's Sons) and in other Counties, who take *Agricultural Pupils*, and theoretically as well as practically instruct and employ them: and for this purpose I beg to recommend, the offer of *Premiums* to those properly qualified Farmers, who shall have thus taken and instructed, the greatest number of Pupils; and the offer of Honorary Medals to those Pupils, who shall have most effectually availed themselves of the instructions received.

almost

almost constantly, either directly or indirectly, for the benefit and accommodation of the others, is as inseparable from civilized Society, as a shadow is from the substance occasioning it.

By this alone it is, that any articles or commodities acquire that general estimation or value, which constitute *Riches* in their possessors; and such Individuals only can be said to be *rich*, or possess property, who can thus command the labour of others, usually denominated *Poor*, in exchange or barter, for a part of their accumulated Riches.

The Laws of every free or well regulated Community, as expressly and fully protecting the *Poor* Man in *his right to labour for whomsoever he pleases*, and on no worse terms than the competition of other labourers and his own necessities may dictate him to agree to, as such Laws do protect the *Rich* Man in the possession of *his Property:* and as the *Rich* can no more feel comfort, or enjoy their Riches, if perchance they might exist, without the service and labour of the *Poor*, than the poor can exist without drawing food, raiment, &c., or money to purchase them, from the stores or pockets of the Rich: it follows, from these *equalities of rights and wants*, in these otherwise unequal classes of Society, that the Poor may, and actually have done so, in all free and wholesomely regulated States, *obtained such prices for their labour, as enabled them to provide for themselves, and bring up their children*, in the same class, generally, with as much comfort, as was consistent with their condition: except in a few cases, comparatively speaking, of permanent disease or lameness, of young orphan children, and of aged and childless persons.

And even, where good-natural abilities were seconded by

by active exertions, economy and prudent forecast, the door has always been left open, for Individuals of the lower class in such States, to accumulate property, and they and their descendants, if alike able and disposed, to permanently establish themselves in the lower ranks at least, of the higher class: while on the other hand, Individuals, weak, indolent, wasteful and improvident, have, as commonly, descended from the higher, and become fixed in the lower class of Society.

The very forcible appeal which individual distress and great suffering, either permanent in a few among the poor, as, already mentioned, or occasional among a greater number of them, during sickness, or particular calamities, like general scarcities of food, &c. make to *the feelings*, which the all-wise Author of Nature implants *in every breast*, as a principal characteristic of our Species; and the powerful aids which nearly all religions, that of the true Christian in particular, have given to this principle of benevolence, in all persons, each in his own sphere, from the highest to the lowest Individual; these (where no Institutions or Laws of the State counteracted), have usually provided, prompt and *sufficient relief* to the suffering Individuals.

I say, *sufficient* relief, from a firm conviction, that more than has been, and will be afforded, solely by the operation of benevolence, or *voluntary* and *discriminating charity*, cannot be given, towards alleviating distress, without counteracting itself, and producing the very evil it might be intended to remove: to think otherwise, appears to me to impeach the goodness or wisdom of Him, who implanted, and fixed the degree of force, of these feelings and principles, so evidently

dently made the only efficient and safe means for the purpose of assuaging human sufferings, whose entire removal or prevention is clearly impossible.

And it is worthy of remark, that this principle is not only more virtuous or praise-worthy, but incomparably *more efficacious*, from one individual towards another *in the same class*, than between the different classes, or, as has often been mistakenly supposed, in its exercise by the *Rich*, towards the Poor. The *Rich*, except on extraordinary occasions, most efficaciously exert their *charity*, each towards those of their own rank, who, through unavoidable misfortunes, suffer, and must otherwise sink to a lower rank, or into the class of the Poor, by giving a preference in wages and employ, to such Individuals of decayed fortunes, about themselves and families. And among the higher ranks, it ought, as formerly was more the case than at present, to be viewed by all, as among *their virtues*, to indulge as much *in state*, and in what of late it has been fashionable to designate vicious *luxuries*, as their properties and incomes will allow; since but few, and those very distinct parts of the luxuries which have been so much railed at in the Rich, will on a just and dispassionate view, be found really vicious, or any way prejudicial to the Community: but on the contrary it will appear, that *wants in the Rich fully commensurate with their means*, and which wants must of course be most of them artificial, and their indulgence sought only, in proportion as general approbation stamps a high value on these indulgencies; are the only means by which the Poor can maintain and avail themselves, of *their rights*, before spoken of, (p. 515) to draw for their full labour and services, voluntarily given to the Rich, a sufficient remuneration, not only for their own and children's support, and comfort suited to their rank, but for enabling the more active and virtuous Individuals of this Class, occasionally to better their condition, and the whole of them, to obey the divine dictates and injunctions, in *administering Charity toward those in their own Class*, more immediately connected with or surrounding them: so promptly, and in such ways and degrees, as to alleviate present distress, without at all relaxing the endeavours, by rigid economy and strenuous exertions in the distressed, to contribute towards, and if possible to earn their own support.

But alas! alas! as far as South Britain is concerned, this is now but an Utopian picture, which I have been drawing of Society: through ages of mental darkness, a proud and insolent Hierachy, who having usurped the places of the first and simple teachers of Christianity, had here, as elsewhere, employed themselves, their Priests, Monks and Nuns, *in preaching Charity, only to the Rich*, and at length, when death approached, enforcing their claims upon this class, by offers of sure passports to seats of bliss, which their past *Lives*, for the most part spent in War, Rapine and Violence, gave little hope of their attaining, *in barter* for portions of their Estates, to the Church, *in trust for the Poor*.

Thus the *Poor* were fatally induced to neglect, and in a degree abandon their own natural right, to *a sufficient remuneration* from the Rich, *in return for their Labour*, to answer all their real wants, and to look to the Priests for assistance, in every case of sickness or occasional distress: and in process of time, the enormous Landed property and Tithes of the Church, thus accumulated, came, so principally to be *depended on by by the Poor*, as greatly to increase their numbers, beyond the demand for labour, by the comparatively impoverished higher class, which should have maintained them, and such, as almost entirely to break and destroy all spirit of independence in this lower class.

In this state of things it was, that Henry the Eighth, having quarrelled with the Pope, lawlessly seized and confiscated the greater part of the Lands and Tithes, which superstition had thus, thro' a long period, extorted from Individuals, as expiatory of their real or imaginary crimes, and on whose produce the Poor, had thus fatally been brought to depend. And this confiscated property, of which the Clergy were Trustees for the Poor, was not only given away, or sold by the Monarch at trifling prices, to purchase acquiescence in his sacrilegious measures, but it should also seem, that *the tenure* of all the remainder of such Church property, as was suffered to remain in *the hands of the new Clergy of his own appointing*, was entirely changed, and the Clerical possessors thereof, as effectually released from any express or implied obligation, *to give to the Poor*, out of the produce of these Lands and Tithes, as the possessors of all other property in the country were.

The class now properly denominated *the Poor*, thus fatally *increased in numbers* [*], *and demoralized*, by the baneful

---

[*] No manner of doubt can remain, on the "principle of population," on which Mr. Malthus has so ably and satisfactorily written, that during several ages which preceded the Reformation, a very considerable *increase took place, in the poorer part of the population, beyond the demand for Labour, by the then state of Agricultural Improvement, and of Civilisation:* and that such *surplus* part of the population depended chiefly, and had indeed been produced by, a dependance on the Lands and Tithes which

baneful and unnatural effects of Eleemosynary distributions by the Clergy, of which they were now suddenly deprived, and left destitute, became in consequence greatly distressed; and for more than half a Century afterwards, the sufferings of this class increased, until a degree of misery was experienced, which has few parallels in history, it is believed, if impartial and full accounts of the state of the Poor in this period, had been suffered to reach us.

It does not appear, during this trying period which succeeded "the Reformation," as it was called, that any proper attempts were seriously made by the Government or the Clergy, towards *retracing the steps* which had so fatally been taken, in the preceding Centuries, with regard to *the dependance of the Poor on Eleemosynary gifts*, from others above them in the ranks of society, rather than on *their own exertions*, or on mutual kind offices amongst themselves.

---

which the Clergy held and administered for them. This dependance being removed, and distress forcing great numbers of the most ingenious and active of this *surplus people*, from being only occasionally employed, in Agricultural or rural Labours, in the vicinities of the Monasteries, principally, to seek more constant employ, in the Trades and occupations carried on in the Towns and Cities:—and to me it appears clear, that *the competition* thus created, among workmen, in the "crafts, mysteries, and occupations" of the Towns, having lowered, and threatening much farther to *lower the Wages earned in these Trades*, &c., was in reality their reason for applying for, and obtaining, that code of *Apprentice Laws* (in the 5th of Elizabeth) which has since brought so many and great evils on the Country, and on this County in particular, as mentioned pp. 502, 508, &c. I cannot view it otherwise than as *a pretence* then set up, (altho' since so often and gravely repeated and argued on) that securing *the perfection of our fabrics*, and guarding *the public from imposition*, were the motives, for desiring and passing these Laws, whose injustice and cruelty towards the surplus population I have mentioned, cannot be denied, or a doubt entertained, but they contributed largely towards the rural misery, which followed their introduction.

The

The propriety, and even the moral obligation on the higher ranks, to afford, in their Houses and Establishments and on their Estates, *sufficient employ and wages* to the industrious and economical Poor around them, *to enable such to support themselves* and their families and dependants, does not seem to have been enforced, from the pulpit or otherwise, and the perfect reasonableness and right, of the virtuous part of the Poor, to expect this at their hands, enforced on the higher class, as ought incessantly to have been done. On the contrary, it should seem, that senseless railings began more than ever to be indulged in, against this class, for spending in luxurious living and great or splendid establishments, what the Clergy contended, or broadly hinted, *thro' them*, as formerly, *they ought to give in Charity*, towards the support of the Poor.

Elizabeth's Government, instead of devising and trying every possible means, to induce and enforce *the spending, in rural and domestic Improvements, and in Luxuries contributing to the employ of the people,* by all the higher class, *as large a portion of their incomes and surplus wealth, as could with prudence be done;* yet leaving to the parties, the perfect liberty of *selecting the Individuals* in the lower class, and apportioning their pay, who were thus intended to be benefited: adopted the fatal and absurd measures, of erecting every Parish into a sort of *Poor Corporation:* assigning the conditions and terms, under which, every person *already poor*, or who might thereafter *by any means become so*, should *be free of*, or as it was termed, *legally settled* in, some one of these new Corporations: the officers of all which Corporations were, with monstrous absurdity, required *to find work* for all the Members of their Corporation, who should apply for it!

it! and, for enabling the Officers to pay for the Labour they were thus *commanded to create,* however uncalled for, or impracticable (from local circumstances) the finding of the same might be, an indefinite portion (even to the whole of them) of the Rents of Lands and Houses within each Corporation, were placed at the disposal of its Officers! who were likewise constituted Guardians, or *Foster Fathers,* to all distressed Orphans, or Bastard *Children,* and charged with their maintenance and Apprenticeship (with only inadequate remedies against the criminal parents of the latter): and which Officers, to crown the absurdity of the whole, were required to execute all these important duties and trusts *without fee or reward!*

Buildings were accordingly erected or hired, and fitted up and stocked with tools and raw Materials, in very numerous Parishes of the Kingdom, at great expences to the occupiers of Lands and Houses, under the denomination of "*Work-houses,*" for *employing* and in some instances lodging and boarding also, the Poor, as above mentioned. Yet notwithstanding the restrictions of *the Apprentice Laws* (which already had so largely contributed to the supposed necessity of these *Poor and Work-house Laws,* as mentioned p. 520 Note) were removed, as far as concerned the free introduction and working at any kind of Trade, in these Parish Work-houses, it was soon found, that positive loss * rather than gain, towards the expences of
mainte-

---

* That this result took place in numerous instances, even before the Poor, owing to the extremity of their previous sufferings, manifested *unwillingness to work,* I have no doubt; and the Labouring in these Houses, being in consequence given up, and food, or money, at first sparingly and by stealth, and at length by permission, and even by compulsion

maintenance, were incurred to the Parishes, in these unnatural attempts at manufacturing *(without skill, or interest in the result,* by the workmen and their directors) Goods, which the markets of the Country did not require!

The Law being imperative, as to providing and

---

pulsion of the Law, given to *the unemployed Poor:* this evil example, soon so paralized the exertions of the 'Poor, in the remaining Work-houses, that similar *losses* occurred in all: and it would I think be found, if the facts had been recorded and preserved, that not a single Parish *Work-house* in the Kingdom has, from its establishment in Elizabeth's reign till now, continued to have *work* performed in it; but all these pretended "Workhouses" have, without exception, at times at least, degenerated into *haunts of idleness, vice and misery.* And such only they would have all permanently remained, under the monstrous and demoralizing system that gave them birth; if at different periods, the pressure of the poorrates had not become intolerable, and more than ordinarily active Parish Officers, had not in consequence introduced, and for a time enforced *work* in the House, much more with the view, *of deterring the idle* and slothful from entering it, than from any other prospect of advantage.

Within the present age, immense thought and pains were bestowed, on the establishment of district *"Houses of Industry,"* in lieu of separate Parish Work-houses, and from which establishments, the most important advantages were anticipated: the inherent evils of the system, seem, however, to prevail over every expedient, for on consulting the Table facing p. 34, in Vol. II. it will appear, from cols. 10, 9, and 8, that in all England and Wales there were in that year (1803), 83,468 persons permanently relieved in Work-houses and Houses of Industry, the gross amount of whose *earnings*, after deducting raw-materials, amounted but to 31,153*l.*, or 7*s.* 5¼*d.* *each person per annum!* In Derbyshire 462 Individuals, earned 432*l.* or 18*s.* 8½*d.* each person, per annum, in its Poor Houses:—most inadequate earnings, these!

Every attempt at supplying the Poor with *work* at their own Houses, have proved alike inefficacious; it is mentioned in the Parliamentary returns, that the Parish Officers of Aston on Trent, were in the habit of purchasing flax for the Poor to spin at home; by which an average *loss* of 10*l.* per annum was sustained, altho' in 1803, the total of these earnings was but 15*l.* 0*s.* 3*d.*!

paying

paying for work, in either food or money, to those who applied for it; rather than incur *loss* thereby, as well as to save the enormous trouble thereby occasioned to the Parish Officers, *money allowances* came soon to be very generally made, *instead of providing work;* and and thus, ere long, even a worse and *more demoralizing system* prevailed, than before the Reformation: because then, however great and mischievous *the dependance* of poor persons may have been, on the usual donations at the Monastery Gates, rather than on their own labour and economy; they well knew, that the Priests distributing these Alms, had *the uncontroled power of withholding them altogether*, at least from immoral, disorderly, idle, or undeserving persons; and were no less sensible, that the whole amount to be distributed (after providing for the Priests, Monks, Nuns, &c.) *was limited*, and of no very great amount, in any particular case.

But now, on the contrary, when *the Law of the Land* directed Parish Officers, *almost without any discretion, to relieve distress, however occasioned! without any limit in total amount, short of the total income from Lands and Houses!!* the dependance placed on *parish relief*, naturally became much greater, and *more mischievous and demoralizing in its effects* on the Poor, than that formerly placed on *Church relief:* Their rate of increase, both of persons *becoming* and of others *born* poor, was proportionably increased, beyond the possible demand for their labour, and the mass of vice and wretchedness among this class was, even in a higher degree increased.

At first, the Law imposed some wholesome restraints on the persons whom its provisions were so directly calculated to increase in numbers, and debase; such as, confinement to *Residence* (except on Certificate granted) within the Parish of their settlement, the wearing of a *Badge*, while receiving relief, &c. But one after another, nearly all these have been removed, principally within the last quarter of a century, in which I have been concerned to observe, scarce a Session' of Parliament pass, without some new Act, either repealing former checks, or giving, according to my views of the subject, greater extent and force, to the overwhelming mischiefs to Society, which the Poor-Law system is hastening to produce, as observed in p. 193.

Of late years nothing has been more common, than to hear persons in all ranks of life, expatiating on *the indefinite right which all the Poor have, to be maintained* (in a certain degree of comfort, and without degradation) *under the Poor-law System: a right* not less sacredly theirs, it is preposterously contended, than that fundamental one of Society, *by which property is secured to its possessors*; or rather, as it ought to have been stated by those persons, *by which that share of property* is secured to those *called* Rich, which will remain to them, after *the indefinite claims of the Poor upon it, are satisfied!*

My views of *the Rights of the Poor*, as explained at the beginning of this Section, have long been very different, and, as I conceive, not only more consistent with justice, but in every moral as well as interested sense, *more advantageous to the Poor themselves:* many of this class, the agricultural and rural part of it in particular, as mentioned towards the bottom of page 193, have been so far deluded by the Poor-law Advocates, as to let go *the substance*, of adequate prices for their labour, and grasp at *the shadow* of parish dependance.

Since

Since it will not be denied, that nothing has been more common, through many years past, in most or every agricultural district of the kingdom, than for the distressed Labourer to make out a case to his Master, of the utter impossibility of subsisting himself and family, on his wages*; but instead of this being followed,

---

* See Mr. William Pitt's calculations on this head, in the Leicester Report, p. 327; and one regarding the Poor in Westmoreland, in 1775 and 1813, in Monthly Mag., Vol. 36, p. 171. In p. 193 I have briefly endeavoured, to decry the practice, of associations of Landlords and Farmers in our Agricultural Societies, offering *Premiums* to those rural Labourers, who, by *extraordinary good fortune* (for seldom with these, could it have been otherwise occasioned) had brought up the *largest families*, without parochial assistance! How plainly do such Premiums admit the facts stated in the text, by signalizing those few Individuals, who may have been able to bear up against *difficulties*, which ought not to have borne on their whole class. To argue, that it is virtue, industry, and economy only, that is intended to be rewarded, is in effect to admit, that the demoralizing effects of the Poor-law system, have spread general and almost universal depravity, among even this Class, the least likely in the whole community, to have been thus deeply contaminated. Besides, what is made *the test of merit* in these cases; but the having *most contributed* to increase the evils of *a redundant poor population!* maugre all that common-sense has dictated and Malthus written on the subject.

I would beg here to suggest, that the Societies alluded to, might do most important service to the Country, by changing these *Premiums*, for others of much larger amounts, to such heads of Poor Families, as kept during a whole year, the most minute, regular and satisfactory *Registers and account*, and delivered them, intelligibly written, to the Society, with proper Certificates of respectable neighbours (who had carefully perused and satisfied themselves of the truth and accuracy of the particulars stated): on the one side, of the appropriation of all *the time of each Individual* of the Family (their names, ages, and descriptions being previously stated) and the precise *wages or remunerations* received for the same, and of *all other incomings*, whether of commodities or money, from charitable persons, or *the Parish*, (this last, not in any degree lessening the claims to a Premium); and on the other side, a minute account of

every

followed, either by an advance of the wages, or the Labourer's seeking some other employ, it has been thus answered by the Master, viz. " I am aware of this,

every outlay for food, fuel, clothing, rent, &c. &c.: exactly distinguishing *the quantity, price and amount of every article purchased or received* into the family store: the size of Garden (if any), and kind and quantity of produce therefrom: the weight of Pork, &c. from any pigs kept and killed for the family use: the quantity of Milk daily, from any cow that may be kept; with notes or particulars, as to the family economy, with regard to using Meat, Flour, Potatoes, Milk, &c. received; viz. whether Broth or Soups were generally made, whether Bread was baked, and Pies, or Puddings, Milk, or other Porridge was usually made; whether Beer, and what quantities were brewed, &c.

In short, as complete a record of the family economy as possible should be required, and explained by written particulars, and forms or specimens, which the Secretary should furnish, to such persons as applied, and came well recommended in writing, as persons able, willing, and likely to compete properly for such Premiums.

Abstracts of these Registers, when received, might be prepared for the use of the Judges appointed to decide the Premiums, and read to the Society, with their award; and then, it is part of my wish, that all the original Registers should be transmitted to the Board of Agriculture, and there preserved: and that on a sufficient number of such being collected, classified abstracts, and average results, of all their more important items, should be printed, with the remarks and reflections they would suggest to the Compiler, as most important documents for Parliament, towards forming an opinion, as to what *can*, and what ought *to be done*, with regard to removing *the necessity of dependance on the Poor-rates*, by that large and important rank of our Poor class, the Agricultural and Rural Labourers and their Families. If this could be done, and steps taken *to check*, rather than increase our *poor population*, particularly by an entire change of our *Bastardy Laws*, and the encouragement of emigration: with a more effectual opening of the door to competition, for the labouring classes, in Corporate Towns: then, I think, would the most important obstacles be removed, towards an entire abolition of the present *Poor-laws;* which must certainly, otherwise, and at no distant period, produce a Revolution in this Island, followed by a night of more than Vandal and Gothic Barbarism!

but

but you must do with it as long as you can, and then *come to the Parish for assistance!*" and which unreasonable and degrading alternative, the Labourer has too generally embraced, without further effort!

Seldom so with the Journeymen of established Trades, who doubly entrenched behind the *Apprentice-laws* and the *Poor-laws*, have, most of them, too commonly and repeatedly *combined*, to take advantage, collectively, of the most urgent periods of their Master's business, for *advancing their wages*, (see instances in p. 508), not only to keep pace with the prices of provisions and necessaries, but also, in too many instances, in great Towns in particular, for continuing, and even increasing the means, of spending one day or more of the week, besides Sunday, in idleness, if not in drunkenness also, and of obtaining the means of various indulgencies throughout the week, to which the Agricultural and Rural Labourers are almost entirely strangers. And yet, if an accurately divided account of the expenditures by Parishes, to relieve *Agricultural and Rural* Persons or their Families, and those who have been connected with the *Trades and Commerce* of the Towns, I doubt not of its appearing, that a much larger burthen, in proportion to the total numbers of the latter, is thrown on the Poor-rates, than by the former, notwithstanding the very superior advantages in point of wages, which the Journeymen and town Labourers have enjoyed. Because, alas! under the demoralizing operation of Poor-law dependance, *high wages*, are only like fuel, added to an already devastating *fire*. But it is time that I now proceed to the particular heads of enquiry prescribed for this Section, in the Plan for these Reports, viz.

1. *The state of the Poor* in Derbyshire appeared to
me,

To face page 529.]

1.
Name and Description of each Parish or Place.

Blackwell, (in Bakewell P.) Township
Alfreton, and Alfreton Outseats, (1) - - - Parish
Castleton - - - - Par
Eaton, Little, (in St. Alkmund P.) - - Township
Breaston, (in Wilne? P.) Libe
Alvaston - - - - Pari
Boulton - - - - Pari
Pinxton - - - - Par

Curbar, (in Bakewell P.) Ham
Pilsley, (in Edensor P.) Towns
Pentrich (4), (in Pentrich P.) Township
Blackwell - - - - Par
Mapperley, (in Kirk Hallam P.) Township
Dale Abbey - - - Paris

Newton Grange (5), (in Ashburne P.) - - - Liberty
Sturston (6), (in Ashburne P.) Hamle
Clifton and Compton, (in Ashburne P.) - - Township
Ashburne, (in Ashburne P.) Township
Hartle, (in Bakewell P.) Towns
Codnor Park and Castle Pari
Denby (7) - - - - Par
Hopwell, (in Draycot? P.) Township

11.		12.	13.	14.	15.	16.	17.	
Number of Children or Persons Relieved permanently out of the House; and of any other Children maintained out of the House; distinguishing,		Number of Persons Relieved occasionally.	Number of Persons (included in the three preceding Columns, 10, 11, 12) above 60 Years of Age, or disabled from Labour by permanent Illness or other Infirmity.	Number of Persons Relieved, not being Parishioners.	Number of Friendly Societies who hold their usual Meetings within each Parish or Place.	Number of Members in the said Societies.	Number of Children in Schools of Industry.	Rated Rental, deduced from Columns 2 and 3.
Under Five Years of Age.	From Five to Fourteen Years of Age.							£
—	1	—	1	3	—	—	—	28
30	22	35	25	10	3	409	—	1,184
3	6	—	10	—	2	291	—	243
11	11	10	10	3	(3) 2	90	—	422
8	8	9	4	—	2	132	—	459
4	5	33	2	—	—	—	—	461
1	—	19	3	—	—	—	—	201
—	—	10	8	4	1	27	—	496
57	53							34,94
4	6	4	3	—	—	—	2	348
—	—	6	4	—	—	—	—	280
7	18	—	6	1	1	57	—	1,250
7	4	7	8	2	—	—	—	1,318
5	4	1	4	—	1	35	—	625
—	—	3	8	—	—	—	—	1,018
23	32							4,839
—	—	—	—	20	—	—	—	524
4	6	7	2	3	—	—	—	4,978
3	1	72	13	—	—	—	—	6,441
11	15	29	18	6	1	61	—	17,874
1	1	2	—	—	—	—	—	806
7	10	13	4	—	—	—	—	6,488
4	4	53	12	—	(8) 4	226	—	14,818
4	7	2	6	—	—	—	—	4,290
34	44							56,271

me, after all the attention and enquiry I could afford to it, to be as good, or perhaps rather better, than in the surrounding Counties, and certainly, I think, in a considerable degree before the average of England and Wales: few able persons seemed to be unemployed, except during temporary stagnations of the Manufactories in which they might be employed, or in the Parishes which had been deluged with a surplus population, from manufacturing Apprentices, as observed, pp. 502, 507.

Going the Rounds, or *House-row*, as the practice was here called, by order of the Overseers, and receiving from him a part of their wages, by those who could not otherwise obtain work, was formerly pretty much practised, it would seem; but its inconveniences, and mischievous operation on the industry and morals of the poor (as Mr. Batchelor has well observed in p. 609 of the Bedfordshire Report), have caused it to be now almost discontinued here.

The state of *Education* among the labouring Poor, appeared to me fully equal to, or better, than among the Poor of the southern Counties; cheap Village *Day-Schools** were seen tolerably distributed thro' the County: several excellent and well-conducted *Sunday-Schools* were in activity. I have since heard, of *Lancastrian Schools* being introduced at Derby, and the same have, I hope, ere this, spread thro' the County, as one

---

* I noted *Boarding Schools* for Boys, in Ashburne, Beighton, Deer-leap, Dronfield, Heath, Norton, and Quarndon.

Free Schools, in Bretby 2, Chesterfield, Etwall, Netherthorp, Newbold, Repton, Risley, West Hallam, Whitfield, and Wirksworth.

The *Schools of Industry*, mentioned in col. 17, facing p. 34 of Vol. II. and their number of children, are stated as follows, viz. Alt-Hucknal 20, Baslow 27, Barrow 10, Bolsover 6, Bredsall 16, Bretby 24, Cole-Aston 6, Cubley 20, Curbar 2, Great Hucklow 51, Ilkeston 37, Killamarsh 26, South-Winfield 2, and Stainsby 20.

of the most efficient means, if seconded by simple and proper *Tracts* and *Books*, put into their hands, for fitting and preparing the rising generation, for emancipation from *the moral slavery of Poor-law dependance*, and its attendant vices and misery.

Bibles, Testaments, and Prayer-books, and what are usually denominated Religious Tracts, however profusely distributed among the Poor, detailing the history and the institutions of Nations and times *altogether different from our own*, the whole couched in such poetic and figurative language, as to be scarcely within their comprehension, can do little or nothing towards this great and desirable end: compared with plain and unadorned statements of the actual and relative conditions and duties of Men in *civilized Society;* not as Society existed under the Mosaic Government, or even among the Apostles and persecuted first Christians, much less in the Feudal, Monkish, and superstitious ages which succeeded, and now, happily, are *here* passed away, but *as things now stand*, among us; and above all, *as they ought and would stand*, and undoubtedly conduce to more general and universal happiness, if human Laws, so mistakenly extolled, did not destroy and counteract in the British *Poor*, those feelings and motives, and remove those just *fears* of evils *immediately to follow* from vice, or from any material disregards of, the divinely excellent dictates afforded by *Reason* and *Conscience*, which the all-bountiful and wise Author of our *Being*, *has seen sufficient*, and made conducive in a superlative degree, to any and all other means, for producing individual and social happiness*.

2. *Annual*

---

* These great and desirable ends, would, I am confident, judging from the workings of my own mind through an active life, and the candid

2. *Annual Receipt and Expenditure* on behalf of the *Poor*. In Sect. V. of Chap. IV. Vol. II. p. 32, I have given some account of the great undertaking, in the year 1803, of collecting and printing a detailed account, of the annual expence of maintaining the Poor of England and Wales, at that time, as also in the years 1784 and 1776. I have there lamented, and cannot help repeating my regret, at not being able to give herein, the whole of these interesting documents which concern the County of Derby †. I gave, however, in a long page to fold out facing p. 34, the totals and averages for each of the seven Hundreds, &c. of the County, and the whole of it, and of England and of Wales, and their totals: and by way of further specimen and illustration,

---

did declarations and confessions of some scores in most ranks in life, whom I have anxiously consulted hereon, and what I have seen in the conduct of others, be almost infinitely more produced, by plain and practical illustrations, essays, and sermons, on the well-known, but not sufficiently understood or enforced text, from the Book of *Reason*, " Honesty is the best policy," than by a course of elaborate *Sermons* (accompanied by the recital of as many strings or rosarys of Prayers) on every fifteen words in the Old and New Testaments, save, these, " Whatsoever ye would that Men should do to you, do ye even so to them ;" or, " As ye would that Men should do to you, do ye also to them likewise." Divine injunctions! which recognize *that principal spring of human actions*, OUR OWN INTEREST, *when rightly and fully understood*, as the test and measure of virtue!

† In page 33 of Vol. II. I omitted to mention in a Note, that besides the sum of 78,219*l.* expended under the *Poor-laws*, in Derbyshire, in the year to Easter 1803, a sum considerably exceeding 5000*l.* it may be presumed, was in that year distributed " in charitable *donations* for the benefit of poor persons" in Derbyshire; the produce of Rent-charges and *Rents* of Lands, which charitable persons had at different modern periods devised, or left for their benefit. The returns to Parliament in 1786 (of which an abstract is given by Mr. Poole in p. 715), shewing, that 1,087*l.* 15*s.* 7*d.* in money and 3,864*l.* 14*s.* 2*d.* in Rents of Lands, had in that year been received and distributed, as above. The total of these annual donations in England and Wales, being then 252,475*l.*

of the subjects contained in the large and important Volume whence they were extracted, I have prepared another long page, to fold out facing page 529.

In this latter page, the titles at top are almost exactly the same as in the volume quoted, and the particulars below them, are exactly so (except the last column, which I have calculated), for 22 Parishes and Places, which have been selected, with regard to the *Pound-rate*, to which, according to these Returns, the expences of the Poor in 1803, severally amounted, viz. 1st, Eight places wherein the *largest* pound-rates are stated in this County, viz. from 20s. 0¼d. to 11s., averaging 15s. 9¼d.; 2d, Six places the nearest to the general *average* rate of the County, viz. from 4s. 2d. to 4s.‡ averaging 4s. 1d., as in Vol. II. p. 34; and 3d, Eight other places, wherein the *smallest* pound-rates in the County are stated, viz. from 8¼d. to 6d., averaging 6¼d. in the pound.

My objects in selecting these Places were, to enquire and shew, as far as these particulars might be capable of shewing it, what are the local and other circumstances, which have occasioned the pound-rate to be extraordinarily high in some places, of an average amount in others, and extraordinarily low in others? But before proceeding to any considerations that arise out of this Table, it will be proper here, to copy the eight explanatory *notes* which occur in the volume, upon some or other of the particulars relating to the places here selected, and for which there was not room on my folding-leaf facing page 529, viz.

---

‡ Several other places, whose Rates are stated to be 4s. in the pound, occur in the Tables; the two that I have selected, are the first that occur, in different Hundreds.

(1.) The

(1.) The Out-seats of Alfreton, include the Hamlets of *Swanwick* and *Green-hill Lane, Somercotes*, Over-Birchwood and Riddings.

(2.) Including 318*l*. 18*s*. 3½*d*. expended towards building and furnishing the Work-house (for Alfreton, situate near Swanwick).

(3.) One of these is a *Female* Friendly Society, consisting of twenty members.

(4.) Pentrich Township, joins in the Work-house at Crich.

(5.) In Newton-grange Liberty, there is only one Farm-house.

(6.) Sturston Hamlet, joins in a Work-house: probably in that of Doveridge.

(7.) Denby parish, joins in the Work-house at Crich.

(8.) One of these is a *Female* Friendly Society, consisting of sixty members.

There must undoubtedly be some mistake, I think, in *Blackwell* in Bakewell, being mentioned in the parliamentary return, as *highest rated to the Poor*, of any place in the County! I heard or saw nothing to favour such a statement, when there in 1808: I think it in the highest degree improbable, that a Rental of 28*l*. only, could have been there rated to the Poor! and that the rate was in 1803, more probably two shillings in the pound than 20s., I venture to think. I regret exceedingly that I had not access to, nor had I indeed seen, this parliamentary volume of Poor Returns, at the time of preparing the sketch for Travelling Memorandums on my Survey, mentioned in the Note on p. 477: because, in such case, the *rates per Pound*, of poor-expences, in these eight *highest* and eight *lowest* mentioned places (and perhaps of more places within these limits of rates), being noted therein, against the names of the several places; on my visiting the places, I should of course have enquired into the correctness of such statements, and into the circumstances which seemed to operate, in producing any really extreme or unexpected Poor-rate.

Unfortunately also, it was not until a few days past

(March

(March 1816), that I selected out the returns of these places from the volume, and when time would not permit, even of my corresponding with persons in or near these places, to ask explanations: and yet, seeing reason, from all which I know or can recollect on the subject, and particularly from considering the last column of my Table, to suspect strongly, that very few if any of the first mentioned eight places therein, are among those which are in reality most *highly rated* to or burthened by the Poor; but, that *the appearance of* such *very high-rates* in the Pound, have in a great measure arisen, from *nominal Rents**, very greatly below the real ones, being used in rating or calculating the poundrates at these places. So also, thinking, that but few of the last mentioned eight places would in reality be found, among *the lowest rated* or least burthened places, by chargeable Poor, but that here, *real Rents* have in general been rated or calculated upon, instead of nominal ones, as in the former cases: in order to throw all the light in my power on this important subject, I have collected out materials, and with some pains and care, have calculated the following: viz.

---

* Since writing the last paragraph in p. 28 of Vol. II., I have seen extracts from a Return made to Parliament from the Tax-Office, of the Rack-Rents on which the Property Tax was paid, in the several Counties, in 1811; whence it appears, that a Rental (including Tithes) of 640,701*l.* was thus assessed in Derbyshire, and that the sum of 378,584*l.* in my last column facing p. 34 Vol. II., is but 59 per cent. thereon! In some other Counties, the nominal Rentals rated to the Poor, are in lower proportion to the real Rents, I believe.

A Table

A Table of Particulars and Deductions, regarding the Employ or Occupation of the Population, and number of the Poor; from the folding Table in Sect. 11. *and those facing* p. 529, *and* p. 34 *of* Vol. II.

1.	2.	3.	4.	5.	6.	7.
Hundreds of the County, and other collected parts of it, and of Great Britain.	Total Population in 1801.	Number of Persons and hundredths, to each Agriculturist in 1808.	Number of Persons and hundredths, to each Manufacturer, &c. in 1801.	Total Number of permanent Paupers in 1803.	Persons and hundredths, to each permanent Pauper, in 1803.	Rated Rental to each permanent Pauper, in 1803.
						l.
Appletree Hundred	28,417	2.95	5.04	1,622	14.44	53
Borough of Derby	10,832	86.66	2.58	735	14.74	20
High Peak Hundred	32,526	8.71	3.84	1,398	23.27	37
Morleston and Litchurch ditto	24,787	4.76	3.59	1,463	16.94	35
Repton and Gresley ditto	13,643	3.17	5.58	973	14.02	46
Scarsdale ditto	37,988	5.27	4.96	1,864	20.38	47
Wirksworth Wapentake	18,394	5.37	3.51	1,082	17.00	40
County of Derby	161,587	5.06	4.08	9,137	17.69	41
Eight places in ditto, highest rated? to the Poor	4,847	7.84	.90	280	17.31	12
Six places ditto, medium ditto	2,133	5.09	4.213	95	22.45	51
Eight places ditto, lowest? ditto	4,267	5.88	2.84	180	23.71	313
Forty English Counties	8,331,434	5.46	4.66	693,289	12.02	34
Twelve Welsh Counties	541,546	2.86	10.06	41,528	13.04	13
All England and Wales	8,872,980	5.18	4.81	734,817	12.08	33

The Population Returns made to Parliament, and printed in 1801, of which an account, and a folding page of abstracts and calculations thereon, will be given in Sect. 11. of this Chapter, contain a column (my 23d) for the total *population* of each Place, and collectively for the Hundreds, Counties, &c. from which col. 2. of the

the aboveTable is taken: another column, (35th) for such of these Persons as were chiefly employed " in *Agriculture*;" and another column (39th) for such of these total persons, as were " chiefly employed in Trades, Manufactures, or Handicraft." By dividing the first mentioned numbers by the second, I have obtained the numbers in col. 3, above; and by dividing the same first numbers by those third mentioned, I have obtained the numbers in col. 4: each to the nearest second place of decimals, or hundredths of persons.

By adding the 4 numbers relating to each place in cols. 10 and 11. of the folding Tables, facing p. 34 of Vol. II. and p. 529 herein, the numbers in col. 5 of the above Table were obtained: by dividing the numbers in col. 2 by those in col. 5, col. 6 was obtained; and lastly, by dividing the numbers in th last cols. of the Tables Vol. II. p. 34, and III. p. 529, by those in col. 5 of the above Table, the last of its columns was obtained.

In consulting this Table (p. 535) it will be material to observe, that *the smaller* the numbers are found, against any place or district, in cols. 3, 4, 6, and 7, *the greater* will *the number of the Individuals* mentioned in the titles of those cols. prove, in proportion to the total numbers in cols. 2 and 5: in other words, the numbers set down in cols. 3, 4, and 6, are *denominators* of fractions, with 1 always as their numerators, considered with regard to those in col. 2*: and the numbers in col. 7, are similar denominators of fractions, with regard to col. 18, Vol. II. p. 34, and Vol. III. p. 529.

---

* Columns 41 and 43 of the folding Table to be given in Sect. 11. of this Chapter, will exhibit the proportionate parts, of the Derbyshire and of the British Population, employed in Agriculture and in Manufactures, &c. in a different form from cols. 3 and 4, in p. 535.

It may be observed with regard to Derbyshire, col. 3, that *the proportion of Agriculturists* to the whole people, is, as might have been expected, *lowest* in the Borough of Derby, where 86⅔ Individuals, reckon only 1 Farmer or Agricultural Labourer amongst them, on the average: and next so, in the High Peak Hundred (see its Places enumerated in Vol. I. pp. 80 to 82, where the proportion is one in 8¼. On the other hand, *Agriculturists most prevail* in Appletree Hundred, where rather less than every three persons, include a Farmer or rural Labourer; and next so in Repton and Gresley Hundred, where they are 1 in rather more than 3; while in Scarsdale Hundred and Wirksworth Wapentake, the proportionate number of Agriculturists are almost equal to each other, and but little less than the general *average* of the County (and of six places mentioned, facing page 529) which follow; viz. 1 Agriculturist to every 5 individuals, very nearly; and which also differs but little from the average of all England and Wales.

From considering the numbers in col. 4, it appears that *the lowest proportion of Manufacturers*, Traders, and Handicrafts, is found in Repton and Gresley, and the next lowest in Appletree Hundred, being 1 in 5¼ and 1 in 5, nearly, respectively. Belper, and some neat places in Duffield Parish, being in the latter Hundred, seem alone to occasion it to shew, a higher proportion of Manufacturers than Repton and Gresley, and occasion this class of persons, and of Agriculturists, in these Hundreds, not to exactly reverse each other, as might have been expected. So in like manner, the Borough of Derby has the *highest* proportion of Manufacturers, &c., or 1 in 2¼ nearly, and the eight places last mentioned in the Table facing p. 529 the next highest,

or

or 1 in $2\frac{2}{10}$ persons; the high proportion of Manufacturing or Trading individuals in this last case, seems entirely occasioned by Ashburne and Compton (which form indeed, but one *Town*) being included therein: High-Peak Hundred would, I think, appear to have more Manufacturers, proportionally, but its proportion of Traders are inferior, and so in a slight degree also with Wirksworth Wapentake and Morleston and Litchurch Hundred. The *average* proportion of the whole County is 1 in 4, and of England and Wales, 1 in $4\frac{3}{10}$, very nearly.

The numbers in col. 6 indicate, that the *highest proportion of Poor persons*, permanently maintained at the public expence, are in Repton and Gresley Hundred, where 1 such Pauper occurs in every 14 persons, very nearly; and the next highest in Appletree Hundred, 1 in $14\frac{4}{5}$. The lowest proportion, occurs in the last-mentioned eight places facing page 529, and in the High-Peak Hundred they average but a trifle more, viz. 1 in $23\frac{2}{10}$ and 1 in $23\frac{1}{11}$; and in Scarsdale Hundred they are 1 in $20\frac{1}{4}$: while on the average of the County, there appears 1 Pauper to $17\frac{2}{11}$ persons, and in all England and Wales, the more considerable average of 1 in 12!; and it is further observable, that this last is a considerably higher ratio of Paupers, among the people, generally, than occurs in any of the divisions of Derbyshire, in my Table, page 535.

Proceeding now to consider, and compare the numbers in col. 7, it appears, that *the highest Rental* assigned in the Poor Rate Returns *to maintain a permanent Pauper*, and consequently the *least burthened places*, occurs in the last mentioned eight places facing page 529, viz. 313*l.* rated Rental to each Pauper; whereas, the Hundred giving the highest Rental, viz. Appletree, has

has but 53*l.* rated to each of such Paupers, and 51*l.* is so rated in six places, of a medium pound-rate. On the other hand, the *lowest* rated Rental to a Pauper, or the *most burthened places*, are made to appear to be, the eight first mentioned places facing p. 529, which thus (but perhaps erroneously, see p. 534) would seem to have only 12*l.* rated to each Pauper!, and next the Borough of Derby, which thus reckons a Pauper to every 20*l.* of rated Rental. While the averages of the County, and of England and Wales, state 41*l.* and 33*l.*, respectively, to be rated to the maintenance of each permanent Pauper.

The discordance observable between the results just above mentioned, amongst themselves, and with all that might have been expected with regard to Rentals, or is indicated by the circumstances known, and mentioned throughout these volumes, all strongly confirm, what is conjectured in p. 534, as to *these Parliamentary Returns being very deficient*, towards shewing *the real proportionate burthen sustained in different Parishes* and places, by the maintenance of the Poor, compared with the *real Rentals*, or the abilities of the places to sustain such burthens.

According to these imperfect means of judging, it would seem to appear, that eight places, *the most highly burthened* by the Poor, in Derbyshire, and eight others *the least so*, differ in no sensible degree, in their proportions of *Manufacturers* (col. 4), compared with the whole population!, contrary to all which is known, and has been mentioned in pages 507, 528: and altho' *the Agriculturists* in these two sets of eight places (col. 3), are by no means equal, but in the proportion of 2 to 3, very nearly.

Again, it would seem, that the proportion of permanent

manent *Paupers* in these two highest and lowest burthened sets of eight places, compared with the whole people therein, are as 1.37 to 1.; and yet compared with the whole rated Rentals therein, they are as 26.08 to 1.; and the average *pound-rates*, calculated in p 532, viz. 15s. 9¼d. and 6¼d., are as 30.36 to 1. That is, if confidence is to be had in the accuracy of these pound-rate statements, of 16 places, made to Parliament, 1⅜ times as many Paupers in the first, as in the last eight, and these places, occasioned 30⅜ times as great a rate of charge on the Rentals, respectively! which is quite incredible, with real Rentals.

I would presume to hope, that what is said above, may contribute to shew the necessity, of Parliament soon requiring a new set of *Returns relative to the Poor* to be made, almost similar to those of 1803, but with some variations and additions, which I beg the liberty here to suggest: the *First* of these is, to enjoin, under a penalty for wilful and material mis-statements, more accuracy in stating the *pound-rate* in col. 3, facing p. 529.

Second, that the *amount of actual Rents*, or the real annual value of all the property *rated*, or which ought by law to have been rated, to the maintenance of the Poor, of each place, should be stated in its return: and in order to obtain as many checks as possible on this very important *datum*, still wanting in the Returns of 1803, it would be proper to require the Returns, for the Year from Easter 1815 to Easter 1816, being (as we should all hope) *the last Year* of the Property or Income Tax.

And that each Parish Officer should be apprised, by a Notice on, or attached to the printed blank Form, sent to him to be filled up, that his answer as to the real

Rental

Rental of his parish or place, would, when received in London, be compared with the Income Returns in the Tax Office, also with calculations made from the total Expenditure and Pound-rate in his Return, and by such various other means of check, as could not fail of detecting any very negligent or wilfully erroneous statements; and in any of which cases, he would not only be troubled with new blank forms, to amend his Return, but with the penalties also of the Act, if he failed in quickly shewing, his first return to have been right, or erroneous through no very blameable cause.

The rated *Rental* for each Return, as soon as received, should be calculated (as is done in the last cols. of my two folding pages of Extracts) and contrasted with the return of *real Rental* (when compared with the Income returns, as above); and in every case, of very material disagreement between these sums, as well as in all the larger deviations from *the mean results*, of the Hundred or district in which a place is situate, whether above or below such mean, the Parish Officer there, should have a set of Queries addressed to him, for the purpose of fully elucidating the cause of all extreme cases, or of correcting such as only appeared so, through error or misconception.

Third, it would be of important consequence, as hinted, p. 528, to direct each Parish Officer, after filling up cols. 10 and 11, to collect into one amount, the four sums therein, in order to obtain the total number of *permanent Paupers* maintained, (as is done in col. 5 of my Table, p. 535), and that he should then account for all these permanent Paupers, in the filling up of *a separate blank Return*, which would contain a line for *every separate Trade, Occupation or Calling*, and two cols. opposite thereto: the *first*, for

for the number of the Men, Women, or Youths, who have at any time actually followed or wrought at, each respective Trade, &c.: the *second,* for the number of Wives, Widows or Children of such working people, who have depended on, but not followed or wrought at the Trades, &c. of their Husbands or Fathers.

Instead of a third col., for all the Paupers not included in the preceding cols., which would be very liable to receive a great and very unequal proportion of persons, in the different Returns, whose former Trades, Callings, or modes of Life, might and ought to be distinguished and known: it would be better, that a sufficient blank should be left at bottom of the printed form, for writing in, any uncommon or omitted Trade or Occupation, and for Cases of permanently unemployed Persons, like insane or crippled Children of Paupers, Beggars, Gipsies, &c. &c.

Fourth, that when, in cols. 10, 8, or 9, any Paupers are maintained, or furnished with materials for work, " in any House of Industry or Work-house," the Return should state, whether such House is situated within their own Parish or place? and if so, whether exclusively maintained and appropriated to their own Poor, or whether other, and what places, join in its expence, and send Paupers thereto? in which latter case, a List of the Parishes or places thus contributing, and their present number of Poor in such House, should be made, and sent with the Return. And in cases of sending out their Poor to general Poor-houses, the places wherein these Poor-houses, Work-houses, or Houses of Industry are situated, should be mentioned.

In preparing these Returns for publication, besides collecting the numbers of Paupers and their dependants, of each *Trade,* &c. in each of the several Hundreds

dreds of the Counties; it would be desirable, to publish the full particulars, regarding a certain number of places, where the burthens of the Poor appeared *greatest*, and where they appeared *least*, in each Hundred, in order to throw every possible light on the very important and laborious enquiry, which should precede any attempts to alter or change the Poor-laws, viz. as to how? and in what degrees, by the different descriptions of persons in the Community? the present burthens of the Poor-rates have been occasioned.

And in aid of such enquiry, it would be desirable, to have a third col. in the blank Forms for Paupers' Trades, &c. which I have suggested in page 541, in which the Parish Officers should be required, according to enquiries, and the best of their knowledge, to insert the average *weekly wages* given, or usually earned, in each and every Trade carried on in their Parish or place, during the year to Easter 1816, in case that year be adopted.

2. *Sums raised by Rates*, for the maintenance of the Poor. On the folding leaf facing p. 529, the full particulars regarding the Poor-rate assessments and expenditures, in 22 Parishes or places of the County, will be seen, for the Year to Easter 1803; and on a similar leaf facing p. 30, in Vol. II. the totals of each of the seven Hundreds, and of the whole County, for the same period, will be found. It would have been very desirable, had I been able, to have enquired into and noted the amounts of Poor-rate expenditures in each place, but this labour, or even that of obtaining the pound-rate of poor assessments, throughout the County, was seen to be more than I could accomplish: I therefore confined my exertions on this head, to the noting of the Rates that are mentioned in p. 35, of Vol. II.; and the rather unusual case of Mr. Thomas Wheeldon,
who

who for 33 acres of Castle Field, opposite the Infirmary in St. Peter, Derby, was paying 65 pounds yearly for Poor-rates!

In 1805, the Gentleman to whom I have alluded at the bottom of p. 185, noted the Poor-rates, as they were then stated by respectable Occupiers, in three places in Repton and Gresley Hundred, on *the actual Rents*; and in 1809 I made similar enquiries of the same Individuals; these particulars I shall give below, for the purpose of shewing the ratios of advance in the Rates; and shall prefix the returns made to Parliament in 1803, from the same places; principally for the purpose of further shewing, as remarked in pages 534 and 539, that the *Pound-rate* Returns of 1803, are, great part of them, calculated on *nominal* and very disproportionate and unequal *Rentals*, viz.

	1803.	1805.	1809.
	s. d.	s. d.	s. d.
Foremarke	3 3	2 0	3 0
Repton	5 1½	1 0	3 0
Walton on Trent	9 0	2 0	4 0

4. *Work-houses.* In page 521, I have endeavoured to depict, the very mistaken and absurd principles, on which Parish Work-houses were first introduced, towards the latter end of Elizabeth's Reign; and mentioned, in the note on page 522, that very few or scarcely any of these, now, answer at all to their *legal name*, having little or no *work* performed within their walls; but they are become the abodes of decrepid age, and helpless childhood, insane, infirm, and diseased adults, and of the idle and vicious, of both sexes and of all ages; the latter and baser characters, in too many instances, forming the bulk of the inmates of Work-houses, at present.

COST OF MAINTENANCE IN WORK-HOUSES.

Far more than half, I think, of the 317 Parishes and places in this County which are separately charged with the maintenance of their own Poor, (see Vol. II. p. 33), have entirely abandoned the plan of keeping up a Work-house of their own, and instead thereof, contribute now, towards the general expences of Rent, Repairs, and Superintendance, of some one of seven or eight general kind of Subscription *Poor-houses*, distributed through the County, and pay a sum weekly for each Pauper, whom they send to be there maintained, as will be more particularly mentioned in the next article of this Section.

I regret much, that neither the Parliamentary Returns, or my Travelling Notes, can furnish a satisfactory list of the *Parish Work-houses*, which still remain exclusively to their own use: but I have ventured below to make one out, from the Returns of 1803, subject to future corrections, by inserting those places omitted, or striking out such as contribute to other Work-houses, instead of keeping one of their own.

Names of Places.	No. of Paupers in House.	Cost of Maintenance, in Pounds.	Cost each, per Ann. £ s.
Alfreton (at Swanwick)	25	175	7 0
Alkmund, St.	13	248	—
Allsaints	50	540	10 16
Ashburne	20	193	9 13
Bonsal	12	111	9 5
Bradwell	6	50	8 6
Chesterfield	26	313	12 1
Dronfield	4	40	10 0
Duffield	3	73	—
Etwall	3	55	—
Horsley Woodhouse	5	49	9 16
Kirk Ireton	3	29	9 13
Lea	3	29	9 13
Measham	7	72	10 6
Mickleover	—	89	—
Norton	—	71	—
Peter, St.	35	440	12 11
Werburgh, St.	32	253	7 18
Winster	14	90	6 8
Wirksworth	40	427	10 13

In the third of the above cols. the nearest pound is set down from the returns col. 4; except, that in Alfreton and St. Peter, I have deducted 19-20ths of the sums stated, in the notes thereon, to have been expended in the year, in *repairing* and furnishing their Work-houses. In St. Alkmund and Etwall, there must, I think, from the unusually high cost of each Pauper, have been similar expenditures not noted, or, their Pauper Returns were perhaps only of such as remained in the House at the end of the year, during part of which, greater numbers had been in it? (the reverse of Bakewell, and 4 other places which will be mentioned presently): and in the case of Duffield, perhaps its high expenditure may have arisen, from expences attending the erection in 1803, of the House of Industry in Belper in this Parish, mentioned in a note thereon, and also, Mickleover and Norton made no return of the Paupers in their House.

On these accounts, the five places last mentioned, are not brought into my calculated cost of each Pauper, in the last col. in which the nearest shilling is set down, as being sufficiently exact for any purpose of comparison.

Perhaps Bakewell, Chapel-en-le-Frith, Litton, Monyash and Snelston, having returned 3 to 14 Paupers as maintained by them *in Houses*, ought to have been included in the above Table, but not certainly knowing, that they have Work-houses, and the very low and disproportioned sums these places have stated, in col. 4 of their Returns, as the cost of their maintenance, shewed such probability of mistake, (perhaps the Paupers mentioned were only a small part of their poor, in some Subscription House?) as occasioned me to omit them.

The Returns to Parliament in 1776, stated this County,

County, to contain 20 Work-houses, capable of holding 1,162 Paupers, which was probably not very different from the number and capacity of Parish *Workhouses*, and of Subscription *Poor-houses*, collectively, in 1803; notwithstanding, that only 462 Paupers were then returned, as inmates of the Houses of this description in the County. In the 2nd head of this article, in order that I may comply with the "Plan" for this Report, the average cost per annum will be stated, and some further remarks made, on the Table of Parish Work-houses above given.

1. *Management* of Parish Work-houses; nothing particularly occurred or was mentioned, for me to note, under this head : as much or more perhaps of order and comfort seemed, or was stated to prevail in these Houses, as exists in them in most Counties, or could probably co-exist, with the idle and depraved habits, and demoralized principles, of the greater part of their inmates.

2. *Expence per head* of maintenance in Parish Work-houses. The particulars stated in the Table in p. 545, of the costs of thus maintaining 282 Paupers, in 1802 and 3, belonging to 15 places, shew an average cost per annum for each, of 9*l*. 19*s*. 4*d*.: those 42 Paupers maintained in the Work-houses of Dronfield, Horsley-Woodhouse, Ashburne, Kirk Ireton, Lea and Measham, coming the nearest in their expence, to this average cost per head. Mr. Thomas Poole, in his "Observations" at the end of the Derbyshire Returns, stated, the whole 462 Paupers in its Work-houses in 1803, to average 11*l*. 13*s*. 3½*d*. for their cost of maintenance; and in like manner he calculated, in p. 716, that the whole 83,468 Paupers maintained in Work-

houses in England and Wales, cost 12*l*. 3*s*. 6¼*d*. each per annum, on the average.

3. *Houses of Industry.* In the Note on p. 523, I have already hinted, at the disappointment of the hopes of those, who a few years ago sanguinely imagined, that District Houses of Industry, established under particular Acts of Parliament, would become general, and not only supersede Parish Work-houses, but very materially, if not entirely counteract the inherent evils of the Poor-Law system, and reform also the vicious habits of its demoralized dupes and victims. But alas, the more precisely and minutely *the Law* interferes in the management and support of the Poor, the greater has, and ever will be the mischiefs, ere long produced: the liberty of *discretion*, and not legal restraint*, in those who have the superintendance of,

---

* When endeavouring, towards the bottom of p. 33 of Vol. II., to account for a higher proportion of charge being incurred in this County, for the removal of Paupers and Suits of Law respecting their *Settlements*, than in England and Wales on the general average; I omitted to mention, in a note, that so far from a *litigious* disposition being indulged in, by the Parish Officers of this County, the increasing and intolerable expences of Quarter-Session Trials, and King's-Bench Appeals on Settlements, had very generally taught them, the propriety and advantages, in cases of dispute on this head, for the Officers of the two Parishes to meet, and after inquiring into each other's case, to amicably settle, and admit their own Poor; or in case of discovering, as very often occurs, when both cases are thus compared, that the Pauper in reality belongs to neither, but to some third Parish, in such case uniting their good offices with such third Parish, to induce the taking home of their Pauper, without that trouble and the vast expence, which the tricking practice of full too many of the legal profession, would spin out into two or three successive Processes. The settlement part of the Quarter-Session business, was represented to me as having greatly declined in consequence, and that Attorneys and Counsellors, &c. were complaining bitterly of the change.

and

and providing for Paupers, can alone prevent the increase of evils, frightful to contemplate.

I am not aware, that a single District "House of Industry" has been established in Derbyshire, or near to its borders, of the kind I have been alluding to; but here, several Subscription *Poor-houses* (as they are more properly called than "Work" or "Industry" Houses) have within the last 50 years been established, and are increasing, and which seem to me replete with so many advantages over, either local Parish Work-houses, under the controul, too often, of illiterate and sordid Parish Officers, annually or half-yearly changed, or of Houses of Industry under Trustees or Directors, too commonly superficially informed and inattentive, and never or but seldom changed, that I trust, the fullest particulars I can give, regarding one of these Poor-houses, will be acceptable to the Board and the Public.

I will first give a List of the eight *Subscription Poor-houses* in the County, adding the number and cost of maintaining, in any of them, the Poor belonging to the Parishes in which they are severally situated, similar to what has been given, regarding separate Work-houses, in p. 545, viz.

Poor-house and places' Name.	No. of own Paupers in House.	Cost of maintenance, in Pounds.	Cost each, per Ann £ s.
Ashover	5	52	10 8
Belper	—	—	— —
Church Sterndale (or Earls)	—	3	— —
Crich	5	50	10 0
Doveridge	7	97	13 17
Heanor	30	154	— —
Ilkeston	9	113	12 11
Rosleston	—	—	— —

And I now proceed to mention, what my Notes and

the Parliamentary Returns furnish, respecting each of these Poor-houses, viz.

Ashover.—About the year 1720, a former Proprietor of the Estate, which now belongs to Sir Joseph Banks, Bart. was induced, by the appearance of a copious *Spring* of clear cold water, which issued from the bottom of the Millstone Grit, at Marsh-green, about 1 m. N of Ashover Church, and the sanguine representations of some Bathing-house speculators, to erect and fit up, a pretty large Building over it, for a public *Bath* and Lodging-house: but which never answering, these Buildings remained shut up several years, until about the year 1767, when all the principal Tenants of its Owner's Estates in Ashover, Matlock, and Darley in the Dale, agreed together, to take these Buildings, and fit them up for a Subscription Poor-house.

This Ashover *Poor-house*, to be under the management of three Directors and a Treasurer, chosen every four years, from among these or other Inhabitants of the three parishes above-mentioned; and the Parish Officers of all the Districts around, were invited, by a printed List of Rules, almost similar to those which I will give under the next head, to *subscribe* and send Paupers, to be lodged, fed, and managed, according to another printed List of Rules, which I will also give; paying quarterly their *quota*, according to the whole number of parishes so subscribing, of the Rent, Salaries, and cost of Utensils and Repairs of the House; and paying monthly, for each Pauper sent, their weekly share of the whole current expences, of maintaining all the Paupers in the House.

This scheme was well approved; and many adjacent Parishes soon subscribed, and they were by degrees followed by others, scattered almost over the whole County, as the following will shew, viz.

A List

A List of the Subscribing Parishes, &c. to the Ash-over POOR-HOUSE, *in* 1804.

Ashley-hay,	Harland,	Shardlow,
Ashover,	Hognaston,	Shirland,
Barlow,	Ireton-wood,	South Normanton,
Baslow,	Kilburne,	Stanton,
Beighton,	Kniveton,	Staveley,
Blackwell,	Little Chester,	Stoney-Middleton,
Bolsover,	Little Eaton,	Stretton,
Brackenfield,	Littleover,	Sutton (in the Dale),
Bradburne,	Longford,	Tansley,
Brampton,	Longston,	Tibshelf,
Brassington,	Matlock,	Trusley,
Brimington,	Newbold,	Tupton,
Callow,	Over-Langwith,	Turnditch,
Calver,	Pilsley in Edensor,	Walton,
Clay-cross,	Pilsley in N. W.	Wensley,
Cole-Aston,	Pinxton,	Weston Underwood,
Curbar,	Plesley,	Williamsthorpe,
Darley,	Ripley,	Wingerworth,
Eckington,	Rodsley,	Yolgrave.
Elton,	Scarcliff,	
Foolow,	Scropton,	

Belper Poor-house was erected in 1803, and I regret to have obtained no information concerning it.

Church Sterndale. The Poor-house here, is subscribed to by the quarters of Hartington, by Mellor, Wormhill, and doubtless by several other places, of which I have not obtained information.

Crich Poor-house, is subscribed to by Denby, Melborne, Mercaston, Pentrich, Wessington, Willington, and several other places, I believe.

Doveridge Poor-house, is subscribed to by Dalbury Lees, Marston-Montgomery, Sturston, Sudbury, and 10 or 11 other places, whose names I have not noted.

Heanor

Heanor Poor-house, is subscribed to by Codnor, Shipley, and many other places.

Ilkeston Poor-house, is subscribed to by Burnaston, Ockbrook, Stanton-by-Dale, and 11 or 12 other places, some of which are in Nottinghamshire.

Rosleston Poor-house, is a plain Brick Building, erected in 1803, and first occupied at Midsummer, 1804, having 12 living or sleeping Rooms, and said to be able to accommodate near 40 Paupers; it is subscribed to by Caldwell, Coton, Croxall, Linton, Stretton in the Fields, and by 18 or 19 other places; Sutton and some others of which, are in the other Counties which near adjoin. I was told before visiting this place in 1809, that this was a regular " House of Industry," under an Act of Parliament, binding 22 Parishes to subscribe for 10 Paupers each; but neither the size of the House, or the information of Mr. *John Taylor*, the Master of it (who formerly had the care of Doveridge Poor-house) in any degree confirmed these statements.

1. *Management* of Subscription Poor-houses. The " Rules, Orders and Regulations, for the better supporting, employing, managing, and governing the Paupers in the *Poor-house* at Ashover, in the County of Derby, made at a General Meeting of the Directors, in the said Poor-house, on the 9th day of March, 1803," and signed by the late Mr. John Twigg, Clerk and Treasurer, which have been mentioned in page 545, are as follows, viz.

I. RULES, &c. *to be observed by the Overseers of the different Parishes or Townships subscribing to the Ashover Poor-house, &c.*

1. That no Parish or Township be permitted to subscribe to this Poorhouse for any shorter space of time than a Year; and that the Overseers of every Parish or Place desirous of withdrawing themselves, shall give
the

the Treasurer or Master one month's notice in writing, previous to the expiration of the then current year.

2. That all Paupers sent to the Poor-house, shall be decently clothed by the Overseers of the place by which they shall be so sent, to the satisfaction of the Master and Mistress of the Poor-house, and that whilst such Paupers continue therein, they shall be kept decently clothed, by or at the expence of their respective Parishes or Townships.

3. That no Paupers shall be sent to the Poor-house during such time as they labour under any infectious disease or complaint.

4. That no unmarried Female Paupers who shall be pregnant, shall be sent to the Poor house, without having previously delivered to the Churchwardens and Overseers of the Township of Ashover, a regular Certificate from the Churchwardens and Overseers of the place to which they belong, acknowledging the Settlements of such Paupers, and of the child or children of which they are pregnant, in the usual manner.

5. That some one of the Overseers of every Parish or Township that subscribes to and has Paupers in the said Poor-house, shall attend at the Poor-house every fourth Monday, for the purpose of paying their respective quotas of the expences incurred, and of inspecting the state of the Monthly Accounts, and also the state and clothing of their Poor in the house. And that some one of the Overseers of every Parish or Township subscribing and not having Poor in the said house at the time, shall by himself, ar agent, attend at the Poor-house once in every Quarter of a Year, for the purpose of paying their respective quotas of the rent, and expences of the utensils of the house.

6. That in case of the non-attendance of the said Overseers at any of the said Monthly or Quarterly Meetings respectively, they shall forfeit and pay to the Treasurer for every such default 2s.; and in case of their non-attendance at the next succeeding Monthly or Quarterly Meetings respectively, then to forfeit and pay the sum of 6s. 8d. for every such repeated default; and in case of their non-attendance at the third succeeding Monthly or Quarterly Meeting, then that every such Parish or Township, whose Overseer shall so neglect attending as aforesaid, shall from that time be considered, as expelled from the Poor-house, and shall not be entitled to derive any benefit or advantage whatsoever from the same, in future.

7. That each Parish Officer attending at the said Monthly or Quarterly Meetings at the Poor-house as aforesaid, shall be allowed a quart of ale, and some bread and cheese only; and shall charge his other expences and journey, to his Parish or Township; and that in future there shall be no Dinners got at the Poor-house for the Officers attending the said Meetings.

II. Rules, &c. *to be observed and obeyed by the Paupers in Ashover Poor-house.*

1. That previous to the admittance of any Paupers into the Poor-house, they shall, if necessary, be stripped and washed clean, and be examined by the Master or Mistress, to see whether they labour under any infectious disease or complaints.

2. That the Paupers on their admittance, shall deliver up to the Master, all such clothes and wearing-apparel as they are in possession of, that the same may be cleaned and aired, and if proper, be worn by the owners; otherwise ticketed and laid by, for their use when discharged from the house.

3. That such of the Paupers as are not prevented by illness or incapacity, shall arise every morning on the ringing of the bell, and set about such work as their several capacities or abilities will admit of, and shall continue to work from eight o'clock in the morning till five o'clock in the evening, from the first day of April till the first day of October; and from nine o'clock in the morning to four o'clock in the evening, from the first day of October to the first day of April, being allowed half an hour at Breakfast and an hour at Dinner, and not working on Sundays, Saturday afternoon, Good-Friday, Christmas-day and the two following days, and on Monday and Tuesday in Easter and Whitsun Weeks.

4. That the Paupers shall be called over every morning before breakfast, by the Master or Mistress, and that such of them as shall not then attend to answer to their names, (unless prevented by illness, or some other sufficient reason, to be allowed of by the Master or Mistress) shall forfeit their breakfast.

5. That Breakfast in Summer be at eight o'clock; in Winter at nine; Dinner all the year at twelve: and Supper at six.

6. That none of the Paupers, except such as are sick, shall be permitted to remain in the Lodging Wards in the morning after seven o'clock in Summer, or half past eight o'clock in Winter, or be permitted to return to these Wards until the hour of going to bed.

7. That all the Paupers (except such of them as may happen to be employed in nursing the Sick) shall be in bed by nine o'clock at night in Summer, and eight o'clock in Winter; and that the fires and candles (except such as shall be necessary for the Sick) shall be put out at those times.

8. That no person (except Magistrates, and Officers of the different Parishes or Townships subscribing to the House) shall be admitted to see the Paupers, without the permission of the Master or Mistress; nor any

of the Paupers permitted to go out, without leave of the Master or Mistress, and that none of them shall be suffered, on any pretence whatsoever, to be out of the House after eight o'clock at night in Summer, and six in Winter.

9. That every Pauper, who having been permitted to go out, shall return at irregular hours, disorderly, or in liquor, shall be restrained that liberty for the space of two months to come, and be fed with bread and water only, the next day.

10. That all Paupers absenting themselves from Prayers or Divine Service on a Sunday, without sufficient cause, to be allowed of by the Master or Mistress,—not finishing their task in due time,—wasting their Provisions,—or guilty of other Offences not before mentioned, shall forfeit their next meal, and for repeated Offences, shall be fed with bread and water only, for two days.

11. That if any of the Paupers shall go out of the House without leave, or shall refuse to work, or shall disobey the Master or Mistress's directions, or shall pretend sickness, or make any false excuse for not working, or shall steal, embezzle, wilfully waste, spoil, or do any damage to any stock, provisions, goods, or furniture belonging to the House, or shall profanely curse, swear, or lie, or be guilty of drunkenness, or of any immodest discourse or behaviour, the Master shall cause the offender or offenders to be confined, and fed on bread and water for such time as he in his discretion shall think fit, not exceeding one day; and that he shall keep a Register, and make a Report of the offence and punishment, at the next meeting of the Directors.

III. Rules, &c. to be observed by the Master and Mistress of Ashover Poor-house.

1. That the Master and Mistress reside within the House, in the apartments appropriated for their use, and that they be not, on any occasion, both absent at the same time, and that neither of them be out, later than 10 o'clock at night, without leave in writing from some of the Directors, or the Treasurer.

2. That the Master and Mistress have a separate Table to themselves; and that they apply themselves wholly to the business of the House, and engage in no other business whatsoever.

3. That the Master or Mistress attend Divine Service at Ashover Church every Sunday morning, in company with all the Paupers, except such as are prevented by sickness or other sufficient reason, to be allowed of by the Master and Mistress.

4. That

4. That the Master do enter into a Book, with proper columns, the Names of all the Poor, their Ages, the times of their admissions, and deaths or discharges, and the names of the Places from whence they were sent.

5. That he call them over every morning, just before breakfast, and see that they perform the several directions prescribed in the Second List of Rules, which has been drawn for their better management.

6. That he keep a regular and fair daily Account, of all provisions, victuals, stores, goods, clothes, materials and things which shall be received into the House, and of all Work-done therein, in such form as he shall be directed; and he is strictly to examine all goods and provisions received, and make a faithful Report to the next Meeting of the Directors, of any deficiency in weight, quantity, or quality.

7. That Prayers be read to the Paupers every evening, half an hour before bed-time, and Grace said before and after each meal, by the Master or Mistress, who are regularly to attend each meal, and see that the Paupers behave themselves decently and orderly.

8. That the Mistress do cause the Beds to be thrown open every morning, by or before nine o'clock, and made by or before 11 o'clock (except those of such of the Paupers as are sick), and every Room swept, and the windows set open by 10 o'clock, every morning, in fair weather; and the floors washed once a-week in Winter, and twice a-week in Summer; and the dishes, platters, trenchers, bowls, and piggins, washed twice a-day: And that the Victuals and Provisions (which shall be as near as possible to what are mentioned in the Bill of Fare fixed and allowed by the Directors,) shall be cleanly and well dressed; and all Fires and Candles except in the Master and Mistress's apartments, and except such as may be necessary for any Paupers who are sick) put out at eight o'clock in the evening.

9. That the Master shall cause all the Rooms in the Poor-house to be white-washed, at least twice in every year, to prevent as far as possible, any infectious Disorder taking place, or happening therein.

10. That the Master and Mistress shall take care that the larder, kitchen, back-kitchen, and other offices, together with the utensils and furniture thereof be kept sweet, clean, and decent; and that the dining-room table, and seats be cleaned immediately after each meal.

11. That all the Paupers be compelled to wash and comb themselves, every morning before breakfast, and in general, to keep themselves clean and decent; and that the Children in the House be all washed, combed, and cleaned every morning before breakfast; and that they be taught to Read, and be catechised, every Sunday afternoon, as soon as they are capable of being instructed.

12. That the Master or Mistress shall, as far as it is practicable, place

RULES OF ASHOVER SUBSCRIPTION POOR-HOUSE.

in the best Apartments, such poor persons (if any) who having been creditable Housekeepers, are reduced by misfortunes, in preference to those who are become poor by profligacy and idleness: and that separate apartments shall, if possible, be provided, for the sick and distempered Poor, and an Apothecary or Surgeon be sent for to attend them, when there shall appear necessity for it, at the expence of the Parish or Place to which such poor persons belong.

13. That when any poor person shall die in the House, the Master shall take care, that the body of such person be immediately removed into some separate apartment, and be decently buried, as soon as may be; and also take care of the clothes of such person, and deliver them to the Overseer of the poor of the Parish or Place to which such person belonged, who is to pay the charges of such funeral.

14. That the funerals of such of the Paupers as die in the Poor-house, shall be as near Eleven o'clock in the forenoon as possible, unless the same happen on a Sunday, and then to be in the afternoon, as soon as Divine Service is over.

Bill of Fare for Ashover Poor-house.

	BREAKFAST.	DINNER.	SUPPER.
Sunday,	Milk-Pottage,	Butcher's Meat, & Garden-Stuff,	Potatoes, or Milk-Pottage.
Monday,	ditto	Suet Dumplings,	Milk-Pottage.
Tuesday,	ditto	Pease-Soup,	Mashed Potatoes, or do.
Wednesday,	ditto	Butcher's Meat, & Garden-Stuff,	Broth, or Milk-Pottage.
Thursday,	ditto	Pease-Soup,	Bread and Cheese.
Friday,	ditto	Suet-Dumplings,	Mashed Potatoes, or Milk-Pottage.
Saturday,	ditto	Stewed Meat, and Garden-Stuff,	Ditto, - - or ditto.

When I examined this *Ashover*-house, in October 1809, Mr. *David Watts* and his wife, had for some years been Master and Mistress thereof, and I saw much reason to approve their orderly and good management of its accounts, concerns and inmates, who were then 38 in number, consisting, as usually has been the case here, of old and infirm Paupers, several silly, idiotic,

or

or slightly insane ones, and of many children, great part of them Bastards, born in the House: at 9 years of age these Children are usually apprenticed out, to the Cotton-Mills, &c. (see p. 501), if not sooner taken home to their respective parishes. Grown Paupers, not seriously insane, were freely suffered to leave the House, whenever by work, or often by begging I believe, they could elsewhere provide for themselves. It is almost unnecessary to add, that no *work*, worth consideration, was performed by these Paupers, merely digging the Garden belonging to the House, gathering *dung* for it from the Roads (see p. 267), and assisting a little to keep the premises clean, were all they did.

The *Ilkeston* Poor-house, in 1809, was under the management of an aged woman as Mistress; about 20 Paupers were in the house, principally old persons and children, most of them Bastards, born in the House, under certificates from the parishes to which their mothers belonged. Some few of the Men worked, occasionally, for the Parish Surveyors, on the Roads, and in cultivating a large Garden for the use of the House: some of the women spun a little flax in the House.

In the *Rosleston* Poor-house there were but 12 Paupers, when I was there (and it never, I believe, had more than 16 at once) old and lame persons, and women lying-in of Bastards, principally: a little flax was occasionally spun by the women.

2. *Expence per head*, of Paupers in *Subscription Poor-houses*. In addition to the particulars contained in the Table, p. 549, I will here extract from the Parliamentary Returns, all the remaining places appearing to have more than one Pauper maintained in any Subscription Poor-house in the County, in 1803, viz.

COST OF MAINTENANCE IN POOR-HOUSES.

Places' Names.	Initials of Poor-houses, (p. 549).	No. of Paupers.	Cost of maintenance.	Cost each, per ann.
Codnor,	H.	5	67 14	13 12
Darley,	A.	2	17 4	8 12
Kilburne,	A.	2	31 14	15 17
Littleover,	A.	3	20 16	7 0
Matlock,	A.	5	32 8	6 8
Mellor,	Ch.	5	41 9	8 4
Newbold,	A.	2	20 16	10 8
Shipley,	H.	2	20 4	10 2
Wensley,	A.	2	27 18	13 19
Weston Underwood,	A.	2	28 6	14 6
Wormhill,	Ch.	2	11 8	5 14

The average annual cost of maintaining each of these 32 Paupers, belonging to 11 places in 3 different Subscription Poor-houses, appears to be 9*l.* 19*s.* 5*d.* which is almost exactly the same as the average cost of maintaining 282 Paupers, calculated in page 547, principally, if not entirely, in Parish Work-houses!

The 26 Paupers belonging to Ashover, Crich, Doveridge, and Ilkeston, maintained in the Subscription Poor-houses in their own Parishes, as mentioned p. 549 are returned, as occasioning an average expence each, of 12*l.* which I am somewhat at a loss to account fully for: as well as for the 30 Paupers belonging to Heanor, costing, in the Subscription Poor-house there, only 5*l.* 2*s.* 8*d.* each per annum, according to the returns to Parliament: perhaps several of them may have been only a part of the year in the House.

Abstracts

Abstracts of Four Years' Accounts, of the Ashover Subscription Poor-House.

TABLE I.

Particulars of Disbursements.	Year 1803-4.						Year 1804-5.						Year 1805-6.						Year 1806-7.						Per Cent. on the Aggregates of the four Years.
	Summer, 28 Weeks.			Winter, 24 Weeks.			Summer, 28 Weeks.			Winter, 24 Weeks.			Summer, 28 Weeks.			Winter, 24 Weeks.			Summer, 28 Weeks.			Winter, 24 Weeks.			
	£	s.	d.	£	s.	d.	£	s.	d.	£	s.	d.	£	s.	d.	£	s.	d.	£	s.	d.	£	s.	d.	£ dec.
Bread, Flour and Oatmeal	23	16	6	45	10	6	38	11	2	47	18	0	41	13	2	47	3	7	65	3	2	56	12	0	21.467
Butcher's Meat and Pork	39	11	2	35	3	8	19	16	2½	43	18	6¼	19	1	10¼	48	6	9½	33	0	10½	76	16	10¼	18.501
Milk - -	27	18	1½	15	5	6	25	15	8	16	13	8	24	9	1½	20	1	11	26	10	10	18	15	4½	10.282
Malt - -	10	12	9	10	0	10	12	17	0	13	9	6	16	4	0	14	4	0	14	4	0	13	11	0	6.161
Groceries -	26	0	1½	27	12	9¼	36	18	3¼	31	18	2¼	28	4	1	32	10	4¼	36	14	0¼	33	12	0	14.852
Coals - -	18	4	0	15	1	0	18	4	0	15	1	0	18	4	0	15	1	0	18	4	0	15	1	0	7.790
Salaries - -	19	3	3	16	7	9	19	3	3	16	16	9	19	13	9	16	16	9	19	13	9	16	16	9	8.472
Rent - -	3	4	9	2	15	3	3	4	9	2	15	3	3	4	9	2	15	3	3	4	9	2	15	3	1.406
Taxes - -	0	19	10½	1	1	11¼	0	17	3	1	3	0¼	1	17	6	0	2	3	2	3	4¼	0	2	3	.491
Utensils & Repairs	16	7	11	3	1	10¼	8	15	0¼	10	1	6½	4	17	1½	8	18	1½	11	9	1½	11	10	1½	4.426
Miscellaneous -	13	18	4	11	17	1	8	10	9	8	3	1¼	11	7	4¼	23	7	9¼	12	18	8¼	14	17	2¼	6.152
Amounts of Disbursements	199	16	9½	183	18	2¾	192	13	4¼	207	18	7½	188	16	9¼	229	7	9½	243	6	8¼	260	9	10¼	100.000
Totals - -	£383 15 0¼						£400 11 11¾						£418 4 6¾						£504 6 6½						£1706 18 1¼

TABLE II.

Endings of the 4 weeks' Accounts.	No. of Paupers in the House.				Weekly Expences of each Pauper's maintenance.				No. of Parishes that had Paupers in the House.				No. of Subscribing Parishes that had no Paupers in the House.				4 Weeks' Charge to each Subscribing Parish for Rent, Utensils, &c.			
	1803.	1804.	1805.	1806	1803. s. d	1804. s. d	1805. s. d.	1806. s. d.	1803.	1804.	1805.	1806.	1803.	1804.	1805.	1806.	1803. s. d.	1804. s. d.	1805. s. d.	1806. s. d.
Summer.																				
April 25, to 21 -	29	29	29	41	3 7¼	2 3	2 2½	3 7¼	22	21	17	24	39	37	37	30	5 9	5 5½	4 4	4 8½
May 23, to 19 -	26	28	30	42	2 5¾	3 4	2 8½	2 6½	21	21	18	25	39	36	35	28	4 0½	4 3¼	4 6	4 4
June 20, to 16 -	25	28	29	42	2 5½	3 6½	2 11	2 5	20	20	18	25	38	35	36	28	5 11	4 5½	4 8¾	3 9¼
July 18, to 14 -	25	25	29	46	3 3¾	2 9	3 3¾	2 7	20	19	19	25	38	36	35	26	4 2½	4 1¾	4 0½	4 7½
August 15, to 11	25	24	27	43	3 1¼	2 5¼	2 3¼	2 6	20	18	17	26	38	37	37	27	3 10¼	3 9¾	3 11¼	4 10
September 12, to 8	25	24	27	38	2 9½	2 8½	2 5	2 3½	19	18	17	25	39	37	37	28	4 2¾	4 9	4 5½	5 4¾
October 10, to 6	23	25	30	38	2 7	2 8¾	2 10¼	2 9½	18	18	21	24	41	37	33	29	4 10½	4 6	4 6¼	6 0½
Winter.																				
November 7, to 8	22	27	33	40	3 9¼	2 5½	2 9¼	3 1¼	17	18	22	25	43	37	32	28	4 11½	5 0	3 10½	4 4¼
December 5, to 1	23	27	36	45	3 10	3 10¼	3 5½	2 11	18	17	24	24	42	37	29	29	3 7¾	4 8¼	4 2¾	3 11¾
Jan. 2, to Dec. 29	26	29	42	46	3 3	3 2¾	3 7½	3 7¼	19	17	24	24	40	37	30	28	4 10½	5 9¾	4 1	5 0¼
January 30, to 26	27	32	39	45	3 6½	3 1¼	3 8	3 0½	18	19	25	27	41	35	29	26	4 6	3 10¼	4 4	7 4¼
February 27, to 23	28	35	35	47	2 11¾	3 4	3 6¼	3 2	19	19	22	28	40	35	32	25	3 10	3 3½	4 8¼	4 6¾
March 26, to 23	29	29	40	51	2 10	4 3¾	3 0	3 11½	20	17	22	25	39	37	32	28	3 6	5 1½	5 10½	4 11¼
Yearly averages	26	28	33	43	3 1¾	3 0¼	2 11¾	2 11½	19	19	20	25	40	36	33	28	4 6½	4 6¾	4 5	4 11
General ditto - -	32				3s. 0¼d.				21				34				4s. 7½d.			

In the *Ashover* Poor-house, in pursuance of the 4th and 6th of its 3d List of Rules, (p. 555), very regular and clear accounts have been kept, and made up, and audited by the Directors and Treasurer, at the end of every 4 weeks; and then, the same have been submitted to the Officers of the subscribing Parishes and Townships, agreeably to the 5th Rule in the 1st List.

And annually, to the Monday nearest to Lady-Day, the total expences of the House, under various heads, for every 4 weeks of the past year, have been collected, and an average 4 weeks' expenditure under each head, calculated and annexed: and to which also have been annexed, an account of the number of Paupers in the House every 4 weeks, and from how many parishes, the expence of each Pauper's maintenance weekly, and the month's *quota* of each subscribing Parish, for the general expences of the House: which abstracts of the accounts, have annually been printed, and sent to each subscribing Parish.

Mr. Watts, the Master of the House, having furnished me with these printed accounts, for the years to L. D. 1804, 5, 6, and 7, I trust that the above abstracts of them in pages 560 and 561, will be deemed worthy of being preserved here, and that the facts in them will be found useful *data*, for comparison with the expences of maintaining Paupers, in other modes and places.

The titles of these two Tables, and what is said above, will, I hope, sufficiently explain them: the last column in Tab. 1. I have calculated, in order to place in the readiest point of view, for comparison in other cases, the proportionate expences, under each of the eleven heads mentioned in the first column.

The second Table, shews the number of Paupers maintained in this Poor-house, to have varied in these

4 years,

4 years, from 22 to 51, and to average 32. The weekly expence of food and necessaries for each of them varied from 2s. 2¼d. to 4s. 3¼d., averaging 3s. 0¼d. weekly, or 7l. 17s. 6d. per annum: the reason, in great part, I conceive, of such considerable variations in the weekly expences, has been owing, to less or more provisions being brought into the Mistress' store, for use beyond the current month.

The number of subscribing Parishes, who at the same time have had Paupers in the House, in these 4 years, has varied from 17 to 28, average 21; and the number of subscribing Parishes, not availing themselves of their privilege, to constantly have Paupers in the House, has at times varied from 25 to 43: the total subscribing Parishes and places (see page 551), being from 51 to 61, at different periods of these four years, averaging 55: so that more than half as many more Parishes subscribe constantly, to the general expences of this House (about 3l. each annually, as will be presently shewn), than constantly have Paupers in it: in order that they may always have a place, ready, for the proper reception of pregnant single women, or infirm or deranged persons, who may fall suddenly, and be likely to continue long chargeable, on such subscribing Parishes.

The monthly amount of these Subscriptions for general expences of the House, have in these 4 years varied, from 3s. 3¼d. to 7s. 4¼d. according as none, or a more considerable degree of reparation to the Premises, or renewing of worn-out Utensils, Linen, &c. have been wanted; the average per Month being 4s. 7½d. or 3l. 0s. 3¼d. per annum.

If therefore, we add the annual average expence of maintenance, in the Ashover House, viz. 7l. 17s. 6d. (above) to the *quota* for general Expences, viz. 3l. 0s. 3¼d.

we shall have 10*l*. 17*s*. 9½*d*. as the average annual expence per head, of maintaining Paupers in this Poorhouse, in the four years which followed Easter 1803; from the Parliamentary Returns for the year that preceded this Easter, 9*l*. 19*s*. 4½*d*. (see pp. 547 and 559) was calculated as the average House Expences of 314 Derbyshire Paupers; and the difference of 18*s*. 5*d*. in these two periods, may, I think, fairly be ascribed to the *general advance* of all articles in price, or to the depreciation of Money, through excessive and increasing taxation, and a paper-money medium.

6. *Box Clubs.*—" Advantages and disadvantages of them ?" To this enquiry, directed by the " Plan," I have given some attention, but previous to any observations directed to this enquiry, I beg to state, and make some calculations on, the numbers of Male and of Female *Friendly Societies*, or Benefit, Sick, or Provident Clubs in the County, and on the proportions their Members bear, to the whole Population of each sex, from the Poor-Returns of 1803 (see Vol. II. p. 34), and the volume of Population Returns of 1801 (see the next Section), &c.

Of the 317 Parishes and Places in the County which made returns respecting their Poor, 133 mention Friendly Societies, of which Societies the total number stated was 267, consisting of 22,681 Members of both sexes; that is, full 85 Members in each, and being 14 in the Hundred, of the whole resident population: whereas in England and Wales, collectively, the 9,675 Societies, which were returned as containing 704,350 Members, averaged less than 75 Members each, and were only 8 per cent. on the whole population.

1. The *Men's* Friendly Societies appear to have been 247 in number, containing 21,505 Members, in Derbyshire,

shire, in 1803, its total male population being 79,401; that is, more than 87 Members in each Society, and composing 27 per cent. or more than 1 in 4, of the *male* population. Of those places which had only one or two Men's Societies held in them, in 1803, the largest numbers of Members to each, were, in Chinley 360, Mellor 315, and Smalley 298; and the smallest numbers of such Members were, in Clown 15, Wessington 15, and Repton 13.

In order to shew more clearly, the extent and distribution of these useful Friendly Societies, in this County, I will extract below, all those places which had, in 1803, three or more Men's Societies, viz.

Places.	No. of Societies.	No. in each.	Per Cent. on the Males.
Alfreton,	3	136	34
Allsaints,	4	52	16
Ashover,	4	103	39
Belper,	7	116	36
Bonsal,	3	69	37
Chesterfield,	13	56	38
Crich,	3	106	44
Denby,	3	69	44
Doveridge,	3	79	68
Eckington,	6	67	30
Glossop,	4	119	34
Heanor,	3	101	54
Ilkeston,	10	78	63
Matlock,	3	91	25
Melborne,	3	185	60
Middleton by W.	3	57	47
Peter, St.	3	48	14
Phoside,	3	120	74
Ripley,	3	81	42
Stoney Middleton	3	54	96*
Tibshelf,	3	77	62
Whittington,	3	54	51
Whitwell,	3	43	35
Wirksworth,	8	63	36

* This extreme result, arises from an error, which I now perceive, in the Return of the Males in this place, in 1801; which, compared with its Return of 1811, shews, that 70 too few Males had been set down on the former occasion: which will reduce the Members of Societies to 68 per cent. on the male population.

Hence it appears, that Chesterfield, Ilkeston, Wirksworth, and Belper, had the greatest number of Men's Friendly Societies, among these 24 places: that the largest Societies in them, were in Melborne, Alfreton, Phoside, and Glossop; and that the highest proportions which their Members bore, to the whole Male population, was found in Phoside, Doveridge, Stoney-Middleton, and Ilkeston. The average number in each of these 104 Societies being 81 Members, and their proportion to the whole Male population where they were held, rather more than 39 per cent. (or something less than one in three) of the Men and Boys; and of course, their proportion to the Men was much greater. In all England and Wales 9215 Men's Societies averaged near 73 Members each; and which formed near 16 per cent. of the whole Male population.

2. *Women's* Friendly Societies. The particulars of these in the County, in 1803, were as follows, viz.

Places.	No. of Societies.	No. in each.	Per Cent. on the Females.
Bakewell,	1	137	18
Belper,	2	39	3
Calver,	1	84	34
Chinley,	1	160	42
Crich,	1	60	9
Cromford,	1	83	15
Denby,	1	20	5
Glossop,	2	23	3
Heanor,	2	35	12
Little Eaton,	1	29	10
Long Eaton,	1	86	34
Matlock,	1	82	6
Mellor,	1	75	9
Muggington,	1	5	3
Sawley,	1	46	12
Spondon,	1	40	9
Wirksworth,	1	96	17

Whence

Whence it appears, that the greatest numbers of these Female Societies in 1803 were, in Belper, Glossop, and Heanor; the largest ones in Chinley, Bakewell, and Wirksworth: and that their Members bore the highest proportion to all the Females, in Chinley, Calver, and Long Eaton. The average number of Members in these 20 Societies was 59, and their Members near 10 per cent. of all the *Female* population of these places. In all England there were then 457 Female Friendly Societies, averaging 76 Members each, and these constituting only $\frac{8}{10}$ per cent. or 8-1000th of the whole Female Population. Whence it appears, that Derbyshire may be said to excel, in the numbers and extent of its Friendly Societies, of both sexes.

I was happy to find, when examining the Northern part of this County, that those excellent institutions, *Female Friendly Societies*, were flourishing, and rapidly increasing: when at Glossop in July 1809, I was shewn the printed Rules, and informed of some particulars, of a Women's Society which held its meetings at the Bull's-Head Inn, and then had increased (from about 23, p. 566) to 175 Members! As the nature and regulations of these *Women's Societies* are little or not at all understood in various parts of the Country, I will give here the Notes which I took, respecting this Society at Glossop.

The objects of the Society are stated at the head of its *Rules*, to be, " For the purpose of support to such Members of this Society as shall at any time be rendered incapable of work, by means of Sickness, Lameness, Old Age, or Casualties, and for the decent Interment of deceased Members, their Husbands and Widowers."

Females from 18 to 25 years of age *eligible* to be admitted,

admitted, without an entrance fine; and others, from 25 to 30 years, to pay as an entrance-fine, 8s. for every year that their age exceeds 25.

New Members, discovered within 2 years, to have any disease likely to occasion their being permanently burthensome, to have their Money returned, and be *excluded*: all Members, after 2 years to be *free* of the Society: but Members entering, or being in any other Benefit Society whatever, are to be *excluded*.

Members eloping from their Husbands, and cohabiting with other Men, or having a Bastard by a Married Man, or three Bastards by any persons, are to be *excluded*: and having a second Bastard by single Men, are to pay a fine of 8s. to the Stock. Conviction for Murder or Felony, is also to *exclude*.

The *Contribution* of each free and junior Member to the Stock, to be 1s. each Quarter, and 4d. *to be spent in Ale*, (in order *to pay for their Meeting-Room*, as the Landlord frankly confessed to me), until 250l. is in hand; after which the Committee may from time to time, either lessen or increase the quarterly contribution, so as to preserve this standing Stock, as near as may be.

Allowance to Sick or disabled Members to be 6s. per week. Lying-in Members, who may be confined to their Room for more than 4 weeks; after their Month, are to be considered as Sick Members.

If a sick or disabled Member is suspected by the Stewards, to be practising any deception on the Society, they may call in a Medical Man to examine her; and if such deception be proved, the Society may *exclude* such Member. Disorders or disabilities brought on by any vicious or unlawful practices of Members, are not to entitle them to any but the Funeral allowance.

The

The *Funeral allowance* of Members is to be 5 Guineas: 2¼ Guineas to be allowed for the Funeral of a Member's Husband; and the widowed Husbands of a Member, continuing single and paying 6*d.* per Quarter to the Stock, after their Wive's decease, to be entitled to 2¼ Guineas for their Funeral expences.

Not a single doubt was, or could I think be started, by any of the many persons of whom I enquired in the County, as to the *unalloyed good, which each of the above descriptions of Friendly Societies produce,* both on *the morals and habits of the Members,* and on many other Individuals of the lower class, among whom they live, and *in keeping down the Poor-rates,* where they prevail, and are *encouraged and held in respect,* as here is generally the case, by the middle and higher classes of the community on the spot: not by the meddling interference, of such of the latter as happen to be Magistrates, under the mischievous powers of the 33d and 35th of the King*, by *enforcing or preventing the*

---

* The sanguine and mistaken expectations, of the framers and promoters of these Acts, that all the Benefit Societies of the Kingdom would eagerly and immediately enrol themselves under them, has been so far disappointed, that of 9228 such Societies in England and Wales (in Counties for which the Clerks of the Peace had made Returns) in 10 years after the passing of the first of these Acts, only 5428 of them had been enrolled, or not 59 per cent. of the whole. Of 267 Derbyshire Societies, 138 only had enrolled themselves, or 52 per cent of them: and in Bedfordshire, in the same period, only 21 per cent. of its Friendly Societies, had made their laws, perhaps after mutilation, irrevocable, and their Funds disposeable, in particular cases, not by the sense of the majority of those, *who raised and owned them,* but, perhaps, by meddling clerical Justices (see Vol. I. p. 93), who ought rather to employ themselves, as recommended in p. 530, in instructing and enlightening the Members of these harmless and truly laudable Societies, as to their dignity, as free Men or Women, compared with Poor-law slaves, and as to their practicable rights and duties, in and towards the community in which they live, and each other.

altering

altering of Rules made for regulating the internal concerns, and distributing the Funds, of such Societies; but leaving *the will of the majority* in such, to regulate every matter and thing concerning them, as on every principle of justice and policy ought to be the case.

I happened, on a Thursday in May 1808, to be at Brassington, on the anniversary of the Men's Friendly Society there, of 120 Members, who had a standing stock of 300*l.* out at interest:—the Members, all clean and in their best clothes, assembled early in the forenoon to transact their business in their Meeting-room, and being there marshalled, each with a tall white wand tipped with blue, walked in procession through the Town, preceded by a band of music and a large handsome blue silk flag, on which was painted the appropriate story of the Good Samaritan (the gift of some patriotic Lady of the district, I believe), to the Church, where all the most respectable Inhabitants of the Town and neighbourhood attended, and heard Service and a highly appropriate Sermon: from which, through streets crowded by approving spectators, the Members returned, in like order, to their Room, to dinner.

I was told, that similar marks of attention and respect, were commonly paid to the Friendly Societies, both of Men and Women, on their anniversaries, and on some other occasions throughout the district, and that the best possible effects were visible therefrom, in *the rising spirit of independence, and spurning at poor-law dependance*, with all its moral degradations, felt by the Members and families, of these very laudable Societies: whose Rules were in general so well contrived, and their management so judicious, that instances were extremely rare, of their Funds being exhausted,

hausted, or even of their being obliged, materially to lessen, their allowances to distressed Members: and *Magistrates never here interfered*, to enforce any payments to them.

An instance, very creditable to the Men's Society in Spondon, (which in 1803, had 319 Members) is related in the " Monthly Magazine," Vol. XXXVII. p. 277, to which Henry Foster was many years ago admitted, and after contributing 7l. 3s., had the misfortune to lose his sight, in which state he continued to receive weekly allowances, and at length the Funeral allowance, which altogether amounted 267l. 2s. 11d.! and many almost similar cases might be found in the Kingdom.

What necessity can there be, for *legal interferences* with Societies thus honourably and liberally disposed? beyond *placing their Funds within the pale of the Law's protection*, as fully so as any private property is placed: which now, to the disgrace of the age in which we live, is probably not the case, with about 40 per cent. in number and value of all these funds*, raised

for

---

* I read in the " Farmer's Journal" of this day (15th April, 1816) among an excellent set of Resolutions entered into by the Cambridgeshire Agricultural Society, one, recommending to the Agricultural Interest, " to support the institution of Benefit Clubs, *on economical plans*, as calculated to diminish the Poor-rates, and *to raise the moral character of the Poor*. And the establishment of " *Saving Banks* " (admirable Institutions! which had been little heard of or tried, at the time of my Survey of the County) as affording a convenient and *secure deposit for the savings* of Individuals, or *of Societies for relief and assistance in sickness and old age*."

It is the last suggestion which I am principally anxious to record here, and earnestly to recommend for adoption, to all the higher Class of Society, for the most useful, honourable and virtuous, of human purposes!

I am fearful that the enactments, *empowering Magisterial interference* with the management of Friendly Societies, *as the price of having legal protection afforded for their Funds*, have operated more to discourage the establishment and increase of those importantly useful Institutions, and to the destruction of many such which did exist, than any other cause: at the same time, that their operation has been, even still more mischievous, in counteracting the natural tendency of these admirable Institutions, " to raise the moral character of the Poor," above the debased state of Poor-law dependance.

No bodies of Men, or their views, possibly, could be more distinct, than those who associated here in 1793 and 1794, with really seditious or treasonable views, and those numerous bodies of profligate *abusers of the Apprentice and Poor Laws*, combined, of whom I have spoken in p. 528, from all the bodies of *persons composing the Friendly Societies of the Kingdom;* except, perhaps, a very few, among those in great and demoralized trading Towns.

In such places, unfortunately, *Benefit Clubs* do but rarely originate in the virtuous impulse and desire of the Members themselves (encouraged and applauded by those above them), for raising their con-

---

ciety, as a preferable mode, perhaps, of securing and improving the Funds of Friendly Societies, to any modification or improvement of the existing Laws regarding such Societies, which ought I think to be repealed: all of their clauses at least, but those which simply go to *the securing of their Funds* to such Societies, from the dishonesty or insolvency of Stewards, Treasurers, or Persons borrowing their Monies.

dition

dition above Poor-law dependance; but where, on the contrary, *the interested views of Publicans*, for bringing, the much too profitable *custom* of these Clubs, to their Houses, aided by low persons, who being ready with their Pens, expect to profit in the office of *Clerks to such Clubs*, originate and set on foot the greater part of them: placarding the walls of such Towns and their vicinities, with *puffs* thereon, as shameless as those common in Recruiting or Lottery Bills, in holding out the promise of *allowances*, so early and greatly disproportioned to the *contributions*, as to be inconsistent with the permanence of any such Institution.

A sufficient number of Members being thus collected, and a set of Rules drawn up, providing for very *frequent Meetings, and Money to be spent at each*, in *the particular Public House*, giving *Name*, usually, to the Club, and these Rules being rendered almost irrevocable, by enrolling them, the Landlord's harvest begins, and lasts, until cases of distress, real or fictitious occur, to occasion a serious call on the Fund, which, backed either by magisterial enforcement, or the threat thereof, continues, at length exhausts the same, and dissolves the Club.

It is, I believe, among the Members of these *Publican's Clubs*, to which, frequently, only one or a few particular trades are invited, that *Combinations of Journeymen*, are engendered and matured, in a very principal degree, and such have, in reality, no connection with the true Benefit or *Friendly Societies*, of the Country at large.

I beg, therefore, as one principal means of encouraging the formation and increase of the latter truly useful and important Societies, that every considerable Village and Town in the Kingdom, should, as soon as may

may be, not by legal interference, which, like keen frost to the early vegetation of the Spring, nips every opening good to the Poor, but by those efforts of private and voluntary beneficence, for which our Island is so famed, be furnished with a large but plain and unexpensive *Room*, near adjacent to, but perfectly *separable from*, a reputable Inn or Public House.

That such *Society Room* should be held in the hands of the Owner of the Inn*, or of the Subscribers' *Trustees*, building it, and *not let with the Inn*, or to the Innkeeper, except specially, for particular occasions, like Wakes, Fairs, or Markets, occurring between the Meeting-times of the Friendly Societies: and, that for very moderate Rents, scarce exceeding nominal ones, the several Societies of the place or vicinity, should have the free use of this Room for their Meetings, on different Evenings and Days.

Then might the clauses in their Rules, *requiring Alehouse spendings* (see pp. 568 and 573), as *the unavoidable means of paying for or procuring their Meeting-rooms in Public Houses* (and where else can they now meet?), be quite abolished: and this solitary circumstance, having any immoral or mischievous tendency, in all the economy of Friendly Societies (when left free from legal interference), be entirely removed.

It must never be lost sight of for a moment, by those in the middle or superior classes, undertaking to promote the establishment of Friendly Societies, that *all the Monies of such Societies must be of their own con-*

---

* In this manner the Freemason's Society of London built their Hall and their Tavern in Great Queen-street, adjacent, and connecting at pleasure, but *do not let them together;* reserving the Hall for every occasion of their own; but specially, on each occasion, letting the use of the Hall to the Tavern keeper, for the accommodation of his Guests.

tributing,

tributing, managed by, and distributed by themselves, according to Rules, alterable at the free will of the majority for the time being: or Poor-law dependance will only change its name, and but slightly abate of its malignant influence: the Poor, in bodily health, must be taught *to respect themselves,* so far as to spurn at charitable gifts in any form: *proper earnings by their labour* (as shewn, I trust, in pages 515, 517, 525, 526, &c.), voluntary contributions therefrom, to provide against adversity, and virtuous and economical habits in spending the remainder of their earnings, on themselves and family connections, of the passing, present and rising ages, seem to me the only means, for again re-organizing and restoring the moral health, of the large and important class of our Society, who have been so cruelly cheated and demoralized, by the Poor-law system.

Saving Banks (as already suggested, see page 571, Note), might prove a very important auxiliary and assistance to Friendly Societies, in realizing *compound* Interest on most of their savings or Fund: and perhaps, for a limited time after their establishment, and until a certain stock be once accumulated, a higher rate of Interest than to others, might with safety and excellent effect be allowed these Societies, from the Saving Bank.

Some assistance might with propriety be given towards the establishment of these Societies, by their Patrons, in providing them, without expences, with proper Account Books, and with a Publication, which is yet *much wanted,* and not likely, I fear, to be furnished but from materials collected by Parliament, in Returns, of abstracted histories and accounts of the original and varied Rules, of as many as possible of such Societies, in all Counties and situations; the ages and

and Trades of all the present Members, and years they have been such; the number of Members, and amounts of Subscriptions for each past Quarter, and of allowances to temporary and to more permanently chargeable Members, separately distinguishing them in each of such Quarters; with remarks on all extraordinary cases of allowances.

From which *data* as many sets of leading *Rules*, and Rates of *Contributions* and *Allowances* (founded on the averages of real cases, rather than on hypothetical calculations), as all the various cases seemed to require, might by a proper person be drawn up, and very fully explained, *as precedents* and guides to the Members, in the very difficult task, in new situations, of forming proper Rules, and Rates of Contribution, and Allowance, sufficiently provident and liberal, for just securing the permanence and utility of the Institution, without the double evil, of keeping out Members by the largeness of Contributions, and of inducing the worst disposed among the Members, *to rely* on the certainty and largeness of Allowances, rather than on their own exertions and economy.

In order to obtain these much wanted *Returns*, a frank and explicit explanation, of the motives for requiring them, and uses to be made thereof, should be addressed to each Society, requesting, that at their next Meeting, the most proper and able person among their Officers, Members, or Neighbours, should be selected and deputed by the Society, to prepare, and make and sign the Return from their Books, agreeably to the printed Queries and Forms or Precedents that would be sent. And that such deputed Person, sending his Return to the nearest Receiver of Land-tax, and after a certain number of days allowed for the Re-

ceiver

selves at the top of the Tables at their Annual Dinner, properly but economically provided, and sit an hour with them afterwards, in familiar conversations, on topics connected with the great and laudable ends in view in these Societies, this would also do good: and if such Visitors, were to contribute somewhat liberally towards

[To face page 577.]

Hundreds, &c.	No. of Parishes and Places, in 1811.	Number Inhabited, in				No. Uninha... 1801.	Emp 181
		1788.	1791.	1801.	1811.	Uninha 1801.	
Appletree Hundred	64	3,431	3,654	4,396	4,904	167	
Borough of Derby	5	1,637	1,754	2,144	2,644	26	1
High Peak Hundred	71	5,582	5,793	6,497	7,173	554	3
Morleston and Litchurch Hundred	49	3,347	3,640	4,617	5,575	111	1
Repton and Gresley Hundred	39	1,931	2,093	2,632	3,047	54	
Scarsdale Hundred	54	6,655	6,905	7,739	8,284	266	
Wirksworth Wapentake	37	3,203	3,358	3,873	4,110	194	1
Totals of Derbyshire	319	25,786	27,197	31,898	35,737	1372	1

Kingdoms.	No. of Counties						
England	40	14,620	—	—	1,472,870	1,678,106	53,965	47,9
Wales	12	1,121	—	—	108,053	119,398	3,511	3,0
South Britain	52	15,741	—	—	1,580,923	1,797,504	57,476	51,0
Scotland	32	1,005	—	—	294,553	304,093	9,537	11,
Great Britain	84	16,746	—	—	1,875,476	2,101,597	67,013	62,

1. 2. 3. 4. 5. 6. 7. 8.

lected and deputed by the Society, to prepare, and make and sign the Return from their Books, agreeably to the printed Queries and Forms or Precedents that would be sent. And that such deputed Person, sending his Return to the nearest Receiver of Land-tax, and after a certain number of days allowed for the Receiver

ceiver to look into it, and see that the Queries and Forms or Precedents have been, with sufficient accuracy and fullness answered and followed, without needless lengthening of the same; calling then on the Receiver, and producing the Resolution of the Society, appointing him to make their Return, the person making (after amendments, if necessary) *a proper Return*, should receive for the same, a reasonable remuneration per folio (of 72 words) in the same, according to an act, which should authorise and direct these payments.

In some places, zealous and well-meaning Patrons of Friendly Societies, among the higher ranks, having enrolled their names in such, make contributions to their Funds, as Members, and even sometimes attend their Meetings for business; this practice I can by no means approve, for reasons before stated. Yet courteous and condescending attentions from the great, to the assembled bodies composing these Societies, on Anniversaries, or occasionally; such as attending their Annual Sermon in the Church (see p. 570), presenting them with Flags, Staves for processioning; with Medals, for the most active and useful in organising and conducting their concerns, and by various other means, calculated *to excite emulation*, and awaken those feelings in the lower class, which Poor-law dependance has well-nigh extinguished, will certainly do much good.

If a few very respectable persons would seat themselves at the top of the Tables at their Annual Dinner, properly but economically provided, and sit an hour with them afterwards, in familiar conversations, on topics connected with the great and laudable ends in view in these Societies, this would also do good: and if such Visitors, were to contribute somewhat liberally towards the Landlord's Bill, it might also be proper; since, assistance towards lessening all necessary expences of establishing, and proper ones of attending the Meetings for managing these Societies, may, apparently, *be given* by others, without the evils attendant on, contributing to, or interfering with the management of their Funds.

Flour Clubs.—During the scarcities of 1795 and 1800, many Associations of working Men were formed in the southern parts of this County and elsewhere, who subscribed to purchase Wheat, and get it ground, selling out the Flour, at prime cost, to their Members: these Societies were several of them found so beneficial, that some of them seemed likely to become permanent. A Society of this kind in Long-Eaton, when I was there in 1809, consisting principally of Stocking-weavers, were in the habit of sending the Wheat they had purchased, to the Grist or Batch Mill at Toton, Notts, paying $3s.$ per quarter to the Miller, who delivered back within 14 lb. the same weight of Flour and Bran, as of the quarter of Wheat he received; by which it was at that time calculated, that $4d.$ to $6d.$ in the stone of flour (14 lb.) was saved to the Members.

No place is assigned for the mention of Charitable Institutions, in the "Plan;" but judging this the most proper place, I will just mention,

Alms-Houses, or Bead-houses, as they are here often called;—the most considerable of these in the County, sometimes called an Hospital, is at Ravenstone, for aged women. Many smaller Alms-houses are scattered thro' the County, which I did not particularly note. I have mentioned in the Note on page 531, the gross sums which have been lately returned to Parliament, as the revenues for supporting Poor Persons in these Alms-houses, and for Donatious to other Poor Persons within the County.

Soup-shops were established, and the Poor greatly benefited by them, during the two great scarcities, in various parts of the County.

I was told at Bakewell, that in 1800, the Mine-owners in some of the Mining Villages, subscribed, and purchased indifferent Wheat, mixed it with Rye and ground it, and had a very coarse kind of *cheap bread* made of it, which they served out to the Men, instead of Wages; in consequence of which, the regular Bakers left off their usual baking of Oat and Wheat Bread, because the Miners had no money to lay out; and that the Men, thus forced to live on bread they were unused to, and that disagreed with them, suffered great hardships therefrom.

SECT. XI.—POPULATION.

ENUMERATIONS of the Houses and People, of districts that have been statistically described, have long been considered as important parts of such details: Mr. James Pilkington, when collecting materials for his "View of Derbyshire," in 1788, was at the pains, to count the Houses in each parish and place in Derbyshire, and which he appears to have done with great care and accuracy, with the exception of Kniveton, which he seems to have forgot[*], and Derby Hills, Edingale, and Ravenstone, omitted on account of their

---

[*] In saying this, I am forced to rely upon the abstract of this enumeration, which Mr. Thomas Brown has printed, at the end of his 4th Report on this County, not having Mr. P.'s work now at hand.

intermixture with other Counties, I believe, making, with these additions, proportionably calculated from later enumerations, a total of 25,786 Houses. Mr. P. also counted the People, inhabiting the 5 Parishes composing the Town of Derby, and of various other Parishes and places distributed through the different Hundreds of the County, to the number of 5,709 Houses, and 27,274 Persons.

In the year 1801 the attention of the Legislature was forcibly called to the consideration, of the actual state and rate of increase of the Population of the Kingdom, in considering the causes and means of averting the recurrence, of such distressing scarcities of Food, as had previously occurred in 1796 and 1800; and an Act was in consequence passed, requiring Returns from the Overseers (or Schoolmasters) of every Parish and place in Great Britain, of the actual state of the population, on the 10th of March 1801; and from every officiating Clergyman, of the registered Marriages, Baptisms, and Burials in every year subsequent to 1779, and in every 10 years of the preceding part of the 18th Century.

These important documents were collected and arranged, and distinct and average results deduced from them, by Mr. *John Rickman*, and were printed by order of the House of Commons, in 1006 folio pages. About 10 years afterwards, a nearly similar Act was passed, requiring like Returns of the state of Population on the 27th of May, 1801, and of Parish Registers for the preceding 10 years; which being collected and prepared in like manner, by Mr. R., were also printed in 742 folio pages.

As far as Derbyshire is concerned in these very curious and important parliamentary details, I was so anxious

to

ACCOUNT OF POPULATION RETURNS. 531

to present the whole of them, and of Mr. Pilkington's enumeration also, properly arranged, for comparison, in each individual Place (319 in number) for which Returns had been made, that I employed one of my Sons, several Months ago, in preparing such materials for this Section of my Report: but finding now, unfortunately, that room will not permit of its insertion, I must, principally, content myself on the present occasion, with giving a long Table to fold out facing page 577, containing the totals of the several Hundreds of the County, and totals of the three Kingdoms; on which Table I have bestowed more than ordinary care and pains, in filling up and calculating its 44 several columns, as I will now proceed to explain.

Mr. Pilkington's enumeration of Houses having been made and arranged, according to the six *Ecclesiastical* divisions or the County, mentioned in p. 93 of Vol. I.*, it became necessary to new arrange his materials, ac-

---

* I omitted in this place to mention, that the Divisions of the County into its 7 HUNDREDS, &c. (I. 78, and II. 1 of Preface) called *Appletree* (Ap H), Borough of *Derby* (B of D), *High Peak* (HPH), *Morleston and Litchurch* (MLH), *Repton and Gresley* (RGH), *Scarsdale* (SH), and *Wirksworth* Wapentake (WW); and its division into 6 DEANERIES, called *Ashburne* (AD), *Castillar* (CaD), *Chesterfield* (ChD), *Derby* (DD), *High Peak* (HPD) and *Reppington* (RD), are scarcely in any way co-extensive; but intermixtures of them occur, in the following manner, viz.

Appletree Hundred will be found to contain parts of AD, CaD and DD: the Borough of *Derby*, of DD: *High Peak* Hundred, of HPD: *Morleston* and *Litchurch* Hundred, of AD, CaD, DD, and RD: *Repton and Gresley* Hundred, of CaD, DD, and RD: *Scarsdale* Hundred of ChD and DD; and *Wirksworth Wapentake*, of AD, ChD, DD, and HPD.

In like manner, *Ashburne* Deanery will be found to contain parts of ApH, MLH, and WW: *Castillar* Deanery, of ApH, MLH, and RGH: *Chesterfield* Deanery, of SH and WW: *Derby* Deanery of APH, B of D: MLH, RGH, SH, and WW: *High Peak* Deanery, of HPH and WW: and *Reppington* Deanery, of MLH and RGH.

cording to the several *Hundreds* thereof, for col. 3 of my Table, and in doing which, I have been careful to collect together, all those small Places which have since been comprized together in the Parliamentary Enumerations and Returns; and on the other hand, to divide, proportionally, all those Enumerations by Mr. P. which have been divided in the subsequent Enumerations and Returns.

Column 12 of my Table came next in order to be filled up, with average results, from Mr. Pilkington's Enumerations of Houses and People, in the several Hundreds; and all of which were found so nearly consistent with each other, and with subsequent averages, as to be adopted without modification, except Appletree Hundred, in which Duffield and Holbrook, comprizing 3-5ths, of all the Houses and People which Mr. P. enumerated in this Hundred, fell then, so much below the ordinary rate of Persons to each House, as to render it advisable, to take another number instead of their average, bearing the same proportion to the result of Mr. P.'s whole enumeration, as the averages of the two subsequent enumerations in these Hundreds, in the succeeding columns, bear to the County results, on these occasions.

The two columns thus obtained from Mr. P. (3 and 12), were then used, in computing col. 21, the estimated numbers of Persons in 1788; which in this mode of separate averages for each of the Hundreds, and adding them, amount to 2071 persons less, than if 4·803 (a general average exclusive of the two places above-mentioned) had been used for calculating the whole County. It will be unnecessary for me to further notice like differences which occur, in cols. 31, 36, 37, and 40, between calculations on Hundred and on County averages, such

such disagreements being unavoidable; in the lower part of the Table they are kept out of sight, by adhering to the casting, and deducing averages from them.

After making in the Parliamentary volume of 1801, corrections, or removes from one Hundred to another, for Brackenfield, Donisthorpe, High-low, Intake, and Stoney-Middleton (see p. 565 Note), cols. 5, 7, 17, 23, 27, 29, 35 and 39 of my Table, were filled up therefrom. There is in this Parliamentary volume another column, entitled, " All other Persons not comprized in the preceding classes ;" that is, of Agriculturists and of Traders, &c. in my cols. 35 and 39, and consequently, by deducting the sum of these last, from col. 23, the numbers of this column *ought* to appear; but owing to mistakes in this part of the Returns, this is often otherwise stated; which, as well as its being of little use to my purpose, occasioned my omitting this column, the last but one in the Parliamentary pages.

In like manner, after making corrections or removes, in the Parliamentary volume of 1811, for Appleby, Grange-Mill (not Iron-Brock-Grange), Ravenstone and Weston Underwood, my columns 2, 6, 8, 9, 18, 24, 28, 30, 34, and 38, were filled up therefrom: the last column but 3 of the volume, entitled, " All other Families not comprized in the two preceding Classes," I have omitted, as not useful to my purpose, and because the sum of cols. 34 and 38, taken from col. 24, will in every case give the numbers set down in this omitted column. In all instances, I have omitted the proportions of Men serving in the Army and Navy, and in the Militia, in cols. 23, 24, 27, and 28.

The next step was, by the calculation and addition of 3-13ths of the increase from 1788 to 1801, to fill up col. 4; shewing the probable number of Houses in each

Hundred, in 1791; that is, 10 years preceding the first, and 20 years preceding the second Enumeration of Houses, in cols. 5 and 6. And in like manner, to calculate the corresponding number of People in 1791, in col. 22, by help of cols. 21 and 23.

I then had materials, for proceeding with the calculations of *average* results, viz. in cols. 10 and 11, of Families in each House, from cols. 17 and 5, and 18 and 6; in cols. 13 and 14, of Persons in each House, from 23 and 5, and 24 and 6: in cols. 15 and 16, of the Rate of increase of Houses in 10 years, from cols. 5 and 4, and 6 and 5: in cols. 19 and 20, of Persons in each Family, from cols. 23 and 17, and 24 and 18: in cols. 25 and 26, of Rate of increase of Persons in 10 years, from cols. 23 and 22, and 24 and 23: in cols. 31 and and 32, of Females to each Male from cols. 29 and 27, and 30 and 28.

It then became necessary to supply, by calculations, the numbers of Families employed in Agriculture and in Manufactures in 1801, in cols. 33 and 37, from cols. 35 and 19, and 39 and 19; and the numbers of Persons employed in Agriculture and in Manufactures in 1811, in cols. 36 and 40, from cols. 34 and 20, and 38 and 20; and then I could proceed to complete my Table, viz. in cols. 41 and 42, of Agriculturists, *per cent*. of the People, from cols. 35 and 23, and 36 and 24: and in cols. 42 and 43 of Manufacturers *per cent*. of the People, from cols. 39 and 23, and 40 and 24.

I will proceed now to some observations, which arise from considering and comparing the several columns and averages in my Table facing p. 577, viz. From calculations on cols. 5 and 6 it appears, that the *number of Houses* in Derbyshire, were, in 1801, nearly 1-46th part of the whole of the Houses in England,

and

and in 1811 they were 1-47th part of the whole: so, from cols. 23 and 24 it appears, that the *Persons* in Derbyshire were, in 1801, 1-51½th part of the whole Population of England, and in 1811 its 1-52d part.

From cols. 7 and 8, and 5 and 6, it appears, that the proportions of *Uninhabited Houses* (including a few new ones, it may be presumed), in Derbyshire, in 1801, was 1 in 23¼ of the whole, and in all Great Britain at the time, 1 in 28: the proportion of *Empty Houses* in 1811 in the County, was 1 in 29½, and in Great Britain, 1 in 33¼: so from cols. 6 and 9 it appears, that at this latter period, the proportion of *new Houses* yet unoccupied in the County, was 1 in 160 of the whole, and in Great Britain 1 in 113.

The most Houses having *double Families* in them, in the Derbyshire Hundreds, occur (cols. 10 and 11) in the Borough of Derby, in 1801 and 1811, yet considerably less so than in Great Britain generally; and those least so incumbered, were in the High-Peak, in both these years.

Cols. 12, 13 and 14, shew the proportions of *Persons in each House* to be, highest in the Borough of Derby in 1788, and to decrease rather rapidly thence through 1801 and 1811 (as the Families also, in the preceding columns, had done): and to be lowest in High-Peak in 1788, in Wirksworth Wapentake in 1801, and in the Borough of Derby in 1811: which last seems rather a curious result, when all the other proportions of Persons to a House in my Table, appear increasing, except Morleston and Repton Hundreds, from 1801 to 1811.

Cols. 15 and 16 shew, the *increase of Houses* to have been greater in Morleston from 1791 to 1801, and in the Borough of Derby from 1801 to 1811: the least increases of Houses were in Scarsdale to 1801 (proving

ing just equal to the average increase in all Great Britain in the next 10 years, and rather exceeding the County average in that period), and in Wirksworth Wapentake to 1811.

The proportionate *number of Persons in each Family*, cols. 19 and 20, appear to have been greatest in Appletree Hundred in 1801, and in Morleston in 1811: and to have been least in the Borough of Derby in 1801 and 1811. The proportion of persons to a House, in general, seems rather increasing : a necessary consequence of increasing population.

Cols. 25 and 26 shew, the average *increase of Inhabitants* to have been greatest in Morleston, in the 10 years to 1801, and in the Borough of Derby, in the succeeding 10 years : the rate of increase evidently lessening, in these periods, except in Derby Town. Fourteen per cent. is very nearly the increase of people in the County, and in England, and in Great Britain also, in the 10 years to 1811 : and consequently, the annual increase of the British People is 1.4 per cent.: that is nearly 1½ per cent. per annum. Further light will be thrown on this important subject, when considering the deduction from Parish Registers, in the next article of this Section.

Cols. 31 and 32 shew, that the resident *Males* are inferior in number to the *Females*, in all instances except in Appletree and Morleston Hundreds, in 1801: and in which particulars there may perhaps be mistakes. In the Borough of Derby, Females most prevailed in proportion to Males, in 1801 and 1811 : the Males most prevailed in Appletree in 1801, and in High Peak in 1811. It seems a little singular, that the proportion of Females seemed increasing, between the enumerations of 1801 and 1811, in every Hundred

of

of Derbyshire; although in each of the 3 Kingdoms it was decreasing: its inland situation may perhaps have occasioned this, aided by its importations of girl Apprentices, and enlistments of boys who have been Apprentices. See pages 501 and 504.

It seems a question of very considerable importance in the political economy of this Country, to ascertain with precision, the proportion of the whole People who are chiefly employed in, or are dependent, like their young children, on *Agriculture*, and also the proportion of Persons who are employed in *Manufactures*, Trades, &c. or are immediately supported therefrom.

The Act of 1801, required Returns of the number *of Persons* in each of the above Classes (col. 35 and 39), and in a third class of unemployed Persons, which I have mentioned in p. 583, but omitted it in my Table. The Sum of which three Classes ought, in every Return, to have agreed with the total Population (col. 23), but which, in so many instances, proved otherwise, that an idea was, as I conceive, too hastily taken up, that no credit was due to the numbers of Agriculturists and of Manufacturers, obtained by these Returns; and accordingly, in framing the Bill for the enumeration of 1811, it was proposed, to require Returns of the employments *of Families*, instead of Persons.

Having myself seen reason to think, that a very *useful approximation*, towards ascertaining the numbers and proportions of the two most important Classes of *employ* for the People had been made, in the Returns of 1801; and that more full and explicit *directions to the Overseers* were only wanting, as to the classing of certain occupations, &c. of a medium or doubtful description, to obtain every practicable degree of accuracy, in a next enumeration; under this impression, I

took

took some pains, when the Bill of 1811 was in the House of Commons, to shew to a few Members, to one in particular, that the proposed alteration, for Returns of *Families*, instead of *Persons*, would only procure *the appearance of correctness*, instead of its reality ; and by which change, an opportunity would be lost, of better defining, or explaining the three Classes of Occupations, by Notes on the printed Forms, and obtaining correspondingly correct and consistent Returns, under each : but unfortunately, I failed in my endeavours on this head ; as also on another, regarding the Towns, to which reference will be presently made.

On a comparison and consideration of cols. 33 and 34, and 37 and 38 ; and of cols. 35 and 36, or 41 and 42, and 39 and 40, or 43 and 44 in my Table facing p. 577, there cannot, I think, remain reason for concluding, with Mr. Rickman, that the answers as to *Families* employed in 1811, "appear to have been made with care and *sufficient distinctness :*"—*with care* they may have been made, as far as, somehow to fit the whole set down as Agricultural, Manufacturing, or Unemployed Families, to the whole Families in a previous column : but when the undefined nature of " a Family" in very numerous cases, Apprentice-Houses and Boarding-Schools, in particular, and the equally numerous uncertainties, as to *which class*, various more or less actively and usefully employed parts of the Community were to be referred, are duly considered, it certainly cannot be maintained, that the late Returns of Family employments, are " sufficient," for any useful purpose whatever.

Whether we reduce the Families into Persons in one Enumeration, or the Persons into Families in the other,

other, by Rules of the parties' own furnishing and acting on (as is done in my columns), and thus make the results of the two Enumerations comparable under these Classes, such great and excessive proportion appears, in the latter Enumeration, of Agriculturists and Manufacturers, as to divest them of all credibility, in my opinion, and therefore I have marked these years with a note of interrogation (?) at the top of my Table.

In Appletree and Morleston Hundreds, for instance, the sum of these two classes in 1811 (cols. 42 and 44) are each 89 per cent. of the whole People therein; leaving only 11 per cent. for the third class of unemployed Persons and their dependants!, which is well known to be much more considerable in these Hundreds, and was, on their average, returned 48 per cent. in 1801!. Throughout the County 19 per cent., and throughout Britain 21 per cent., are the proportions thus assigned to the third Class in 1811; whereas, in 1801 they were returned 56 and 60 per cent. respectively, or very nearly three times as many, as are 10 years afterwards assigned to them, in every case.

Further proofs of the same thing, viz. of *the loose and very different manner in which*, in different Counties and Places, *Employments were assigned to whole Families of persons at once*, in 1811, appear, I think, on the face of a sort of supplemental Sheet, to its Volume of Returns, prepared by Mr. R. and printed in 1816: whose 8th column is similar to my 42nd: in which, altho' *Lancaster, Stafford*, and *Warwick* Counties (in this order), *are ranked very low* in Agriculturists (proportionate numbers of Families and Persons considered, and always excepting the Metropolitan County of Middlesex), as I concluded to be the

case,

case, when writing p. 476, before seeing this Sheet, or making any similar calculations whatever, yet *Durham* and *Northumberland*, are, by Mr. R.'s Table, assigned *lower* Agricultural Rank of the same kind, than *Yorkshire*, or even than *Stafford* or *Warwick* Counties! : and what cannot but be equally inconsistent with all the facts, *Chester, Cumberland, Kent, Leicester*, and *Nottingham* Counties, are assigned *equal rank* (within 1 per cent. only), *with each other and with Derbyshire!* How can Cumberland and Northumberland differ, in their proportionate numbers of Agricultural Families, as 38 to 29?

Sincerely hoping and trusting, that the important subject of Population will again be taken up by the Legislature, in or previous to 1821, and that a new Enumeration will then be made; and so on at the end of every 10 years: I venture here earnestly to recommend, that the former mode of returning *the employ of Persons*, may be again restored, and acted on, under three well explained Classes of Occupations, &c.*

In

---

* The Notes of direction to the several Overseers, to which allusion is made in page 588, might be somewhat like the following, viz.

The 1st, *or Agricultural Class*, to contain, all Farmers, or Occupiers of Land, as a business:—their hired and day Servants and Labourers: —their several Wives, young Children (or older ones at home, not employed out of this class), and Household Servants.—The Bailiffs and Farming Servants, &c. of Gentlemen or Manufacturers, &c. who also occupy Land.—Market Gardeners and their home Work-people, &c.— Drovers and Salesmen of Live Stock. Certain Handicrafts, occasionally employed on dwelling-houses, as Thatchers of Ricks and Hovels; Masons of Fence-walling; Carpenters of Gates, Rail and Pale Fences, &c.

The 2nd, *or Manufacturing Class*, to contain, all Manufacturers, Tradesmen, Dealers, &c.—their Clerks, Overlookers, Journeymen, Apprentices, Workmen, Shopmen, Porters, &c.—their several Wives, young

IMPROVED POPULATION RETURNS, SUGGESTED.

In pages 90 and 91 of Vol. I., I have briefly described the intermixtures and want of co-extension in the Counties, Hundreds, and Parishes on the southern borders of Derbyshire, and Pinxton on its eastern border, see p. 19 of Preface to Vol. II. and in various parts of its interior, all which will, I hope, be attended to, in preparing and distributing, and filling up the Forms, for the next general Enumeration, so that *sufficiently separated Returns may be made*, and with proper information for their subsequent collection, that the *Enumerations* may exactly tally with the divisions of *Parish Registers*, of Marriages, Christenings and Burials, which are to be treated of under the next article of this Section.

---

young Children, (or others at home, not employed out of this class):—Converters of, Dealers in, or Hawkers of Provisions:—Merchants, Traders and their Clerks, Warehousemen, &c.—Bankers, Stock-brokers and their Clerks, &c.—Auctioneers, Brokers, Valuers, &c.—Professional Men and Artists, &c.—Teachers, Schoolmasters, &c.—Miners, Colliers, Quarriers, Pit-men, &c.—Canal-diggers, Managers, Boat and Railway-Men, &c.—Fishermen, Watermen, Ferrymen, Pilots, &c.—Road-makers and menders, Toll-men, &c.—Scavengers, Dustmen, Chimney Sweepers, &c.—Stage Coach and Waggon Masters, Coachmen, Waggoners, Guards, Book-keepers, Porters, &c.—Post-Chaise, Hackney Coach and Chair Men, &c.—Inn, Tavern, Coffee-house, Public-House, Lodging-house, or Bathing-house Keepers, Waiters, Servants, Hostlers, &c.—The Post-Office Keepers, Clerks, Letter-Carriers, &c.

The 3d, or mostly Unemployed Class, to contain, Rich or unemployed Persons;—their Families (except any individuals of them at home, that may be employed in either of the preceding Classes, or be commissioned or enrolled in the Army or Navy), and Household Servants and Attendants:—Game-keepers, Huntsmen, &c.—Military and Naval Officers, &c. or Men, unemployed or unpaid by the State, or Foreigners not in any Trade or Profession:—Clerical Men, of all degrees:—Government and Tax Officers, Clerks, &c.:—Excise and Custom Officers:—Showmen, Comedians, Female Prostitutes, &c. &c. Poor Widows and aged Persons:—Orphan, or Soldiers' and Sailors' young or unemployed Children:—permanently Lame or diseased Persons, Paupers, Beggars, &c.

If

If we take England and Wales to contain 37,267,000 Acres* (see the General Report of the Board, on "En-

---

* Mr. Rickman's Table, on a Sheet, comparing the " Area, Fertility, and Population" of the Kingdom (already adverted to in p. 589), shews, as the result of a very accurate mode of measuring the surface of Mr. Arrowsmith's magnificent Map of England and Wales (mentioned in the Note on p. 206), that England and Wales contain 37,094,400 *Statute Acres*, England 32,342,400 Acres, and Derbyshire 656,640 Acres. From other sources, that 31,830,101*l*., 30,084,348*l*., and 640,701*l*. are the respective annual values of *Rents and Tithes* in these places, on which the Income Tax was assessed in 1811; that 10,150,615 persons, 9,538,822, and 185,487 persons (see my col. 24, and p. 577), respectively, were the whole *Resident Population* in that year; that 2,142,147 Families, 2,012,391 Families, and 37,440 Families (col. 18) respectively, were the *whole Number of Families:* and that 770,199 Families, 697,353 Families, and 14,283 Families (col. 34), respectively, were the numbers of *Families chiefly employed in, and maintained by Agriculture*, in that year.

From which *data*, Mr. R. shews, that 549*l*., 595*l*. and 624*l*. respectively, were *the annual value per square Mile* of the surface, of England and Wales, of England, and of Derbyshire, respectively; or *per Acre*, 17*s*. 2*d*., 18*s*. 7*d*., and 19*s*. 6*d*. respectively, yielded by the Tithe-free and Tithable, cultivated and waste Lands, Waters, Roads, &c. altogether. The number of *Persons on a square Mile*, Mr. R. calculates at 175, 189, and 181, respectively; and, that Land and Tithe Owners drew annually, on the average, as *Rent from each Agricultural Family*, 41*l*. in England and Wales, 43*l*. in England, and 45*l*. in Derbyshire: the highest of these County Agricultural Family Rents being in Northumberland 89*l*. per annum, and the lowest in Anglesea, 14*l*. and in Carnarvon, adjoining it, 15*l*. per annum.

Can it however be so?, that *Anglesea*, none of whose surface is so sterile, high, or uneven, as full one-quarter, or perhaps one-third of Northumberland is, on its western side, yields not one-sixth of the Rent and Tithe from a family of Agriculturists, on the average, as Northumberland does? any more, than that the latter Families yield *twice* the Rent and Tithes each, that Agricultural Families do, thro' so rich a County as Derbyshire? I fully believe not; but think, that the *irregular and uncertain Returns* made, as to the numbers of Agricultural *Families*, as observed in p. 589, is the source of these, and multitudes of other inconsistencies.

closures,"

closures," p. 143), the same had in 1801, 23¼ Acres to each inhabited House, and 4⅐ Acres to each Individual; while Derbyshire, if containing 622,080 Acres (as mentioned in pages 76 and 312 of Vol. I.), gave a House to each 19½ Acres, nearly, and a Person to less than every 3⅘ Acres of its surface: much as Writers had said to its disparagement, in the National Scale, (as observed, Vol. I. p. 95), on the score of its (formerly) bleak and sterile Wastes, and the alledged inhospitable climate of its Alpine Mountains.

The *Towns* of a Kingdom, as having an Economy peculiarly their own, and differing materially in their political and moral effects, according to their extent or number of their Inhabitants, it has long been desirable with Political Economists, to obtain a correct and perfect enumeration of all the larger British Towns: their expectations were, however, considerably disappointed, after the publication of the Returns of 1801; from finding few more than the Corporate Towns, separately and clearly distinguished therein, while an inadvertent abbreviation of " Township" into *Town*, in other parts of the printed volume of Returns, led to the commission of various and egregious errors[*], by those who attempted to form therefrom, more extensive Lists of the population of British Towns, and

---

[*] The most striking of these consisted, in representing *Hundersfield*, constituting indeed four Townships instead of one "Township," of scattered Villages and Houses in Rochdale Parish: and *Spotland*, two other "Townships" in the same Parish, in Lancashire, each as Towns, of 10,671 and 9,031 Inhabitants, respectively!; and the "Township" of *Quick* (and Sadleworth), of the like scattered kind, in the same Parish, but in the West Riding of Yorkshire, as another Town, of 6,713 Inhabitants!!

which erroneous Lists have, unfortunately, in the interim, obtained a wide circulation in various works.

When commencing the Section on Canals in this Volume, I saw the necessity, in order to mark the connection of each of such Establishments with the Population of its district, of making a very careful research into the Population Volume of 1811, in order, as far as I found practicable, to extract all the larger British *Towns*, separately from the country parts of the several Parishes, Townships, &c. in which they are situated, and to mark their order as to magnitude: which List I have mentioned in the Note on p. 298, and promised its insertion here, in Alphabetical order, as follows, viz.

An Alphabetical List of 700 of the largest British Towns*, shewing their Number, in the order of Magnitude, and Numbers of People.

Order or No.	Towns' Names.	No. of Persons.	Order or No.	Towns' Names.	No. of Persons.
144	Aberbrothick	5,280	497	Alcester	1,862
23	Aberdeen	21,639	153	Aldstone Moor	5,079
328	Abergavenny	2,815	678	Alford	1,169
478	Abergeley	1,944	253	Alfreton	3,396
428	Aberystwith	2,264	150	Alloa	5,096
167	Abingdon	4,801	137	Alnwick	5,426
230	Alban's, St.	3,653	417	Alton	2,316

* The erroneous statements mentioned in the last Note, and others, as to *Towns*, in the 13th volume of the "Monthly Magazine," pp. 405 and 479, having been wide spread, I judged it proper, in order, as far as possible, to counteract these errors, to send a copy of this List (but differently arranged, viz. in the order of the Population), for insertion in that Work, and where it is printed in Vol. 40, p. 487, preceded by several observations and suggestions, for further correcting this List, to which obervations I beg to refer.

462 Altrin-

POPULATION OF BRITISH TOWNS.

Order or No.	Towns' Names.	No. of Persons.	Order or No.	Towns' Names.	No. of Persons.
462	Altrincham	2,032	598	Beaconsfield	1,461
430	Amersham	2,259	425	Beaminster	2,290
653	Ampthill	1,299	515	Beaumaris	1,810
263	Andover	3,295	308	Beccles	2,979
192	Andrews, St.	4,311	602	Bedale	1,078
260	Annan	3,341	178	Bedford	4,605
379	Ardrossan	2,526	585	Beer-Alston	1,504
440	Arundel	2,188	129	Belper	5,778
580	Asaph, St.	1,520	475	Berkhamstead	1,963
451	Ashburne	2,112	532	Berwick, North,	1,727
299	Ashburton	3,053	89	Berwick-on-Tweed	7,746
284	Ashby-de-la-Zouch	3,141	106	Beverley	6,737
377	Ashford	2,532	248	Bewdley	3,454
26	Ashton, Underline, and Staley Bridge	19,052	487	Bicester	1,921
			269	Bideford	3,244
318	Atherstone	2,921	631	Biggar	1,376
618	Attleburgh	1,413	491	Biggleswade	1,895
404	Axminster	2,387	645	Bingham	1,326
249	Aylesbury	3,447	168	Bingley	4,782
524	Aylsham	1,760	6	Birmingham	80,753
116	Ayr	6,291	517	Bishop Aukland	1,807
591	Bakewell	1,485	635	——— Castle	1,367
606	Baldock	1,438	358	——— Stortford	2,630
612	Bampton	1,422	508	——— Waltham, Hants	1,830
327	Banbury	2,841	97	——— Wearmouth	7,060
237	Banff	3,603	38	Blackburn	15,083
409	Bangor	2,383	393	Blandford Forum	2,425
396	Barking	2,421	579	Blyth, Northumb.	1,522
615	Barmouth, Merio.	1,417	458	Bodmin	2,050
307	Barnard Castle	2,986	20	Bolton, Great & Little,	24,149
564	Barnet	1,579	83	Boston	8,180
158	Barnsley	5,014	560	Bourn	1,591
204	Barnstaple	4,019	305	Bradford, Wilts,	2,989
669	Barton	1,228	88	Bradford, Yorksh.	7,767
354	Basingstoke (g)	2,656	647	Bradnich	1,321
16	Bath	31,496	422	Braintree	2,298
378	Battle	2,531	459	Brampton, Cumb.	2,043

637 Bran-

Order or No.	Towns' Names.	No. of Persons.	Order or No.	Towns' Names.	No. of Persons.
637	Brandon	1,360	389	Cardiff	2,437
135	Brechin	5,559	446	Cardigan	2,129
278	Brecon	3,196	48	Carlisle	12,531
668	Brent, South,	1,230	94	Carmarthen	7,275
140	Brentford	5,361	620	Castle Cary	1,406
664	Brentwood	1,238	696	Cawood	1,053
188	Bridgenorth	4,386	353	Chapel-en-le-Frith (g)	2,667
160	Bridgewater	4,911	543	Chard	1,688
218	Bridlington	3,741		Chatham, see *Rochester.*	
238	Bridport	3,567	279	Cheadle, Staff.	3,191
636	Brigg, Glanford,	1,361	175	Chelmsford	4,694
51	Brighton	12,012	78	Cheltenham	8,325
7	Bristol	71,297	368	Chepstow	2,531
311	Bromley	2,965	233	Chertsey	3,629
101	Bromsgrove	6,932	456	Chesham	2,071
686	Bromyard	1,101	37	Chester	16,140
164	Broseley	4,850	183	Chesterfield	4,476
482	Bruntisland	1,934	113	Chichester	6,425
576	Bruton	1,536	251	Chippenham	3,410
306	Buckingham	2,987	471	Chipping Norton	1,975
531	Bungay	1,730	146	Chorley	5,182
641	Burford	1,342	572	Christchurch	1,553
187	Burnley	4,386	507	Chudleigh	1,832
319	Burrowstowness	2,919	643	Chumleigh	1,340
207	Burton-on-Trent	3,979	182	Cirencester	4,540
74	Bury, Lanc.	8,762	235	Clackmannan	3,605
84	Bury St. Edmunds	7,986	563	Cleobury Mortimer	1,582
	Buxton, unknown.		523	Clitheroe	1,767
180	Caernarvon	4,595	312	Cockermouth	2,964
697	Caistor	1,051	387	Coggeshall	2,471
619	Calder, Mid,	1,408	47	Colchester	12,544
239	Calne	3,547	407	Coldstream	2,384
85	Cambletown	7,807	573	Coleford	1,551
54	Cambridge	11,108	552	Coleshill	1,639
687	Camelford	1,100		Colnbrook, unknown.	
672	Campden, Chipping,	1,214	141	Colne	5,336
59	Canterbury	10,200	457	Columb, St. Major	2,070

177 Con-

POPULATION OF BRITISH TOWNS.

Order or No.	Towns' Names.	No. of Persons.	Order or No.	Towns' Names.	No. of Persons.
177	Congleton	4,616	374	Dorchester	2,546
81	Coventry	17,923	267	Dorking	3,259
559	Crail	1,600	70	Dover	9,074
308	Cranbrook	2,994	522	Downham	1,771
502	Crediton	1,846	359	Downton	2,624
300	Crewkerne	3,021	255	Drayton	3,370
509	Crich (g)	1,828	498	Driffield	1,857
262	Crieff	3,300	454	Droitwich	2,079
398	Cromarty	2,413	640	Dronfield	1,343
658	Cromford	1,259	41	Dudley	13,925
535	Crowland	1,713	288	Dumbarton	3,121
611	Crowle	1,424	67	Dumfries	9,262
86	Croydon	7,801	208	Dunbar	3,965
452	Cuckfield	2,068	340	Dunblane	2783
320	Cullumpton	2,917	17	Dundee	29,616
557	Culross	1,611	53	Dunfermline	11,649
627	Cumnock, New,	1,381	638	Dunkeld	1,360
468	Cumnock, Old,	1,991	466	Dunmow, Great,	2,015
169	Cupar	4,758	293	Dunse	3,082
173	Dalkeith	4,709	556	Dunstable	1,616
155	Darlington	5,059	105	Durham	6,763
281	Dartford	3,177	370	Dursley	2,580
529	Dartmouth	1,734	566	Dysart	1,578
334	Daventry	2,758	567	Easingwold	1,576
92	Deal	7,351	360	Eastbourne	2,623
654	Deddington	1,296	329	East Grinstead	2,804
322	Deerham, East,	2,888	5	Edinburgh	82,244
343	Denbigh	2,714	571	Egremont	1,556
	Deptford, see *Greenwich.*		179	Elgin	4,602
45	Derby	13,043	134	Ellesmere	5,689
216	Devizes	3,750	513	Eltham	1,813
156	Dewsbury	5,059	194	Ely	4,249
588	Dingwall	1,500	298	Enfield	3,055
366	Diss	2,590	597	Epping	1,473
295	Dolgelly	3,064	294	Evesham	3,068
100	Doncaster	6,935	684	Ewell	1,135
420	Donnington, Castle,	2,308	28	Exeter	18,896
			492	Eye	

Order or No.	Towns' Names.	No. of Persons.	Order or No.	Towns' Names.	No. of Persons.
492	Eye	1,893	505	Guisborough	1,834
602	Fairford	1,444	189	Haddington	4,370
625	Fakenham	1,382	365	Hadleigh	2,592
62	Falkirk	9,929	102	Hales Owen	6,888
416	Falkland	2,317	516	Halesworth	1,810
209	Falmouth	8,933	69	Halifax	9,159
261	Fareham	3,325	265	Halstead	3,279
321	Farnham	2,911	112	Hamilton	6,453
504	Farringdon, Great,	1,843	541	Harlow	1,695
229	Faversham	3,655	562	Harrowgate	1,583
568	Fishguard	1,572	530	Hartland	1,734
608	Flint	1,433	698	Hartlepool	1,047
225	Folkstone	3,697	220	Harwich	3,732
431	Fordingbridge	2,259	148	Haslingden	5,127
132	Forfar	5,652	512	Hastings	1,823
316	Forres	2,925	351	Hatfield	2,677
648	Fowey	1,319	628	Hatherleigh	1,380
473	Framlingham	1,965	510	Havant	1,824
639	Frodsham	1,349	292	Haverford West	3,093
65	Frome	9,493	185	Hawarden	4,436
147	Gainsborough	5,172	226	Hawick	3,688
73	Gateshead	8,782	688	Hay	1,099
445	Germans, St. Cornw.	2,139	616	Helmsley	1,415
2	Glasgow	100,749	423	Helston	2,297
480	Glastonbury	1,937	273	Hemel Hempstead	3,240
79	Gloucester	8,280	291	Henley	3,117
241	Godalming	3,543	695	Henley in Arden	1,055
679	Godstone	1,156	93	Hereford	7,306
87	Gosport	7,788	210	Hertford	3,900
453	Goudhurst	2,082	384	High Wycomb	2,490
231	Grantham	3,646	122	Hinckley	6,058
289	Gravesend	3,119	234	Hitchin	3,608
694	Grays Thurrock	1,055	662	Hoddesdon	1,249
27	Greenock	19,042	313	Holbeach	2,962
12	Greenwich & Deptford	36,780	674	Holsworthy	1,206
335	Grimsby, Great,	2,747	302	Holyhead	3,005
310	Guilford	2,974	114	Holywell	6,394

333 Ho-

POPULATION OF BRITISH TOWNS.

Order or No.	Towns' Names.	No. of Persons.	Order or No.	Towns' Names.	No. of Persons.
338	Honiton	2,735	435	Kinross	2,214
361	Horncastle	2,622	217	Kirkaldy	3,747
317	Horseley	2,925	634	Kirkby-Lonsdale	1,368
534	Horsham	1,714	548	Kirkby-Moorside	1,673
	Hounslow, unknown.		665	Kirkby-Stephen	1,235
514	Howden	1,812	333	Kirkcudbright	2,763
69	Huddersfield	9,671	436	Kirkham	2,214
19	Hull	26,792	219	Kirkintilloch	3,740
401	Huntingdon	2,397	533	Kirkwall	1,715
332	Huntley	2,764	195	Knaresborough	4,234
415	Hythe	2,318	448	Knutsford	2,114
483	Ilfracombe	1,934	131	Lanark	5,667
441	Ilminster	2,160	68	Lancaster	9,247
400	Inverkeithing	2,400	356	Langholm	2,636
57	Inverness	10,757	528	Lauder	1,742
44	Ipswich	13,670	570	Laugharne	1,561
130	Irvine	5,750	525	Launceston	1,758
264	Ives, St. Cornw.	3,281	536	Lavenham	1,711
392	Ives, St. Hunts	2,426	285	Ledbury	3,136
184	Jedburgh	4,454	13	Leeds	35,951
	Johnstown, unknown.		223	Leek	3,703
103	Keighley	6,864	21	Leicester	23,146
259	Keith	3,352	449	Leighton-Buzzard	2,114
186	Kelso	4,408	24	Leith	20,363
91	Kendal	7,505	581	Lenham	1,509
546	Keswick	1,683	274	Leominster	3,238
272	Kettering	3,242	117	Lewes	6,221
526	Keynsham	1,748	110	Lewisham	6,625
123	Kidderminster	6,057	75	Lincoln	8,599
60	Kilmarnock	10,148	372	Linlithgow	2,557
276	Kilsyth	3,206	632	Linton	1,373
621	Kimbolton	1,400	472	Liskeard	1,975
439	Kinghorn	2,204	157	Litchfield	5,022
663	Kingsbridge	1,242	4	Liverpool	94,376
496	Kingsclere	1,863	405	Llandiloes	2,386
199	Kingston	4,144	685	Llandilo-fawr	1,103
555	Kington	1,617	603	Llandovery	1,442

Qq4   499 Llan-

Order or No.	Towns' Names.	No. of Persons.	Order or No.	Towns' Names.	No. of Persons.
499	Llanfair	1,855	330	Marlow, Great,	2,799
474	Llangadoch	1,964	617	Marshfield	1,415
211	Llangelly, Carnar.	3,891	286	Maryport	3,134
382	Llanrwst	2,509	553	Mawes, St.	1,639
447	Llantrisaint	2,122	578	Measham	1,525
402	Lochmaben	2,392		Melcombe Regis, see Weymouth.	
1	London, Westminster and Southwark	1,009,546	549	Meldrum, Old,	1,655
			200	Melksham	4,110
565	Longtown, Cumb.	1,579	287	Melrose	3,132
138	Loughborough	5,400	444	Melton Mowbray	2,145
171	Louth	4,728	438	Mere	2,211
280	Lowestoft	3,189	55	Merthyr Tidvel	11,104
198	Ludlow	4,150	667	Middlewich	1,232
221	Luton (g)	3,716	659	Midhurst	1,256
503	Lutterworth	1,845	383	Mildenhall	2,493
586	Lydd	1,504	527	Milton, Kent	1,746
484	Lyme Regis	1,925	554	Milverton	1,637
355	Lymington	2,641	268	Minchinhampton	3,246
58	Lynn (King's)	10,259	699	Minehead	1,037
49	Macclesfield	12,299	493	Modbury	1,890
43	Machars, Old, Aberd.	13,731	511	Moffat	1,824
661	Machynlleth	1,252	193	Monk-Wearmouth	4,264
154	Madeley	5,076	243	Monmouth	3,503
277	Maidenhead	3,203	72	Montrose	8,955
66	Maidstone	9,443	550	Moreton Hampstead	1,653
350	Maldon	2,679	270	Morpeth	3,244
680	Malling, West,	1,154	601	Mottram	1,446
222	Malton, New,	3,713	151	Mould	5,083
3	Manchester	98,573	337	Moulton, South,	2,739
693	Manningtree	1,075	587	Mountsorrel	1,502
104	Mansfield	6,816	115	Musselburgh	6,393
700	Marazion	1,022	381	Nairn	2,504
120	Margate	6,126	205	Nantwich	3,999
539	Market Harboro'	1,704	336	Neath	2,740
582	Market Weighton	1,508	469	Neots, St.	1,968
371	Marlborough	2,579	95	Newark	7,236

477 New-

POPULATION OF BRITISH TOWNS.

Order or No.	Towns' Names.	No. of Persons.	Order or No.	Towns' Names.	No. of Persons.
477	Newburgh	1,951	675	Penrith	1,191
161	Newbury	4,898	344	Penryn	2,713
118	Newcastle, Underline	6,175	203	Penzance	4,022
18	Newcastle, upon Tyne	27,587	629	Pershore (Holycross)	1,378
375	Newent	2,538	35	Perth	16,948
488	Newmarket	1,917	228	Peterborough	3,674
213	Newport, Hants	3,855	174	Peterhead	4,707
412	Newport, Monmouth.	2,346	656	Petersfield	1,280
380	Newport-Pagnel	2,515	388	Petworth	2,459
609	Newport, Pembro.	1,433	413	Pickering	2,332
450	Newport, Salop	2,114	8	Plymouth	56,060
489	New Radnor	1,917	479	Plymouth Dock	1,941
	New Sarum, see *Salisbury*.		575	Pocklington	1,589
			236	Pontefract	3,605
561	Newton, Lanc.	1,589	394	Pontypool (g)	2,423
461	Newton, Montgo.	2,025	166	Poole	4,816
433	Northallerton	2,234	149	Port Glasgow	5,116
77	Northampton	8,427	651	Port-Patrick	1,303
626	Northwich	1,382	9	{ Portsea / Portsmouth }	40,657
11	Norwich	37,256			
15	Nottingham	34,253	681	Potton	1,154
159	Nuneaton	4,947	227	Prescot	3,678
657	Oakham	1,266	33	Preston	17,065
604	Oakhampton	1,410	551	Prince Risborough	1,644
36	Oldham	16,690	624	Pwllheli	1,383
427	Olney	2,268	403	Ramsey	2,390
296	Ormskirk	3,064	197	Ramsgate	4,221
644	Orton	1,333	56	Reading	10,788
244	Oswestry	3,479	125	Redruth	5,903
363	Otley	2,602	421	Renfrew	2,305
323	Ottery St. Mary	2,880	463	Retford, East,	2,030
506	Oundle	1,833	145	Richmond, Surry	5,219
46	Oxford	12,931	297	Richmond, Yorksh.	3,056
589	Padstow	1,498	275	Rickmansworth	3,230
10	Paisley	37,722	266	Ringwood	3,269
385	Peebles	2,485	232	Rippon	3,633
397	Pembroke	2,415	108	Rochdale	6,723

22 Ro-

Order or No.	Towns' Names.	No. of Persons.	Order or No.	Towns' Names.	No. of Persons.
22	Rochester, Chatham, and Stroud	21,722	29	Shrewsbury	18,543
			544	Sidmouth	1,688
673	Rochford	1,214	325	Skipton	2,868
271	Romford (g)	3,244	521	Sleaford	1,781
547	Romsey	1,681	666	Sodbury, Chipping,	1,235
429	Ross	2,261	369	Solihull	2,581
315	Rotherham	2,950	596	Somerton	1,478
240	Rothsay	3,544	64	Southampton	9,617
600	Rothwell	1,451	574	Southend	1,541
649	Royston	1,309	352	Southwell	2,674
437	Rudgley	2,213	633	Southwold	1,369
518	Rugby	1,805	191	Spalding	4,330
242	Rutherglen	3,529	162	Stafford	4,868
655	Ruthin	1,292	690	Staindrop	1,087
558	Ryde	1,601	460	Stains	2,042
349	Rye	2,681		Staley-Bridge, see *Ashton*.	
252	Saffron Walden	3,403	181	Stamford	4,582
80	Salisbury	8,243	494	Standon	1,889
595	Saltash	1,478	652	Stevenage	1,302
419	Sandbach	2,311	127	Stirling	5,820
339	Sandwich	2,735	32	Stockport	17,545
346	Sanquhar	2,709	196	Stockton-on-Tees	4,229
96	Scarborough	7,067	605	Stokesley	1,439
519	Sedberg	1,805	418	Stone	2,314
650	Sedgfield	1,307	495	Stonehaven	1,886
258	Selby	3,363	590	Stoney Stratford	1,488
395	Selkirk	2,422	201	Stourbridge	4,072
486	Sevenoaks	1,922	676	Stow, on the Wold	1,183
337	Shaftsbury	2,635	467	Stowmarket	2,006
545	Sheerness	1,685	485	Stranraer	1,923
14	Sheffield	35,840	326	Stratford, on Avon	2,842
176	Shepton Mallet	4,638	391	Strathaven	2,439
256	Sherborne	3,370	689	Stratton	1,094
90	Shields, North,	7,699	424	Stromness	2,297
71	Shields, South,	9,001	142	Stroud, Glouc.	5,321
202	Shiffnall	4,061		Stroud, Kent, see *Rochester*.	
630	Shipston	1,377			

599 Stur-

Order or No.	Towns' Names.	No. of Persons.	Order or No.	Towns' Names.	No. of Persons.
599	Sturminster Newton	1,461	254	Ulverstone	3,378
246	Sudbury	3,471	592	Uppingham	1,484
50	Sunderland	12,289	465	Upton	2,023
314	Sutton Coldfield	2,959	283	Uttoxeter	3,155
411	Swaffham	2,350	399	Uxbridge	2,411
82	Swansea	8,196	76	Wakefield	8,593
642	Swindon	1,341	490	Wallingford	1,901
593	Tadcaster	1,483	136	Walsall	5,541
403	Tain	2,384	461	Walsham, Norf.	2,035
304	Tamworth	2,991	426	Waltham Abbey (g)	2,287
172	Tavistock	4,723	214	Walthamstow	3,777
98	Taunton	6,997	133	Wandsworth	5,644
569	Tenbury	1,562	406	Wantage	2,386
677	Tenby	1,176	257	Ware	3,369
331	Tenterden	2,786	538	Wareham	1,709
376	Tetbury	2,533	153	Warminster	4,866
165	Tewkesbury	4,820	52	Warrington	11,738
414	Thame	2,328	111	Warwick	6,497
390	Thetford	2,450	362	Watford	2,603
442	Thirsk	2,155	682	Watlington	1,150
691	Thornbury	1,083	139	Wednesbury	5,372
345	Thorne	2,713	206	Wellingborough	3,999
247	Thurso	3,462	81	Wellington, Salop	8,213
583	Tickhill	1,508	212	Wellington, Somerset.	3,874
671	Tideswell	1,219	348	Wells, Norfolk	2,683
107	Tiverton	6,732	364	Wells, Somerset.	2,594
324	Topsham	2,871	250	Welsh Pool	3,440
443	Torrington, Great,	2,151	622	Wem	1,395
341	Totness	2,725	594	Wendover	1,481
432	Towcester	2,245	455	Wenlock, Much,	2,079
501	Tring	1,847	520	Westbury	1,799
121	Trowbridge	6,075	607	Westerham	1,437
386	Truro	2,482	683	Wetherby	1,140
124	Tunbridge	5,932	170	Weymouth, and M. L.	4,732
434	Turreff	2,227	542	Whitburn	1,693
215	Twickenham	3,757	99	Whitby	6,969
126	Tynemouth	5,834	367	Whitchurch	2,589

148 White-

Order or No.	Towns' Names	No. of Persons	Order or No.	Towns' Names	No. of Persons
61	Whitehaven	10,106	613	Wokingham	1,419
148	Whitehorn	1,935	470	Wolsingham	1,983
152	Wick	5,080	39	Wolverhampton	14,836
40	Wigan	14,060	196	Woodbridge	4,332
309	Wigton, Cumb.	2,977	614	Woodstock	1,419
537	Wigtown	1,711	540	Wooler	1,704
476	Wilton	1,963	34	Woolwich	17,054
282	Wimborn Minster	3,158	623	Wooton Basset	1,390
500	Wincanton	1,850	577	Wooton Underedge	1,527
660	Winchcombe	1,256	42	Worcester	13,814
109	Winchester	6,705	128	Workington	5,807
119	Windsor	6,155	224	Worksop	3,703
670	Winslow	1,222	347	Worthing	2,692
245	Wirksworth	3,474	301	Wrexham	3,006
143	Wisbeach	5,309	646	Wye	1,322
410	Witham	2,352	610	Yarm	1,431
342	Witney	2,722	30	Yarmouth, Great,	17,977
373	Wivelscombe	2,550	290	Yeovil	3,118
584	Woburn	1,506	25	York	19,099

The above List of British *Towns*, notwithstanding it will be found, I trust, not less comparatively accurate, than it is also more extensive, than any previous one that has been collected, is yet, I am sensible, *very far from being sufficiently accurate*; " owing (as I have elsewhere observed) to very few, if any, of our large Towns, being co-extensive with the several Parishes and Townships in which they lie," and whence it appears (I beg to repeat), that " no accurate or useful Returns of the actual Population of such Towns could be obtained, without requiring, as would be perfectly easy to do, a separation of the Return, from every Parish or Township, which lies *partly within a Town* (above a certain size, say 500 Inhabitants), and *partly without it*."

The

The errors in this List, occasioned as above-mentioned, are not all of *excess*, in number of people, although perhaps the most considerable in number and amount, are so: but many *parts of Towns*, therein mentioned, are at present collected, without notice or remark, into other Townships, Parishes, Hundreds, and even Counties, in several instances; and consequently, all such places appear in *defect*, unless the country parts, of their other Parishes or Townships, happen to compensate the same.

It rarely happens that a considerable Town stands in the middle of, or wholly within a single Township, or even so, with respect to a Parish, Hundred, or County, as we contemplate larger and larger Towns; because of the circumstance, of all these divisions being far more ancient, than the present magnitudes of any of our Towns: most of which are situated on both banks of a Brook, or River, which, in very numerous of the instances, separate the Townships, Parishes, Hundreds, or Counties, or some or all of these.

Having been myself alike unsuccessful, in my wishes and endeavours, to see the last Population Bill improved, on this head, as on the other head mentioned in p. 588, I will not yet despair, if I should so long live, of seeing, through the sanction of the Board of Agriculture, these Improvements carried into effect, in 1821, the next regular period for an Enumeration; when we may hope to see a just comparative view of the extent and consequence, of all the larger and middling British Towns.

1. *Tables of Births, Burials, and Marriages.* I had some time ago made preparation, as is mentioned in p. 581, for giving here the full particulars regarding Derbyshire, from the two population volumes, of the

Marriages,

Marriages, Baptisms, and Burials, for every 10 years from 1700 to 1780, and for every year thence to 1810, and the Marriages for every year from 1754 to 1810, for each Hundred of the County, and for the whole collectively, but room will not permit of the insertion, except of the last Table, somewhat modified, and a Table of Abstracts of Parish Registers, with a few observations thereon.

A Summary of Parish Registers in Derbyshire.

Years.	Marriages.		Baptisms.				Burials.		
	No.	Average of 5 Years, to	Males.	Females.	Total.	Average of 5 Years, to	Males.	Females.	Total
1700	—	—	1367	1247	2614	—	1153	1205	2358
1710	—	—	1199	1178	2377	—	989	909	1898
1720	—	—	1299	1328	2627	—	1042	989	2031
1730	—	—	1399	1315	2714	—	1313	1250	2563
1740	—	—	1504	1479	2983	—	990	1003	1993
1750	—	—	1549	1500	3049	—	1001	967	1968
1760	849	811	1807	1639	3446	—	1070	1128	2198
1770	970	911	1928	1925	3853	—	1214	1327	2541
1780	984	986	2091	2021	4112	—	1381	1382	2769
1781	953	988	2085	2059	4144	—	1292	1340	2632
1782	1010	990	2092	2090	4182	—	1273	1384	2657
1783	1064	999	2075	2062	4137	—	1244	1200	2444
1784	1068	1016	2168	2075	4243	4164	1323	1373	2696
1785	1087	1036	2266	2064	4330	4207	1505	1490	2995
1786	1004	1047	2115	2025	4140	4206	1297	1362	2659
1787	1029	1050	2240	2068	4308	4232	1364	1231	2595
1788	1042	1046	2317	2067	4384	4281	1347	1451	2798
1789	1127	1058	2309	2164	4473	4327	1403	1568	2971
1790	1050	1050	2376	2076	4452	4351	1479	1543	3022
1791	1109	1071	2403	2173	4576	4439	1428	1485	2913
1792	1223	1110	2392	2071	4463	4470	1415	1512	2927
1793	1149	1132	2373	2234	4607	4514	1569	1559	3128
1794	1092	1125	2234	2196	4430	4505	1464	1598	3062
1795	1113	1137	2294	2239	4533	4522	1609	1658	3267
1796	1240	1163	2314	2221	4535	4514	1395	1521	2916
1797	1185	1156	2490	2376	4866	4594	1593	1696	3289
1798	1185	1163	2377	2288	4665	4606	1384	1507	2891
1799	1132	1171	2361	2355	4716	4663	1516	1572	3088
1800	1100	1168	2225	2176	4401	4637	1546	1631	3177
1801	1134	1147	2205	2121	4326	4595	1587	1703	3290
1802	1463	1203	2609	2524	5133	4648	1534	1571	3105
1803	1516	1269	2870	2653	5523	4820	1874	1940	3814
1804	1395	1322	2880	2667	5547	4986	1425	1491	2916
1805	1276	1357	2818	2606	5424	5191	1449	1457	2906
1806	1330	1396	2686	2652	5338	5393	1386	1449	2835
1807	1357	1375	2807	2782	5589	5484	1500	1588	3088
1808	1300	1332	2793	2796	5589	5497	1360	1396	2756
1809	1325	1318	2758	2693	5451	5478	1502	1475	2977
1810	1383	1339	2682	2699	5381	5470	1790	1856	3646

Abstracts of the Parish Registers of the Hundreds of Derbyshire, and of the Kingdom at large.

Hundreds, &c.	No. of Registers.	Marriages.						Baptisms.				Burials.
		Numbers, on average of Five Years, to			Persons to one Marriage. 1810.	Children to one Marriage. 1810.	Numbers, on average of Five Years, to			Persons to one Baptism. 1810.	Persons to one, in 1810.
		1788.	1791.	1801.	1811.			1788.	1791.	1801.	1811.		
Appletree Hund.	33	128	137	168	188	140	3·4	578	561	539	640	40	76
Borough of Derby.	5	101	105	143	179	73	2·5	319	333	366	449	29	47
High Peak Hund.	29	250	242	243	262	141	4·4	986	1043	1035	1149	32	56
Morleston and Lit. Hund.	35	133	145	145	176	168	4·5	522	522	670	800	37	66
Repton and Gres. Hund.	25	66	68	92	90	157	4·4	275	278	298	432	36	74
Scarsdale Hund.	35	229	232	247	282	147	5·0	1026	1061	1188	1402	30	53
Wirksworth Wap.	19	146	133	119	154	132	3·9	529	551	504	598	34	61
Derbyshire, - -	181	1050	1060	1168	1330	138	4·1	4232	4351	4637	5470	34	60
England, - - -	10,313	64,275	65,719	69,927	78,408	122	3·0	217,261	228,686	242,470	282,109	34	51
South Britain, - -	11,159	67,886	69,368	73,527	82,953	122	3·6	229,499	241,567	255,427	297,428	34	52
1.	2.	3.	4.	5.	6.	7.	8.	9.	10.	11.	12.	13.	14.

The Table in p. 607, is extracted from the Population Volumes of 1801 and 1811, and needs no further description; except as to its 3rd and 7th columns, which I have added, to facilitate the calculation of the number of Inhabitants in the County in any past year, (according to rules which I will mention presently), whether for comparison with the Enumerations which have actually taken place, or for supplying their want in other years, &c.

In order to render Mr. Pilkington's labour in 1788 (see p. 579), of enumerating the Houses in the County, and a portion of its People, as useful as possible, I have in the Table in p. 608, added cols. 3 and 5, carefully calculated from the volume of Population Returns *; cols. 4 and 10 have been added, for the convenience of those who may wish to calculate the Population of each Hundred in 1791, for comparison with the numbers in col. 22 of my Table facing page 577, deduced from quite other considerations than these Parish Registers. Cols. 5, 6, 11, and 12, are in like manner deduced from the volumes of Returns.

I have calculated col. 8 from cols. 12 and 6. Col. 7 is deduced from col. 6 and 24 in my Table p. 577 from which col. 24 and col. 12 here, col. 13 is obtained; and from col. 24 and the 5-year averages of Burials, taken from the volume of Returns, col. 14 is calculated.

---

* The Parish Registers being made up to *the ends of the years*, instead of the days of Enumeration, which were in March and May, my 5-year averages in this Table, are those for 1787, 1790, 1800, and 1810, and 4 years preceding each of them, although entitled 1788, 1791, 1801, and 1811.

An average of 5 years, seem to have been adopted by Mr. Rickman in his calculations, in the introductory parts of the volumes of Returns, because the Births of Children within that period after Marriage, form so considerable a proportion of the whole number of Baptisms from Marriages, on the average, that nearly similar results may be expected to be obtained by calculation in a direct simple proportion of the population of one period from those of another period, when the five year averages of Marriages and of Baptism in each are known, by using either of these Registered *data*.

Thus, for example, in Appletree Hundred, if 188 Marriages in 1811, occurred amongst 26,350 persons, 168 Marriages in 1801 would occur amongst 23,546 persons. Again, if 640 Baptisms in 1811 occurred amongst 26,350 persons, 539 Baptisms would occur amongst 22,192 persons, in 1801. But the number of persons on the actual Enumeration of that year, proved to be 23,417; shewing the result *by Marriages*, to be, in this particular case, much the nearest to the truth of the *two;* altho' generally, Mr. R. concludes the contrary to be the case, and has preferred calculating, by means of the Baptisms. Those of my Readers who are curious in these matters, may thus try the consistency of their calculations, and the enumerations, by all the various modes of check, which these Tables furnish.

On considering col. 7, it appears, that the greatest proportion of Marriages out of the whole people, in any of the Hundreds in Derbyshire, occurs in the Borough of Derby, where 1 in 73 were annually married in 1810, and the lowest proportion in Morleston, where 1 in 168 only married yearly; the average of the County being 1 in 138, and of England and Wales, 1 in 122.

In col. 8 it appears, that the lowest proportion of Children to a Marriage was in Derby 1 in 2½, and the highest in Scarsdale, double the former: the County average being $4\frac{1}{10}$, and that of South-Britain, 1 in $3\frac{6}{10}$.

In col. 13 it appears, that the greatest proportion of Births out of the whole people, occurs in the Borough of Derby, 1 in 29, and the lowest in Appletree, 1 in 40; the County and South British averages being 1 in 34; or nearly mid-way between the above extremes.

In col. 14, the Deaths appear to occur in the greatest proportion in the Borough of Derby, 1 in 47 persons: the Hundred of the least mortality was Appletree, 1 in 76: the mean of the County being 1 in 60, and of South Britain, 1 in 52.

Before quitting the subject of these Parish Registers, I beg to remark, that several rather material instances remain, of inaccuracies, owing to the Enumerations treated of in the last article, not being co-extensive with these Parish Registers, of which I am now speaking, viz. 1st, in places where *the Counties are intermixed*, as observed in p. 591; 2nd, where *the Parishes and Hundreds are not co-extensive*, or intermix; and 3d, where *the Townships, Parishes, or Hundreds are intermixed*, in like manner.

In order to remedy all these defects, in future Population Returns, I beg to suggest, that the Printed Forms sent to the Parish Officers, should contain queries to the following effect, viz. 1st, What is the name of the *Church or Chapel of Ease*, at which the Inhabitants of the Parish, Township, &c. for which you will make Returns, usually Marry, Christen and Bury? and the name and residence of the present *officiating Minister* there?—2nd, In case of different parts or divisions of the Parish, &c. for which you will make

Return, generally using different Churches or Chapels respectively, for their Marriages, Christenings, and Burials, mention the name of each, and the name and address of its Minister; and let *separate Returns*, in Form corresponding with the printed one, be made for the Population of each of such divisions: so that Enumerations may be capable of being made up, exactly to tally with the Parish Registers, in every case: which last, would be of considerably greater consequence, than rigidly adhering to the divisions of Hundreds, in making up the Returns: but both of these arrangements might easily be attended to and shewn, in the printed volume.

2. *Has the progress of Population depended solely on Food?* or on the permission and facility of raising *Cottages?* In order to answer this important inquiry by the Board, I have, on a very careful collation of the three Enumerations of Houses in the County in 1788, 1801, and 1811 (which are mentioned in p. 580), extracted all those places which shew a regular and very notable increase of Houses; and in another List, all those places which have decreased or fallen off, since the first period, viz.

A List of Places much increased in Numbers of Houses.

Places.	Houses.			Observations.
	1788.	1801.	1811.	
Alkmund, St.	244	411	576	Silk-Mills and other Factories, see pp. 479 to 496; and Trade; Canal-wharfs, pp. 355 to 360, &c.
Allestry	21	70	74	Vicinity to Derby; the late Mr. Mundy's Improvements.
Belper	433	831	1023	Inclosed in 1789, see II. 76.— Messrs. Strutt's Manufactories and Improvements, see II. p. 21.
				Brackenfield

A LIST OF INCREASED PLACES.

Places.	Houses. 1798.	1801.	1811.	Observations.
Brackenfield and Wooley-Moor	30	50	63	Inclosed, with Morton.
Bretby	26	48	55	The late Earl of Chesterfield's Improvements, see II. 21.
Buxton (part of)	89	170	180	The resort of Company to the Baths.
Chilcote	14	32	36	Inclosed.
Church Broughton	48	76	80	Ditto.
Codnor and Loscoe	56	218	307	Ditto, in 1791; Stocking-weaving (omitted p. 487).
Cole-Aston	46	43	94	Ditto, in 1808: Cutlery—perhaps the new E end of Dronfield Town, is included, in 1811?
Crich, Coddington, Wheatcroft	175	272	361	Ditto, in 1786: Lead-Mines, and Bull-bridge Wharf, p. 338; Limestone-Quarries, II. 421, Manufactures, &c.
Cromford (part of)	120	207	230	Ditto, with Wirksworth; Mr. Arkwright's Cotton Mills and Improvements, II. 21; Canal-Wharf, p. 337: Lead-Mines, &c.
Darley Abbey	47	92	116	Cotton-Mills, and Factories: Wharf, p. 356.
Dore	35	83	84	
Ferneylee and Whaley-bridge	39	69	69	Inclosed in 1772: Cotton-Mill, Factories: Canal-Wharf, p. 402.
Glossop, parish	1121	1547	1911	Manufactories surprisingly increased here, see p. 498: Canal and Rail-way Wharfs: Mr. Oldknow's Improvements, II. 21; new Roads, &c. &c.
Hartshorn	76	112	140	Inclosed: Factories: Major Hassal's Improvements.
Hassop	13	27	31	Lead-Mines: Mr. Eyre's Improvements.
Haslewood-Lane	20	55	68	Inclosed with Duffield.
Heanor	58	125	341	Inclosed: Collieries, Canal-Wharf, see p. 364: Stocking-weaving, &c.
Hulland, and Ward and Intake	39	75	97	Inclosed, about 1773.

A LIST OF INCREASED PLACES.

Places.	Houses. 1788.	1801.	1811.	Observations.
Ible and Grange-Mill	11	16	27	Inclosed. *Note,* Grange-Mill (Ivenbrook G) is improperly entered "Iron-Brock" G., and in the wrong Hundred, in 1811.
Ilkeston	354	487	599	Ditto, in 1794: Collieries; Canal-Wharf, p. 364: Stocking-weaving, &c.
Kilburne	50	78	90	Stocking-weaving, &c.
Kirk-Hallam	8	13	16	Canal by it, p. 400.
Lea and Dethick	44	108	116	Inclosed in 1776: Canal-Wharf, p. 311; various Factories.
Little Eaton	39	69	84	Canal and Rail-way Wharf, p. 357; Stone-Quarries; Stocking-weaving, &c.
Measham	120	210	254	Canal-Wharfs; Manufactories; and the late Mr. Wilkes's Improvements, see vol. II. p. 362.
Middleton by Wirk.	64	154	189	Inclosed, with Wirksworth: Lead-Mines.
Ockbrook	81	161	143	Inclosed, in 1772: Moravian Establishment: Stocking-weaving.
Sawley	78	145	169	Ditto, in 1787: Canal-Wharfs, pp. 364 and 423; Stocking-weaving, &c.
Shardlow and Great Wilne	62	103	142	Canal-Wharfs and large Warehouses, pp. 433 and 447; Boat-building, Malting, Lace-weaving, &c. &c.
Shipton and Heanor-wood	39	83	91	Collieries: Canal and Rail-way Wharf, p. 401: Mr. Mundy's Improvements; Stocking-weaving, &c.
Shirland and Higham	176	227	263	Inclosed, in 1777: Bleaching, Factories, Stocking-weaving, &c.: Miss Willoughby's Improvements.
Stanton-Ward and Newhall	49	160	190	Collieries.
Stapenhill	81	100	99	Inclosed; Frame Lace-working, &c.

Tansley

Places.	Houses. 1788. 1801. 1811.			Observations.
Tansley	29	81	84	Cotton, Bump and other Mills, Dyeing, &c.
Totley	21	48	48	Paper, Lead, Grinding and other Mills, Scythes, &c.
Turnditch	30	46	64	Lime-Quarries, II. 431; Linen-weaving, &c.
Werburgh, St.	398	564	784	Various Trades and Manufactures, see pp. 479 to 496.
Wessington	29	113	70	
West-Hallam	66	95	118	Collieries; Canal Wharf, p. 401.
Willington	36	52	68	Inclosed, in 1779: Canal Wharf, p. 433.
Yeldersley	6	32	40	

The above 44 places, contained 4541 Houses in 1788, which, in 23 years afterwards, had increased to 9634, or 2⅕ times as many as formerly!

The "Observations" which I have annexed, as to the most noted circumstances attending each place in the above List, pretty plainly point out, *Inclosures, Canal Establishments, successful Manufactures and Rural Improvements,* to have been the most operative causes, in the rapid increases of Houses and People, which are here shewn.

Nearly all other places in the County, except the few in the List that follows, shewed a moderate increase, and pretty uniformly so, with some few exceptions, of Houses, and People, within the period above mentioned, as is indeed evident, from comparing cols. 3, 5, and 6, and 21, 23, and 24, in the Table facing page 577.

A List of Depopulated Places.

A List of Places, decreasing, or nearly stationary, in numbers of Houses.

Places.	Houses. 1788. 1801. 1811.			Observations.
Abney, and A. Grange	30	28	25	In a high and rather inaccessible situation for Roads.
Alsop and Cold-Eaton	11	8	7	
Beeley	58	54	52	Wanted Roads, and the Inclosure; since effected.
Birchover	70	29	20	Probably a part of Stanton (stated to have 64, 158, and 737 Ho.) was included in 1788?: decline of the Lead-Mines in Winster and Wensley.
Calke	13	12	8	Removed, from Sir H. Crewe's Park.
Darley in the Dale	231	210	204	Inclosed in 1766: failure since of Lead-Mines in Wensley; and the decline since of the Toad-hole Cotton-Mills, &c.
Darwent-Chapel	26	23	20	Wants Roads, see I. 236; and III. 226.
Foolow	94	58	58	Exhaustion, nearly, of its Lead-Mines.
Grindlow	30	23	24	Ditto.
Middleton by Yolg.	45	39	42	Ditto.
Monyash	53	66	60	Ditto, inclosed.
Rowland	29	25	25	Ditto.
Sheldon	33	28	32	Ditto, inclosed.
Tideswell	254	284	271	Ditto, Cotton-Mill, Fustian, Calico, and Stocking-weaving, &c.
Wardlow	40	32	37	Ditto, Calico-weaving. Inclosed, in 1808.
Wensley	150	139	135	Ditto.
Wheston	13	10	11	A Road through it wanted, see p. 225, Note: inclosed in 1807.
Winster	218	190	188	Inclosed about 1765: complete exhaustion of its Lead-Mines, see I. 338, 379; the Town is in a miserably decaying state.
Yolgrave	136	154	172	Its Lead-Mines had quite declined, but some recovered a little, about 1805, see I. 339.

Among

Among the above places, Monyash, Tideswell, and Yolgrave, had not actually decreased in their numbers of Houses, but are inserted here, on account of the stagnation and highly injurious effects, visible in them, from ceasing to be considerable Mining Towns, as they were formerly. The remaining 16 places had in 1788, collectively, 1091 Houses, which in 23 years were reduced to 888, or almost one-fifth, on the average! *Failing Mines, and Factories, and the want of Roads and of Inclosures,* seem plainly, the operative causes, in the decline of these particular places: " Food" was certainly not less plentifully produced in the district around them, than formerly, but *profitable employ decreasing,* necessity compelled the removal of a part of the people, to more fortunate places.

A superficial observer of the many Cottages erected of late years by the Roads' sides, in some newly Inclosed Parishes, mentioned at the bottom of p. 76, Vol II. might be led to suppose, that " permission to erect Cottages," was alone wanting, to increase their numbers: but to me, more obvious reasons presented themselves, in almost every instance, in the *increased demand for labour,* in the improved cultivation of the soil (besides the produce of *food* consequent thereon), and in the increasing employ of the people in home Manufactures, like Stocking or other Weaving, Lace-working, &c. or in some of the various Manufactories of the place or Neighbourhood, which are enumerated in pages 479 to 496.

The poorer class in Derbyshire are, as observed in p. 21 of Vol. II. fully and very comfortably provided with Dwellings; but I cannot conceive, that this circumstance has, in any material degree, contributed to the very high and striking degree of increase in its population,

pulation, which the details in this Section establish. The *spirit and energy of the people possessed of property*, in promoting and judiciously carrying into effect Inclosures and Agricultural Improvements, Roads, Canals, Rail-ways, &c.; establishing, and spiritedly and judiciously conducting, Factories of so many kinds, opening Collieries, Quarries, &c. &c; these have given opportunity for the no-less commendable *energy and industry of the labouring class*, to operate so much to their Country's and their own benefit and happiness; and the increase of their numbers has been the necessary consequence.

3. *Is the County over or under-peopled?* and at what price of Wheat? Certainly the County *is not over-peopled*, by *the industrious* part forming the vast majority of its great and increasing mass of Inhabitants: to whom, any " price of Wheat" cannot, more than temporarily, be an object, or any way influence their numbers: if Corn is dear, more labour will be turned to its cultivation, and this evil will be remedied, before the people are lessened, now that famines have happily ceased to desolate the Country.

But here, unfortunately, as every where else under the demoralizing effects of the Poor-law system, idle and vicious Persons have too much increased; because, among them and the poorer of the People, almost in general, the " preventive check," which *moral restraint* and virtue should impose, on an over-abundant increase of People (on which Malthus, Matthews, and many others have ably written), has little or no *operation:* but the other and lamentable alternative, vice and misery, is principally operative, in preventing an excessive increase: and because it is only, with the thinking and virtuous better part of the Poor class,

and

and with such in the middle and upper ranks of Society, that increase is in any considerable degree moderated, by moral restraint, from early marriages, or from illicit practices, which last, in too numerous instances, no less to the disgrace of morals, than to the injury of the community, are almost every where become prevalent.

Bastardy*, as will be seen by the details regarding the Poor in the last Section (p. 522 and 527, N.) and the marriages between Paupers, and between the idle and dissolute Poor, contribute now, in no inconsiderable degree, to *an increase of the People, of the very wrong sort:* instead of a more healthy increase of virtuous and independent middle Men, as Mr. Matthews expresses himself.

That most absurd of all provisions of the Poor-law

---

* Mr. *William Matthews*, in his excellent Essay on Population, in the "Letters and Papers" of the Bath Society, Vol. XII. p. 310, Note, after remarking on the evil effects of licentious intercourse between the Sexes, by persons in the higher stations in life, on the manners of those below them, says, that supposing their conduct had been otherwise, "They would then say, with a rational authority, among young Men of inferior stations, that illicit intercourse is equally base and fraudulent, and that connubial enjoyments should be the privileges only of the sober and industrious: that the *lazy Poor* have no social or moral right to be concerned in the propagation of their species: that no man, however circumstanced, who is not honourably and industriously disposed, and who has not a rational prospect of maintaining a child, has any right in Society, to become the Father of one: that any Man who disregards and violates such plain rules of conduct, should be considered as a depredator on the social rights of others, and as a scandal in Society.

"This principle (continues Mr. Matthews) I hold to be applicable even to *married persons;* however severe it may appear, at first sight, to any of our readers, I must contend for its moral fitness, as a general rule; and the adoption and maintenance of it, as such, as far as may be practicable, to be worthy of the exemplary *Master,* the *Magistrate,* the *Philosopher,* and the *Divine.*"

system,

system, which enables, and enjoins on an Overseer the monstrous task, of compelling a vicious or thoughtless indigent young pair, who having committed *a crime against Society* as well as Morals, in viciously adding to the population, *to marry*, and add to their former one, many other scarcely less flagrant *crimes against Society, in the birth of every Child which they are unable, and above all, are unmindful to provide for!* This absurd law and practice, I say, wants immediately to be abolished; and on the contrary, no legal provision could so effectually aid the cause of female virtue, or benefit the public on this head, as a similar Law, to those often and wisely provided between adulterous parties in higher life, *forbidding marriage altogether*, between those who had anticipated or violated its sacred rites and usages.

It is well known, that 9 out of 10 at least, of every violation of female chastity, is effected under the real or feigned pretence and expectation of *an early subsequent Marriage*, and under the notion, which the law most absurdly countenances, that *a marriage ceremony performed*, at any time before the birth of an illicit offspring, cures all the civil and moral defects of its origin!—can any thing be a greater mockery of a sacred Institution, or more mischievious in its effects on Society?

Instead of reviving the ineffectual mummery of penance in White Sheets, which is disused here and every where else in England, I believe, or any similar expiatory measures; *the Mothers of Bastard Children should be made entirely chargeable with their maintenance* (the contrary of Elizabeth's policy, see p. 522), and under the strong coercive lash of the Law, *should be made*, not only to labour for bringing up
their

their Children, but for the re-payment of all charges, which a Public Bastardy House might incur, during their lying-in, or for their maintenance while unable to work, on account of their Bastard; the operation of which, would far more deter young women from easy compliance, than *the power* of sending them to Bridewell for a short time, which is rarely if ever acted on here, or elsewhere, I believe. The Fathers of Bastards, when convicted, not by the mere and unsupported Oath of the Woman (in the absence of the accused! as at present), but by the verdict of a Jury, should certainly *be imprisoned*, and perhaps fined also, in a given proportion to their property or means, by way of atonement to that community, against whose most importantly sacred institutions they had sinned.

4. *Healthiness of the County.* My observations and inquiries, shewed Derbyshire to be in general very healthy: the ordinary *diseases* of England were none of them unusually prevalent or malignant here; while the few unusual or peculiar disorders, which I will mention presently, were too local and rare, to affect in any material degree the general happiness or numbers of the people.

The *Ague* was very prevalent here, as in Bedfordshire and numerous other places, 40 years ago; but disappeared almost entirely in this County, about or before the end of the last century: the last place which I heard of its occurrence, was at Ash.

At Ashover I was told, that an efficacious remedy was there used for the *Hooping-cough* in children, by merely causing them to play or remain some time daily, in the fumes of burning Tar, in an out-house.

At Killamarsh, it was mentioned to me, that great
case

ease from the pains of *Stone in the Bladder* had been obtained, by the use of Tea made of black Currant Leaves, dried in the Autumn.

For the medical assistance and custody of *Insane Persons*, there is a respectable Mad-house at Calow, 1 m. ESE of Chesterfield. Lunaticks among the poorer class, are many of them kept in the Subscription Poor-houses, as mentioned in p. 558.

The Derbyshire *Thick Throat*, monstrous Craw, or Bronchocele, prevails less or more in Ashover (Hillside or Rattle), Belper, Cromford, Duffield, Farlowgreen, Holbrook, Overton and Wirksworth, and is occasionally seen thro' other places, about the centre of the County: it is an unsightly enlargement of the fore-part of the neck of Women, and very rarely, or in a trifling degree attacking Men: it is rarely either painful or discoloured, but of a soft fleshy consistence, unattended by danger, I believe, altho' asthmatic women, who have Throats much thickened and enlarged all round, state it to increase their difficulty of breathing.

Considerable differences of opinion seem to prevail, as to the cause and progress of this disease: Mr. George Nuttall, who seemed to have paid considerable attention to the subject, considered it, as decidedly hereditary, in particular families: others of the same way of thinking stated, that it commonly first appears in Girls of 8 to 12 years of age, and usually the enlargement of their necks continued increasing, for 4 or 5 years afterwards. Mr. John Milnes, on the contrary, considered it a disorder, engendered or aggravated on the bleak sides of Hills, and to which married Women were peculiarly subject; 19-20ths of the cases within his knowledge, being those of Child-bearing,

poor

poor Women. Mr. Cooper of Eyam was said to have reputation in allaying the symptoms, or in curing this very unfortunate female disease.

Sore Eyes, have been occasioned to a few Miners and Colliers, whose works are subject to corrosive Water; as in Mullet-hill Lead Mine, Alton, Berley-moor, Oakerthorpe, Simondley, and Troway Collieries, and some others, I believe, occasioned by the splashing of such water, it is said.

Belland, Flight, Mill-reak, or Lead-colic, which has been mentioned in p. 392 of Vol. I., not unfrequently attacks those Persons who work in the Lead Slag-mills, and those who work or live in or near to some of the Lead-smelting Cupolas, where such are not well constructed and situated, for harmlessly venting their deleterious fumes: Lead Ore or Slag, in almost ever so small quantities, coming in contact with the food, are productive of this terrible disease. Its chief symptoms are, obstinate costiveness accompanied by great pains of the Bowels.

The use of much fat Bacon, Lard, Butter, &c. is said to be beneficial, in protecting the Lead-smelters from Belland; and some assert, that strong Coltsfoot Tea, often taken, has operated its cure.

In p. 86, I have mentioned the disorder in Cows, Calves, &c. occasioned by drinking stagnant water, from the Ore-dressers' Buddle-ponds, about the Lead Mines; and in pp. 145 and 161, should have noticed the same thing, with regard to Sheep and Horses, and p. 179 with regard to Poultry; which last soon die, if suffered to roam or feed on Mine-hillocks, or on Yards, Roads, or Paths, strewed or mended with Mine rubbish, or with Lead Slag; dogs, and other small animals are

frequently

frequently killed in the same way. About Baldmere and other Mines, and at Barber-fields and some other Cupolas and Slag-mills, many Cattle are said to have died, at different times, from drinking Bellamded or Buddlers' Water: the injury of the same to Fish, is mentioned in Vol. I. p. 377. About Lea-wood and other Cupolas, small particles of slug, and of ore perhaps, have proved fatal to Poultry, Dogs, &c.

Asthmas, are so prevalent among the Grinders of Scythes and Cutlery in the northern part of the County (see p. 489), and so liable to bring on Consumptions, owing to the spicula of metal and dust of stone, which they inhale, that few of these men attain 50 years of age: the Grinders of cast-iron goods, like laundry Smoothing-irons, are particularly affected. Dreadfully fractured foreheads, and broken limbs, from the breaking of unsound or over-driven Grind-stones, are also extremely common, around Sheffield; more so than they need be, if greater care was used, as Mr. Joseph Hutton of Ridgeway informed me; in whose Blade-mill, only one accident, of a broken thigh, had happened to his Grinders, in the space of 30 years.

5. *Food of the People, and mode of Living.* The Tradesmen and Farmers of Derbyshire, appeared to me to live very comfortably, and to be very hospitably inclined towards strangers: the comforts enjoyed by the Labouring people, seemed to me greater than I have been used to witness in the southern parts of the Kingdom; notwithstanding, that nearly all their *Bread*, in all the northern parts of the County in particular, was of the thin, *soured*, soft kind of Oat Cake (like a Pancake, except as to greasiness, as described in Vol. II. p. 30), which to most strangers, and even to those fond

of

of the very thin, unsoured, hard Haver-cake of the West Riding of Yorkshire, is at first, a very unpalatable kind of food.

Very *fat* New Leicester *Mutton*, was perhaps formerly more generally relished among the manufacturing Labourers, than at the time of my Survey, see p. 128. On the mode of supplying *Milk* for the larger Towns, see pp. 30 and 40; and a recommendation to Farmers, more generally to supply their Labourers with this wholesome and important food of Children, in p. 195.

6. *Customs, Opinions, Amusements, &c. of the People.* An ancient custom still prevails in Chapel-en-le-Frith, Glossop, Hayfield, Mellor, Peak-Forest, and other places in the north of the County, I believe, of keeping the floor of the Church, and Pews therein, constantly strewed or littered, with dried Rushes: the process of renewing which annually, is called the *Rush-bearing*, and is usually accompanied by much ceremony. The Rush-bearing in Peak-Forest, is held on Midsummer Eve, in each year.

In Chapel-en-le-Frith, I was informed, that their Rush-bearing usually takes place in the latter end of August, on public notice from the Churchwardens, of the Rushes being mown and properly dried, in some Marshy part of the Parish, where the young people assemble; and having loaded the Rushes on Carts, decorate the same with Flowers and Ribbons, and attend them in procession to the Church; many of them huzzaing and cracking Whips by the side of the Rush Carts, on their way thither; and where every one present lends a hand, in carrying in and spreading the Rushes. In Whitwell, instead of Rushes, the Hay of a piece of grass land called the Church-close, is annually,

nually, on Midsummer Eve, carted to and spread in the Church.

The *Well*, or fine running *Spring* of Water in Tissington Town (which is mentioned Vol. I. p. 506, and II. 268), is annually, on Holy Thursday, finely decked out with Flowers, by the Inhabitants: a similar custom formerly prevailed in Bakewell, and in other places of the County, I believe.

Instead of the Harvest-home feasts of most other Districts, what is called (I don't know why), the " Hare-getting" or " Hare-supper," is given, by the Corn Farmers of some parts of Derbyshire, to their Reapers, or Shearers as they are here called, and their Wives, on finishing the cutting of their Corn; and it is usual for such Shearers, " to give three cheers on first setting to," to shear or mow, and " three wheeps and a hallow," on finishing the cutting.

The superstitious and unfounded *opinions* formerly held by many in the Peak Hundreds, that the motions of Hazel-sticks held in the hand, or of Meteors in the air, could point out the situations of Veins of Lead Ore!; but which Ore, the whistling of a Miner could again drive away!, have already been mentioned in p. 316 of Vol. I.; and so has the silly notion, that the blooming of Pease could occasion Fire-damp in Mines, been mentioned in p. 336; and that the castrating of Colts, Calves, &c. would prove fatal, when the Moon was in certain signs of the Zodiac!, in p. 87 of the present volume; with a suitable admonition thereupon, to the London Company, enriching itself, by the perpetuation of Astrological humbug, among the unthinking part of the English Population: who, but for the sordid and unpatriotic conduct of this

Body,

Body, would ere this, have almost entirely forgotten such fooleries; which, if they are to continue to be circulated, as at present, the Minister would do well, I think, to double the Stamp-duty, on all such Almanacks as contained Astrological details or predictions.

I have been told, that the young Children of Matlock, are yet often made to stare and tremble, at the relations by their more childish Nurses and Grand-Mothers, of the devouring feats of a former Dragon of that place!. Satyrs, or imaginary Wild Men, were confidently said, formerly, to inhabit Hobsthirst Rocks, on the N side of Fin Cop Hill, mentioned in Vol. 1. p. 33; and I was myself gravely told, in Tansley, that Fairy elves are still frequently heard to squeak !, in the damp cavities of the Rocks, over which the water-fall in Lumsdale is projected, which is mentioned Vol. I. 489.

Many persons in Castleton are said to believe, that the Sun appears to dance up and down at its rising on Easter Sunday Morning, when viewed from the top of the Castle Hill adjacent; and that numbers repair thither, almost annually, in expectation of seeing it ! By others, the Sun is said to illuminate more of the surface of that deep valley, in the shortest days, than it did some years previously !

My reason for putting these stories upon record, is, the hope, that better-informed Persons, residing in or visiting the places mentioned, or others, where similar absurdities are believed, will be induced to laugh such credulity out of countenance, and inculcate juster beliefs and notions of things: and by no means do I wish to shew hereby, that the common people of Derbyshire are more prone to superstition and credulity, than those of the Neighbouring Counties, because, I believe, the reverse in some cases to be the fact; and compared as to these weaknesses in the common people, of most of those of Scotland and of Wales, they have wonderfully the advantage.

The Village *Wakes* or Feasts, are prominent seasons of Festivity and *amusement* to the lower ranks of Society, in almost every part of the County; they begin on a Sunday, and continue through most, or perhaps all of the ensuing week: Mr. James Pilkington, in his " View," Vol. II. p. 55, seems rather too severe in condemning those Wakes, which, except their Sunday commencement, and the cruel Sports which are yet too commonly indulged in thereat, are thought by many well-informed persons, with whom I conversed on the subject, to be rather beneficial than otherwise. A thorough cleaning of the Cottage, and mostly a white-washing of its Rooms, annually precede the Wakes; the Children and their Parents are then, if possible, new clothed: previous economy is exercised by most, for accumulating the means of providing Meat, Beer, &c, and various exertions are made on these recurring occasions, which tend to keep alive feelings and principles, which otherwise the Poor-law system might utterly extinguish.

The disgraceful *Sports* to which I have alluded, as fit subjects of immediate and entire suppression are, *Bull-baiting*, by means of Dogs, who tear and lacerate the harmless and confined Animal, as was still the case in Ashover, and various other places in the County. In Tutbury, on the County bounds, the still more inhuman and dangerous practice, of goading a loose Bull, about the Streets of the Town, was only a few years ago put down by the Civil Authority.

Badger-baiting is common at the Wakes; and sometimes *Bear-baiting*: near to Buxton, in June 1810,

1810, a young Bear was thus tortured by Dogs, and money collected from the foolish and brutal spectators, for the more inhuman Brute, who was conducting this poor inoffensive Animal about the Country, *for gain* by daily repetitions of these tortures! In October of the same year, Joshua Jebb and W. A. Lord, Esqrs., two respectable Magistrates near Chesterfield, committed one John Smith to prison, as a vagabond *Bearward*. Let all High Authorities cease to practise or countenance similar cruelties (p. 200), and the existing laws be properly enforced, and these disgraces of an enlightened Christian Age, must soon cease.

Throwing at Cocks, tied to a stake, which still, I fear, disgraces the Shrove Tuesdays of particular districts in most parts of England, has happily disappeared some time from Derbyshire: I wish I could say the same also, with regard to the silly and brutal practice of *Cock-fighting*, p. 180. The profanation of Christmas Day, by the *Squirrel-hunt* of Stanton, has been mentioned p. 244: I remember, when a youth, seeing a similar but more general hunt, of small Animals, and even Game, by the lower class of people, about Wetherby in Yorkshire, on the 5th of November.

On this latter occasion, *Bonfires* are, as in every other part of England I believe, still lighted by the Boys in the Evening; but always in Derbyshire, I believe, this is done without the Towns and Villages, on some elevated spot. In 1814 I happened to come thro' Coventry, in the Mail Coach, on the Bonfire Evening, and have since thought it one of the fortunate events of my life, to have escaped an over-turn, owng, to not less I think than 20 very large Fires in the narrow Streets of that City, around which the thoughtless rabble were at the time throwing fire-brands, and discharging pistols, guns, &c.; many of which Bonfires we were necessitated to pass so near, that the wheels of the Coach actually passed over the skirts of some of these Fires! If the Magistrates of this City have not sense or spirit enough, to repress these proceedings, the Solicitor to the Post-Office ought to do it, in justice to travellers by their Coaches, and the Public. Similar disorders in Lewes, were a few years ago repressed, after repeated Quarter Sessions trials and imprisonments.

Desperate *Foot-ball* contests, or rather cudgelling fights, under this pretence, were some years ago common at the Ashover and other Wakes; but such contests were come to be very harmlessly conducted, at the time of my Survey; except in the Streets of Allsaints, in the Town of Derby, whose Foot-ball outrages on Shrove Tuesday were mentioned, as the most disgraceful remains of former Barbarism in the County. Throwing of *Quoits* seemed a very prevalent amusement of the lower and more idle part of the manufacturing People, at the Ale-house Doors, in the north of the County, about Sheffield in particular: often so near to the Road and Path-ways, as to endanger passers by.

The afternoons of particular Fairs, at most of the places mentioned in pp. 457 and 458, are more or less devoted to amusement and jollity, among the young folks: if this be the main purpose of the day, it is called a *Gig-fair*. Strolling vagabonds, under the name of *Mountebanks*, often attend these Fairs, and are improperly allowed also, to ramble at other times thro' the Villages, not dispensing quack Medicines, as formerly, by which the Revenue might be benefited,

whatever

whatever became of the people's healths, but for openly gambling, or holding " Little Goes !"; and shewing tumbling feats, only to collect together their thoughtless dupes: surely these mischievous strollers ought to be suppressed; *Gipsies* also, whose Camps, or plundering Quarters, I saw pitched in Butterley in Ashover, and in Kirk-Hallam, while on my Survey, ought no longer to be endured: the female wretches of these nefarious gangs, do inconceivable mischief, as *Fortune-tellers*, among the daughters and servants of Farmers and others, whose pilferings for rewarding these miscreants, too often lead to more serious stealings, for themselves.

Sitting in the Parish or Township *Stocks*, a summary and wholesome mode of punishment, for the less heinous offences against good morals, seems here, and almost every where else, growing into entire disuse: altho', ridiculously enough, every country place continues religiously to uphold its Stocks: on a great many occasions, when seeing them repairing or new ones erecting, or such as lately had been renewed, I have inquired, whether any one in their Place remembered a single instance of the Stocks being *used*; but have almost invariably, except by very old persons, been answered in the negative.

The Parish *Pounds*, for stray or trespassing cattle, in the stony districts of the County, are not, as is too common in similar situations elsewhere, inclosed by such high and close *walls* and doors, that poor Animals, from not being *seen* therein by passers by, may suffer greatly by hunger and thirst, before their owners get to hear of their situation: but here, a proper portion of the side next the public Road, is enclosed by open

Rails, for admitting such view by Travellers, of the imprisoned Animals.

Many Villages in Derbyshire have their small local *Prison*, Watch-house, Cage, or Lobby, as they are here sometimes called, built of brick or stone, in a circular or square form, and arched and topped conically or pyramidically with the same materials, in a very durable and secure manner.

The Lord of the Manor of Ilkeston, maintains constantly *a Gallows* (on which I could not learn that any one was ever hanged) near the Ernwash River and county bounds, on the East of the Town: in order, as is said, that the Inhabitants of the Manor may avail themselves of a Charter, by John of Gaunt, for paying only half of the usual Tolls in any of the Markets or Fairs of the Kingdom.

CHAP. XVII.

OBSTACLES TO IMPROVEMENTS.

SECT. I.—RELATIVE TO CAPITAL.

RICH Farmers are rare in this County, as observed in p. 40 of Vol. II. and rather scarce in every other, I believe; it is not here only, but in every County in Britain, which I have visited, that the want of more *Capital*, for stocking and managing their Farms, in a more perfect and spirited manner, by the bulk of the Farmers, is experienced as an evil, of considerable magnitude, and a great draw-back on the produce, which ought to be available to the Public: here and every where else, the major part of the Farms are too small (see Vol. II. p. 26) for the most perfect management, or the most abundant or cheap produce *being brought to market.* The disposition, laudable in itself, which English Land-owners have, for preserving the same Tenants or Families on their Estates, seems carried to a mischievous excess, in great numbers of instances. The Tenants in several places, who are alluded to in p. 36 of Vol. II., are said to leave the occupation of their Farms to their Widows or Children, &c. by Will, with as much confidence as if it were their own Estate, or held on a long Lease; and that a custom of long standing prevails with the Noble Owner and his Agents, of entirely respecting such
‘ bequeaths,

bequeaths, except on very particular and rare occasions.

Many Gentlemen, influenced perhaps too much, by the paltry and fallacious considerations, of preserving their Game, and the facilities for their field Sports, have studiously kept on their lands, the same Persons or Families, altho' not hesitating to embrace every opportunity, of advanced prices of Agricultural produce, to raise their *Rents*, far beyond what *the small disposeable produce of such small occupations*, as mostly prevail on these occasions, would bear.

Landlords, inconsiderately relying upon *seizures* for their Rents, and Farmers, *not dealing upon credit*, occasion many poor Tenants to keep possession of, and linger on upon small Farms, which they are unable properly to cultivate, to the great injury of the Country, and of themselves also, in reality; since most of this description of Men, and their Families, would enjoy more comforts, and their labour be vastly more efficient and beneficial to the Public, in Trades or Manufactures, or even as Bailiffs and Servants to more opulent and larger Farmers, whom it is, in almost every district, desirable to settle, on several of these small occupancies united into one: under liberal and *assignable* Leases, which would not fail of remedying this material defect, of the want of sufficient Capitals, or of proportionate ones to those embarked in all our trading and other concerns, for the best possible cultivation and improvement of the Soil.

SECT. II.—OBSTACLES RELATIVE TO PRICES.

Notwithstanding the check which an enormous importation of Foreign Corn, has at times, imposed on the prices of the Farmer's chief produce, in the age now passing, no very serious obstacle to improvement, of the Farms sufficiently large, and in hands sufficiently opulent, has, I think, been experienced : but with the small and the poor Farmers, the case has in very numerous instances been different: compelled by the calls of their Landlords, the Tax-gatherers, and their Tradesmen, to bring their produce early, and in disproportionate quantities, to market, the prices during these periods, have been usually depressed, below the amount necessary to support or increase the little Capital, remaining in the seller's hands, for future cultivation : to the consequent and inordinate enrichment of *Middle-Men* and Speculators, thro' the aid of Paper Money (see p. 510), in the traffic of the necessaries of Life, thus prematurely taken from the Barns and Stores, of the most proper people to hold them, the growers ; who scattered, unused to speculation, and limited as to property or credit, could not have made the Public pay those high prices, towards the latter ends of many Seasons, which they have often had to deplore : and which temporary high prices, as hinted in the last Section, have been made the pretence, by many unthinking Landlords, for raising the Rents of small and poor Farmers, who had sold nearly all their produce, before the periods when such prices obtained.

Gentlemen,

Gentlemen, who will continue to have very numerous Tenantries, in states approaching the vassalage of past ages, and of course, with capitals and means, utterly disproportioned to those of dealers and speculators, must be content to draw lesser amounts of Rent, than they otherwise might do, and to see also, the Public less beneficially served, than they might be, by larger and more equal Farms, in the hands of men, whose object would be, increased and *large produce,* rather than *enhanced prices,* of a limited stock on hand.

Derbyshire, from the distribution of its Fairs and Markets, p. 457 and 459, its great and general facilities for Commerce and Trade, as mentioned p. 509, and the number, activity, and means of its Inhabitants, has had less reason to complain of the want of adequate prices for agricultural produce, generally, than several others in the Kingdom.

SECT. III.—OBSTACLES RELATIVE TO EXPENCES.

It will be gathered from what is said in pp. 27 and 28 of Vol. II. and pp. 534 and 592 N. of the present, that Derbyshire Farmers, generally, have no great occasion to complain of *Rents,* except perhaps the smaller and poorer ones, as observed in the two last Sections: and as to expences of cultivation, such seem in no instances peculiar or inordinate, or such as to check improvements, in any material degree.

The expences of *Roads,* to the Farmers, might, under

under a better system of Laws and proceedings on the subject, be materially diminished, as observed in pp. 236, 240, &c. and by a more equal assessment of the Funds for the Roads, upon Property in general. Poor-rates and Tithes will be more particularly mentioned presently.

SECT. IV.—OBSTACLES RELATIVE TO WANT OF POWER TO INCLOSE.

IN pages 77 and 341, Lists of the uninclosed common Fields and open Commons are given, by which, and the remarks subjoined, the extent of this impediment to further improvements in the County, may be guessed at. A General Act, unexpensively giving powers and facilities for dividing, exchanging, and inclosing of common, intermixed and open Lands, with powers to exonerate such from Tithes, either for Land or Corn-Rent equivalents (if, unhappily, more general powers or means for such exonerations, are not soon given), would prove very beneficial to this County; but more so to others, where a larger proportion of uninclosed Lands yet remain.

The obstacles which *Mineral Rights* and the Mining Laws of the Peak-Hundreds present, in some instances, and the modes of commutation on Inclosures, have been mentioned in Vol. I. pp. 356, 359, 363; and II. p. 81.

Obstacles relative to want of Leases, from their univer-

universality and importance, ought to have had a Section in the "Plan," devoted to their mention. Derbyshire, as I have said in p. 36 of Vol. II., suffers a good deal from the want of Leases, but not more so, I believe, than is now almost general, owing, principally, to the uncertain and fluctuating value of our nominal *currency**, occasioned by the late unparalleled increase of Paper Money, as Wars and Taxes have fatally progressed, which required and gave it birth and circulation. *Restrictive Covenants* on the Farmer's management, where Leases exist, or in the express or implied conditions of Lettings without Leases, seem less numerous or mischievous in their operation here (page 39 of Vol. II.), than in several other Counties they are said to be. The most vexatious and injurious covenants in Leases, which I have ever met with, were on an Estate 2 or 3 miles to the S of Horncastle in Lincolnshire.

SECT. V.—OBSTACLES RELATIVE TO TITHES.

TITHES, an impost, increasing with the outlay and exertion of the Farmer, as the Income Tax did upon the personal labours of Professional Men, must needs

---

* In the "Farmer's Journal," No. 382, Vol. 8, p. 19, I have endeavoured to shew the necessity of, and the means of, obviating this grand objection to Leases, by means of *Corn-Rents:* and in No. 436, Vol. 9, p. 35, I endeavoured to shew, that great good to the Community would have resulted, from applying the same mode of regulation of the value of the Currency, to the dividends in the Public Funds, Interest, Pensions, &c.

prove

prove very discouraging, and injurious to the cause of Improvement, and their entire commutation*, for Lands, or for Corn Rents, seems now imperiously called for. I believe it to be generally true, as mentioned in page 29 of Vol. II. that Lay Impropriators (who can certainly plead *no divine right* to protect them from a commutation) are considerably more severe in exacting their Tithes, than the Clergy, collectively; and I believe it not less certainly true, that the non-resident part of the Clergy, are vastly more strict and oppressive in this respect, than that most useful and respectable part of the beneficed Clergy, the resident and officiating Ministers. The latter circumstance is the more offensive, galling, and disheartening to the Farmers who pay the Tithes, from knowing, how very shabbily in general, such pluralists, and other non-resident Clergy, pay their poor *Curates*, for performing all the duties, for which any peculiarity or sacredness in the right of Tithes, can with justice be maintained.

It is well known as a general fact, that the Clergy, of the greater part, of the more valuable Livings, employ Curates; rarely, or but occasionally residing, and yet, from a Return made to Parliament, which is preserved in p. 118 of Vol. 36 of the " Monthly Magazine," it appears, that of 3730 Curates, employed by non-resident Clergymen in England and Wales, in 1811, one only of these received 150*l*. a year; 56 only of them had 100*l*. or upwards, averaging less than 103*l*. each; while 1931 of these Curates, or considerably more than half of the whole, had stipends of only

---

* See some excellent observations on the nature of *Tithe* claims, and on the propriety of their commutation, in the " Farmer's Journal," No. 454, Vol. 9, p. 184.

44*l*. 4*s*. per annum, 17*s*. per week, or 2*s*. 5*d*. per day, on the average! Labourers' pay!:—33 of them even, had no more than 10*l*. per annum each!! The living of Sawley in this County, is said to be worth 2400*l*. per annum, out of which, during 20 years, only 40*l*. per annum was paid the Curate!; who being at the end of this time, unable to procure an advance of stipend, petitioned Parliament, on the hardship of his situation. An Act has since passed, for compelling more adequate stipends to Curates.

The amount of Tithes in the County, according to the Property Tax returns, was in 1811, only 7¼*d*. and a trifle more in the Pound, upon the total Rental of the County; so great a part of this Rental being, on exonerated Lands, on Houses, &c. In Lancashire the Tithe is 8½*d*.; on the average of England and Wales it is 1*s*. 7*d*.; in Sussex 3*s*. 10*d*., and in Hampshire 3*s*. 10*d*.: occasioned more, by the proportional extent of exonerated Lands, than by any other causes of variation, I believe.

SECT. VI.—OBSTACLES RELATIVE TO POOR-RATES.

To these most serious and increasing impediments to agricultural prosperity and Improvement, I have already given considerable attention, in the last Chapter; and have shewn, in pages 502, 504, 507, 518, &c. the peculiar hardships to which the Farmers of particular parts of this County are subjected, in maintaining a surplus and vitiated Population, collected, raised, and fixed there, by the Manufacturers. I must not therefore enlarge further here on the subject: except to mention, that since that Chapter was printed off, a

Gentleman

Gentleman of Derby has informed me, that two of its Parishes, a few years ago, since my Survey, I believe, availed themselves of a general Act of Parliament, which passed for encouraging *Houses of Industry*, by which, any one or more Parishes may, singly or jointly, incorporate themselves; and in a great measure, get rid of the Powers, which Magistrates, too frequently, through the artifices and deceptions of Paupers, misuse, in the ordering them specific relief from the Parish; and that being thus enabled to confine relief to the old, disabled, and really distressed, and to offer *work*, and *plain food* only, in pay, to the idle and improvident, the latter had been driven away, in such numbers, as very materially to have lessened the Poor-rates, in these two Parishes: I was very anxious to have learnt, during this communication, that the manners and habits of the Poor-law applicants, above alluded to, had been reformed, rather than the scene of their deceptive arts, transferred to other Places (see p. 523, Note), less able to detect or counteract them. The Gentleman alluded to, has promised to send me further details respecting the above, and which I hope may arrive in time, to be recorded in the Appendix to this Volume.

SECT. VII.—OBSTACLES RELATIVE TO A WANT OF DISSEMINATED KNOWLEDGE.

I do not think that the progress of Improvements is in any very material degree impeded from this cause, in Derbyshire.

1. *Agricultural Libraries.* I did not happen to learn, whether any of the Agricultural Societies which will be mentioned in Sect. 1. of the next Chapter, have attached

attached a Library of Books on the subject, to their Establishment: I fear not, as such could not fail of doing good. Of *Reading Societies*, of a general nature, there is a respectable one at Chesterfield, and others at Derby and the other principal Towns, I believe.

2. *Cheap Publications*. It has happened, unfortunately for the Country, I think, that the Funds allowed to the Board of Agriculture have been so very limited, that except frequent Letters, some Papers, and a few very short Tracts, which have been printed and issued gratuitously, they have been able to do nothing towards *the cheap circulation of Agricultural Books*, to the British Farmers generally: and such being the case, neither the Board, or its Members individually, in the Provincial Societies, to some of which most of them belong, have seen the propriety of recommending to, or of supporting propositions in such Provincial Societies, for the appropriation of any parts of their funds to the same object, desirable as it is, and useful as it could not fail of proving, if judiciously carried into effect.

Agricultural Books, generally, have perhaps as large a sale in Derbyshire, as in most other Counties. The " Farmer's Magazine," a respectable Scotch quarterly Publication, has, I believe, a few subscribers in the County: the " Agricultural Magazine," a London monthly Publication, has considerably more readers therein; but the most generally read, and perhaps the most useful, of publications on this important subject, is the " *Farmer's Journal*," a weekly Publication, by Messrs. Evans and Ruffy, of Budge-row, London, one quarter, and often near half of which is devoted to, and filled with original Correspondence and Essays, on Agricultural, Rural, and Politico-Economical subjects,

princi-

principally by able and practical Men, scattered through the three united Kingdoms; and not unfrequently they are obliged to issue an extra Sheet, wholly *filled* with such subjects: the remainder of its usual columns, embrace the topics of a common Newspaper: on which account, its sheets are on all occasions subjected to the Newspaper *Stamp* duty. If the Legislature were to allow this, and all similar publications, of a price to be limited, when either wholly, or one-fourth part or more, filled with Agricultural Essays and Correspondence, instead of News, a drawback of half the stamp duty, and at the same time to go free of postage, as at present, important benefits to the cause of Improvement, might, I think, result therefrom.

SECT. VIII.—ENEMIES TO IMPROVEMENT.

1. *Red*, or *Wire-worm*. In pages 108, 123, and 129, of Vol. II. I have extracted all which was mentioned to me, on the depredations of this Insect or Grub, on the Pastures, the Wheat, and the Oat Crops of the County. In the " Farmer's Journal" of the 23d May, 1814, p. 267, some useful observations on this subject, by Mr. Robert Paul, are inserted.

2. *Slug*, Grub or other Insects. In the " Farmer's Journal" of 4th June, 1813, the expedient is mentioned, of strewing Cabbage leaves over some Lands of young Wheat, on which the *Slug* was preying, in order to entice them off, and take and destroy them. The *Grubs* of the *Cockchafer* and of a Tipula, in Foremarke, Ingleby, Losco, Repton, and other places in the southern parts of the County (see Vol. II. pp. 108 and 123), attack the roots of the Carex, Nardus, and

other coarse grasses, which occasions the plants to wither, and this proves a signal to the Rooks and Crows, to pull up such plants of Grass, and take the Grubs at their Roots. Such an enormous number of Cockchafers are said to have bred in Blythe, in Notts, in 1788, that the Women and Children employed for the purpose, collected 3,743 pecks of them!, which were burnt. See Far. Jour. 3d Jan. 1814, p. 107.

In Ingleby, the ley Crops of Corn are often greatly injured by a small white *Grub* with a yellow head, and by the scratchings of the Crows after them: on first breaking up old Leys, the Grubs often lay on the surface, as thick as sown and unharrowed Corn should do: after a proper time is allowed the Rooks and Crows to destroy them, the Land is again stirred, and time given, 2 or 3 successive times, to those useful Grub-destroyers; without which precaution, on some lands there, no Corn could be raised, it is said. The manner of destroying *Worms* which injure pastures by their Casts, is mentioned in p. 178 of Vol. II.

When crossing the high heathy Moor Lands, near Moscar House, on the Borders of Yorkshire, towards the latter end of June 1809, I was much struck by seeing, almost every top branch of the heath plants, surrounded by a large froth, called *Cuckoo-spittle*, in the centre of each of which, I found a small green Grub (Cicada Spumaria), which the Persons I inquired of near the spot told me, would turn to a swarm of Horse *Flies*. See the "Monthly Magazine," Vol. 30, p. 97.

Caterpillars, very commonly breed on, and destroy crab Hedges, as observed p. 89 of Vol. II.: in Ingleby, the white-thorn Hedges are often greatly injured by the same cause: on the new Inclosure of Ashby Wolds, it appeared, as though thorn hedges could not be raised,

owing

owing to the depredations of these insects, until Mr. Henry Smith of Norris Hill, on repeated occasions, employed women to cut off and collect and burn, all the pockets or nests of cob-web, in which the eggs had been laid, by the Fly occasioning these Caterpillars; and by that means nearly extirpated them. In p. 214 of Vol. II. the *Insects* affecting Raspberry bushes are mentioned: and in p. 137, the *Fly* affecting Turnips is treated of.

At Little Haddon, NE of Haddon Park, I saw some *Ant-hills*, which seem to be rather rare in Derbyshire, as observed p. 403 N. Vol. II. *Crickets* are often very troublesome, about houses and places where large Fires are kept: Mr. Pilkington, in p. 364, Vol. I. of his " View," speaks of powder of dried Carrots, mixed with arsenic, and laid about, being effectual in destroying them.

Mr. Francis Blakie described to me the mode, by which those mischievous insects *Wasps*, are destroyed, about Bretby: the nests being sought and discovered, they are attacked *in the day-time*, when a great part of the wasps are abroad, collecting their wax and honey; by first stupifying those in the nest, by introducing a fusee, or large straw, full of lighted gunpowder, into the entrance hole; the nests are then dug out and effectually destroyed, and, as quickly as may be, the earth is replaced over its scite, around the neck of a wine-bottle, half filled with small-beer and wasp's comb, and sunk therein; which the wasps, on returning, mistaking for their nests, enter and are drowned. The danger of being *stung* in these operations, was described as being very small, compared with less effectual night attacks on them: if the hungry and active wasps are thus stupified in their nests, because the loaded and tired wasps, returning home, are seldom found to sting, unless buffeted on their approach to the nest.

3. *Rats, Mice,* and other small animals, called Vermin: the precaution of cutting the lower part of the sides and ends of Corn Ricks *very smooth*, to prevent Rats and Mice gaining a first lodgment in the loose outside straws, is mentioned in pages 67 and 113 of Vol. II. Dwarf Elder Leaves are said by Mr. Pilkington (View, 1. 372) to drive away Mice, from where they are strewed. The occasional injuries done by *Squirrels*, are mentioned p 244 of Vol. II. *Moles* are not particularly troublesome by their burrowings in the Fields in Derbyshire, since the subject was not, I believe, once mentioned to me, nor did it force itself any where on my observation. In the Farmer's Journal, 9th May, 1814, p. 250, some useful observations on this subject will be found.

4. *Sparrows,* and other Birds. The mode in which Corn-ricks are made up to render them less accessible to Sparrows, &c., is mentioned p. 112 of Vol. II. In the fields about Brailsford, as well as in the Gardens more generally, I saw numerous *Bird-Clacks* erected on poles, to be turned by the wind, for frightening away Birds from the Crops: these form a very proper first exercise of mechanical ingenuity in Boys, to contrive and make these Bird wind-mills: they were among my own first efforts, and happening some years afterwards, while still a youth, to have spent some time in contriving on paper, and making out rather detailed drawings, of all the different constructions of them, which I had then seen, or could think of; these, falling into my eldest Son's hands (see 1. 397) while a mere Child, with others of my juvenile attempts in this

way,

way, seemed the visible means, of awakening and calling into action, a curiosity and talent, for drawing and for mechanical exercises and pursuits.

Tom-tits are found mischievous little Birds, to the Buds of Fruit-trees, and more especially so by their arts, in alarming and bringing Bees to the holes of their Hives, in winter time, if such are not then stopped up, or the hole of a mouse-trap set before it, in which little Tommy may be caught by the neck, when reaching in, and thinking to pick up and eat a poor Bee. The depredations of *Wood-Pigeons* on Turnips, are mentioned p. 148 of Vol. II.

Rooks, were formerly more prevalent in the County than at present; I noticed large Rookeries only in Eggington, Etwall, Foston, Radburne, Stanesby, and Sutton on the Hill: these Birds, and Crows, certainly do occasional mischief to the Corn, and to Turnips in the winter season: but it seems the opinion of most competent persons, that as Grubs and insects are almost their entire food, they are beneficial rather than otherwise, provided the district is not greatly overstocked by them, as sometimes is the case.

Swallows are very useful Birds, by living wholly on flies: in a Building in Osmaston-Cottage Garden, I saw in August 1809 a nest, and second brood in the same year, of these pretty Birds, fledged and ready to fly, and they and the old ones, were much tamer than I had ever before witnessed. A single pair of *Ravens*, and never more, it is said, have annually built, for time past immemorial, in the Cliff called Raven's-Nest Tor, in Overton: other Rocks or Cliffs, bearing the names of Raven's Tor, in Alport, Milnhouse-dale, Wirksworth, &c. would seem to indicate, that these large and mischievous Birds, formerly bred in such places.

"Other

"Other Vermin," under the denomination of *Game*, have been enumerated in pages 199 to 202:—and the "Means of preventing" mischiefs, from all the above Enemies to Improvement, have been mentioned, as far as I heard or noted them.

CHAP. XVIII.

MISCELLANEOUS ARTICLES.

SECT. I.—AGRICULTURAL SOCIETIES.

1. *Those established in the County.*—At *Derby*, a respectable and rather numerous Agricultural Society has existed since 1794, of which some mention has already been made in pages 109 and 129: its *Shows of Live Stock*, at the time of Derby Easter Fair, and in the 2d week of July, annually, have been well supported by competitors, and are esteemed to have done much good, in promoting the improvement of the chief domestic Animals: other important objects of Rural Improvement, to which this and other similar Societies might with advantage turn their attention, have been suggested in different parts of these Volumes, and will be recapitulated in the " Conclusion" of this Report.

In *Repton*, a Society has been held for several years past; confining its objects to a Show and Premiums for *fat Wether Sheep*, annually, of which mention is made in pp. 132 and 133. It has, I believe, been principally composed of New Leicester breeders, disposed, many of them, for retracing some of the steps of excessive refinement, in the more fashionable flocks of this breed of Sheep; and in this respect, was judged by several practical Farmers around, to have done much good.

In *Hayfield*, a Society of Mountain or Woodland *Shepherds* and Sheep-keepers, in Edale, Glossop, Hayfield,

field, and Woodland Liberties, of which mention has been made in page 91, was established on the 23d of July, 1790:—when it was agreed, and since acted on, to hold Meetings on the 23d of July and 7th of November, annually, or on the next days when Sundays occurred; every Member to attend such Meetings, or send 1s. towards the expences at the Inn, of those who do attend:—that a Shepherd (Samuel Garside) do attend the Meetings of the Saltersbrook Shepherd's Society (on the 20th of July and 5th of Nov. or following day, if Sundays) to collect and bring to these Meetings, such stray Sheep as may be brought there, belonging to any Members of this Society, and be allowed 4s. for his wages and expences in so doing: —and that any Member, causing a quarrel, or dispute to arise in any of these Meetings, is to forfeit 2s. 6d. or be excluded. Through the exertions of Mr. George Greaves of Rowlie, the Rules of this Society, as above, and the full description of all the several *modes of marking their Sheep*, adhered to by its Members, was first printed, in a small Book, in 1798; which has since been reprinted from time to time, as occasion required.

The Saltersbrook Society holds its meetings on the days above mentioned, at the Inn of that name in Yorkshire, just beyond the bounds of this County, and of Cheshire also; for the Liberties of Bradfield, Glossop, Holmfirth, Longdendale, Penistone, and Woodland: one of whose printed Rules is, that any Person bringing strayed Sheep to their Meetings, the owner or owners of such Sheep shall pay or cause to be paid, reasonable expences for taking up, keeping, and bring them to the Meetings.

A General List of South-British Agricultural Societies.—Thro' several years past, I have been careful, when

GENERAL LIST OF AGRICULTURAL SOCIETIES.

when reading Agricultural Books or Intelligence, to refer to and enter in a List, all the *Names, Places of Meetings*, Names of *Presidents*, and of *Secretaries*, with the addresses of the latter, of all the Agricultural Societies I have seen mentioned, or have heard of, in England and Wales. The Presidents, of so many of these Societies, are changed annually, and their Secretaries are likewise so often changed, that it would be of little use to give that part of my List; but the other parts, will not I trust, be deemed intrusive here, as a record of the greater part, if not all of the Agricultural and Live-Stock Societies, existing in South-Britain in 1816, viz.

Societies in England and Wales.

Abbey-holm, Farmer's Club (Cumberland)	Abbey-holm.
Anglesea Isle (North Wales)	Llangefni.
Barnard Castle (Durham)	Barnard Castle.
Bath and West of England (Som.)	Bath.
Bawtry, Farmers' Club (Yorks. W R)	Bawtry.
Bedfordshire	Bedford.
Berkshire	East-Ilsley.
Board of Agriculture	No. 32, Sackville-street, Piccadilly, London.
Border, of England and Scotland, 1815	Kelsoe.
Boston (Lincolnshire)	Boston.
Bramber, Rape (in Sussex)	Horsham, and Steyning.
Brecknockshire (South Wales), 1815	Brecon.
Caermarthenshire (South Wales)	Caermarthen.
Caernarvonshire (North Wales)	Caernarvon.
Cambridgeshire, 1812	Cambridge, and Newmarket.
Cardiganshire (South Wales)	Aberystwith, and Cardigan.
Christchurch (Hampshire), 1794	Christchurch.
Cleveland (Yorks. N R)	Guisborough?

Cornwall

Cornwall	Bodmin, and Helston.
Craven (Yorks. W R)	Skipton.
Derbyshire, see p. 649	Derby.
Derby, West Hundred (in Lancashire)	Prescot.
Devon, North,	Barnstaple.
Devon, South,	Ivy-Bridge, and Totness.
Doncaster (Yorks. W R)	Doncaster.
Dorsetshire	Blandford, and Dorchester.
Drayton (Shropshire)	Drayton.
Durham	Darlington, and Durham.
Essex	Chelmsford.
Farmers' Club	Thatch'd House, St. James's-street, London.
Fenwick Farmers' Society (Durham?)	Fenwick.
Glamorganshire (South Wales)	Cowbridge.
Hampshire	Winchester.
Hayfield, Shepherd's Society (Derby), see p. 650	Hayfield.
...fordshire	Hereford, and Leominster.
...rtfordshire	Great Berkhamstead, and Hertford.
Holderness (Yorks. E R)	Headon.
Howden (Yorks. E R)	Howden.
Kendal (Westmorland)	Kendal.
Kent	Canterbury.
Kent Association	Ashford.
Kimbolton (Huntingdonshire)	Kimbolton, and St. Neots.
Lamport (Northamptonshire)	Lamport.
Lancashire	Lancaster.
Leicester and Rutland	Eaton, Leicester, Melton Mowbray, and Oakham.
London, Society of Arts,	John-Street, Adelphi, Strand.
Manchester (Lancashire), 1767	Altringham, and Manchester.
Melford (Suffolk)	Clare.
Merino Society	Freemason's Tavern, London.
Merionethshire (North Wales)	Endernion, and Penrhyn.
Newcastle (Northumberland)	Newcastle-on-Tyne.

GENERAL LIST OF AGRICULTURAL SOCIETIES. 653

Newcastle Cattle Society, 1815	Newcastle-on-Tyne.
Newcastle and Pottery (Staffordshire)	Newcastle Underline.
Norfolk	Blowfield near Norwich, East Dereham, Lynn, and Swaffham.
Northamptonshire	Northampton.
North-Lonsdale (Lancashire)	
Nottinghamshire	Newark, and Southwell.
Oswestry (Shropshire)	Oswestry.
Oxfordshire	Dorchester, and Woodstock.
Pembrokeshire (South Wales)	Haverfordwest, and Narbeth.
Penwith, Hundred (in Cornwall)	Penzance.
Peterborough (Northamptonshire)	Peterborough.
Petersfield (Hampshire)	Petersfield.
Preston (Lancashire)	Preston.
Radnorshire (South Wales)	Penybont, and Presteign.
Repton Sheep Society (Derbyshire, (see p. 649	Repton.
Rushy-ford (Durham)	Rushyford, near Darlington.
Saltersbrook Shepherd's Society, (Yorks. W R) see p. 650	Saltersbrook Inn, near Penistone.
Seamer (Yorks. N R)	Seamer.
Shiffnal (Shropshire)	Shiffnal.
Shiney Row (Durham)	Shiney Row, near Chester-le-Street.
Shropshire, 1810	Shrewsbury.
Smithfield Club	Freemason's Tavern, London.
Staffordshire	Litchfield, and Stafford.
Surrey	Dorking, Epsom, Guilford, and Ryegate.
Sussex	Lewes.
Sutton (Nottinghamshire)	Sutton in Ashfield.
Torrington (Devonshire)	Great Torrington.
Tyne Syde (in Durham and Northumb.)	Newcastle, and Ovingham.
Wellingborough (Northamptonshire)	Wellingborough.
Whaley (Lancashire)	Whaley.
Whardale (Yorks. W R)	Otley.

Whitby-

Whithy-Strand and Pickering-Lythe (Yorks. N R)	Hackness.
Wiltshire	Amesbury, Deptord-Inn, Devizes, East Lavington, Swindon, and Wilton.
Wirral and Western Cheshire (Cheshire)	Cartlet Ferry-House.
Wolsingham (Durham)	Wolsingham.
Worcestershire, 1816	Worcester.
Workington (Cumberland) 1804	Ireby, Keswick, and Workington.
Wortley (Yorks. W R)	Penistone, and Wortley-Chapel.
Wrexham (Denbighshire)	Wrexham.
Wyvelscombe (Somersetshire)	Wyvelscombe.
Yorkshire	York.
York, East Riding	Driffield.

Besides the above, I have collected the Names and some particulars, of a great number of Societies held in Scotland and in Ireland, all I have met with: but in the former of these Kingdoms, many of their Societies or Clubs, consist of small bodies of practical Farmers, often assuming the name of the small or obscure place or Inn where they meet; and which has not been sufficiently described in the printed accounts of their Cattle Shows and Meetings, for enabling me to refer them with any certainty, to the County and District to which they severally belong: otherwise, I should have wished to make the above, a more complete List of these Patriotic and useful Bodies in the United Kingdom: a *desideratum* which I hope some abler hand will ere long supply.

The late *Earl of Chesterfield*, for some years previous to his death, gave several *Premiums* annually, to his Tenants about Bretby, the liberal and judicious arrangements of which, are said to have had a very beneficial effect, in stimulating them to improved management:

nagement: the particulars of the distribution of these useful Premiums, in two of these years, will be found recorded in the " Farmer's Journal," of the 27th Dec. 1813, and the 15th of Jan. 1816.

Mr. W. B. Thomas's annual *Sheep-shearing*, and exhibition of Merino Sheep, Wool, and Cloth, has been mentioned in p. 118.

In *Derby*, a *Philosophical Society* has been some years established, and done considerable good, in diffusing Scientific and useful Information: some of the Members occasionally deliver Lectures on the subject of their particular studies or pursuits.

Another such Society was several years ago formed, and held its meetings successively at Chesterfield, Bakewell, and Sheffield; but before the time of my Survey, the latter meetings had been discontinued, and the Society met at *Chesterfield* only; sometimes Agricultural subjects are, I believe, occasionally introduced, and discussed at its meetings, but no Premiums are offered, or publications issued by it. Chesterfield seems a very eligible situation, between the Lead-Mining and the Coal and Iron Mining districts, for one of those provincial *Geological and Mineralogical* Societies, of which I have spoken in recommendation, in pages 217, 271, 298, and 302 of Vol. I.: intending, that such Societies should employ themselves, in useful *matters of fact*, rather than in bewildering Theories.

The *Coal-masters*, or occupiers of Coal-pits (see Vol. I. p. 188) on the Eastern Border of Derbyshire, with those of the Western border of Notts., are associated, as is mentioned p. 183 of Vol. I.; which Society holds its meetings at Eastwood, Notts. and I believe at Alfreton also.

The *Iron-Masters*, or occupiers of Iron Furnaces (see Vol.

Vol. I. p. 397), in Derbyshire and Yorkshire, adjoining, in March 1800 associated themselves, and have since held Meetings, at Barnsley, in March, at Bradford in June, Sheffield in September, and Rotherham in December, on the Wednesdays nearest to the full Moons. At several of the early Meetings of this Society, practicable and very valuable Papers were read by Joseph Dawson, Esq. and other Members, on the processes and economy of Iron smelting. Some of these Papers, and the Rules of the Society, are recorded in Clennel's " New Agricultural Magazine," Vol. II. p. 295.

2. *Where Agricultural Societies are wanting.*—It has appeared to me, that if such Societies held Meetings, in Chapel-en-le-Frith or Castleton, in Chesterfield, and in Measham, they might do considerable good to the cause of general Improvement: especially, if, deviating from the beaten track, their efforts were directed to accomplish some of the desirable objects, which I will endeavour to enumerate in the concluding Chapter.

SECT. II.—PROVINCIAL TERMS.

ALL the provincial terms and names of things and operations, or the Dialect, which had become familiar to me in early youth in Bedfordshire, and which I had been led to suppose were, great part of them, peculiar to that County and its vicinity, I found, during my sojournments in Derbyshire, were well known to many persons there, and were most of them in common use, mixed with other terms, unknown I believe, in the more southern and eastern parts of England.

Mr.

Mr. W. Bainbridge, a learned Gentleman of Alfreton, with whom I was not so fortunate as to form an acquaintance while in that neighbourhood, seems, from two Papers which he has published in the "Monthly Magazine," Vol. 40, p. 297, and Vol. 41, p. 311, to have correctly studied the language of Derbyshire, and has therein ably communicated a great deal concerning its peculiarities, to which I must beg to refer.

The forming of a *Glossary* of all the *terms* for things and operations, used by the several practical Men or others, with whom I have communicated, as well as in all the Books and Writings which I have perused, relating to Mineral concerns and processes, and Rural ones any way connected therewith, has, since entering on the Survey of this County, in 1807, in a particular manner engaged my attention: and my Glossary has in consequence swollen, to a size, not to be given here for want of room; but which it is my intention, at the first convenient opportunity, to revise and complete, and to publish it in a separate Tract; in which form I cannot doubt of its proving useful to other Mineral Surveyors, to practical Miners, &c. when they travel, and to Geological and other inquirers, towards enabling them to remove the chief, and indeed the only material impediments there are, to acquiring most important and valuable Information, from nearly every one of the practical Colliers, Miners, Quarriers, &c. &c.; but who are as naturally *shy of communicating* with those *who have not learnt the terms of their art*, much more so with those who by gestures or otherwise, appear to shew their contempt of them, as any of us should be, at *giving information* to a Foreigner, who might superciliously come among us, extremely inquisitive as to all our particular concerns,

cerns, but willing, only, to receive them *in his own language*, on account of his declared contempt, of the gothic barbarism of ours!

Such as may think I am thus over-rating the importance of acquiring a very extensive *knowledge of the terms*, used by the most common as well as the more informed of Mineral and Rural Persons, by all those who may wish to acquire and profit by the local and practical knowledge such Men possess, I would beg to assure them, that in almost every part of England, and in various parts of Wales and Scotland, I have always found this habit of studying and acquiring local terms, and *using them*, as early and constantly as is practicable, while communicating with the practical Men I have had intercourse with, to be the best and *most material qualification*, for profiting by, a free, full, and correct communication, of the valuable knowledge such persons mostly possess.

And I would beg to correct here, a too common mistake, in supposing, that the peculiar *terms* used by local Miners, and other similar and Rural bodies of Men, are loosely and capriciously applied, and therefore incapable of conveying correct information: because, I have uniformly found, that in every particular District, *one set of terms only are in use*, however small the District may be, ere another set of local terms will be found to commence, in part or in whole; and that altho' such terms, and the definition the users of them may be able to give thereof, may fall considerably short of scientific precision or truth, in frequent instances, I will venture to say, that even the mistakes of these practical Men, at the same time that they are easily discoverable by an attentive Observer on the spot, are in far greater and more extensive degrees excusable,

THE IMPORTANCE OF LEARNING LOCAL TERMS. 659

cusable, by *their accordance with appearances* of the things they are applied to, than many of the fancifully theoretic Terms and Distinctions, part German, part Latin, part English, &c. &c. part true and more false (see Preface of Vol. I. pp. viii, x, &c.) which the fashionable School of Geognosts and Wernerian Mineralogists*, seem of late years to have *determined*,

to

---

* I hope I shall be excused for adverting here, to the unhandsome and uncalled-for attack on the 1st Volume of this Report, and on its Author, by a highly respectable and able *Chemist*, a tolerable Mineralogist of the School alluded to, but considered by many as a dogmatical *pretender* to most other species of knowledge. Could I or the readers of my account of the Strata and Minerals of Derbyshire expect? that more than four years after its publication, any Journalist could be found to assert, that *the Inhabitants of the County only*, can reap benefit from my 1st Volume, because, as he is pleased falsely to assert (having so many scientific as well as other synonyms and popular descriptions before him, if he ever looked into the Book?), my names are *all local, and can convey no precise information!*—probably they will not, to blinded partizans of the Werno-Jamesonian jargon, of loose and falsely *theoretic* terms and distinctions, who will not, as far as in them lies, *suffer the publication* (see the account of their rejection of the Paper, Map and Section mentioned in the last paragraph of p. ix, of Vol. II. and adverted to in p. vi, of the Preface hereto, &c.), or *make mention of*, or reference to (unless it be to asperse and libel, see Phil. Mag. Vol. 45, p. 339, &c. &c.) the Geological writings *by any other persons*, but those of the Wernerian, or perchance the Huttonian, Schools, or which *are not couched in the terms* of the former of these Theories.

It is not true, that I treat " *scientific Names* with ridicule and contempt," as this Journalist has unblushingly asserted:—truly scientific Names, must be such, as *clearly, fully*, and *truly* describe th ethings to which they are applied, and such I ever have and shall continue to study, to respect, and to use *on all suitable occasions*: but falsely called " scientific Names," involving, if not wholly founded on the most palpable and visible *contrarieties to truth and nature*, I mean (as they well know) not as to the nomenclature or descriptions of *hand Specimens*

to have made, *the only medium of communicating*, on these subjects, which form my professional pursuits, and the principal subject of my writings.

Throughout the copious Index to each of these three Volumes, and through their pages, I have endeavoured to introduce *synonyms*, for explaining every technical or local term, or unusual word used, which appeared to require it; so that I may hope, no material difficulties will arise to my Readers, from the omission here, of a Derbyshire Glossary, which want of room, as well as the intentions above explained, compel me to omit in this place.

---

of Minerals, but as to *the strata*, " assemblages of Strata" (Vol. I. p. ix, &c.), or " Rocks," *mineralogically* NAMED, and as to the terms for their relative *positions* in the Earth's Mass, their *forms, changes* they have undergone, &c. &c. all which last, have really very little or *nothing to do, with technical Mineralogy*, or the knowledge and habits acquired, by study in the Cabinet Room or the Laboratory, as I have lately endeavoured to shew, in the Phil. Mag. Vol. 47, page 355, Note.

CONCLUSION.

MEANS OF IMPROVEMENT;
AND MEASURES CALCULATED FOR THAT PURPOSE.

IT now only remains, preparatory to taking leave of my Readers, to recapitulate or refer to those several hints and suggestions for Improvements, which my many kind Contributors, who are enumerated in the Prefaces to the former Volumes, or myself, have been enabled to give, on the various subjects treated of in this Report, and to add a few other hints which may occur to me while so doing.

I have observed, that most of the Gentlemen whom the Board have employed to Survey and to prepare *County Reports,* have in this concluding Chapter confined themselves to two or three, or but to a few subjects, which seem to have appeared to them, as the most strikingly important Objects of Improvement wanting in the County, or to the Agricultural Public, generally; entering, some of them, at considerable length, into arguments in this place, in favour of their adoption.

But the strict attention which I have all along paid to the very full and sufficiently well-arranged "Plan," containing 1031 heads or divisions,* of the subjects to be

---

* As appears, from a series of Numbers affixed to my travelling interleaved Copy of the "Plan," before commencing my Derbyshire Survey,

be treated of by the Board's Surveyors, and the care I have bestowed on interposing a great number of *other heads*, omitted in different parts of this "Plan," in the most consistent way therewith, that I have been able:—these, have already occasioned the mention or suggestion (preceded or accompanied by the facts and remarks which gave rise to, and appeared best calculated to enforce them) of so many subjects, on which the County, Parochial, Social, or Personal concerns of Derbyshire-men (often in common with those of the Kingdom at large) *might be altered for the better*, or Improved; that on a review now of these suggestions, collectively, I shrink from the task *of selecting* any small number of these suggestions, or subjects of Improvement, on which, in particular, to dilate in this place, as of primary or chief importance; but have chosen rather, and which I hope my Readers will approve, in as brief a manner as possible, merely to mention or enumerate the different points of suggested Improvements, and means for their accomplishment,

---

for the purpose of reference thereto: and in which Copy, on the other hand, references were afterwards made to the pages, in a continued Series, of the several Pocket *Memorandum-Books*, (opening the long way, and interleaved with soft blotting paper), in which the details of my Observations and Answers to enquiries were written, with a Pencil, on the spot and at the time—which general *Memorandums* of Observations, Remarks, short Extracts, &c. since I adopted this mode, are now so extended, as to occupy near 5400 pages; which Memorandums are for the most part Indexed, alphabetically, under the very various heads to which they relate: and I have taken the liberty here, and in p. 477 and its Note, to record so much of the methods pursued, in collecting and arranging the subjects and items (perhaps unprecedented in number) of local particulars and facts, which these Volumes contain: with the hope, that they may stimulate, and perhaps in some instances assist others, to commence and further perfect similar investigations and details, regarding several, if not all of the other British Counties.

in

in the same order as they have arisen, in the Chapters, Sections, and Heads, or divisions of this Report, with references to its pages, for the more ample details and arguments which will there be found.

I will presume to hope, that such a collected view, of Improvements *which may be made*, will stimulate numbers to the adoption of such of them, as appear most applicable to their own cases; that some other Persons will be induced to give, in consequence, a more close and systematic attention to this important subject, for correcting any mistaken views of things which may have been taken herein, and for supplying the many defects which will doubtless appear therein, on such an examination. To the Committees and Members of Agricultural and other Societies for " Internal Improvement," they may also prove useful, towards suggesting and selecting new subjects for their Premiums, or discussions.

A brief Enumeration of IMPROVEMENTS, *or Means for such, which have been already suggested, and a few others.*

IN VOLUME I.

That Parliamentary Commissioners should be appointed, for making exchanges between the different *Counties, whose borders intermix*, so as not to divide Parishes, Townships, &c., as at present, p. 91, (and 89, 90; and III. 611, &c).

That Lord Lieutenants of Counties, would, (in imitation of the late Dukes of Devonshire, Richmond, and Norfolk, and others), not place Clerical Men in the *Commission of the Peace*, p. 93.

That very large and populous Parishes should be divided into smaller ones, with more, and more roomy *Churches*, and resident Clergymen, for preventing, the otherwise necessary, increase of Dissenters, p. 94.

That Noblemen and Gentlemen would provide the apparatus, and order their Gardeners to keep regular *Meteorological Registers*, or Journals of Rain, Heat, Pressure, &c. p. 103.

That Quarriers, Lime-burners, Agriculturists, Masons, Road-makers, &c. should separately try, and ascertain the precise qualities of *each of the several beds of stone*, in the Rocks they excavate; as to setting or fertilizing qualities of the Limes they produce, the durability of the Walls, Roads, &c. they make, &c. p. 115, 157; II 411, &c.

That proper *Trials for Coals* (by general Subscription) should be made on the S and S E of Ashburne, p. 159.

That Coal-owners, or the Legislature, would look into and prevent the *waste of soft and small Coals*, p. 185, 346.

That justice to the Metropolis and E and S E of England, &c. and the preventing of the waste last mentioned, should be alike consulted, in a thorough revision of the *Duties on Coals* carried coastways, p. 186.

That Coal-owners and Lessees would cause, in every practicable case, their *Coals to be worked the long way*, as is successfully practised between Wollaton, Notts. and Sheffield, Yorks. &c. p. 188, 344, 350.

That Coal-owners, Masters, &c. and Agents, would very carefully preserve in Books, the full accounts or Journals of all *Boreings, Sinkings, and under-ground Workings* for Coal, or other Minerals, p. 216, 234.

That

That *Mineral and Geological Societies*, and Collections of local Specimens should be formed, p. 217, 271, 298, 302; III. 655.

That Mineral and Geological Observers, should far more minutely Note on the spot, and describe *the localities of Specimens of Minerals*, and observations, than is yet usual, p. 298, 302.

That practical Miners, Colliers, Agents, &c. in different Districts, would more freely and fully communicate, on the accidents from *Fire Damp*, and others, and the means of preventing them, p. 334; and Phil. Mag. Vol. 45, p. 436, 446.

That Colliers would search for, ascertain, and draw to the surface, in the progress of their Works, those substances which occasion the *spontaneous firing* of some Coal Works and Heaps, p. 349, and Phil. Mag. Vol. 45, p. 418.

That Owners of Coals, Ironstone, &c. would cause the top or *vegetable soil* to be removed and stacked, before beginning to sink any Pits, or to form heaps of rubbish on the surface; and fill up level and return such soil upon them, when disused, 351*, Phil. Mag. Vol. 35, p. 413.

That a public Law should authorise Colliers and Miners, under proper regulations, to *drain the surface* of their Works, and old hollows above their level, and stop water from descending into them, p. 351, 501, Phil. Mag. Vol. 45, p. 445.

That *Cannel Coal* should be wrought, and not left

---

* At Adelphi Furnace, the stacking of the vegetable Soil, levelling the open and Bell Pits, from whence the Ironstone had been wrought, and evenly spreading the top-soil thereon, cost in 1808, about 3*s*. per Rood (of 49 square yards), or near 2½ per cent. on the value of the Ironstone obtained.

or wasted, as is too common, on account of its harmless, although rather unpleasant crackling, when laid on the fire, p. 352.

That the *Mining Laws* of the Peak Hundreds of Derbyshire, should be entirely revised and altered, to cause the *consolidation of Titles* to Veins, where Ore remains unwrought, to enable (on making proper compensation) the *driving of Soughs*, erecting Engines, &c. to unwater the same, &c. p. 364, 370, 381, Note. And so, that Owners and Occupiers of the Lands should be able to prevent the devastation of their fields, and poisoning of their waters, by *Cavers and Buddlers* of old Mine-Hillocks, p. 363, 377; III. 623.

That Iron-masters and Coke-burners would abandon the burning of open heaps of Coals, and adopt some of the *close ways of Coke-making*, p. 399.

That Colliers should search and examine their Binds and other quickly perishing strata, to ascertain such as can be wrought and advantageously used in the *Marling of Lands*, p. 446; II. 407.

That the Legislature should remove all remaining *Taxes on Bricks and Tiles used in draining*, or in building or repairing *Farm Out-houses*, Walls, &c. p. 455; II. 395.

That the occupiers of rocky, dry Pastures should cause *Cattle-ponds*, on improved principles, to be made therein, p. 494; and Drinking-places, in fields in general, p. 495.

VOL. II.

That Land-owners should discontinue the employment of *Attorneys* as their Land Stewards, p. 241.

That Surveys and impartial Valuations of *all the properties*

properties in Parishes, should be made, for justly *equalizing the Poor-rates* in them, p. 3: and that the Legislature should enact a more equal distribution of them on property of all kinds, than separate parochial assessments admit of, p. 34; III. 540, &c.

That the erection of *Wooden Buildings* in or near *stony* districts, should be discontinued, p. 12: and the use of *Straw for Thatching* all permanent Buildings, as much as possible laid aside, p. 14.

That Stone or Brick *paved Thrashing Floors*, should be substituted for planked wooded ones, p. 17.

That Land-owners, while Tithes continue uncommuted (for Lands or Corn Rents, III. 639), should take *leases of Tithes* from their Lay or Clerical owners, and let their lands *tithe-free*; and any tithes of other Person's Estates to their several Occupiers, p. 29.

That *White-strawed Crops* of Grain, should not immediately succeed each other; or any other kind of Crop be too often repeated, p. 39, 103, 175, 401, 406, &c.

That Farmers, as well as the Bailiffs of Gentlemen Agriculturists, should keep and preserve very full and methodical *Farm Accounts*; which desirable ends, the ruled and titled blank Account Books, printed annually by Mr. J. Harding of St. James's-street, London, are well calculated to promote, p. 42.

That Agricultural Societies, instead of spending so very large a proportion of all their attention and Funds on Live Stock, now that Animals are so much improved, should offer liberal *Premiums* for the best new, or most efficiently *improved Implements for Culture*, when properly recommended, by their practical users, p. 43.

That

That *Drill-rollers*, for breaking down obdurate Clods in *Fallows*, should be more used, p. 46.

That the *Drilling for hire*, of Corn and Turnips, Horse-hoeing, &c. should be undertaken, and generally encouraged, p. 48.

That Magistrates, Gentlemen, and Persons generally, should discourage the rising and mistaken clamours of Poor Persons, in some districts, against *Thrashing Machines:* and that encouragement should be given to the use of Portable Machines, in *Thrashing for hire*, as preferable to erecting so many Machines, p. 55, 124.

That the *Wheels* of Carriages should be made with *cylindrical rims;* and one and two-horse Carts, more generally used, p. 60.

That the use of correct *Weighing Machines* should be much extended, p. 65; III. 245.

That encouragement should be given to the use of *Chimney-sweeping Apparatus*, as improved by Mr. George Smart, of Ordnance Wharf, Westminster Bridge, for dispensing with climbing Boys, p. 69.

That if a general Act, so much desired, for *Inclosing* of open and intermixed Lands, cannot be obtained: the *Fees* on Acts for Inclosing small tracts of Land, certainly ought to be taken off, that their improvement may not be entirely delayed by so improper a cause, p. 78, 343, 637.

That the use of *two-horse Ploughs* should be more generally adopted, p. 95.

That the *drilling of Crops*, and the *Dibbling* of others, should be more practised, p. 96, 98.

That neatness and security in the forming, thatching, and protecting the *Roofs and sides of Ricks* of Corn and of Hay, should be more attended to: and for which end, *Premiums* given by Societies, to the Farm Servants who excel in performing these operations in the best and most expeditious manner, would have good effects, p. 112, 180.

That the Agricultural Societies (lessening, if necessary, their Live-Stock Premiums) should give more attention and *Premiums*, for the better and more advantageous culture of *arable Crops*, of Corn and Pulse, of other Roots besides Turnips, and of useful plants in general, p. 113, 140, 149, &c.

That the *pareing of Stubbles* and carrying off the Straw and Weeds, and slightly burying the Seeds, immediately after harvest, should be more extensively tried, p. 124; III. 182.

That the *manuring of Grass Lands* should be performed in warm and moist weather, instead of frosts, p. 185, 457.

That a better system of *stocking grazing Land*, so as to prevent the growing up of useless seed-stalks, would be worthy the attention and *Premiums* of Societies, p. 189.

That the Legislature should require Road-Surveyors, both Turnpike and Parish, carefully to *destroy Thistles* and *Docks* by their Roads' sides, in or before their flowering time; and render it criminal in any one, to maliciously suffer Weeds to grow up on his Lands to annoy others, p. 193.

That more attention should be given in seeding down, and in Weeding Grass Lands, to exterminate *broad-leaved Plants*, for improving the quantity and quality of Hay, p. 196.

That the true *Fiorin Grass* of Dr. Wm. Richardson, should be cultivated on Bogs, p. 202, 346, 467.

That the valuable liquid Manure of the *Sewers of Towns*,

Towns, should not be wasted, as at present, p. 212, 453.

That *Apple Orchards* should be more generally planted on the Red Marl Lands, where not too tenacious and wet in Winter, p. 214; and that *Walnut Trees* should be more planted, p. 215.

That the *planting of precipitous Lands*, and very rocky ones, impracticable to the Plough, should be encouraged by *Premiums*, and on no others; but on the contrary, *Premiums* for *clearing and cultivating* lands adapted to the Plough, but occupied by Woods or Plantations, will be more proper, p. 226, 236, 237, 261.

That much higher prices, and other indulgencies should be granted, to the Timber-growers who supply the Royal Dock-yards with *large Ship Timber*, p. 227, 316, 321.

That the earliest practicable clearing of *Spring Woods*, or those growing Underwood and Trees, and shutting of them up, after the falls in them, should be attended to, p. 228.

That *Ivy and Moss* should be carefully cleared from the bodies of Timber Trees, p. 231.

That Timber Fallers, should *dish the stools of fallen Trees*, rather than waste Timber and labour, in leaving them round-top't, p. 232.

That all vacant places in Woods should be carefully filled up by planting, p. 233.

That Young Men, heirs to Estates, should seriously undertake and persevere in, *planting, pruning*, and keeping minute *accounts* of the progress and produce of Plantations and Woods, p. 239.

That Plantations should be made rather *thick of Trees*, and be pruned and thinned, p. 243.

That

That *Larch Trees* should be a good deal increased, p. 252; and *Oaks*, on strong soils, on Ironstone Shales, in particular, p. 254 (I. 395).

That Land-owners, stipulating for their *Lessees to plant* and preserve hedge-row Trees, should allow them compensations for so doing, p. 258.

That *Hedge-row Timber Trees* should be selected or planted, and carefully pruned and trained, at the expence of the Landlord, p. 259.

That *Spanish Chesnut Trees* should be more cultivated, p. 266.

That high and exposed Waste Lands, should be *sheltered* by belts of Planting, or the whole planted, and fields cleared out therein, some years after, to protect cultivation on them, p. 268 (I. 382; II. 243).

That careful and persevering attention to *Pruning Forest Trees*, on the improved principles (described p. 296) should be given, from the Nursery upwards, as being one of the greatest, as well as most wanted, of Rural Improvements, p. 282, 296.

That *Premiums* should be given to those *operative pruners*, who excel in perfection and dispatch of their business, p. 304 N. 306.

That the utmost attention and thought should be given, to the timely and rightly conducting *the thinning of Plantations*, as being the most difficult part of their management, p. 311.

That the system under which the greater part of the *Royal Forests* have been managed, should be entirely changed, p. 315, 353.

That Oak Trees, Spires and Underwood, should always be *felled before peeling the Bark*; and the peeling, drying and loading, &c. should be let *by weight*, p. 331, 333.

That

That the Owners of Woods and Timber, should frequently, if not always, *house or stack the Bark*, on their premises, and sell it out there, by cubic measure or weight, p. 336 (333).

That the extreme *twigs and buds of Oak* should be collected, for sale to the Tanners, p. 339.

That the *pruning* of neglected Oak Plantations and Woods (as well as their thinning), and the *lopping* of Oak Pollards (until such are exterminated, p. 260), should be done in the Spring, and *the branches peeled*, p. 339.

That the valuable tract of Land called *Synfin Moor* or Fen, should be more effectually drained and improved, p. 350.

That Land-owners should turn their anxious attention to permanent *Improvements*, which may be effected on their Estates, and become immediately profitable, and wherever the Tenants don't adopt them, to direct and bear the cost of such, and charge Interest on the out-lay, as additional Rent, p. 360.

That *Premiums* should be given to successful *Professional Drainers*, and undertakers of the work *by measure:* and it would be well to stimulate the taking, and theoretically and practically instructing *Draining Pupils*, p. 383 (361).

That thoroughly *drained Bogs*, should be immediately pulverized on the surface, to prevent deep cracking, and sterility, p. 396.

That *deep fence ditches* should surround all wet lands, and be kept clear of obstructions and stagnant water, p. 397.

That the "Commissioners of Sewers" (which any County or district may have appointed), or the enactment of a new Law, if necessary, should compel the

entire

entire removal of all Trees and large *stems of aquatic Woods*, growing in the Banks, and projecting in the stream in ordinary or flood times, of Rivers and Brooks, in all flat and occasionally inundated parts, p. 398.

That the projecting points in the sides of *Brooks and Rivers* should be removed, low or sloping weir-hedges constructed, to stop the further wear of hollow places, and the sides of such streams reduced to easy and regular grassy slopes, p. 398, 400.

That the *periodical* and systematical *pareing and burning* of the surface soil should be discouraged or prohibited: altho' the advantages of this mode of reclaiming Waste Lands, are great, p. 400.

That the *Marling* of Lands, in the Red Marl districts (which so prevail in the middle parts of England, p. 146, 147, Phil. Mag. Vol. 39, p. 28), where naturally rather dry, or made so by under-draining, should be more practised than at present, p. 407 (446).

That frequent and careful experiments on the effects of *Liming Lands*, should be made, as well in those districts, where it has long been practised, as where it has not been used, p. 408, 412, &c.

That the *dust of Limestone Roads*, should be collected and kept and used *dry*, instead of Lime, in manuring, p. 445.

That the *Sanding of clayey Lands* should be carefully and more frequently tried, p. 447.

That *Gypsum*, pounded and ground to a fine powder, *without previously heating it*, should be carefully and more extensively tried, as a Manure, p. 448.

That *Peat-Ashes* and dust, should be very carefully and more generally tried, as Manure, p. 448: paying attention to *the nature and situation of the*

Peat whence it is made: see Farmer's Jour. No. 409, 24th July, 1815, p. 234.

That *Coal-Ashes* should be carefully tried, on different soils, and situations, in all those parts of the County, and elsewhere, where they are neglected, as the valuable manure they universally prove, I believe, 449, (186, 414).

That *Bones* of every kind, *cores of Horns* at the Tanners (p. 452), &c. should be carefully collected (perhaps boiled, p. 450) and *ground**, and applied as manure, p. 449.

That more extensive and careful experiments should be made, than heretofore, on the *raising of crops for, and the ploughing them in green, as manure*, p. 452.

That the offensiveness of Privies in the Country, should be more generally prevented, by frequently throwing down small quantities of loose earth (or lime?) to be dug out for manure, p. 454.

That *Dung-holes and Yards* should be improved, by under-drains, by excluding too much water, by Cesspools, Pumps, Landers, &c. p. 455 (453).

That the use of Drags, instead of Pummels, to Dung, Lime, Marl or Stone *Carts*, should be tried, p. 456.

That more care and attention should be paid to the increasing, making and applying of *Compost Manures*, p. 457 (184, 400, &c.).

---

* Mr. *John Charlton* of Calow (Vol. I. p. 163 Note), makes improved Mills for this purpose, wholly of Iron, and delivered them, in 1815, at the Canal-Wharf at Chesterfield (p. 317), at 50/. each. It occurs to me, that a much *finer grinding of the Bones*, by a subsequent operation, to the excellent Mills I allude to (and which would be easier performed on boiled Bones), would make them a more economical and effective manure?.

That *Irrigation* should be introduced and extended, wherever water is plenty; taking care to give very ample *slope* to the surface, in every part, p. 463.

That no opportunities should be neglected, of washing or conveying *liquid manures*, from Cattle-Yards, Sewers, Streets, Roads, &c. on to adjacent Grass, Garden, or Arable Lands, p. 465, (212, 453, &c.).

That *Premiums* should be given to successful *Professional Irrigators*, and undertakers of the work *by measure:* and it would be well to stimulate the taking, and theoretically and practically instructing *Irrigating Pupils*, p. 467, 481.

That Land-owners should construct proper Irrigation works for their Tenants, or grant Leases, on condition of such being done by them, p. 477 (361).

That *Inclosure Acts* should contain Clauses, providing for Irrigation, wherever practicable, p. 482, 491.

That Steam-Engine Mills, and Wind-mills (p. 492), should be erected, in or near to Towns, where they are most wanted, in lieu of all the *Water-mills* which most injure the adjoining low lands, or impede the Drainage or Irrigation of others, p. 489, (462).

That certain lengths or districts of *Valleys* should be improved (under Acts of Parliament, where various properties occur) by side *Canals* nearly or quite level, from which over-shot instead of under-shot Water-wheels, might be supplied, and the Drainage and Irrigation of any parts of the lands below them, be effected, p. 490.

That the low Meadows, or occasionally Inundated Lands, by the sides of the Dove, Trent, Derwent, and other Rivers, should be *Embanked;* with provision for *warping* or mudding such parts as may lie below the level of very *thick flood waters*, p. 494.

VOL. III.

That *Milk-Farming*, or the keeping of a succession of Cows in Milk, the year through, for the nearly equable and full supply of the *Inhabitants of Towns and Villages* with Milk, should be more generally adopted, p 30.

That *Cows*, the best carcassed, or most disposed to increase and *fatten*, and at the same time good and large *Milkers*, and proper Bulls, should be more generally selected by Farmers, to breed from, p. 32.

That the very expensive and silly practice of *colouring Cheese* should be discountenanced, p. 58.

That the *Cheese Fairs* should be more generally attended for business, and no Cheese delivered into the Factor's Warehouses, before absolute bargains for the prices, p. 62.

That the selling of all fat Animals by their *weight alive*, should be encouraged; as is done with regard to Hogs in Ireland, in Anglesea, &c. &c. p. 74, 131, 173.

That *Sheep*, the best carcassed or most disposed to increase and *fatten*, bearing the most valuable *Wool*, quantity and quality considered, and being at the same time prolific and hardy, should be selected to breed from, p. 97, 114, &c.

That *Wool-fairs* should be established in every district, and the Growers make as great a point as they can, of attending and selling thereat, 137.

That breaking of Colts, and *Training of Nags*, should be made a separate business from Farming, p. 155.

That

That vigilant attention should be paid by the Breeders of Improved *Pigs*, to timely crossings with other's Stock, p. 164.

That Pigs should have the gristle of their Noses cropped when very young, and be *tethered* by the neck, and fed on natural and artificial Grasses, much more extensively than at present, p. 169.

That *Premiums* should be offered to Rural Labourers who have well performed the greatest quantity of *job or measured Work*, at fair and usual prices, p. 193.

That *Premiums* should be offered for Essays on the best principles and *modes of letting and conducting Rural Works*, with proper Prices of each, &c. p. 194.

That *Gas-Lights* should be substituted for Tallow or Oil Lights, in Towns and large Factories; and an adequate *increase of duty on imported Tallow*, compensate the British Farmers and Whale-Fishermen, p. 197.

That the *Game Laws* should speedily undergo a full and liberal discussion, and be altered, so as to less injure the Farmer than at present; and that Gentlemen would yield their Grouse Moors, to planting and cultivation, in many instances, p. 198, 199, 202.

That the County would be benefited by the making of some *new lines of Road*, for avoiding Hills *, &c. p. 225.

That accurate Surveys of districts of the *public Roads* should be made, for suggesting Improvements therein, p. 231.

---

* I had the pleasure of learning, a few days before writing this, from Richard Arkwright, Esq. that part of the suggested improvements, between Cromford and Great Rowsley (p. 236), had been carried into effect.

That *permanent Road Acts*, in districts, or a general one, should be passed, instead of the present expensive and inadequate system, p. 235.

That *a Board of Roads and Bridges* should be appointed, for Auditing Road Accounts, and advancing or lowering Tolls, when requisite, p. 235, 240.

That Roads would be easily kept in a much better state, if Carriages had their *wheels* at different *widths apart*, p. 241, 242.

That a progressive scale of Tolls, should induce the carrying of *less Weights*, generally, than at present; and the mischievous exemptions in favour of very *broad wheels*, be entirely abolished, p. 144.

That Spirit and Wine Licenses to detached Inns on the Roads, should be altered (if not already so done in the Act of this Sessions?), p. 247.

That *Mail-Coaches* should pay Tolls, p. 248, (II. p. 317).

That all the largest *round smooth Pebbles* laid on Roads, *should be broken*, in order to allow of their wedging in, and becoming fast, and the Road smooth, p. 250, 281.

That *the hardest quarry stones*, only, should be laid on the Roads, when broken almost uniformly small, p. 254.

That Land-owners should employ the Poor, in winter, or when Work is most scarce, in systematically digging, sifting, and *preparing Gravel for the Roads;* and prevent Road-Surveyors devastating their Lands and Commons, p. 252, (238).

That much more of the *Labour on Roads* should be Let or contracted for, than at present, p. 260.

That the *Statute-duty* Laws should be greatly altered, for placing at the disposal of more competent Road

Road Surveyors, instead of Team-work in kind, Money, raised by a far more equal Rate, *on property in general*, p. 262, 637.

That *Premiums* should be given to *Road Surveyors* who manage the best, considering expence, situation as to materials, traffic on, and state of their Roads, throughout given periods.

That Surveyors should remove from the Roads and Lanes, all *Nuisances*, as shelves of Rock, Bye-sets, Ploddings, Scotching-stones, large Stones, either fast in the Road or loose, the breaking of Stones, and heaps of such, Timber, Bushes, or Dung, Thistles, loose Cattle, &c. &c. p. 263 to 267.

That all *unnecessary widths of Roads* should be reduced; but not by the nuisance of *belts of planting*, 268, 269.

That *Hedges* by the Roads should be *without Trees* (except a few tall *pruned* ones, II. 259), be cut low, and kept *clipped;* and such be planted by the sides of the open Roads, with *Gates*, on the best construction, only, when such are suffered to remain, across the less frequented Roads, p. 272, (II. 92.)

That *Foot-paths* should be provided by the side of the Roads; and a great many of *those across the fields stopped*, and all of these be so altered, as to follow the Fences, or such altered to suit the Paths, p. 273.

That an Act should be passed, allowing and providing for, the making of *Rail-way branches to Canals*, more extensively than at present, p. 287.

That Government should not claim *exemptions from Tolls on Canals, Rail-ways or Roads* made at private expence; or set *limits to the profits*, of the Proprietors of the former establishments, beyond fixing the maximum of their Tolls, p. 248, 291, (II. 317).

That Canal Companies should allow, the *stacking of Coals* on the Wharfs of great Towns, in store for part of the Winter's consumption, and take *Bonds* for the Tonnage due thereon, until sold off, p. 293.

That Canal and Rail-way Companies should be authorised, to cause the legal adjustment, and compensation to be made, between the Workers of Coal-Pits, Mines, Pits, Quarries, &c. who dispute about *draining* one another's Works: and the more extensive measure be adopted, of appointing *Commissioners of Mineral drainage and ventilation*, p. 293 (I. 351, 501), Phil. Mag. Vol. 45, p. 445.

That Canal Companies should well consider and attend to suggestions, as to the operations of their *Bye-laws*, regulating the Trade, &c. on their Canals, p. 326.

That further *junctions of the Canals* in and near Derbyshire should be made, viz. either from near Lea-wood to near Bugsworth, p. 369, or from Horninglow to Uttoxeter, and from Endon to Marple, p. 388 (386). Also from near Rotherham to near Killamarsh, and from Chesterfield to Pentrich-lane (with a nearly level Rail-way to Ashover Lime-works), p. 393 to 395. And from near Nutbrook-mouth to Trowell-heath, p. 400.

That a speedy and entire reformation of our national *Weights and Measures* should take place, on the principles concisely stated in the Note*, p. 464.

That

---

* Since the Section of my Report, here referred to, was printed off, Earl Stanhope and other noble Peers, have done themselves great honour, by causing the rejection of Sir George Clerke's very lame and inefficient Bill on this subject, and towards appointing *a Commission*, for more thoroughly inquiring into the subject. The discovery which has been made, of great numbers of *Avoirdupoise Weights*, having

within

That *Re-Surveys* of the principal *Manufacturing Counties* should be made for the Board of Agriculture; and, that the whole of the Counties should be again surveyed, is perhaps desirable, p. 477.

That *Agricultural and Rural Labourers*, should be more adequately paid, 499, 526 (193).

That the practice of employing constant successions of *Children*, in particular Manufactures, should be discouraged, p. 501, 506 (II. 33).

That *Premiums* should be given to encourage the taking, instructing, and employing of *Agricultural Pupils*, by practical Men;—and others, to the most industrious and qualified of such Pupils, p. 514, Note.

That *Premiums* should be given to those heads of

---

within some years past, been shamelessly *stamped* in Westminster, *without being weighed!* (thus distributing throughout the Kingdom, *doubtful or false Weights* of this kind), has almost entirely changed the ground, on which I have for years been arguing, in favour of the Avoirdupoise Pound, as our standard decimal of new and *unit* Weights; and now, after a long communication with Earl Stanhope, I am happy in entirely concurring with his Lordship, in favour of the *Metre* and *Gramme* of France, as our future *Units* of Measure and Weight, and our own *Pound Sterling*, as that of Money; and I beg sincerely to wish and hope, that the enlarged views of the Commission to be appointed, and of the Legislature, in consequence, may lead them to the same conclusions.

The call now (June 1816) so strongly expressed, for a Coinage of *Pound-pieces* of Gold, instead of Guineas, which once prevailed amongst us, comes most opportunely for the adoption above mentioned; and I would beg earnestly to press on the attention of His Majesty's Ministers, the propriety and great advantages which will result, from accompanying the issue of these new *Pound-pieces*, by moderate numbers of *Tenth-pieces*, *Hundredth-pieces*, and *Thousandth-pieces*, of a Pound, for circulation *along with the present (or new) Silver and Copper Coins*, as recommended in p. 465, until the Public are fully habituated to their relations, and see by use, the utility of a *decimal scale of Money*, agreeing with that of our numeration and arithmetic, as ere long, I hope, it will also do, with all our weights and measures.

Poor

Poor Families, who kept and produced regular and minute Journals and *Accounts of the Labour, Earnings, and Expenditure of their Families*, p. 526, Note.

That the Laws regarding *Bastardy* and Pauper Marriages, should be entirely altered; and the *Emigration* facilitated, of those who here cannot obtain employ, p. 527 Note, 619.

That the powers of *Corporate Towns*, to exclude poor Persons from labouring at various of the common Employments, within their liberties (now that Apprenticeship exclusions are elsewhere abolished, p. 520, Note, 508), should be considerably modified, if not abolished altogether, p. 527, Note.

That a full enquiry should be made into the origin, progress, and Funds for *Combinations amongst Apprenticed Journeymen*, and other bodies of Workmen, with a view to more effectual measures or Laws, for their suppression; and equally so, *Subscriptions* of combined Masters, for depressing Wages below the results of fair *competition*, among all properly qualified Workmen, whether of the apprenticed class or not, p. 528, 573.

That increased encouragement should be given, to spread and increase *Lancasterian, and Sunday Schools*; and to the distribution among their Scholars, and the poor generally, of *Tracts* and plain and easily intelligible Books, calculated in a more practicable way than at present, to inform and better them, 529, 530, 569 Note.

That Parliament should require new and improved *Returns from Parishes* and Townships, of the real burthens *per pound*, which the *Poor* in each occasion, the Trades or modes of life Paupers have severally followed,

followed, the Wages usually earned in such Trades, &c. p. 540 to 543.

That *Parish Officers* would less refer to, or be led by Lawyers, or to the Sessions, for adjusting doubtful Settlement Cases, but amicably settle such amongst themselves, p. 548, Note.

That *Subscription Poor-houses*, would in many situations be preferable, to erecting or keeping up small separate Parish Workhouses, p. 549, 563.

That the Laws regarding *Friendly Societies* should be reconsidered, and much altered, p. 569, 572, Note.

That encouragement should be given, to increase and extend *Friendly Societies*, by very cheaply letting them *Rooms*, procuring them *Information*, Books, &c.; but not by subscribing to, or interfering with their *Funds*, p. 574, 578.

That Parliament, in 1821, if not sooner, should require new *Population Returns*, in which the numbers of *Agriculturists* and of *Manufacturers*, &c. should be far more accurately ascertained, than in the last Returns, p. 590, (528).

That the Returns, above alluded to, should be so separated and described, as to admit of more accurate matching of the *Enumerations*, with the *Parish Registers*, than at present: and so, as to admit of vastly more correctly stating, the actual *Population of the Towns* of these Kingdoms, than is yet done, p. 591, (611), 604.

That the Legislature should require double *Stamps* to be put on all Almanacks containing *Astrological predictions*, or Calendars adapted to the uses of its knavish or foolish Pretenders or Dupes, p. 627 (87).

That more effectual measures or Laws should be resorted to, for suppressing *Bear-baiting*, *Bonfires* in
the

the streets of Towns, *gambling Mountebanks, Gipsies* and *Fortune-tellers*, which disgrace, annoy, and injure the Community, p. 629, 630, 631.

That Land-owners should embrace favourable opportunities, of consolidating the lands of the smaller and poorer of their Tenants, into more *adequate Farms*, and grant *Leases* of such on liberal terms (and if assignable the better) to persons of sufficient capital and skill, to carry every practicable Improvement into effect thereon, p. 633, 636, 638, (II. 26).

That the Legislature should contribute efficiently, towards *the cheaper circulation of Agricultural Books*, and those relating to Rural Statistics, towards which, Provincial Societies should also lend their aid; and half *Stamp* Duty, only, should be paid by Agricultural Newspapers, p. 642, 643.

That on the appearance of many Grubs on the newly ploughed Lands, or webs or pockets of Caterpillars, &c. on Hedges, Women and Children should be employed to pick, and burn them, p. 644, 645.

That Proprietors of Estates, should more generally imitate the praise-worthy conduct of the late Earl of Chesterfield, and others, in giving *Premiums* among their own *Tenants*, or those of the immediate vicinity, for plain and practicable Improvements, effected by them, p. 654.

That new *Agricultural Societies* should be established, in some others of the principal Towns of the County; or that its general Society should hold Meetings in these places, in rotation, p. 656.

APPENDIX.

Eleven Years' Abstracts and Averages of Meteorological Observations made at Derby, Lat. 52° 58' N, Lon. 1° 32' W: about 170 feet above the level of the Sea: by the late Mr. Thomas Swanwick, of the Commercial Academy, who died in March 1814.

Years.	Thermometer.						Barometer.					Pluviometer.
	Hottest Days.	Degrees Farh. on those Days.	Wind ditto.	Coldest Nights.	Degrees Farh. on those Nights.	Wind ditto.	Annual means, Degrees and Tenths.	Highest observations.	Inches, and Decimals, then.	Wind ditto.	Lowest observations.	Inches, and Decimals, then.	Wind ditto.	Totals of Rain, in Inches and Decimals.
1803	July 19	80	W	Dec. 8	18	N W	46·8	Nov. 30	30·40	N	Nov. 11	28·50	S W	—
1804	June 25	83	N E	Dec. 16	14	N	44·5	Feb. 21	30·50	N W	Jan. 28	28·90	S E	—
1805	August 11	76	S	Dec. 13	22	N W	47·6	Nov. 15	30·79	N	Feb. 5	28·81	N	—
1806	June 10	82	S	Jan. 30	25	W	44·2	May 19	30·61	S	Jan. 7	28·72	W	—
1807	July 10	82	W	Jan. 17	17	N W	45·4	Mar. 1	30·75	N E	Nov. 20	28·85	N W	—
1808	July 13	90	S W	Jan. 22	12	N W	43·7	Feb. 25	30·98	N	Dec. 2	28·71	W	—
1809	July 27	78	S W	Jan. 19	17	E	45·5	Mar. 8	30·40	W	Dec. 18	28·15	S E	28·51
1810	July 7	76	S W	Feb. 21	15	S W	47·8	Feb. 21	30·36	S W	Nov. 10	28·85	E	29·59
1811	May 13	78	S.	Jan. 30	10	E	48·0	Mar. 12	30·38	N E	Oct. 26	28·51	S W	24·36
1812	July 9	78	S E	Dec. 9	19	N W	43·7	Dec. 7	30·48	N E	Oct. 19	28·34	S	28·79
1813	July 30	80.	S	Jan 29	17	S W	46·3	Nov. 4	30·51	N	Oct. 17	28·93	N	20·34
Averages	July 8	80	S S W	Jan. 11	16	N W	45·8	Jan. 30	30·51	N	Dec. 1	28·66	—	24·32

In pages 96 and 97 of Vol. I. I had occasion to lament, the not having met with any continued registers of the *Thermometer* or *Barometer*, kept in the County; in consequence of which, the late Mr. Swanwick (I. 104), while suffering under the illness, which a few months afterwards, terminated his useful life, kindly sent me the Table in p. 685, except its last line, which I have calculated.

Whence it appears, that the *hottest* day, or greatest height of Fahrenheit's Thermometer, and *coldest* night, on the average of 11 years, occurred within 4 days of half a year apart, or 17 days and 21 days after the Summer and Winter Solstices, respectively; the Spring Interval, of increasing heat, being the shortest, by that number of days, and the whole of this increase of heat, being 64° annually, on the average.

The *greatest heat* of these years, July 13, 1808, exceeded the hottest day of 1810 (July 7) by 14°, the former exceeding the average by 10°, but the latter falling below it only 4°. The *greatest cold* of these years, Jan. 30, 1811, exceeded the coldest night of 1806 (Jan. 30th) by 15°, the former exceeding the average by 6°, and the latter being warmer by 9°. The extreme range of temperature, between July 13, 1808, and Jan. 30, 1811, having been 80°, or just the same as the average hottest day, annually, of the whole period.

The lowest average heat of any of these 11 years was in 1808 and 1812, and the highest in 1811, their difference being $4°\frac{1}{10}$; the former exceeding the mean of the whole period by $2°\frac{1}{10}$, and the latter falling short of it by $2°\frac{3}{10}$.

The greatest heat of 1808, exceeded the general mean of the whole 11 years (or $45°\frac{1}{10}$) by $44°\frac{2}{10}$, and the greatest cold of 1811, fell below the general mean $35° \tfrac{5}{10}$.

The *lightest state of the Air*, (least Atmospheric pressure, or lowest level of the Mercury in the Barometer tube) and the *heaviest* state of the Air (&c.) on the average of 11 years, occurred only 60 days apart, at the same interval before and after the commencement of the year, or 20 days before and 40 days after the Winter Solstice, respectively; the average variation annually, being $1\tfrac{54}{100}$ Inches.

The *greatest pressure* of the Atmosphere in these 11 years, Feb. 25th, 1808, exceeded that of the heaviest day as to pressure, in 1810 (Feb. 21) by $\tfrac{62}{100}$ Inch, the former exceeding the average by $\tfrac{42}{100}$, but the latter being below it only $\tfrac{11}{100}$ of an Inch. The *least pressure* of these years, Dec. 18, 1809, fell short of the lightest pressure in 1813 (Oct. 17) by $\tfrac{71}{100}$ Inches, the former being below the average $\tfrac{11}{100}$, and the latter above it $\tfrac{27}{100}$ of an inch. The extreme range of pressure, between Feb. 25, 1808, and Dec. 18, 1809, having been $2\tfrac{13}{100}$ Inches: but how these extremes were related, either to the annual or general mean pressures, Mr. Swanwick has, unfortunately, not given me the means of knowing. In case any of Mr. S.'s family or friends have it in their power to furnish me with the *mean pressure* for each of these 11 years, I should be much obliged thereby.

The total depth of *Rain*, annually, in 5 years, was greatest in 1810 (see I. 104) and least in 1813; their difference being $9\tfrac{1}{4}$ Inches; the former exceeded the general average by $5\tfrac{37}{100}$ Inches, and the latter fell short of it by $3\tfrac{84}{100}$ Inches.

In

FORM OF AGREEMENT FOR A COLLIERY COTTAGE.

In p. 194 of this Volume, I intimated my intention of speaking in Sect. 8 of Chap. XVI., on *Cottages attached to Manufactories,* in order to shew the kind of Agreement, which is in some instances made, for letting a Cottage (with its Garden perhaps) to a Workman, so long as he continues in the Owner's employ, and no longer: but having afterwards forgot the same, I beg here to preserve, *A Form of Agreement* for Letting a Cottage on the Estate of a Coal-Master, to one of his Colliers, which I met with in use, viz.

" Memorandum of Agreement made the —— day of ——, in the year of our Lord, —— between T. U. on behalf of V. W. and Y. Z. as follows, that is to say, the said T. U. for and in consideration of the said Y. Z. having undertaken and agreed to work and employ himself, at the Colliery of the said V. W. in the capacity of a Collier, upon such terms and conditions as are specified in the *Contract* entered into for that purpose, doth hereby undertake and agree, that the said V. W. shall and will, find and provide, a Dwelling-house or Tenement, with other appurtenances thereto, situate at ——, for the residence and habitation of the said Y. Z. and his family, as Lodger, for and during such time or term only as the said Y. Z. shall continue to work and be employed at the Colliery of the said V. W. he, the said Y. Z. paying the weekly rent or sum of —— for the same; which said rent or weekly sum of —— he, the said Y. Z. doth hereby undertake, promise and agree to pay unto the said V. W. his Agent, or Steward, accordingly; and also, that he will not do, or permit to be done or committed, any hurt, injury, or damage whatsoever to the said Dwelling-house or Tenement, or other appurtenances thereto belonging, during the time that he and his family shall reside therein; and that, upon the determination of the aforesaid Contract, and immediately upon his ceasing to work at the said Colliery, he the said Y. Z. and his family, shall and will remove out of and quit the said Dwelling-house or Tenement, with —— thereto belonging, or, in default thereof, that it shall and may be lawful to and for the said V. W. his Heirs and Assigns, and his and their Agents, Servants, Bailiffs, Workmen and others, to enter into and upon the said Premises, and to remove the Goods, Chattels and Things of him, the said Y. Z. therefrom, as fully and effectually, to all intents and purposes, as he or they might do by virtue of a Writ of Possession.

" As

"As witness the hands of the said parties, the day and the year first above written."

T. U.
Y. Z.

Witness, A. B.

———

The *Work-people* retained on Wages at the large Cotton-Mills (p. 502, II. 208, &c.) and in some other Factories, are most of them *bound*, for a certain time, and I have thought it would be right to preserve here, *a Form of Agreement* for such purpose, which was in use at one of the Cotton-Mills, which I visited.

"This Agreement, made the —— day of —— in the year —— between V. W. of the one part, and Y. Z. of the other part, as follows, that is to say, the said Y. Z. for the consideration here underwritten, doth promise and agree to and with the said V. W. that —— the said Y. Z.* —— shall and will become Servant unto, and diligently serve, abide, and continue with the said V. W. from the date of these presents, for, during, and unto the full end and term of —— weeks now next ensuing, as his hired Servant, and diligently and faithfully, according to the best of —— power, skill and knowledge, exercise and employ —— in, and shall and will, during the said term, do and perform all such service and business whatever, as well relating to the trade of a Cotton Manufacturer, which the said V. W. now useth, as in and about any other business, matter or thing whatsoever, as he the said V. W. shall, from time to time, order, direct and appoint, to and for the most profit and advantage of the said V. W. that —— can, and shall and will keep the secrets of the said V. W., and likewise shall and will be just, true and faithful to him in all matters and things, and in nowise wrongfully destroy, embezzle, or purloin any monies, goods or things whatsoever, belonging to the said V. W., and shall make and give up fair and true

———

* Sometimes Y. Z. signing this Agreement, is a Husband and Father, or a Mother, and agrees also for his Wife, and one or more young Children, or a Mother for her Children, in which cases their names are here inserted, and the *plural* used, where these "Servants" are afterwards mentioned or spoken of herein.

accounts of all —— dealings whatsoever in —— said employment, without fraud or delay, when and so often as —— shall be thereunto required, and shall not absent —— from work without leave; and also find and provide wearing apparel, and also meat, drink, washing, and lodging. And in consideration of the premises, and of the several matters and things by the said —— to be performed as aforesaid, the said V. W. doth promise and agree to and with the said —— that he the said V. W. shall and will, well and truly, pay unto the said —— the sum of —— shillings and —— pence per fortnight†, as and for the hire and wages of the said ——. Witness the hands of the said parties, the day and year abovesaid."

V. W.
Y. Z.

Signed in the presence of { A. B.
C. D.

---

† In case of more than one Person being included in this Hiring, the Names and Wages of each, fortnightly, are here specified.

INDEX AND GLOSSARY.

ACRE, a long measure, 28 or 32 yards, page 469.
——— a square measure, 4840 yards, 461.
——— statute, proposed, in 1795, as a standard of Measures, 461.
Adelphi Canal, described, 294.
Adzes, Axes, Hatchets, &c. where made, 493.
Agricultural Books, are pretty well circulated in the County, 642.
——————— Implements and Tools, where made, 493.
——————— Labourers, are worse paid than Manufacturers, 499, 526.
——————— ——— less burthen the Parishes, than Traders, &c. 528.
——————— Libraries, more of them, would be useful, 641.
——————— Pupils, premiums to, recommended, 514 Note.
——————— Societies, a general List of, in England and Wales, 650.
——————— ——— established in the County; at Derby, Repton, and Hayfield, 649.
——————— ——— where others are wanting, 658.
Agriculture, how affected by Commerce, 512.
——————— ——— by Manufactures, 500.
Agriculturists, their proportion to the whole People, in different parts, 536.
Ague, almost unknown now, as a disease, 621.
Alabaster Pits, Gypsum, where situated, Vol. I. p. 149.
Alms Houses, or Bead-houses, 578.
American Turkeys, of a large kind, 178.
Amusements, Sports, &c. of the People, 628.
Angling, mode of permitting it to the Public, in private Waters, 208.
Animal products, of the County, Trades, &c. therein, 479.
——————— imported, ditto, ditto, 481.
Ankerbold and Lings Rail-way, described, 295.
Ants given as a medicine, to Cows having the Blood-water, 80.
——— their hills are rare in the County, 645.
Apprentice Laws, their gross abuse, 508.
——————— their origin, 528.
Apprentices taken by Manufacturers, in large numbers, 501, 508.
Aqueduct Bridges, on Canals, 308, 406.
Articles of chief export or sale from the County, 497.
Artificial *wants* in the Rich, are highly beneficial to the Poor, 517.

Ashby-de-la-Zouch Canal, &c. described, 297.
Ashler, Building-Freestone, where dug, Vol. I. p. 416.
Ashover and Chesterfield, proposed Canal, 304.
Ashover Subscription Poor-House, its history, management, and accounts, 549 to 564.
Ash Timber, where grown, Vol. II. p. 245.
Asses, Stallions of such, where kept, 163.
—— used for draught, in the Coal-pits, &c. 161.
Associations, for the improvement of Roads, recommended, 230.
Asthmas and Consumptions, occasioned by grinding Tools, 624.
Astrological Impositions, ought to be discontinued, 87, 627, and 629.
Auctioneers, Names of some, 460.
Auction Sales, mode of conducting them, 460.
Auditing of Turnpike Accounts, by a Board, recommended, 235.
Author, his intended Glossary of *terms*, relating in any way to Mineral Concerns, 657.
—————————— History of Canals, 284, 456.
—— his scheme in 1795, for decimal Weights, Measures, and Monies, 462—his modifications thereof, in 1816, 681 Note.
—— his Survey of the County was made, Sept. 1807, to Dec. 1809, Vol. II. p. viii. 469.
—— Survey of Turnpike Roads and Rivers, published in A. Arrowsmith's large Map, January 1815, 207 Note.
—— his travelling Notes, (p. 662 N.) kept during the Survey, are preserved, and should be arranged and printed, 477 Note.
—— his various suggestions, for improving the County and Kingdom, 663.
Axes, Hatchets, Adzes, &c. where made, 493.

BACON, where cured, in a large way, 173.
Badger-baiting, still sometimes practised, 628.
Bakestones, of Grit and Shale, where made, Vol. I. pp. 431 and 444.
Bankers in the County, a List of, 511.
Bank-notes, less desired than those of Country Bankers, and why, 510.
Banks, Sir Joseph, Bart. the Author's chief Patron, in his minute Mineral and Agricultural Survey of this County, 160.
Baptisms, see *Births*.
Bark of Timber, measures and weights for selling it, 472.
Barley, where grown, Vol. I. p. 125.
Barnsley Canal, described, 306.
Barometrical Observations at Derby, thro' 11 years, 685.
Barytes, sulphate of, or Cawk, where dug, Vol. I. p. 461.

Baskets and Whiskets, where made, Vol. II. p. 262, III. 482.
Baslow and Brimington, proposed Canal, 308.
——— and Chesterfield, ditto, ditto, 311.
Bastardy, the Laws relating to it, require entire alteration, 620.
——— very mischievously increases *the poor* population, 522, 527, 619.
Bay, a measure, of slating, 500 square feet, 469.
Beans, where grown, Vol. II. 132.
Bear-baiting, not quite discontinued, 628—but should be suppressed, 683.
Beef, and Veal, where produced, 1 to 76.
Bees, a good many kept, 183.
Beesams and Brooms, where made, Vol. II. 234, III. 482.
Belland, a disorder of Cows and Horses, Sheep, Poultry, &c. 86, 623.
——————— of Lead-smelters and workers, 623.
Belper and Morley-park Rail-way, described, 313.
——— excessive Poor-rates there, occasioned by manufacturing abuses, 508.
——— proposed Canal, 312.
Benefit Societies, see *Friendly Societies.*
Bible Societies, 530.
Bills, Hatchets, Axes, Adzes, where made, 493.
Bird-clacks, common in gardens and fields, 646.
Birds, considered as Game by Sportsmen, 198.
———, those injurious to Agriculture, 646.
Birmingham and Fazeley Canal, described, 313.
Births, Marriages, and Deaths, Parish Registers of in the County, 605 —in the Hundreds and the Kingdom at large, 608.
Black-game, on the Moors, 199.
Black-leg, Hyon, Spade, &c. a dreadful disease of Cows, 76.
Blade-grinding Mills, where situated, 489.
Blanket-weaving, where carried on, 479.
Bleaching, where carried on, 483.
Blood-Horses, Breeders of such, 155.
Blood-water, Foul-water, &c. a disease of Cows, 80.
Boarding-Schools, Boys', a List of in the County, 529 Note.
Board, of Roads and Bridges, recommended by the Author, 235.
Boars, of improved Breeds, where kept, 165.
Boat, or Barge-building, where carried on, 482.
Boilers, of wrought iron, where made, 488.
Bone-crushing Mills, where situate, Vol. II. p. 442.
Bonfires, are dangerous in Towns, 629.

Book-clubs, Reading Societies, 642.
Bout, a cubic measure, 24 Dishes of Ore, 475.
Box Clubs, or *Friendly Societies*, 564.
Bran, a substitute for, to Horses, 157.
Brasses, or Copperas-stones, where dug, Vol. I. p. 219.
Brass Foundry, where situate, 493.
Breaking of Stones, on the Roads, dangerous to Passengers, 265.
Breedon Rail-way, proposed, 316.
Breweries, public, where situate, Vol. II. p. 127.
Brick-clay Pits, where situate, Vol. I. pp. 445, 447, 451.
Bricks, Building, where made, Vol. I. pp. 445 and 452.
—— Draining, where made, Vol. I. p. 453.
—— Fire, or Furnace, where made, Vol. I. p. 451.
—— for mending Roads, where used, 257.
Bridgewater, the Duke of, his Canal, 316.
Bridle-bits, and Buckles, where made, 493.
Broad-wheel, or rolling, Carriages, introduced on mistaken principles, 233.
———— crush the materials, laid on Roads, 233.
———— their exemptions from Tolls, injurious to Roads, 233.
Brooks, and Rivers, of Derbyshire, correct Maps of, 206 Note.
Brown-Ger, a scouring disease of Sheep, 145.
Buckles and Bridle bits, where made, 493.
Bud, a disease in Lambs, 149.
Buddlers, do unnecessary injuries to the River Fisheries, 203.
Buddling, and washing, of Ores, where carried on, Vol. I. pp. 363, 378.
Building-Bricks, where made, Vol. I. p. 445 and 451.
—— Stone, Ashler, Freestone, where dug, Vol. I. p. 418.
——, the measures used in estimating it, 469.
—— Tiles, plane, &c. where made, Vol. I. p. 451.
Bull-baiting, still sometimes practised, 628.
Bull-Houses, construction of, 72.
Bull-Letters, of the improved or New Long-horn kind, 6.
Bullocks, or Oxen, on the fattening of, 27.
Bulls, Cows, Beasts or Neat Cattle, different breeds of, 1.
—— on the fattening of, 29.
Bump, or Candle-wick, where made, 484.
Burials, see *Deaths*.
Burnt Stones, used on the Roads, 256.
Bushel, a measure, 22 or 34, 35, 35½, 36, 37, or 38 quarts, 470.
—— a weight, 33 or 60, or 90 lbs. 470, 471.
Butter, and Cheese, exist separate in new Milk, 51 Note.

Butter,

Butter, how it is prevented tasting of Turnips, Willow-leaves, &c. 64.
——— making, the processes of, 64.
——— tasting of wild Garlick, is not disliked in some Towns, 66.
——— the comparative quantities from different breeds of Cows, 26.
——— weights for selling it, 471.
——— where made, 1 to 66.
Button-moulds, of Bone and Horn, where made, 479.
Bye-sets or Gutters, obliquely across the Roads, are dangerous, 264.

CABBAGES, given to Deer, in Snows, 201.
Calf-fatting House, a complete one, 73.
Calico-printing, where carried on, 484.
Calico-weaving, ditto, 484.
Calves, fat, proportions of their carcass and offals, 74.
——— on the fattening of, 23.
Calving-House, a complete one, 73.
Cambrick-weaving, where carried on, 484.
Canals, Rail-ways, &c. a general Work on, intended by the Author, 284, 456.
——— and their Branch-cuts and Rail-ways, &c. described, 290.
——— a Map of them, in and near the County, Plate III. p. 193, 212.
——— and Rail-ways, their effects in increasing Population, 298 Note, 615.
——— part of a System of Improved Communication, 206.
Candles, are more, and Lamps less used, than is proper, 197.
——— of Tallow, where made, 479.
Candle-wick, Bomp or Bump, where made, 484.
Cank-stones, used on the Roads, 255.
Cannon Balls, and Shells, where cast, 488.
——— where cast and bored, 488.
Capital, or Property, employed in Agriculture, is less than should be, 633
Carp, castrated, for fattening more speedily, 204.
——— the growth of some in a year, 204.
Carpet-weaving, where carried on, 479.
Carriage, by Stage Waggons, regulated by Law, 275.
——— prices of, in Carts and Waggons, 274.
Carriages adapted to Rail-ways, and to common Roads, 295.
Casting, or Founder's Sand, where dug, Vol. I. p. 463.
Cast-Iron Cutlery, where made, 494.
——— Nails, where made, 490.
Castrating of Deer, in a particular way, 201.
——— of Fish, for fattening, 204.

Y y 4

Caterpillars,

Caterpillars, very destructive to Hedges, 644.
Cattle-doctors, Veterinary Surgeons, a list of, 76, 160.
——— Life-Insurance Company, 87.
——— Neat, Beast, Cow or Bull-stock, described, 1.
——— proportionate live and dead weights of, 73.
Chaff, steamed, a valuable food for Cows, 23 Note.
Chains, of Iron, where made, 488.
Chair-bottoms, of straw, where made, 483.
Chaldron, a cubic measure, 32 heapt Bushels, 473.
Chamomile Flowers, where grown, Vol. II. p. 169.
Charcoal-burning, where carried on, 482.
——— grinding, ditto, 482.
——— measures used in selling it, 472.
Charity must be exercised by the Poor as well as the Rich, 517.
——— voluntary and discriminating, can alone effectually relieve distress, 516.
Cheap bread, sold to the Poor, 579.
——— publications on Agricultural topics, are wanted, 642.
Checkt-linen Weaving, where carried on, 486.
Cheese and Butter, exist separately in new Milk, 51 Note.
——— Chambers, for drying and keeping it, 60.
——— Factors, who buy Cheese of the Dairy-men, 62.
——— making, on the Lands most proper for it, Vol. II. p. 191, III. 43.
——————— the process of, in different Dairies, 46.
——————— where carried on, 1 to 63.
——————— whether more profitable than Grazing? 32, 39.
——— on the modes of selling it, 61.
——— presses, on the construction of, 59.
——— the annual weight made per Cow, in different Dairies, 44.
——— the comparative quantities of, from different breeds of Cows, 33, 35, 37.
——— weights used in selling it, 61.
Chert-stone, used for the Roads, 254.
Chesterfield and Swarkestone Canal, proposed, 328.
——————— Canal, described, 317.
——————— has a Philosophical Society, and wants a Mineralogical one, 655.
——————— the late Earl of, his Leases, II. 38—Farm Accounts, II. 40—liberal Premiums, III. 654.
Children are principally employed in Cotton-Mills, 501.
Chimney-pots, where made, Vol. I. p. 450.
China-clay Pits, where situated, Vol. I. pp. 299, 447.

China

China stone, or white Chert, where dug, Vol. I. pp. 273, 415.
—— ware, where made, Vol. I. 447.
Chisels, gouges, &c. where made, 493.
Church Property was formerly held in trust for the Poor, principally, 519.
Churning, the proper heat for it, 68.
Churns, for butter, where made, Vol. II. p. 68.
Circulating medium of the County, principally paper, 510.
Cisterns, and Troughs of Stone, where made, Vol. I. p. 432.
Clasp, or Carpenter's Nails, where made, 490.
Clay-burning, for Manure, by Mr. Wilkes, II. 406.
Clay, burnt for Road-making, 257.
—— pits, Brick, China, Fire, Pipe, Potter's, Tile, &c. where situated, Vol. I. pp. 299, 445 to 452.
Clipt Hedges, next the Roads, very desirable, 270.
Clock, and Watch-making, where carried on, 494.
Coal-masters' Society at Eastwood, &c. 655.
Coal-pits, where situated, Vol. I. p. 188.
Coals, are the chief fuel of the County, 196.
—— wastefully used on Roads, 258.
—— Weights and Measures, used in selling them, 472.
—— worked under Canals, 321, 347, 364, 399.
Cockchafers, the Grubs of them very injurious, 648.
Cock fighting, still exists in the County, 180, 629.
—— throwing, has been abolished, 629.
Coke-burning, where carried on, 489.
—— valuable, after Gas for lights is extracted, 197.
Coldest nights, at Derby, thro' 11 years, 685.
Colour-grinding or Paint-mills, where situate, 494.
Colouring of Cheese, very expensive and useless, 58.
Commerce, its effects on Agriculture, 512.
——— of the County, not very extensive, 509.
Commercial Canal, proposed, 329.
Commons and Waste Lands, destroyed for Road-materials, 253.
Concave Roads, or hollow in the middle, condemned, 275.
Conclusion of the Report, 661.
Congleton Rail-way, proposed, 331.
Consumptions prove fatal to the grinders of Tools, 624.
Convex Roads, the best form for them, 276.
Coping-stones, for Walls, where made, Vol. I. pp. 423, 432.
Copperas-stone, where dug, Vol. I. p. 219.
——— works, where situated, Vol. I. p. 218.
Copper Coins, of the Soho Factory, in general circulation, 512.

Cord,

Cord, a cubic measure, 128 or 155, or 162½ feet, 472.
—— a long measure, 29 yards, 474.
Cords, Halters, Ropes, &c. where made, 487.
Corn-bags, or sacks, where made, 487.
—— markets, proposed to be all held on Saturday only, 459.
—— Measures, and Weights, for selling it, 470.
—— Mills, Flour, Meal, Vol. II. p. 492.
—— Rents, for Lands, Tithes, Interest, Salaries, &c. recommended, II. 31, and 638 Note.
—— Scives, or Riddles, where made, 483.
Corve, a cubic measure, 2¼ level bushels, 472.
—— a weight, 240 pounds, 473.
Cottagers, accounts of their receipts and expenditures much wanted, 526, Note.
—————— keeping Cows, and renting Land, 194.
Cottages, attached to Farms, 189, 194.
—————————— to Mines, Manufactories, &c. 638.
Cotting, or housing of Sheep in Winter, 127.
Cotton-mill Apprentices, an injurious system, 501.
—— ropes, where made, 487.
—— spinning Mills, where situate, 485.
—— thread, measure by which it is sold, 475.
Coventry Canal, described, 331.
Country Banks, a List of those in the County, 512.
—————— their Notes, preferred to any others, 510.
County boundaries, described, Vol. I. pp. 2, 375.
—— Surveys, suggestions as to repeating them, 477 and 681.
Cow Doctors, professional, a List of, 76.
—— houses, complete ones, 71.
—— loggers, for those apt to break Pasture, 72.
—— tyes, for fastening them to the Stalls, 72.
Cows, Bulls, Beasts, or neat Cattle, the different breeds of, 1.
—— fat, the proportions of their Carcass and Offals, 74. Note.
—— for Dairying, which are the most profitable breeds? 32.
—— kept for Cottagers, 187, 194.
—— on the fattening of them, 25.
Crane, a simple one for Canal Wharfs, 318, 336.
—— for tippling or emptying Trams of Limestone, 405.
Cream, from Milk and from Whey, the modes of collecting it, 65.
—— the comparative quantities of, from different breeds of Cows, 36.
Crickets in houses, how destroyed, 645.

INDEX.

Cromford and Bakewell, proposed Canal, 352.
——— Canal, described, 336.
Crossing of Sheep, a Table of the progress made yearly, 120.
Crowstone, Ganister, or Galliard Pits, where situated, Vol. I. p. 180.
——— used for the Roads, 255.
Cruel sports, are not quite laid aside in the County, 628.
Cuckoo-spittle or froth, on plants, incloses insects, 644.
Cudgelling or single-stick sports, formerly practised, 630.
Cupolas or Lead-smelting Furnaces, where situate, Vol. II. p. 385.
Curates of non-resident Clergymen, are too poorly paid, 639.
Currency depreciated, through Paper Money, &c., has abolished Leases, 638.
Curriers, or Leather-dressers, where resident, 489.
Currying of Cows, where practised, 23.
Customary Acres, different from the Standard, disused, 461.
Customs, curious ones yet remaining, 625.
Cutlers and Nailers, great abuses of the Apprentice Laws by them, 508 Note.
Cutlery, Knives, Forks, Scissars, where made, 494.
Cutters of Calves, Colts, &c. used to consult Moore's Almanack, and the Moon, 87.
Cutting of Hedges, prices, 189.
Cylinder, casting and boreing for Engines, where carried on, 489.

DAIRY Houses, account of some, 68.
Dairying, Cheese and Butter-making, &c. where carried on, 30.
——— whether more profitable than Grazing? 32.
——— which breed of Cows are best for it? 32.
Dairies, or herds of Cows, of the Devon breed of Cows, 14.
——————————————— Devon and French Cows, 20.
——————————————— French Cows, 16.
——————————————— Hereford Cows, 15.
——————————————— long and short-horn Cows, 17.
——————————————— long-horn and Devon Cows, 18.
——————————————— New long-horn Cows, 4.
——————————————— old long-horn Cows, 1.
——————————————— old and new long-horn Cows, 18.
——————————————— Scotch Cows, 16.
——————————————— Scotch and White Cows, 20.
——————————————— Short-horn Cows, 1.
——————————————— Short-horn and Devon Cows, 19.
——————————————— and French Cows, 19.
——————————————— and Lincoln Cows, 20.

Dairies,

INDEX.

Dairies, or herds of Cows, Short-horn and White Cows, 20.
——————————— Welsh Cows, 16.
——————————— White Cows, 16.
Day-labour, prices of, 186.
Dead-weights, and live-weights, of fat Cattle, 73.
——————————— Sheep, 129.
——————————— Swine, 172.
Deaneries of the County, their relations to the Hundreds, 581 Note.
Dearne and Dove Canal, described, 353.
Deaths, Marriages, and Births, Parish Registers of in the County, 605.
—in the Hundreds, and the Kingdom, 606.
Deer, an account of their consumption of Hay, 201.
—— castrated in a particular way, 201.
—— fallow, and Red, where kept, 200.
—— Parks, a List of, in the County, 200.
——————— some kept in an unimproved state, 201.
Derby Cake, of prepared Arnatto for colouring Cheese, 55, 59.
—— Canal, described, 355.
—— Agricultural Society, 109, 129, 649.
—— Philosophical Society, 655.
Derbyshire Thick-throat, a disease of Women and Girls, 622.
Derwent River, Navigation on, formerly, 360.
Devon breed of Cows, where kept, 14.
—— and French breed of Cows, 20.
Dialect of Derbyshire, on its peculiarities, 656.
Dibbling of Corn, prices for, 189.
Dilhorn proposed Canal, 360.
Diseases of Cattle, Sheep, &c. 76, 140, 159, 176, 623.
—— of the people of the County, 621.
Dish, a cubic measure, 672 cubic inches, 474.
——————————— 14 or 16 pints, 474.
—— a weight, 58 pounds, 475.
Distempers of Hogs or Swine, 176.
——————— of Horses and Colts, 159, 623.
——————— of Neat Cattle and Calves, 76.
——————— of Poultry, 623.
——————— of Sheep and Lambs, 140, 623.
Distressed Poor Persons, are most effectually relieved by others of their own Class, 517.
Dogs, often destroyed by Lead Ore or Slag, 623.
Don and Chesterfield Junction Canal, proposed, 393.
—— and high Peak Junction Canal, proposed, 420.

Donations

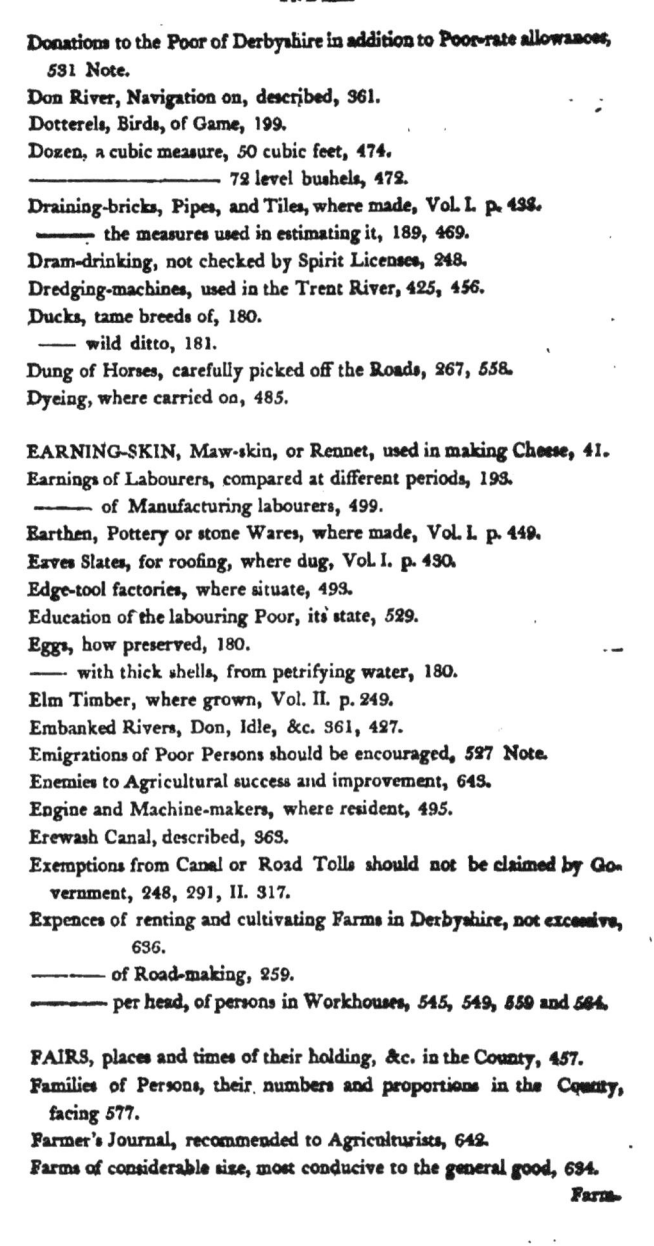

Donations to the Poor of Derbyshire in addition to Poor-rate allowances, 531 Note.
Don River, Navigation on, described, 361.
Dotterels, Birds, of Game, 199.
Dozen, a cubic measure, 50 cubic feet, 474.
——————————— 72 level bushels, 472.
Draining-bricks, Pipes, and Tiles, where made, Vol. I. p. 438.
———— the measures used in estimating it, 189, 469.
Dram-drinking, not checked by Spirit Licenses, 248.
Dredging-machines, used in the Trent River, 425, 456.
Ducks, tame breeds of, 180.
—— wild ditto, 181.
Dung of Horses, carefully picked off the Roads, 267, 558.
Dyeing, where carried on, 485.

EARNING-SKIN, Maw-skin, or Rennet, used in making Cheese, 41.
Earnings of Labourers, compared at different periods, 193.
———— of Manufacturing labourers, 499.
Earthen, Pottery or stone Wares, where made, Vol. I. p. 449.
Eaves Slates, for roofing, where dug, Vol. I. p. 430.
Edge-tool factories, where situate, 493.
Education of the labouring Poor, its state, 529.
Eggs, how preserved, 180.
—— with thick shells, from petrifying water, 180.
Elm Timber, where grown, Vol. II. p. 249.
Embanked Rivers, Don, Idle, &c. 361, 427.
Emigrations of Poor Persons should be encouraged, 527 Note.
Enemies to Agricultural success and improvement, 643.
Engine and Machine-makers, where resident, 495.
Erewash Canal, described, 363.
Exemptions from Canal or Road Tolls should not be claimed by Government, 248, 291, II. 317.
Expences of renting and cultivating Farms in Derbyshire, not excessive, 636.
———— of Road-making, 259.
———— per head, of persons in Workhouses, 545, 549, 559 and 564.

FAIRS, places and times of their holding, &c. in the County, 457.
Families of Persons, their numbers and proportions in the County, facing 577.
Farmer's Journal, recommended to Agriculturists, 642.
Farms of considerable size, most conducive to the general good, 634.

Farm-ways, or Field Roads, 271.
Fattening of Neat Cattle, particulars of, 23, 27.
——— of Sheep, particulars of, 128.
Fat Mutton, not always objected to by Manufacturers, 128, 107, 623.
Fatting-house, for Calves, 73.
Feasts, or Village Wakes, 628.
Felmongers, where resident, 480.
Female Friendly Societies, a List of, their Rules, &c. 566, 567.
Fencing, the measures used in estimating it, 469.
Ferrys, over the Trent River, 282, 429.
File-making, where carried on, 494.
Filtering Cisterns, where made, Vol. I. p. 434.
Fiorin Grass, should be cultivated on the Mosses and Bogs, 91 Note.
Fire Clay Pits, where situated, Vol. I. p. 450.
——— or Furnace Bricks, where made, Vol. I. p. 451.
——— or Furnace Stone, where dug, Vol. I. pp. 221, 228, 431.
Fir Timber, where grown, Vol. II. pp. 246 to 267.
Fish, a List of those common in the County, 203.
——— castrated, for fattening, 204.
——— destroyed by the Lead-miners' Water, 624.
——— preserved as Game, 203.
——— ponds, management of some, 204.
——————— vegetables cultivated on their sides, for the Fish, 204.
Flags or Pavier Stones, where dug, Vol. I. pp. 424, 427.
Flat Roads, preferred by some persons, 277.
Flax Spinning Mills, where situate, 483.
Flies, injurious to Agriculture, &c. 645.
Flint-grinding Mill, for the Potteries, where situate, 494.
Floods in Autumn, of great importance to Water-meadows, 143.
Flour-clubs, for providing it cheaply, 578.
——— or Meal-mills, where situate, Vol. II. p. 492.
——————— weights for selling it, 470.
Flukes, are sometimes found in the Livers of sound, as well as rotten Sheep, 140.
Fluor-spar Mines, where situate, Vol. I. p. 460.
——————— or Petrifaction Workers, where resident, Vol. I. p. 461.
Fly-blown, or Maggotty Sheep, 148.
Fodder, a Weight, 2184, 2240, 2340, 2408, or 2820 lbs. p. 475.
Folding of Sheep, little practised in the County, 127.
Food, and modes of living, of the Inhabitants, 624.
——— for Hogs or Pigs, 168.
——— ——— Horses, 157.

Food

Food for Neat Cattle in Winter, 91.
———— Sheep, 126.
———— the comparative quantities eaten, by different breeds of Cows, 32.
Foot-ball, played in Towns, a great nuisance, 630.
Foot-paths across Fields, are great nuisances, 273.
———————— by Roads' sides, 272.
———— Rot, or Foul, a disease of Cows, 85.
———— a disease of Sheep, 115, 147.
Fords, across the Derwent, Trent, and Dove Rivers, 262.
Forest (of Sherwood), Breed of Sheep, 96.
Forged Bank-Notes, said to be common, 510.
Forges, or Bar Iron Factories, where situated, Vol. I. pp. 397, 403.
Forms of Agreements by which Cottages are attached to Works, 688.
———————————— by which Work-people are bound to Manufacturers, 689.
Fortune-telling Gipsies, should be suppressed, 631.
Foul-water, Blood-water, &c. a disease of Cows, 80.
Founder's, or Casting Sand, where dug, Vol. I. p. 463.
Foundries for Brass and Iron, 493, and Vol. I. p. 404.
Fowls, breeds of domestic ones, 179.
Foxes, are useless vermin, 202.
Fox-hunting, not much practised in the County, 202.
Frame-knitting, of Lace and Stockings, where carried on, 486, 487.
———— Smiths, or Stocking-Loom Makers, where resident, 494.
Freemartin, a sort of barren Cow, 87.
Free Schools, a List of in the County, 529 Note.
Freestone, Building, or Ashler, &c. where dug, Vol. I. p. 416.
French, or Alderney breeds of Cows, 16.
Friendly Societies, Benefit, Box, Sick, or Provident Clubs, 564.
———————————— of Men only, where held, &c. 564.
———————————— of Women only, where held, &c. 566.
———————————— the Rules of one, 568.
———————————— the Act relating to them requires great alteration, 569.
———————————— their abuse, in some great Towns, 572.
———————————— their Members make Procession on their Anniversaries, in Derbyshire, 570.
Frying-pans, where made, 489.
Fuel of the County, chiefly Coals, 196.
———— sold cheap to the Poor, by the late Duke of Bedford, 30.
Fullers Earth, for the Clothiers, where dug, Vol. I. p. 465.
Fulling Mills for Cloth and Leather, where situate, 480.

Furnace

Furnace Cinders, or Slag, as Road-materials, 257.
Fustian-weaving, where carried on, 485.

GABLE-STONES, for roofs, where made, Vol. I. p. 432.
Gadding of Cows, when fly-bitten, very injurious, 22.
Gallon, a cubic measure, 231 or 268¾, or 282 cubic Inches, 471.
Gallows, one maintained by Charter in Ilkeston, 632.
Game, Animals, 200—Birds, 198—Fish, 203.
——— fowls kept, for brutal Cock-fighters, 180.
——— laws, require amendment, 198.
Garden-Pots, where made, Vol. I. p. 450.
Gardens, of sufficient size, important to Cottagers, 195.
Garget, or Gargle, a disease of Cows, 25, 84.
Gas lights, the produce of, from 6lbs. of Coals, 197.
——— used, and the apparatus made, 197.
Gate-fastenings, of different sorts, 271.
Gates across Roads, are inconvenient, 270.
Gauges, or Rings, for Road-Stone breakers, 254.
Geese, tame, have much decreased in the County, 179.
——— wild, 199.
Geological and Mineralogical Societies are wanted, at Chesterfield, &c. 655.
Giddiness, a disease of Cows, 85.
——— or turn, a disease of Sheep, 146.
Gig, or holiday Fairs, at Newhaven and other places, 458, 630.
Gipsies, are an intolerable nuisance, 631, II. 308 Note.
Glass-making, where carried on, 494.
Glossary or Vocabulary of Terms, see the Indexes to these 3 volumes, and p. 656.
——— an enlarged one intended for publication, by the Author, 567.
Glossop Parish, its high Poor-rates, 507.
——— ——— its numerous Manufacturers, 498.
Glue-making, where carried on, 480.
Goats, few now kept in the County, 149.
Gorse, or furzen tops, bruised for Horses, 159.
Gouges, and Chisels, where made, 493.
Government, exempting itself from Tolls on Turnpike Roads and Canals, made at private expence, is improper and injurious, 248, 291.
Grains, Brewer's, used in fattening Oxen and Cows, 96.
Grantham Canal, described, 366.
Gravel, cost of digging and levelling Pits, 259 Note.
Gravelling of a Road, how best performed, 236, 250.

Gravel,

Gravel, often wants breaking for the Roads, 250.
———— should be dug and sifted for Roads, by Land Owners, 238, 251.
Grazing, whether more profitable than dairying? 32.
Green Food, cut for Cows, &c. in stalls, 22.
——————————— Hogs, 168.
——————————— Horses, 158.
Gresley's Canal, described, 368.
Grinding-mills, for Cutlery, Tools, &c. where situate, 489.
———— Stones, where dug, Vol. I. p. 435.
Gritstone Sheep, a crossed breed, formerly common, 123.
———— used on the Roads, 255.
Grouping, a mode of catching Trout, &c. 206.
Grouse, Game, on the Moors, 199.
Grubs or Slugs, injuries by, 643.
Gunpowder-making, where carried on, 494.
Gutter-Tiles for Roofs, where made, Vol. I. pp. 451, 453.
Gypsum, Alabaster, or Plaster Pits, where situated, Vol. I. p. 149.
———— Measures and Weights, used in selling it, 474.
———— striated, used on the Roads, 258.

HAMMER, tilt or skelper Mills, where situated, 490.
Hams, of Mutton, cured, 129.
———— of Pork, ditto, 174.
Handicrafts, Traders and Manufacturers, their proportionate numbers, in different parts, 536, 577.
Hank, a long measure, 840 yards, 475.
Hare-getting, a curious custom, 626.
Hares, are too much preserved, in some places, 202.
Hatchets, Axes, Adzes, &c. where made, 493.
Hat-making, where carried on, 481.
Hay and Straw, weights for selling them, 471.
——— the chief winter food of Cows, 21.
Healthiness of the County, 621.
Heaps of stones, left unspread on the Roads, 265.
Hedge-wood and lop, carelessly left in the Roads, 262.
Heifers, spayed, worked in a team, 70.
Hereford, or middle horn breed of Cows, 15.
High-Peak Junction Canal, proposed, 369.
Hills, on a Road, ascents of some, per yard, 224.
Hip-tiles for Roofs, where made, Vol. I. p. 451.
Hired Servants, wages of, &c. 184.
Hoeing of Corn and Turnips, prices of, 189.

Hoes, where made, 494.
Hogs, food for them, 168.
—— live and dead weights of some, 172.
—— on the breeds of, 164.
—— should be kept by Cottagers, 195.
Holiday or Gig Fairs, in the County, 458, 630.
Hollow Roads, not found in Clay Strata, 221, 263.
Hones, Whetstones, where dug, Vol. I. p. 440.
Hooping-cough, a remedy for it, 621.
Hoops of Iron, where made, 490,—of wood, 485.
Hop-bags, where made, 485.
Horn-Tips, for Cows, of brass, 74.
Horses, on the breeds of, &c. 150.
Horse-shoe Nails, where made, 491.
Hottest days at Derby, thro' 11 years, 685.
Hounds, a List of Packs kept in the County, 202.
House-floor Sand, where dug or made, Vol. I. p. 463.
House-row, or going the Rounds for work, 529.
Houses of Industry, or District Work-houses, 523 Note, 548, 641.
Houses Inhabited, New, and Uninhabited, their numbers and proportions in the County, 580.
Hoven, swelled, or risen-on, a Disease of Cattle, 85.
Huddersfield Canal, described, 371.
Hundreds of the County, their relations to its Deaneries, 581 Note.
Hundred Weight (Cwt.) 112 or 120, or 128 pounds, 471, 473, 475.
Hunting of Deer, a cruelty, discontinued in the County, 200, 629.
Hurdles, of wood and iron, where made, Vol. II. p. 234.
Hyon, Spade, Black-leg, &c. a dreadful disease of Cows, 76.

IDLE River, Navigation on, described, 375.
Implements of Agriculture, where made, 495.
Importation of Corn, injurious to the Farmer, II. 174, 261, III. 635.
Improved communication, by Roads, Rail-ways, and Canals, 206, 283.
Improvements, obstacles to, 633.
—————— recapitulation of the suggestions for such in this Report, 663.
Inclined Planes, or steep Rail-ways, 289, 403, 436.
Inclosing, cannot be done in some places, on account of expence, 637.
—————— is a great source of increased Population, 615.
Inclosure Act, a general one much wanted, 637.
Increase of the Poor, occasioned by dependance on Alms, 520, 524.
Indictments of Parishes, for Toll Roads, a hardship, 239.

Industry

Industry Schools for Poor Children, a List of, 529 Note.
Inflammation of the Lungs of Horses, is common, 160.
Inns, new, or large ones, on the Roads, 246.
—— small ones, oppressed by the spirit Licenses, 247.
Insane persons, where kept, 622.
Insects, injurious to Agriculture, 645.
Iron-Forges, and Furnaces, where situate, Vol. I. pp. 397, 403.
—— Foundries, where situate, Vol. I. p. 404.
—— Masters' Society, at Sheffield, Rotherham, &c. 655.
—— Rail-ways, for Carriages, 283.
—— Slag, or Furnace Cinders, for Roads, 257.
Ironstone, measures used in estimating it, 474.
———— or Ore, Pits, where it is dug, Vol. I. p. 217.
Job, or Piece-work, 189.
——————— its great advantages, 191.
——————— premium for, suggested, 192.

KNIVES and Forks, of Cast Iron, where made, 494.
Knowledge on Agricultural subjects, more wanted, 513 Note, 641.

LABOURERS, rules for managing them, 191.
——————— Rural, their pay inadequate, 193.
Labour, on the prices of, &c. 184.
Lace-weaving, and working, where carried on, 486.
Lamb and Mutton, where produced, 88 to 133.
Lambing-fold, description of a temporary one, 127; and *Plate* I. p. 128.
Lambs, fatted, 129.
——— shearing of them practised, 136, 144.
Lamps, should be more used, instead of Candles, 197.
Lancasterian Schools, where established, 529.
Land-measures of the County, are statute, 461.
Language or Dialect of Derbyshire, 656.
Larch Timber, where grown, Vol. II. p. 252.
Lead Mines, where situate, Vol. I. p. 252.
—— Ore, measures and weights used in selling it, 474.
—— Pipe-making, where carried on, 491.
—— Slag, used on the Roads, 258.
—— Smelting Cupolas, where situated, Vol. I. p. 385.
—— the weights used in selling it, 474.
Leases, not being granted, retards improvement, 637.
Legislative Authority, in matters of Rural Economy, 198.
Leicester and Melton-Mowbray Navigation, described, 381.

Leicester

Leicester Navigation, described, 376.
Leicestershire and Northamptonshire Union Canal, described, 382.
Leather Mill, where situated, 480.
Letters, hours of receiving and sending, at different Towns, 249.
Letting of all Rural Works, by the piece, recommended, 191, 194.
——————————— Essays on, wanted, 194.
Ley, a long measure, 120 yards, 475.
Licenses to small Inns, oppress them, 247.
Lime-burning in Rushall, Staff. particulars of, 453 Note.
Lime-kilns, where situated, Vol. II. pp. 415, 433.
Lime, measures and weights used in selling it, 473.
Limestone, broken small for Roads, 252.
—————— Quarries, where situated, Vol. I. p. 408.
Linen-weaving, where carried on, 486.
———— Yarn-spinning, where carried on, 485.
Liquids, measures used in selling them, 471.
Littering, or bedding of Hogs, 172.
————————————— Horses, 159.
Live and Dead Weights, of Fat Cattle, 73.
———————————————— Hogs, 172.
———————————————— Sheep, 129.
Live Stock, on the breeds and management of, 1.
Load, a cubic measure, $3\frac{1}{2}$, or 40 or 46 cubic feet, p. 469, 472, 473, 474.
——————————— 3 or 20 heapt Bushels, p. 470, 474.
——————————— $2\frac{1}{2}$ or 3, or 144 level Bushels, p. 472, 473.
——————————— 126 or 144 pints, p. 474.
—— a long Measure, 70 Yards, p. 472.
—— a Weight, 200 lbs. or 5040 lbs. 472, 474.
Lobby, a Parish Prison, Cage, or Watch-house, 632.
Local terms and Names of things and processes, are more fixed and definite than most persons suppose, 658.
Long and short horned mixed Breed of Cows, 17.
—— horned and Devon, mixed breed of Cows, 18.
Loose Horses or Cattle in Roads, a nuisance, 267.
Loughborough Navigation, described, 384.
Lump, Pye or Pot Stones, for the Forges, where made, Vol. I. p. 431.
Luxurious living by the Rich, affords employ and comforts to the Poor, 517.

MACCLESFIELD proposed Canal, described, 386.
Machine and Engine-makers, where resident, 495.
Mad-houses, one at Calow, 622.
Maggotty or fly-blown Sheep, 148.

Mail-coach exemptions from Tolls, injurious to the Roads, 248.
——— routes, across or near Derbyshire, 210.
Malt-kiln plates, of Iron and Pottery, where made, 490.
——— making, where carried on, Vol. II. p. 127.
——— Mills, of steel, where made, 595.
Manchester, Ashton, and Oldham Canal, described, 388.
Mangles for Clothes, where made, 595.
Man Traps, for Gardens, &c. 170.
Manufactured Goods, exported from the County, 496.
Manufacturers, and Traders, often make good Farmers, 513.
——————— Traders, and Handicrafts, their proportionate numbers in different parts, 536, 577.
——————— purchase Land, at high prices, 505.
Manufactures, for home trade, are most beneficial, 500.
——————— increase the Poor's Rates, 500.
——————— their consequence in the County, 476.
——————— their wonderful increase in Glossop Parish, 498.
Manufacturing Labourers, are better paid than Agricultural ones, 499, 528.
Map of Turnpike-roads, Rail-ways, and Canals, facing p. 193, 206.
——— with Rivers and Brooks, a correct one by Arrowsmith, 206 Note.
Marble Quarries, where situated, Vol. I. p. 412.
——— works, sawing and polishing, Vol. I. p. 412.
Mares, breeding, and Stallions, where kept, 153.
Market Boat, from Swarkestone to Derby, 357.
——— Towns and Days, in the County, 457.
Marl Pits, for Manuring Land, where situate, Vol. I. p. 456.
Marriages, Births, and Deaths, Parish Registers of in the County, 605—in the Hundreds, and in the Kingdom, 608.
Mason's Mortar Sand, where dug or made, Vol. I. p. 463.
Materials for making and repairing Roads, 250.
Mattresses, of Straw, for Beds, where made, 483.
Means of Improvement, suggested, 661.
Measures and Weights, the Author's scheme for entirely reforming them, 462, 681 Note.
Meat, Butcher's, where produced, 1 to 176.
Mechanists, Engineers, &c. where resident, 495.
Meer, a long measure, 29 yards, 474.
——— a square measure, 14 square yards, 474.
Merino and Ryeland, crossed Sheep, 125.
——— and South Down, ditto, 125.
——— and Woodland, ditto, 125.

Merino or Travelling Spanish breed of Sheep, 113.
Mersey and Irwell Navigation, mentioned, 291.
Meteorological observations at Derby, through 11 years, 685—results of ditto, 686.
Mice and Rats, 646.
—— often injure young Plantations, 202 Note.
Middle men, or Speculators in Agricultural produce, 635.
Milestones on Roads, are neglected and defaced, 245.
Milk, comparative quantities of, given by different breeds of Cows, 36.
—— for the supply of the Towns, 30, 40.
—— should be sold cheap by Farmers to their Labourers, 195.
Milled or rolled Lead, where made, 492.
Millstones, where dug, Vol. I. p. 221, 272.
Millwrights, where resident, Vol. II. p. 493.
Mineral Map and History of the County, proposed, 206 Note.
——— Products, imported, Traders, &c. therein, 493.
——————— of the County, Trades, &c., therein, 488.
——— rights, injurious to cultivation, 637.
Mites, in Cheeses, how destroyed, 61.
Moles, are not very troublesome in the County, 646.
Money, a proposal for its decimal division, 465, 681 Note.
Moor Game, on heathy Hills, 199.
Mountebanks, are an intolerable nuisance, 630.
Mowing, of Grass and Corn, prices of, 189.
Muddy taste of Fish, how removed, 205.
Mules, bred in the County, 163.
Mundy, F. N. C. Esq. the Catalogue and Prices of his Sale of Cow Stock, 10.
Muslin weaving, where carried on, 486.
Mutton and Lamb, where produced, 88, 134.
——— Hams, of New Leicester Sheep, 129.
——— of Merino or Spanish Sheep, much approved, 119.
——— of Woodland Sheep, 92, 95—Forest Sheep, 96—Portland Sheep, 112.
——— very fat, of New Leicester Sheep, preferred by some Manufacturers, 128, 625.

NAG-HORSES, the training of them, made a business, 155.
Nailers and Cutlers, their great abuses of the Apprentice Laws, 508.
Nail-making, Wrought and Cast, where carried on, 491.
—— rods, where rolled and slit, Vol. I. pp. 403 and 404.
Navigations and Canals, the Map of, facing p. 193, described, 212.
Needle-making, where carried on, 595.

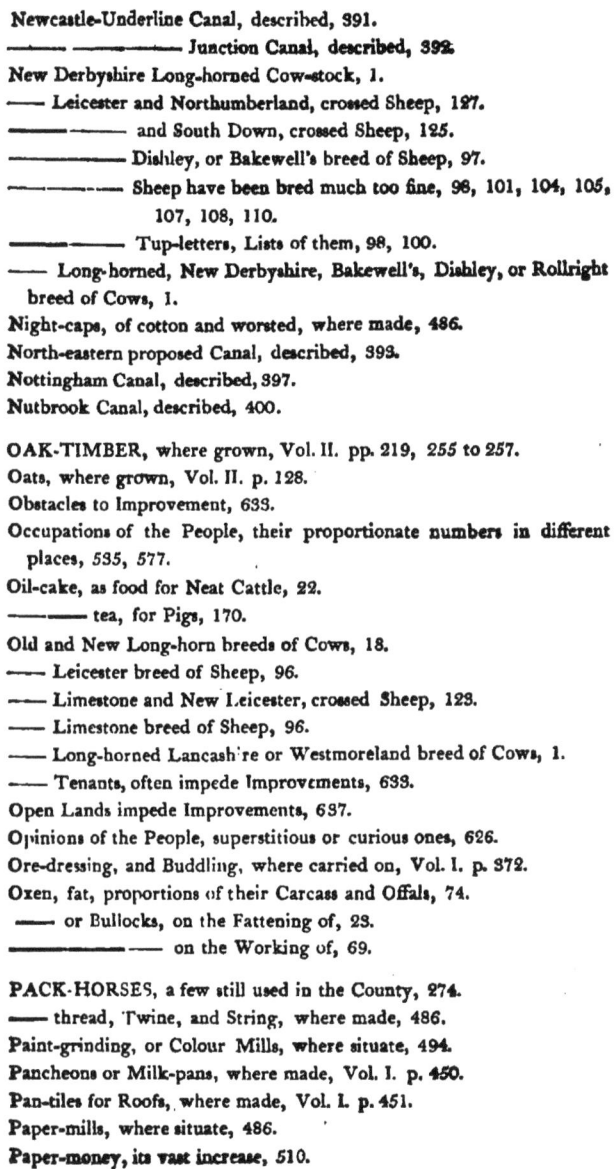

Newcastle-Underline Canal, described, 391.
────── ────── Junction Canal, described, 392.
New Derbyshire Long-horned Cow-stock, 1.
──── Leicester and Northumberland, crossed Sheep, 127.
────── and South Down, crossed Sheep, 125.
────── Dishley, or Bakewell's breed of Sheep, 97.
────── Sheep have been bred much too fine, 98, 101, 104, 105, 107, 108, 110.
────── Tup-letters, Lists of them, 98, 100.
──── Long-horned, New Derbyshire, Bakewell's, Dishley, or Rollright breed of Cows, 1.
Night-caps, of cotton and worsted, where made, 486.
North-eastern proposed Canal, described, 393.
Nottingham Canal, described, 397.
Nutbrook Canal, described, 400.

OAK-TIMBER, where grown, Vol. II. pp. 219, 255 to 257.
Oats, where grown, Vol. II. p. 128.
Obstacles to Improvement, 633.
Occupations of the People, their proportionate numbers in different places, 535, 577.
Oil-cake, as food for Neat Cattle, 22.
────── tea, for Pigs, 170.
Old and New Long-horn breeds of Cows, 18.
──── Leicester breed of Sheep, 96.
──── Limestone and New Leicester, crossed Sheep, 123.
──── Limestone breed of Sheep, 96.
──── Long-horned Lancashire or Westmoreland breed of Cows, 1.
──── Tenants, often impede Improvements, 633.
Open Lands impede Improvements, 637.
Opinions of the People, superstitious or curious ones, 626.
Ore-dressing, and Buddling, where carried on, Vol. I. p. 372.
Oxen, fat, proportions of their Carcass and Offals, 74.
──── or Bullocks, on the Fattening of, 23.
────────── on the Working of, 69.

PACK-HORSES, a few still used in the County, 274.
──── thread, Twine, and String, where made, 486.
Paint-grinding, or Colour Mills, where situate, 494.
Pancheons or Milk-pans, where made, Vol. I. p. 450.
Pan-tiles for Roofs, where made, Vol. I. p. 451.
Paper-mills, where situate, 486.
Paper-money, its vast increase, 510.

Paring-shovels, where made, 494.
Parishes are often indicted for neglect, &c. of the Trustees of Roads, 237.
Parish Officers, should have and exercise more discretion, regarding the maintenance of the Poor, 548.
Parish or Township Roads, 262.
──── or Town Stocks, Pounds, Prisons, Cages, Lobbies, 631.
──── Registers of Births, Deaths, and Marriages in the County, 605 —in the Hundreds, and the Kingdom, 608.
──── Roads, are deprived of Statute-Duty by the Toll Roads, 236.
Parks (Deer), a List of, 200.
Partridges, Birds of Game, 199.
Patten-rings, or Clog-irons, where made, 491.
Paupers, or Poor permanently maintained, their proportionate numbers in different places, 535, 538.
──── the annual cost of their maintenance in Work-houses, 543, 549, 559, 564.
──── their former Trades, &c. desirable to be known, 541.
Paving bricks, where made, Vol. I. p. 451.
──── stones or Flags, where dug, Vol. I. pp. 424, 427.
Peak-Forest Canal, described, 402.
Pease, where grown, Vol. II. p. 132.
Peat, for fuel, mode of digging it, 197.
──────── was formerly more used than at present, 196.
──── pits, turf, where situated, Vol. II. p. 349.
Perch, or Pole, a square measure, $30\frac{1}{4}$ yards, 461.
Persons, Female and Male, their numbers and proportions in the County, 577.
Petrifaction or Spar-workers, where resident, Vol. I. pp. 150, 459, 461.
Pheasants, Birds of Game, 198.
Philosophical Societies, in the County, 655.
Piece, a weight, $176\frac{1}{4}$ pounds, 475.
──── work, its great advantages, 191.
──────── prices of, 189.
──────── rewards should be given for excelling in it, 192.
Pig, a weight, $352\frac{1}{2}$ pounds, 475.
──── see Hogs.
Pigeons, are much decreased, 181, 199.
──── wild or Wood, 200.
Pike, are most voracious Fish, 204.
──── preparation of, for the Table, 205.
Pint, a cubic measure, $28\frac{7}{8}$, or $35\frac{2}{7}$, or $35\frac{1}{4}$, or 48 cubic inches, 471, 474.

Pipes

Pipe-bricks, where made, Vol. I. p. 456.
—— clay-pits, where situated, Vol. I. p. 448.
—— makers, (Tobacco), where resident, Vol. I. p. 448.
Pipes for water, of Earthenware, where made, Vol. I. p. 449.
—————— of Lead, where made, p. 506.
—————— of Zink, ditto, p. 506.
Plane-irons, Carpenters', where made, 493.
—— tiles, Building, Roof, where made, Vol. I. p. 451.
Planishing, or Tilt Mills, where situated, 490.
Plantations, and Belts by the Roads sides, are improper, 269.
————— destroyed by Field Mice, 202 Note.
Plaster of Paris, where prepared, Vol. I. p. 150.
Plate-iron, Rolling-Mills, where situate, Vol. I. p. 403.
Ploddings, or Causeways across Roads, are dangerous, 264.
Poisoned, a disease of Sheep, so called, 149.
Pole, or Perch, a square measure, 30¼ yards, 461.
Political Economy, 198.
Poor-houses, by subscription of several Parishes, for their Paupers, 549 to 563.
—— Laws, are replete with evil consequences, 193, 195, 526 Note.
————— important preliminary steps to their abolition, 527 Note, 618.
————— the causes which led to their enactment, 518.
————— the modern alterations of them, have increased their evils, 193, 524.
—— Persons, necessarily form a large class in every Community, 515.
—————— or permanent Paupers, expences yearly per head, of their maintenance in different places, 545, 547, 549, 459.
—————— their former Trades, &c. should be specified, in new Returns, 541.
—————— their proportionate numbers in different places, 535.
—— Rates, amounts of them in different places, Vol. II. 32; III. 532, 544.
————— increased by Manufactures, 507, 528.
————— in the Pound, are very unfairly stated, in many cases, 533, 538, 543.
———————— a Table of, in several places, 532.
————— prevent Improvements, 640.
————— Returns to Parliament, should be *repeated*, with some improvements, 538.

Poor

Poor Tenants, and Occupiers, unable to improve their Lands, 633, 635
—— the state of this class of the People, 528.
Population, mode of calculating it, from Parish Registers, 609.
—————— not excessive in the County, 618.
—————— of 700 Towns, 298 Note; a List of them, 594.
—————— of particular Towns, has declined, 616.
—————— on what has its progress depended? 612 to 618.
—————— Returns, defects in them, and suggestions for their improvement in future, 583, 586, 590, 604, 611.
—————— statements, averages, &c. concerning it, 535, 579 to 594.
—————— the increase of in 44 places, 612.
Pork and Bacon, where produced, 164 to 176.
Portland breed of Sheep, where kept, 112.
Post Towns, and Offices, a List of them in the County, 249.
Potatoes, weights for selling them, 471.
Pot-stones, Pye, Lump, for Iron Forges, where made, Vol. I. p. 431.
Potteries, Earthen and Stone Wares, where situated, Vol. I. p. 449.
Potters' Clay Pits, where situate, Vol. I. p. 448.
Poverty, or the necessity for the many to *Labour*, is unavoidable in civilized society, 514.
Poultry, house, a complete one, 178.
—————— particulars concerning, 177.
Pound, a weight, 16 or 17 ounces, 471.
Pounds, Parish, for trespassing Cattle, &c. 621.
Premiums, given by Earl Chesterfield, 654.
—————— suggested for accounts of Cottagers' receipts and expenditures, 526 Note—for Essays on Letting Works to Labourers, 194—for job or piece-work, 192, &c. &c. see pp. 667 to 684—for taking Agricultural Pupils, 514 Note.
Prices, and expences of products, 475.
—————— of agricultural produce, has been too low for small Farmers, 635.
Prisons, Parochial, Cages, Lobbies, 632.
Private, and Coach Roads, leading to Mansions, 271.
Products of the County, which are chiefly exported therefrom, 497.
Professional Bull-Letters, of improved Breeds, 6.
—————— Road-makers, the names of, 261.
—————— Tup or Ram-Letters, of improved Long-woolled Breeds, 98, 100.
—————— Stallion-Letters, the names of, 151.
Profits of Manufacturers, are greater than those of Agriculturists, 506.
Property, or Riches, cannot long remain equally divided in Society, 514.

Props for Coal Pits, Puncheons, where made, 491.
Provident Clubs, see FRIENDLY SOCIETIES.
Provincial Terms of the County, some remarks on, 656.
Provisions, of different kinds, cost of in Ashover Poor-house, 557.
────── prices of, 195.
Public-houses, by Canals, are improper, 327.
Pumping of Coal Mines, disputes about it, how to settle, 295, 380.
Puncheons, or Props for Coal-pits, where made, 491.
Purging of Cattle, before turning them to graze, 25.
Pye-stones, Pot or Lump, for the Iron Forges, where made, Vol. I. p. 431.

QUART, a cubic measure, $57\frac{1}{4}$ or $67\frac{1}{7}$, or $70\frac{1}{4}$ cubic inches, 471.
Quarter, 8 or 9 heapt Bushels, 472, 473.
────── a cubic measure, 8 level Bushels, 470, 473.
────── a long measure, 7 yards, 472.
Quarter-Cord, a long measure, $7\frac{1}{4}$ yards, 474.
Quoits, a common Ale-house Game, 630.

RABBITS, Tame and Wild, 176, 202.
────── Warrens, none remain in the County, 177.
Race-horses, Breeders of them, 155.
Railway-branches, of any length may be made, to the Cromford Canal, 287.
Railways, a Map of them in and near the County, *Plate* III. p. 193, 287.
────── cost of laying Iron ones, 289, 300.
────── not applicable to general merchandize, like Canals, 352.
────── of Wood, formerly, 288.
────── part of a System of Improved Communication, 206, 284.
────── public ones, much wanted in Durham and Northumberland, 285.
────── Waggons, drawn by Steam-engines, 339.
Rain-guage, observations in Derby, through 5 years, 685.
Rasps and Files, where made, 494.
Rats and Mice, modes of preventing their mischiefs, 646.
Ravens, breed in some of the Cliffs in the County, 648.
Reading Societies and Book-Clubs, 642.
Reaping Hooks and Sickles, where made, 494, 496.
────── of Corn, how estimated, 471.
────── of Corn, prices of, 189.
Reason and Conscience, the excellence of their dictates, 530.
Recapitulation of the various suggestions for Improvements made in this Report, 661.

Red

Red Lead, Minion, where made, 491.
——— Water, a disorder of Cattle, 80.
——————————— of Sheep, 145.
Reeves and Ruffs, Birds of Game, 200.
Rennet, Maw-skin, Earning-skin, &c. used in Cheese-making, 57.
Rentals rated to each permanent Pauper, in different places, 534.
——————— to the Poor, in various places, facing Vol. II. p. 34 and III. 529, last cols.
—————————————— should be better ascertained, in new Returns, 540.
Rents are at first advanced, by Manufacturers settling, 498.
——— of small Farms, are in many instances too high, 634.
Reptiles, Vermin, &c. 644.
Repton Sheep Society, 132, 133, 649.
Reservoirs, large ones, for Canals, &c. where situated, Vol. I. p. 496, 399, 441.
Restrictive Covenants in Leases, often injurious, 638.
Restrictions, of injudicious and unfair kinds, imposed on Canal Companies, 322, 424.
Rewards proposed, for greatest earnings by Job-work, 192.
Riches or Property, derive their value from the Labour they can command, 514.
Rich Person's Wants, ought to secure competence to the industrious Poor, 515, 517.
Rickets, Warfar or Evil, in Sheep, 147.
Rider, and hard vein-stuff, used on the Roads, 254.
Ridging-stones, for Roofs, where made, Vol. I. 431.
Rights of the Poor, wherein they chiefly consist, 515, 525.
Rise of Rents, occasioned by Manufacturers, 498.
Rivers and Brooks, in and near Derbyshire, a correct Map of them, 206 Note.
Rivets, for Coopers, &c. where made, 492.
Road-Engineers were unknown in England until lately, 229.
——— makers, a List of the Names of, 261.
——— making, measures used in estimating it, 469.
———————, principles modernly adopted therein, 278.
———————, principles that anciently prevailed therein, 219.
Roads, accounts of them, 216.
——— a String Level for setting them out, 278.
——— bad and miry, places noted for them, 263.
——— general Survey of practicable improvements therein, should be made and recorded, 231.

Roads,

Roads, large Stones are nuisances on them, 266.
───── Maps and Sections of new ones, required by Parliament, 229.
───── materials, for making and repairing them, 250.
───── on the different forms of, 275.
───── rocky, and uneven, are dangerous, 263.
───── some new lines of, suggested, 225.
───── the expences of them oppress Farmers, 636.
───── unfenced, with gates across them, are inconvenient, 270.
───── were made unnecessarily wide, at times, 268.
───── with paved foot-paths by them, 272.
Roarer, or common Buller, a diseased Cow, 25, 87.
Roasting Pigs, 168.
Rochdale Canal, described, 410.
Rolling, or flattening and slitting Mills, where situated, Vol. I. p. 408.
Roman Roads, across or near Derbyshire, 216.
Rood, a long measure, 7 or 8 yards, 469, 472.
───── a square measure, 7, or 44, or 1210 yards, 461, 469.
Rooks, in moderate quantities, are serviceable, 647.
Rooting of Hogs, how prevented, 169.
Roots, as food for Neat Cattle, 21.
Rope-making, Cord, &c. where carried on, 487.
Rot in Sheep, causes of assigned, and attempts to cure it, 140.
Rottenstone, where dug, Vol. I. p. 231.
Rouen, or Rowen, Ducks, a fine sort, 180.
Rounds or House-row, gone by Poor Persons for Work, 529.
Ruck, a cubic measure, $5\frac{3}{4}$ yards, 472.
Ruffs and Reeves, Birds of Game, 200.
Rural Economy, Chapter on, 184.
Rush-bearing, a curious Custom, 625.
Ruts in Roads, avoided, by different lengths of axletrees, 241.
Ryeland, or Ross, breed of Sheep, 112.

SACKING, or Corn-bags, where made, 487.
Sail-cloth, where made, 487.
Sale of Mr. Mundy's Cow Stock, particulars and prices, 10.
Sales by Tickets, the mode of conducting them, 460.
Salmon passes and traps, in the Derwent River, 205.
Salt, exuding from stone, &c. attracts Pigeons, 181 Note.
───── is not given here to Sheep, 126.
Salting of Butter, not with Brine, 68.
Sand Pits, for various uses, where situate, Vol. I. p. 463.

Saving

Saving Banks, are excellent institutions, 571 Note, 575.
Sawing-Mills, for Stone, and for Wood, Vol I. p. 428, and II. p. 235.
Scab, a disease of Cows, common formerly, 85.
———— of Sheep, 148.
Schools, Boarding, Day, Free, Industry, Lancasterian, and Sunday, in the County, 529.
Scissars, where made, 495.
Scives, of iron and wood, where made, 483, 493.
Score, a cubic measure, 20 or 22 heapt bushels, 473.
Scotch and White, mixed breed of Cows, 20.
———— or Highland, breed of Cows, 16.
Scotching, or Scoiting Stones, dangerous on Roads, 264.
Scouring, a disease of Sheep and Lambs, 145.
———— sand, where dug, Vol. I. pp. 279 and 463.
————, Shooting-out, &c. a disease of Cows and Calves, 82.
Screws, for Carpenters, where made, 492.
Scythe-sharpening Sticks, and Stones, where made, Vol. I. pp. 439, 437.
———— smiths, where resident, 495.
Seed-weeds, in Stubbles, how destroyed, 182.
Self-interest, rightly considered and acted on, not a base or unworthy motive, 530 Note.
Shammy Leather, where made, 480.
Shear-hogs, sheared Lambs, 136.
Shearing of Lambs, particulars concerning, 136, 140.
———— of Sheep, ditto, 136.
Shed, Stalls, Yards, &c. for Cattle, 70.
Sheep-Dung, collected at the Lees, on the Moors, 91.
———— injured by the Frost, if not ploughed in as soon as Turnips are eaten off, 126 Note.
———— four native Derbyshire Breeds of them, 88.
———— Houses, and lambing Folds, 127.
———— how prevented from leaping Wall-fences, 128.
———— Lees, or shelter Walls, on the Moors, 127.
———— live and dead weights of, when fat, 129.
———— on the breeding and treatment of them, 88.
———— on the food of, 126.
———— 17 Breeds and Crosses of them, now kept in the County, 89.
———— Shearing, Mr. W. B. Thomas's public one, 118.
————————, particulars concerning, 136.
————, Wash, description of one, 134; and Plate II. p. 144.
Sheeting weaving, where carried on, 486.
Sheet-Lead, where made, 492.

Shelling,

INDEX.   719

Shelling, Shilling, or Oat-Meal Mills, where situate, Vol. II. pp. 129, 457.
Shepherds' Societies, on the Moors, 91, 650.
Shock of Corn, 12 Sheaves, half a Thrave, 471.
Shoe-factory, where situate, 480.
—— maker's Nails, where made, 491.
Shooting-out, scouring, &c. a disease of Cows, 80.
Short-horn, and Devon, mixed breed of Cows, 19.
————— and French, mixed breed of Cows, 19.
————— and Lincoln, mixed breed of Cows, 20.
————— and White, mixed breed of Cows, 20.
—————, Holderness, Yorkshire, or Durham, mixt breed of Cows, 1.
Shot, of Iron and Lead, where made, 488, 492.
Shovels, and Spades, where made, 496.
Sick-Club, see *Friendly Societies*.
Sickles, for Reaping, where made, 496.
Silk-spinning Mills, where situate, 482.
—— stockings weaving, where carried on, 482.
Silver Tokens, much circulated, 512.
Skelping or Planishing Hammer Mills, where situate, 490.
Skimming of Milk, for Cream, how performed, 64.
Skinners, or Leather-dressers, where resident, 480.
Slag of Lead Ore, Mills for grinding it, &c. Vol. I. p. 383, 391.
Slates, for Roofs, where dug, Vol. I. p. 429.
Slipping, or Picking, of Calves, before their time, 86.
Slitting and rolling, Iron Mills, where situate, Vol. I. p. 403.
Slugs, or Grubs, injuries by, 643.
Small Cows, advantages of them, 34.
—— Farms, are a National loss, 634, 635.
Snipes, as Birds of Game, 200.
Snuffers, where made, 496.
Soap-makers, where resident, 480.
Soda-water, where made, 496.
Soiling, or giving green food in stalls, to Horses, 158.
———————————————————— to Neat Cattle, 22.
Sore Eyes, occasioned by corrosive Water, 229.
Soup-shops, for the Poor, 579.
South-down Sheep, account of the Flocks of, 110.
Spade, Hyon, Black-leg, &c. a dreadful disease of Cows, 76.
Spades and Shovels, where made, 496.
Spanish, spotted, coarse-woolled Sheep, 123.
—— travelling or Merino fine-woolled Sheep, 113.

Spar,

Spar, or Petrifaction-workers, where resident, Vol. I. p. 150, 459, 461.

Sparrows, their depredations on Corn-ricks, &c. 646.

Spinning Mills, see *Cotton, Flax, Silk, Wool.*

Sports and Pastimes, of the People, 628.

Spreading of Dung, Lime, &c. 190.

Springs and Wells of Water, decorated with Flowers, 626.

Spurs, Bridle-bits, &c. where made, 496.

Square, a square measure, 100 feet, 469.

Squirrel-hunting, yet practised in the County, 244, 629.

Squirrels, occasionally injure Fir Trees, 646.

Stack-pots and Caps, for Hovels, where made, Vol. I. p. 432.

Stafford Rail-way, described, 415.

Staffordshire and Worcestershire Canal, described, 415.

Stag-hunting, a premeditated and repeated cruelty, disused in the County, 206, 629.

Stallion-letters, a List of them, 151.

——— Pastures, or Horse Fields, 156.

——— Show, and Sale of them, 156.

Stalls, Yards, Sheds, &c. for Cattle, 70.

Statute-duty, an improper system for Roads, 260.

Steam-Engines, applied to draw Railway Carriages, 329.

——————————— to wind up or draw Coals, &c. I. 338.

——————— makers of them, where resident, Vol. II. p. 493.

Steel Mills, for Malt, Pepper, &c. where made, 495.

Sties, or Yards, &c. for Hogs, 171.

Stilton Cheeses, some made in Derbyshire, 57.

Stirrup-irons, where made, 496.

Stockingers, and other Manufacturers, dreaded by some Landlords, 504.

Stockings-weaving, of Cotton, Silk, Wool, where carried on, 480, 487.

Stocks, or other Parochial Punishments, now rarely used, 631.

Stone, in the Bladder, how relieved, 692.

—— digging, prices of, 190.

—— a weight, 14 pounds, 471, 475.

—— measures and weights used in selling it, 474.

—— Mills, for sawing and polishing it, where situated, Vol. I. pp. 423, 427.

—— Wares, Pottery, Delph-wares, where made, Vol. I. p. 449.

Subscription Poor-houses, for Paupers of various Parishes, details regarding them, 548 to 564.

Sulphur-works, at the Lead Furnaces, where situated, Vol. I. p. 468.

Summer food, and soiling, of Horses, 158.
—————————— of Neat Stock, 22.
Sunday-Schools, several in the County, 629.
Superstitious opinions and customs, yet remaining, 626.
Surveyors of Parish Roads, absurdly chosen, 262.
Swadlingcote and Newhall, proposed Railway, 418.
Swallows, are serviceable in eating Flies, 647.
Swanwick, the late Mr. Thomas's Meteorological Table, 685.
Swine, see *Hogs*.

TANNING of Leather, where carried on, 481.
Tape-weaving, where carried on, 438.
Tawers, Codders, or White Leather makers, 481.
Taxes, high ones impede Improvements, 638.
Tench, castrated, for fattening, 204.
Tenter-hooks, where made, 492.
Terms and Names of things and operations, by local Men, are important to be known, 657—are not so arbitrarily used, as many suppose, 658.
Ternbridge and Winsford, proposed Canal, 419.
Tethering of Hogs, recommended, 169.
Thermometrical observations at Derby, through 11 years, 685.
Thick-set and Fustian Weaving, where carried on, 485.
Thick-throat, a disease of Women, in the County, 622.
Thistles, are often nuisances by Road-sides, 267.
Thomas, W. B. his public Sheep-shearing, 655.
Thrashing of Corn, prices of, 190.
Thrave of Corn, 2 shocks, 24 sheaves, 471.
Thread-spinning, sewing Cotton and Linen, where carried on, 488.
Three-quarter Stack, a cubic measure, 105 feet, 473.
Throwing at Cocks, a brutal custom, 629.
Ticket-sales of Articles, how conducted, 460.
Ticks, insects infesting Sheep, 149.
Tile-clay Pits, and Kilns, where situated, Vol. I. p. 451.
—— stones, or grey Slates, where dug, Vol. I. p. 429.
Tilt Hammer, and Skelper Mills, where situated, 490.
Tin'd-plate workers, where resident, 496.
Tire, for wheels, where made, 493.
Tithes, are impediments to Improvements, 638.
—————— their amounts compared with Rents, 640.
Toasting Cheese, Derbyshire, made at Shottle, Aldwark, &c. 57.
Tobacco-pipe making, where carried on, Vol. I. p. 448.
Tod, a weight, 28 pounds. 475.

722   INDEX.

Tokens of Silver, in circulation instead of Shillings, 512.
Toll-bars, often are improperly placed and managed, 242, 512.
——— very unequally affect different Towns and Places, 242.
Tolls on Roads, should progressively increase, with the weight carried, 244.
——————— remitted to Carriages of new Widths, 242.
Tom-tits, mischievous little Birds, 647.
Ton, a cubic measure, 15 or 23 feet, 469.
——————— 20 level Bushels, p. 469.
—— a weight, 2240 or 2400 Pounds, 471, 473, 474, 475.
Tool-makers, and Engineers, where resident, 495.
Towns, their importance in Political Economy, 593.
——— a List of 700 British ones, with their Population, 594.
——— their Population, is at present very inaccurately stated, 604.
Tracts, and plain useful Books, wanted among the Poor, 530.
Traders, Manufacturers, and Handicrafts, their proportionate numbers, in different parts, 535, 581.
Training of Nag Horses, made a profession, 155.
Travelling Notes, arranged by *Places*, are of important use, 477, 662 Notes.
Trenching, or digging, measures used in estimating it, 190, 469.
Trent and Mersey Canal, described, 430.
—— River (lower) Navigation, described, 421.
——————— Upper Navigation, formerly, 428.
Troughs and Cisterns of stone, where made, Vol. I. p. 432.
——— of stone, for feeding Hogs, 171.
Trout, growth of some, &c. 206.
Trowels, Paring-shovels, &c. where made, 494.
Trustees of Turnpike-roads, sometimes act improperly, 237.
Tub, a cubic measure, 2¼ Bushels, 472.
Tunnels for Canals, described, 200, 319, 321, 343.
Tup, or Ram-Letters, Lists of them, 98, 100.
Turkies, of a large kind, 178.
Turning-mills, for Wood, where situate, 483.
Turnips, their taste in Butter, how prevented, 64.
Turn, Turney-headed, or giddy Sheep, 146.
Turnpike Acts, are on temporary and imperfect principles, 232, 235.
——— Roads, Accounts of the Trustees, require Auditing, 235.
——————— a Map of them, in and near the County, Plate III. p. 193, 207.
——————— are part of a system of Improved Communication, 206.

                                                         Turnpike

Turnpike Roads, compensated by Canal Companies, for loss of Tolls, 359.
──────── imperfect principles on which the first were constructed, 221, 229.
──────── modes of managing them, 259.
──────── one taken by a Parish, from the Trustees, 239.
──────── should be placed under a Special Board, 235.
Twine and Packthread-spinning, where carried on, 486.

VEAL and Beef, where produced, 173.
Vegetable productions, of the County, Trades, &c. therein, 483.
──────── imported, ditto, 482.
Venison, Buck, Doe, Havier, 200.
Vermin, or mischievous small Animals, &c. 645.
Veterinary Surgeons, names of, 159.

WAKES, or Village Feasts, 628.
Wall-Fences, destroyed by Carters on hilly Roads, 264.
──────── how secured against wild Sheep, 128.
Walling, fence-walling, prices of it, 190.
Warm, steam-cooked Food, for Milking Cows, 32.
Warping, or mudding of Lands, in Yorkshire and Lincolnshire, 428.
Warts, or excrescences on Cow's bellies, 85.
Washing Machines, where made, 496.
──────── of Sheep, an improved mode of, Plate II. p. 144.
──────── before shearing, 134.
Wasps, a mode of destroying them, 645.
Waste Land, by Roads, should be inclosed, but not planted, 369.
Watch and Clock-making, where carried on, 494.
Watering of Roads, with intent to improve them, 280.
Water Meadows, if properly made, probably won't rot Sheep, 143 and 144.
Water-Mill Owners, oppose and obstruct Canals, 322, 358.
Wavy, or Undulating Roads, 279.
Way-leaves, for private Rail-ways, near Newcastle, 285.
──── Posts, or finger boards, are much neglected, 246.
Weeds, the seeds of in stubbles, how best destroyed, 182.
Weighing-Engines, on Roads, 235, 245.
──────── should weigh Goods for hire, 245.
──────── where made, Vol. II. p. 65.
──────── Houses, for Boats on Canals, Vol. I. p. 184.
Weights and Measures, in use in the County, 460.
──────── the Author's schemes for entirely reforming them, 462, 681 Note.

Weir-hedges, for straightening Rivers, 427.
Wells and Springs of Water, decorated with Flowers, 626.
Welsh breed of Cows, where kept, 16.
Wheat, where grown, Vol. II. p. 113.
Whetstones, or Rubbers, where dug and made, Vol. I. pp. 437, 440.
Whey-Butter, the different modes of making it, 66.
—— Cream, given to fatten Calves, 24.
——————— the mode of obtaining it, 65.
—— the principal liquid food of Hogs, in the County, 170.
Whimsey Steam-engines, for drawing or winding up Coals, &c. 340.
Whiskets, or Baskets, where made, Vol. II. p. 262; III. 482.
White breed of Cows, with black Ears, where kept, 16.
White-lead making, and grinding, where carried on, 493.
Wild Ducks, 199.
—— Geese, 199.
—— or Wood Pigeons, 181, 200.
Wilden and King's Bromley, proposed Canal, 449.
Willow Timber, where grown, Vol. II. p. 267.
Wingerworth and Woodthorp Rail-way, described, 451.
Winters, are sometimes long, in the Peak Hundreds, 21.
Wire-drawing, where carried on, 493.
—— working, safes, scives, skreens, &c. where carried on, 493.
—— worms, injurious to Corn, 648.
Women's Benefit or Friendly Societies, 566.
Woodeaves Canal, described, 451.
Wooden Rail-ways, where still in use, 288.
Wood and Timber, measures used in selling them, 472.
——— brush or Kid, used in the foundations of Roads, 258.
——— cocks, Birds of Game, 200.
———, is but little used for fuel in the County, 169.
——— land or Moor-land breed of Sheep, a List of the Breeders of them, 92.
——— ———————————————, their treatment, 88.
——— Pigeons, injure Turnip Crops, 647.
Woollen Factories, Spinning, Weaving, Dressing, &c. where established, 481.
Wool-bags, where made, 485.
—— Chambers, 136.
—— Fairs, recommended, 137, 676.
——, Weights and Prices of Fleeces, 93, 139.
——, Weights by which it is sold, 137, 475.
Work-houses, or Parish Poor-houses, 544.

Work-

Work-houses, the origin of the Laws for establishing them, 522.
Work-people, Forms of Agreement for attaching them to Works, 689.
Working-Horses, in Teams, &c. 156.
——— of Oxen and Heifers, in farm business, 69.
Worms, in the ground, are injurious, 644.
Worsted Machine Makers, where resident, 496.
——— Spinning Mills, where situate, 481.
Wyrley and Essington Canal, described, 451.

YARDS, Stalls, Sheds, &c. for Cattle, 70.
Yearly hired Servants, Wages, &c. 185.

ZINK Mines, where situated, Vol. I. p. 406.
Zink-Works, Plate, Wire, Pipes, &c. where situate, 493.

LIST OF PUBLICATIONS

OF

THE BOARD OF AGRICULTURE,

Which may be had of the Publishers of this Volume.

ENGLISH AND WELSH REPORTS.

	£	s.	d.
Bedfordshire, with Map and Plates, by Mr. Batchelor	0	15	0
Berkshire, ditto, by Dr. Mavor	0	18	0
Buckinghamshire, ditto, by the Rev. St. John Priest	0	12	6
Cambridgeshire, Map, by the Rev. Mr. Gooch	0	9	0
Cheshire, by H. Holland, Esq. Map and Plates	0	10	0
Cornwall, Map, by Mr. Worgan	0	12	0
Derbyshire, Vol. I. Map and Plates, by Mr. Farey, sen.	1	1	0
———— Vol. II. Plates, by ditto	0	15	0
———— Vol. III. plates, ditto			
Devonshire, Map and Plates, by C. Vancouver, Esq.	0	18	6
Dorset, Map, by Mr. Stevenson	0	12	0
Durham, Map and Plates, by Mr. Bailey	0	10	0
Essex, 2 vols. ditto, by A. Young, Esq.	1	4	0
Gloucestershire, ditto, by T. Rudge	0	9	0
Hampshire, ditto, by C. Vancouver, Esq.	0	16	0
Herefordshire, ditto, by J. Duncombe, A.M.	0	7	0
Hertfordshire, ditto, by A. Young, Esq.	0	8	0
Huntingdonshire, Map and Plate, by Mr. Parkinson	0	9	0
Kent, ditto, by Mr. Boys	0	8	0
Lancashire, Map, by R. W. Dickson; revised and prepared for the press by W. Stevenson	0	14	0
Leicestershire and Rutland, Map and Plates, by Messrs. Pitt and Parkinson	0	15	6
Lincolnshire, Map and Plates, by A. Young, Esq. 2d edit.	0	12	0
Middlesex, Map, by J. Middleton, Esq. 2d edit.	0	15	0
Monmouthshire, Map, by Mr. Hassall	0	7	0
Norfolk, ditto, by A. Young, Esq.	0	12	0
———— ditto, by Mr. Kent	0	6	0
Northamptonshire, ditto, by W. Pitt, Esq.	0	8	0
Northumberland, Cumberland, and Westmorland, Map and Plates, by Messrs. Bailey, Culley, and Pringle	0	9	0
North Wales; containing the Counties of Anglesey, Caernarvon, Denbigh, Flint, Merionydd, and Montgomery, ditto, by Walter Davies, A.M.	0	12	0
Nottinghamshire, Map, by Robert Lowe, Esq.	0	5	0
Oxfordshire, Map and Plates, by A. Young, Esq.	0	12	0
Shropshire, ditto, by the Rev. J. Plymley, A.M.	0	9	0
South Wales; containing the Counties of Brecon, Caermarthen, Cardigan, Glamorgan, Pembroke, and Radnor, ditto, by Walter Davies, A.M. in 2 vols.	1	4	0
Staffordshire, ditto, by Mr. Pitt	0	9	0
Suffolk, ditto, by A. Young, Esq.	0	10	6

Publications of the Board of Agriculture.

	£	s.	d.
Surrey, Map, by Mr. Stevenson	0	15	0
Sussex, Map and Plates, by the Rev. A. Young	0	15	0
Warwick, Map and Plate, by Mr. Murray	0	8	0
Wiltshire, Map and Plates, by Mr. Davis	0	9	0
Worcestershire, ditto, by Mr. Pitt	0	10	6
Yorkshire, (East Riding), ditto, by Mr. Strickland	0	12	0

SCOTCH REPORTS.

	£	s.	d.
Argyleshire, with Map and Plates, by Dr. J. Smith	0	9	0
Berwickshire, ditto, by Mr. Kerr	0	14	0
Caithness, ditto, by Capt. Henderson	0	15	0
Clydesdale, Map, by Mr. Naismith	0	7	0
Dumfries, Map and Plates, by Dr. Singer	0	18	0
East Lothian, Map, by Mr. Somerville	0	6	0
Galloway, Map and Plates, by the Rev. S. Smith	0	9	0
Hebrides, Maps and Plate, by J. Macdonald, A. M.	1	1	0
Inverness-shire, Map and Plate, by the Rev. Dr. Robertson	0	14	0
Kincardineshire, Map, by Mr. Robertson	0	12	0
Nairn and Moray, ditto, by the Rev. W. Leslie	0	14	0
Peebles, Map and Plates, by the Rev. C. Findlater	0	10	6
Ross and Cromarty, Map, by Sir G. S. Mackenzie	0	9	0
Roxburgh and Selkirk, Maps and Plates, by the Rev. Dr. Douglas	0	9	0
Sutherland, Map and Plates, by Capt. Henderson	0	12	0
West Lothian, ditto, by Mr. Trotter	0	9	0

Communications to the Board of Agriculture,	Vol. I.	1	1	0
Ditto,	Vol. II.	1	1	0
Ditto,	Vol. III.	0	18	0
Ditto,	Vol. IV.	0	18	0
Ditto,	Vol. V.	1	1	0
Ditto,	Vol. VI.	1	10	0
Ditto,	Vol. VII.	1	11	6

A Sheep-Walk used in Bedfordshire.

A Temporary Lambing Fold, made of Hurdles & Rough Poles.

Milton Keynes UK
Ingram Content Group UK Ltd.
UKHW022103150124
436101UK00005B/168